TRANSMISSION PIPELINE CALCULATIONS AND SIMULATIONS MANUAL

TRANSMISSION PIPELINE CALCULATIONS AND SIMULATIONS MANUAL

E. SHASHI MENON
Vice President
SYSTEK Technologies, Inc.
USA

Amsterdam • Boston • Heidelberg • London
New York • Oxford • Paris • San Diego
San Francisco • Singapore • Sydney • Tokyo

Gulf Professional Publishing is an imprint of Elsevier

Gulf Professional Publishing is an imprint of Elsevier
225 Wyman Street, Waltham, MA 02451, USA
The Boulevard, Langford Lane, Kidlington, Oxford, OX5 1GB, UK

Copyright © 2015 E. Shashi Menon. Published by Elsevier Inc. All rights reserved.

No part of this publication may be reproduced or transmitted in any form or by any means, electronic or mechanical, including photocopying, recording, or any information storage and retrieval system, without permission in writing from the publisher. Details on how to seek permission, further information about the Publisher's permissions policies and our arrangement with organizations such as the Copyright Clearance Center and the Copyright Licensing Agency, can be found at our website: www.elsevier.com/permissions

This book and the individual contributions contained in it are protected under copyright by the Publisher (other than as may be noted herein).

Notices
Knowledge and best practice in this field are constantly changing. As new research and experience broaden our understanding, changes in research methods, professional practices, or medical treatment may become necessary.

Practitioners and researchers must always rely on their own experience and knowledge in evaluating and using any information, methods, compounds, or experiments described herein. In using such information or methods they should be mindful of their own safety and the safety of others, including parties for whom they have a professional responsibility.

To the fullest extent of the law, neither the Publisher nor the authors, contributors, or editors, assume any liability for any injury and/or damage to persons or property as a matter of products liability, negligence or otherwise, or from any use or operation of any methods, products, instructions, or ideas contained in the material herein.

ISBN: 978-1-85617-830-3

Library of Congress Cataloging-in-Publication Data
A catalog record for this book is available from the Library of Congress

British Library Cataloguing in Publication Data
A catalogue record for this book is available from the British Library

For information on all Gulf Professional Publishing
visit our website at http://store.elsevier.com/

CONTENTS

Preface — xi

1. **Introduction to Transmission Pipelines** — 1
   1. Trans-Alaska Pipeline (North America) — 5
   2. Tennessee Gas Pipeline (North America) — 7
   3. Rockies Express Pipeline (North America) — 7
   4. TransCanada Pipeline (North America) — 8
   5. The Bolivia–Brazil Pipeline (South America) — 8
   6. GasAndes Pipeline (South America) — 9
   7. Balgzand Bacton Pipeline (Europe) — 9
   8. Trans-Mediterranean Natural Gas Pipeline (Europe–Africa) — 9
   9. Yamal–Europe Pipeline (Europe–Asia) — 10
   10. South Caucasus Pipeline (Asia) — 11
   11. West-East Natural Gas Pipeline Project (China–Asia) — 11
   12. The Caspian Pipeline (Russia–Asia) — 12
   Reference — 13

2. **Standards and Codes** — 15
   1. Codes, Standards, and Regulations — 15
   2. Boiler and Pressure Vessel Code — 19
   3. Federal and State Laws — 20
   4. ASME Council for Codes and Standards — 21
   5. API Standards and Recommended Practices — 22
   6. Manufacturers Standardization Society — 23
   7. Pipe Fabrication Institute Standards — 24
   8. American Institute of Steel Construction — 24
   9. American Concrete Institute — 25
   10. National Association of Corrosion Engineers — 26
   11. Fluid Control Institute Standards — 26
   12. Hydraulics Institute Pump Standards — 26

3. **Physical Properties** — 29
   1. Properties of Liquids and Gases — 29
   2. Units of Measurement — 30
   3. Mass, Volume, Density, and Specific Weight — 34
   4. Specific Gravity and API Gravity — 37

5.	Viscosity	41
6.	Vapor Pressure	51
7.	Bulk Modulus	51
8.	Fundamental Concepts of Fluid Flow	53
9.	Gas Properties	56
10.	Mass	56
11.	Volume	57
12.	Density and Specific Weight	58
13.	Specific Gravity	58
14.	Viscosity	59
15.	Ideal Gases	64
16.	Real Gases	69
17.	Natural Gas Mixtures	69
18.	Pseudo Critical Properties from Gravity	71
19.	Adjustment for Sour Gas and Nonhydrocarbon Components	72
20.	Compressibility Factor	72
21.	Heating Value	79
22.	Summary	81
23.	Problems	81

4. Pipeline Stress Design — 83

1.	Allowable Operating Pressure and Hydrostatic Test Pressure	83
2.	Barlow's Equation for Internal Pressure	85
3.	Gas Transmission Pipeline: Class Location	89
4.	Line Fill Volume and Batches	93
5.	Gas Pipelines	95
6.	Barlow's Equation	96
7.	Thick Wall Pipes	97
8.	Derivation of Barlow's Equation	99
9.	Pipe Material and Grade	101
10.	Internal Design Pressure Equation	102
11.	Mainline Valves	103
12.	Hydrostatic Test Pressure	104
13.	Blowdown Calculations	144
14.	Determining Pipe Tonnage	145
15.	Summary	148

5. Fluid Flow in Pipes — 149

1.	Liquid Pressure	149
2.	Liquid: Velocity	154

3.	Liquid: Reynolds Number	156
4.	Flow Regimes	158
5.	Friction Factor	159
6.	Pressure Drop from Friction	165
7.	Colebrook–White Equation	167
8.	Hazen–Williams Equation	168
9.	Shell-MIT Equation	170
10.	Miller Equation	172
11.	T.R. Aude Equation	173
12.	Minor Losses	175
13.	Internally Coated Pipes and Drag Reduction	179
14.	Fluid Flow in Gas Pipelines	181
15.	Flow Equations	183
16.	General Flow Equation	184
17.	Effect of Pipe Elevations	187
18.	Average Pipe Segment Pressure	188
19.	Velocity of Gas in a Pipeline	189
20.	Erosional Velocity	192
21.	Reynolds Number of Flow	194
22.	Friction Factor	197
23.	Colebrook–White Equation	198
24.	Transmission Factor	202
25.	Modified Colebrook–White Equation	206
26.	AGA Equation	209
27.	Weymouth Equation	213
28.	Panhandle A Equation	216
29.	Panhandle B Equation	219
30.	Institute of Gas Technology Equation	222
31.	Spitzglass Equation	225
32.	Mueller Equation	227
33.	Fritzsche Equation	228
34.	Effect of Pipe Roughness	229
35.	Comparison of Flow Equations	231
36.	Summary	233

6. Pressure Required to Transport — 235

1.	Total Pressure Drop Required to Pump a Given Volume of Fluid through a Pipeline	236
2.	Frictional Component	237
3.	Effect of Pipeline Elevation	237

	4.	Effect of Changing Pipe Delivery Pressure	241
	5.	Pipeline with Intermediate Injections and Deliveries	242
	6.	System Head Curves: Liquid Pipelines	255
	7.	Hydraulic Pressure Gradient: Liquid Pipeline	258
	8.	Transporting High Vapor Pressure Liquids	263
	9.	Hydraulic Pressure Gradient: Gas Pipeline	264
	10.	Pressure Regulators and Relief Valves	268
	11.	Summary	271

7. Thermal Hydraulics — 273

1. Temperature-Dependent Flow — 273
2. Formulas for Thermal Hydraulics: Liquid Pipelines — 277
3. Isothermal versus Thermal Hydraulics: Gas Pipelines — 289
4. Temperature Variation and Gas Pipeline Modeling — 292
5. Review of Simulation Model Reports — 294
6. Summary — 315
7. Practice Problems — 316

8. Power Required to Transport — 317

1. Horsepower Required — 317
2. Effect of Gravity and Viscosity — 321
3. Gas: Horsepower — 322
4. Summary — 327

9. Pump Stations — 329

1. Introduction — 329
2. Liquid-Pump Stations — 329
3. Summary — 367

10. Compressor Stations — 369

1. Introduction — 369
2. Compressor Station Locations — 369
3. Hydraulic Balance — 376
4. Isothermal Compression — 376
5. Adiabatic Compression — 378
6. Polytropic Compression — 381
7. Discharge Temperature of Compressed Gas — 382
8. Compression Power Required — 383
9. Optimum Compressor Locations — 387
10. Compressors in Series and Parallel — 393

	11. Types of Compressors: Centrifugal and Positive Displacement	397
	12. Compressor Performance Curves	398
	13. Compressor Head and Gas Flow Rate	400
	14. Compressor Station Piping Losses	401
	15. Compressor Station Schematic	404
	16. Summary	404
11.	**Series and Parallel Piping**	**405**
	1. Series Piping	405
	2. Parallel Piping	415
	3. Locating Pipe Loop: Gas Pipelines	429
12.	**Meters and Valves**	**431**
	1. History	431
	2. Flow Meters	432
	3. Venturi Meter	433
	4. Flow Nozzle	436
	5. Orifice Meter	437
	6. Turbine Meter	439
	7. Positive Displacement Meter	440
	8. Purpose of Valves	443
	9. Types of Valves	444
	10. Material of Construction	446
	11. Codes for Design and Construction	447
	12. Gate Valve	448
	13. Ball Valve	449
	14. Plug Valve	450
	15. Butterfly Valve	450
	16. Globe Valve	452
	17. Check Valve	452
	18. Pressure Control Valve	453
	19. Pressure Regulator	453
	20. Pressure Relief Valve	455
	21. Flow Measurement	455
	22. Flow Meters	456
	23. Venturi Meter	467
	24. Flow Nozzle	469
	25. Summary	470

13. Pipeline Economics — 473

1. Economic Analysis — 473
2. Capital Costs — 475
3. Operating Costs — 480
4. Feasibility Studies and Economic Pipe Size — 480
5. Gas Pipeline — 487
6. Capital Costs — 489
7. Operating Costs — 495
8. Determining Economic Pipe Size — 499
9. Summary — 514
10. Problems — 516

14. Case Studies — 519

1. Introduction — 519
2. Case Study 1: Refined Products Pipeline (Isothermal Flow) Phoenix to Las Vegas Pipeline — 519
3. Case Study 2: Heavy Crude Oil Pipeline 2 Miles Long without Heaters — 527
4. Case Study 3: Heavy Crude Oil Pipeline from Joplin to Beaumont (Thermal Flow with Heaters and no Batching) — 537
5. Case Study 4: Heavy Crude Oil Pipeline (Thermal Flow with Heaters and DRA) — 543
6. Case Study 5: Water Pipeline from Page to Las Cruces — 546
7. Case Study 6: Gas Pipeline with Multiple Compressor Stations from Taylor to Jenks — 549
8. Case Study 7: Gas Pipeline Hydraulics with Injections and Deliveries — 558
9. Case Study 8: Gas Pipeline with Two Compressor Stations and Two Pipe Branches — 562
10. Sample Problem 9: A Pipeline with Two Compressor Stations, Two Pipe Branches, and a Pipe Loop in the Second Segment of the Pipeline to Handle an Increase in Flow — 567
11. Sample Problem 10: San Jose to Portas Pipeline with Injection and Delivery in SI Units — 571

Appendix — 577
References — 587
Index — 589

PREFACE

This book was written to provide guidance on the design of liquid and gas pipelines for both practicing engineers as well as graduate engineers entering the pipeline field as their first employment.

We assume the engineer is familiar with basic fluid mechanics including the Bernoulli's equation. Some knowledge of pumps and compressors is also assumed.

This book covers pipeline hydraulics as it applies to transportation of liquids and gases through pipelines in a single phase steady state environment. It will serve as a practical handbook for engineers, technicians, and others involved in the design and operation of pipelines transporting liquids and gases. Currently, existing books on the subject are mathematically rigorous, theoretical, and lack practical applications. Using this book, engineers can better understand and apply the principles of hydraulics to their daily work in the pipeline industry without resorting to complicated formulas and theorems. Numerous examples from the author's real life experiences are included to illustrate the application of pipeline hydraulics.

The application of hydraulics to liquid and gas pipelines involve understanding of various properties of fluids, concept of pressure, friction and calculation of the energy required to transport fluids from point A to point B through a pipeline. You will not find rigorous mathematical derivation of formulas in this book. The formulas necessary for calculations are presented and described without using calculus or complex mathematical methods. If the reader is interested in how the formulas and equations are derived, he should refer to any of the books and publications listed under the Reference section toward the end of this book.

This book covers liquid and gas properties that affect flow through pipelines, calculation of pressure drop due to friction, horsepower required, and the number of pumps or compressor stations required for transporting the fluid through a pipeline. Topics covered include basic equations necessary for pipeline design, commonly used formulas to calculate frictional pressure drop and necessary horsepower, the feasibility of improving an existing pipeline performance using drag reduction additives (for liquid lines), and power optimization studies. The use of pumps, compressors, and valves in pipelines are addressed along with modifications necessary to improve pipeline throughput. Economic analysis and transportation tariff calculations are also included. This book can be used for the analysis of both pipeline gathering systems, plant or terminal piping, as well as long distance trunk lines. The primary audience for the book is engineers and technicians working in the petroleum, water, and process industry. This book could also be used as a textbook for a college level course in pipeline hydraulics.

We are indebted to Ken McCombs of Elsevier for encouraging us to write this book and also for waiting patiently for two years for us to complete this book while Shashi was recuperating from a quintuple heart bypass and sepsis. We would also like to acknowledge our sincere appreciation to Katie Hammon and Kattie Washington who were both very instrumental in getting the book in print. Finally, I would like to dedicate this book to my father and mother, who always believed I could write a technical book, but unfortunately did not live long enough to see it completed.

We invite comments and suggestions for improvements of the book from readers of the book and to point out any errors and omissions they feel. We sincerely hope this book will be an excellent addition to the Pipeline Engineer's library.

E. Shashi Menon, PhD, PE

Pramila S. Menon, MBA

Lake Havasu City, AZ

CHAPTER ONE

Introduction to Transmission Pipelines

Pipelines are used to transport liquids or gases from point of origin to point of consumption of liquids or gases. Transmission pipelines may be small diameter such as 4 in or the average size may range from 24 to 32 in or more in diameter. Over the course of several years, much larger pipelines have been built in the United States and abroad ranging from 48 to 60 in or larger diameter. These pipelines may be short lines, such as gathering lines ranging from a few feet to as much as a couple of miles. They may also be long trunk lines a few thousand miles long. In addition to providing the necessary pipe material, we must also provide the necessary pressure in terms of pumping equipment and drivers as well as other related appurtenances such as valves, regulators, and scraper traps. The Trans-Alaska Pipeline is a well-known large-diameter pipeline built in the United States during the past 25 years at a cost of more than $8 (US Billion) dollars.

In this book, we will concentrate on transmission pipelines used to transport liquids such as water, refined petroleum products as well as natural gas or compressible fluids such as propane and ethane. More sophisticated pipelines have also been built to transport exotic gases and liquids such as ethylene or compressed high-density carbon dioxide (CO_2). The latter pipelines require extensive hydraulic simulation or modeling taking into account the thermodynamic properties of CO_2 including liquid vapor diagrams as well as the complex formulas that define the behavior of high density CO_2.

Starting with 1866 in Pennsylvania, United States, when the first practical pipeline was constructed by the entrepreneur and scientist Edwin Drake, the United States set the stage for the proliferation of practical utilization of pipelines ranging from a few miles to tens of thousands of miles all over the world.

It must be noted that although the US pioneered pipeline efforts in the 1800s, credit must be given to engineers, technicians, and scientists that paved the way for progress in transporting "black gold" to satisfy the twentieth century requirements of mankind, which has reached a level unimaginable particularly during the past few decades. Considering that oil was available for about $20 per barrel (bbl) in the 1800s, we are now experiencing a tremendous price increase of $100 to $150 bbl in recent years. There does not seem to be a let up in the consumption of crude oil and petroleum products despite the fact that the industrialized nations have spent enormous amounts of research and development efforts in replacing oil with a more renewable energy sources such as solar and wind power. The largest consumption by the public for crude oil is the application of diesel and gasoline for motor vehicles. Despite the enormous progress made with electric cars and non–crude oil–based fuels such as compressed natural gas, liquified natural gas, and hydrogen gas, for a long time to come crude oil and their derivatives will remain a major portion of the energy source for worldwide use. For comparison, consider the cost of crude oil today at $100–120 per bbl versus electricity at $0.15 per KWH compared with natural gas cost of $8–10 per MCF. Of course these are only approximations and can vary from country to country depending on Organization of Petroleum Exporting Countries, and other natural gas and crude oil price regulating organizations.

The most important oil well ever drilled in the United States was in the middle of quiet farm country in northwestern Pennsylvania in a town called Titusville. In 1859, the newly formed Seneca Oil Company hired retired railroad conductor Edwin L. Drake to investigate suspected oil deposits. Drake used an old steam engine to drill a well that began the first large-scale commercial extraction of petroleum. This was one of the first successful oil wells drilled for the sole purpose of finding oil. This was known as the Drake Well. By the early 1860s, western Pennsylvania had been transformed by the oil boom. This started an international search for petroleum, and in many ways eventually changed the way we live.

The reason Drake chose Titusville as the spot to drill for oil was the many active oil seeps in the region. As it turns out, there had already been wells drilled that had struck oil in the region. The only problem was, they were not drilling for oil. Instead, they were looking for salt water or drinking water. When they struck oil, they considered it a nuisance and abandoned the well. At the time, no one really knew how valuable oil was.

Later on, they hoped that "rock oil" could be recovered from the ground in large enough quantities to be used commercially as a fuel for lamps. Oil had already been used, refined, and sold commercially for one of its byproducts: kerosene. Along came a gentleman named Bissell who would try to extract the rock oil from the ground by drilling, using the same techniques as had been used in salt wells. Bissell was simply looking for a better, more reliable, and plentiful source.

Table 1.1 shows a list of long-distance pipelines being used around the world to transport gas, crude oil, and products from the fields to areas of

Table 1.1 Various Transmission Pipelines in North America

Project Name	Start Point	End Point	Diameter (inches)	Length (Km)	Capacity (1000 b/d or bn cm)
—	Bakersfield	Los Angeles	—	—	—
—	Chicago	Cushing	$2 \times 12, 22$	—	—
—	Clearbrook	Minneapolis	16	—	—
—	Clearbrook	Bismark	10	—	—
—	Cushing	Wood River	22	703	275
—	Dallas	Lima	20	—	—
—	Guernsey	Chicago	8, 12, 20, 24	—	—
—	Los Angeles	San Juan	16	—	—
—	Los Angeles	San Francisco	34	—	—
—	Louisiana	Lima	22	—	—
—	Midland	Corpus Christi	10, 12	—	—
—	Midland	Cushing	2×16	—	—
—	Midland	Borger	12	—	—
—	Midland	Houston	1, 24	742	310
—	Minneapolis	St. Louis	20	—	—
—	Minneapolis	St. Louis	24	—	—
—	New Mexico	Cushing	20, 24	832	350
—	Port Arthur	Midland	10	—	—
—	Prudhoe Bay, Alaska	Valdez	34	—	—
—	San Juan	Houston	12, 16	—	—
—	Santa Barbara	Houston	10	—	—
—	Saint James	Patoka	40	1068	1175
—	Wichita	Kansas City	34	—	—
Portland natural gas transmission	Westbrook	Colebrook	—	—	—
—	Hugoton	Denver	2×20	—	—
—	Los Angeles	San Diego	36	—	—
—	Los Angeles	Houston	36	—	—
—	Louisiana	Pittsburgh	—	—	—
—	Louisiana	Detroit	—	—	—
—	Mountain Home	Reno	16	—	—
—	New Orleans	Portland	42	—	—
—	Salt Lake City	Pendleton	22	—	—
—	Sal Lake City	Bakersfield	-	—	—
—	San Juan	Bakersfield	24, 30	—	—
—	San Juan	El Paso	2×30	—	—
—	C33	Raleigh	—	—	—

(*Continued*)

Table 1.1 Various Transmission Pipelines in North America—Cont'd

Project Name	Start Point	End Point	Diameter (inches)	Length (Km)	Capacity (1000 b/d or bn cm)
—	Amarillo	El Paso	6	—	—
—	Baton Rouge	Washington, DC	6, 30	5081	550
—	Billings	Minot	8	—	—
—	Billings	Casper	6, 12	1097	100
—	Bismark	Detroit	10	—	—
—	Casper	Rapid City	12	—	—
—	Chicago	Green Bay	10, 16	516	166
—	Chicago	New York	—	—	—
—	Chicago	Saint Louis	—	—	—
—	Denver	Wichita	—	—	—
—	Denver	Houston	6, 8	—	—
—	Denver	Sinclair	—	—	—
—	Des Moines	Cushing	—	—	—
—	El Paso	Midland	2 × 8	—	—
—	Ferndale	Eugene	16	—	—
—	Houston	Port Isabel	—	—	—
—	Houston	Philadelphia	36, 40	—	—
—	Joliet	Toledo	8, 18	990	300
—	Kansas	Detroit	10, 16	516	166
—	Lake Charles	Hammond	8, 28	2248	283
—	Los Angeles	San Diego	10	—	—
—	Los Angeles	El Paso	12	—	—
—	Midland	Rock Springs	8	—	—
—	Midland	Houston	10	—	—
—	Minneapolis/ St. Paul	Midland basin	8, 10	—	—
—	Minneapolis/ St. Paul	Tulsa	8, 12, 20	—	—
—	Mount Belvieu	Raleigh	6, 12	2097	100
—	Omaha	Chicago	—	—	—
—	Omaha	New Orleans	10	—	—
—	Port Arthur	Abilene	12	—	—
—	Port Arthur	Albany	2 × 16, 20	—	—
—	Salt Lake City	Spokane	8	—	—
—	San Bernardino	Las Vegas	8, 14	—	—
Sacramento line	San Francisco	Bakersfield	8, 12, 14	—	—
—	Spokane	Billings	10	—	—
—	Tulsa	Detroit	2 × 10, 12, 14	566, 493	75, 100

Table 1.1 Various Transmission Pipelines in North America—Cont'd

Project Name	Start Point	End Point	Diameter (inches)	Length (Km)	Capacity (1000 b/d or bn cm)
Canada/United States					
—	Edmonton	Puget Sound	2	—	92
—	Edmonton	Guernsey	8, 12	—	—
—	Edmonton	Detroit	20, 24, 34	—	1400
—	Montreal	Chicago	30	—	—
—	Regina	Guernsey	6, 16	600	85
—	Calgary	Barstow	2 × 34, 36	—	—
—	Fort Nelson	Melford	10, 30, 36	—	—
Maritimes and Northeast	Goldsboro	Westbrook	30	56	—
—	Edmonton	Windsor	12	3170	92
—	Portland	Montreal	24	770	109
—	Portland	Montreal	18	770	—
Alaska Gas Pipeline	Prudhoe Bay	Edmonton	—	—	—

use. Sometimes these fields are located in one country or continent and then transported by pipeline for distribution through several countries.

1. TRANS-ALASKA PIPELINE (NORTH AMERICA)

This 48-in-diameter steel pipeline zigzags across the frozen tundra of Alaska for 800 miles. It stretches from Prudhoe Bay, on Alaska's North Slope, to the northernmost ice-free port at Valdez, Alaska, on Prince William Sound. Along the way, it must travel over three mountain ranges, cross more than 500 rivers and streams, over three unstable earthquakes faults, and through the migration paths of the caribou and moose. The construction of the Trans-Alaska Pipeline (the most expensive private undertaking) cost $8 billion. The completed, 48-inch -diameter pipeline was opened for business in 1977.

The pipeline was purposely built in a zigzag configuration to allow the pipe to move more easily from side to side and lengthwise in cases of earthquakes or temperature-related fluctuations. The effectiveness of this design was proven in 2002 when the pipeline survived a 7.9-magnitude earthquake. Where it runs over fault lines, the pipeline rests on perpendicular so-called "slider supports," which are long rails that will allow the pipeline to slide with the ground movement. Approximately 420 miles of the pipeline was built above ground because of the unstable soil conditions from the thaw sensitive permafrost and 380 miles below ground. To keep the oil flowing, there are 11 pumping stations along the length of the pipeline, each containing four motor-driven pumps. Of these 44 pumps, however, only around 28 are operating at any one time, depending on flow.

The total includes such items as $2.2 million for an archaeological survey and $1.4 billion for the Valdez terminal. Also included are the pump stations, 13 bridges, 225 access roads, the three "pig" launching/receiving facilities, more than 100,000 lengths of 40-ft pipe, 14 temporary airfields, and salaries for the total number of construction workers and employees over the life of the construction project.

One of the most significant innovations built into the system are heat exchangers. Because the temperature of the oil flowing through the pipe can reach more than 120° F, the heat could be transferred from the pipe through the specially designed supports and could melt the permafrost. This would cause the pipeline to sink into the melted permafrost, causing catastrophic damage and spillage. To prevent this scenario, heat exchangers were placed on top of the pipes. The heat is transferred from the base through pipes containing ammonia to the heat exchangers that are then is cooled by convection to the surrounding air.

Monitoring of the pipeline is accomplished by several methods. Aerial surveillance is performed several times a day, a task that can take 2 h or more. Another is by sending inspection gauges, called "pigs," through the line on a regular basis that can relay radar scans and fluid measurements back to the launching facility as they travel within the line.

The total oil production since June 1977 is reported to be well in excess of 500 billionbbl. Although the best production year was 1988 when approximately 745 million bbl of crude was shipped, since then the yearly yield has steadily declined to a low of only 270 million bbl in 2007.

In the late 1980s, when oil flow through the pipeline climbed, the higher throughput (a 30% increase) was made possible by the injections of a drag-reducing agent (DRA), thus avoiding additional construction costs such as adding pipes or pumps. The DRA is a poly-alpha-olefin, or nonsaturated carbon with very large long-chain molecules composed of hydrogen and carbon atoms. A rough estimate of the cost savings in facility construction from DRA is approximately $300 million.

DRA was first injected into the pipeline on July 1, 1979. One of the disadvantages of DRA is it would lose its desirable properties once it passed through a pump station. Thus, batches of the agent had to be injected in the pipeline at regular intervals to keep the oil flowing smoothly.

The advantage of DRA is that it reduces turbulence flow in the crude and creates laminar flow. It also does not cause any degradation in the crude oil pumped through the pipeline and does not coat the pipeline wall. Another key development was the success DRA manufacturers had in converting the substance into water-based slurry products that are freeze-protected and can be more easily transported, injected, and cleaned up than the original gel.

Today, the pipeline is still using DRA to cut back on energy costs. Injection of DRA allows the shut down or a scale-back power usage at pump stations 7 and 9, thus saving on electricity consumption. As long as usage of DRA is cheaper than fuel, station maintenance, and manpower, it is practical to continue using the product.

Nevertheless, the importance of the Trans-Alaska Pipeline cannot be overstated. With nearly 40 million gallons of crude flowing through its line each day, and with more still hidden underground, the United States can look to Alaska to provide this necessary commodity in the future.

2. TENNESSEE GAS PIPELINE (NORTH AMERICA)

The Tennessee Gas Pipeline is a set of natural gas pipelines that run from the Gulf of Mexico coast in Texas and Louisiana through Arkansas, Mississippi, Alabama, Tennessee, Kentucky, Ohio, and Pennsylvania to deliver natural gas in West Virginia, New Jersey, New York, and New England [1]. The pipelines were constructed by Tennessee Gas Transmission Company beginning in 1943 and are now owned by Kinder Morgan. It is one of the largest pipeline systems in the United States.

The pipeline is 14,000 miles long and 32 in in diameter, providing natural gas to the eastern seaboard of the United States.

3. ROCKIES EXPRESS PIPELINE (NORTH AMERICA)

The Rockies Express is a 1679-km-long pipeline that runs between the Rocky Mountains in Colorado and Eastern Ohio. One of the largest pipelines ever constructed in the United States, the Rockies Express cost $5.6 billion to complete and has the capacity to supply about 16.5 billion cubic meters (bcm) of natural gas a year. The project was completed in three sections. The 528-km REX Entrega section runs between the

Meeker Hub in Rio Blanco County, Colorado, and the Cheyenne Hub in Weld Country, Colorado. The REX West section, which is divided into seven spreads, runs 1147 km in a 1070-mm pipe from Weld County to Audrain County in Missouri, near St Louis. There is also an 8-km, 610-mm branch connecting to the Williams Energy–owned Echo Springs Processing Plant in Wyoming. The final section of the pipeline, REX East, is a 1027-km, 1070-mm pipeline running from Audrain County, Missouri, to Clarington in Monroe County, Ohio. This section was completed in November 2009.

4. TRANSCANADA PIPELINE (NORTH AMERICA)

The TransCanada pipeline is a system of natural gas pipelines, up to 48 Inches in diameter that carries gas through Alberta, Saskatchewan, Manitoba, Ontario, and Quebec. It is maintained by TransCanada Pipelines, LP. It is the longest pipeline in Canada.

The completion of this project was a spectacular technological achievement21 In the first 3 years of construction (1956–58), workers installed 3500 km (2188 mi) of pipe, stretching from the Alberta-Saskatchewan border to Toronto and Montreal. Gas service to Regina and Winnipeg commenced in 1957 and the line reached the Lakehead before the end of that year.

Building the Canadian Shield leg required continual blasting. For one 320-m (1050-ft) stretch, the construction crew drilled 2.4-m (7.9-ft) holes into the rock, three abreast, at 56-cm intervals. Dynamite broke up other stretches, 305 m (1001 ft) at a time.

On October 10, 1958, a final weld completed the line, and on October 27, 1958, the first Alberta gas entered Toronto. For more than two decades, the Trans-Canada pipeline was the longest in the world. Only in the early 1980s was its length finally exceeded by a Soviet pipeline from Siberia to Western Europe, an approximately 4196-km (2607-mi)-long pipeline.

5. THE BOLIVIA–BRAZIL PIPELINE (SOUTH AMERICA)

The Bolivia–Brazil pipeline is the longest natural gas pipeline in South America. The 3150-km (1960-mi) pipeline connects Bolivia's gas sources with the southeast regions of Brazil.

The pipeline was built in two stages. The first 1418 km (881 mi) long stretch, with a diameter varying from 24 to 32 in (610 to 810 mm), started operation in June 1999. It runs from Rio Grande to Corumbá in Mato Grosso do Sul, reaches Campinas in the state of São Paulo, and continues to Guararema, where it is connected with the Brazilian network. The second 1165-km (724-mi)-long stretch, with a diameter varying from 16 to 24 in (410 to 610 mm), links Campinas to Canoas, near Porto Alegre in Rio Grande do Sul, was completed in March 2000.

The maximum capacity of the pipeline is 11 billion cubic meters per year (390 billion cubic feet per year) of natural gas. The total cost of the pipeline was US$2.15 billion, of which US$1.72 billion was spent on the Brazilian section and US$435 million on the Bolivian section.

6. GASANDES PIPELINE (SOUTH AMERICA)

The GasAndes Pipeline is a 463-km (288-mi)-long natural gas pipeline from La Mora, Mendoza, in Argentina to San Bernardo on the outskirts of Santiago, Chile. The diameter of the pipeline is 610 mm (24 in) and the annual capacity is 3.3 bcm. It is supplied mainly from the of the Neuquén gas fields. Total investment in the project was US$1.46 billion.

7. BALGZAND BACTON PIPELINE (EUROPE)

The Balgzand Bacton Pipeline is the first natural gas pipeline between the Netherlands and the United Kingdom. The overall length of pipeline is 235 km (146 mi), of which around 230 km (140 mi) is offshore. The pipeline's diameter is 36 in (910 mm) and working pressure is 135 standard atmospheres (13,700 kPa). The initial capacity is 16 bcm per year, which will be increased to 19.2 bcm by the end of 2010 by installing a fourth compressor at the compressor station at Anna Paulowna. The direction of gas flow is from the Netherlands to the United Kingdom. The overall cost of the project was around €500 million.

8. TRANS-MEDITERRANEAN NATURAL GAS PIPELINE (EUROPE–AFRICA)

The Trans-Mediterranean is a 2475-km-long natural gas pipeline built to transport natural gas from Algeria to Italy via Tunisia and Sicily. Built in 1983, it is one of the longest international gas pipeline systems and has the capacity to deliver 30.2 bcm/y of natural gas. The Trans-Mediterranean pipeline begins in Algeria and runs 550 km to Tunisian border. From Tunisia, the line passes 370 km to El Haouaria in the Cap Bon province and then crosses the 155-km-wide Sicilian section. Passing through Mazara del Vallo in Sicily, the pipeline moves a further 155 km from Sicily to the Strait of Messina and 1055 km in the Italian mainland to northern Italy with a branch to Slovenia. The pipeline consists of nine compressor stations, including one in the Algerian section, three in the Tunisian section, one in Sicily, and four in the Italian section.

9. YAMAL–EUROPE PIPELINE (EUROPE–ASIA)

The Yamal–Europe natural gas pipeline is a 4196-km (2607-mi)-long pipeline connecting natural gas fields in Western Siberia and in the future on the Yamal peninsula, Russia, with Germany.

The planning of the Yamal–Europe pipeline started in 1992. Intergovernmental agreements between Russia, Belarus, and Poland were signed in 1993. In 1994, Wingas, the joint venture of Gazprom and Wintershall, a subsidiary of BASF, started building the German section of the pipeline. The first gas was delivered to Germany through the Belarus-Polish corridor in 1997. The Belarusian and Polish sections were completed in September 1999 and the pipeline reached its rated annual capacity of about 33 bcm of natural gas in 2005, after completion of all 31 compressor stations.

The pipeline includes around 3000 km (1900 mi) in Russia, 575 km (357 mi) in Belarus, and 680 km (420 mi) in Poland. The German gas system is connected to the Yamal–Europe pipeline through the Jamal-Gas-Anbindungsleitung pipeline. The pipeline is initially supplied by gas fields in the Nadym Pur Taz District of the Tyumen Oblast and eventually will be supplied from the Bovanenkovo field of Yamal peninsula after construction of the 1100-km (700-mi)-long Bovanenkovo-Ukhta pipeline, a part of the Yamal project.

The capacity of the pipeline is 33 bcm of natural gas per annum. The diameter of the pipeline is 1420 mm (56 in). The pressure in the pipeline is secured by 31 compressor stations with a total rated capacity of 2399 MW.

The Russian section of the pipeline is owned and operated by Gazprom. The Belarusian section is owned by Gazprom and operated by Beltransgaz. The Polish section is owned and operated by EuRoPol Gaz S.A., a joint venture of the Polish PGNiG, Russian Gazprom (both 48% of shares), and Polish Gas-Trading S.A. (4% of shares).

10. SOUTH CAUCASUS PIPELINE (ASIA)

South Caucasus Pipeline (also known as: **Baku–Tbilisi–Erzurum Pipeline**, **BTE pipeline**, or **Shah Deniz Pipeline**) is a natural gas pipeline from the Shah Deniz gas field in the Azerbaijan sector of the Caspian Sea to Turkey.

The 42-in (1070-mm)-diameter gas pipeline runs in the same corridor as the Baku–Tbilisi–Ceyhan pipeline. It is 692 km (430 mi) long, of which 442 km (275 mi) is laid in Azerbaijan and 248 km (154 mi) in Georgia. The initial capacity of the pipeline is 8.8 bcm (310 billion cubic feet) of gas per year, and after 2012 its capacity could be expanded to 20 bcm (710 billion cubic feet) per year. The pipeline has a potential of being connected to Turkmen and Kazakh producers through the planned Trans-Caspian Gas Pipeline. Azerbaijan has proposed to expand its capacity up to 60 bcm (2.1 trillion cubic feet) by building a second line of the pipeline.

11. WEST-EAST NATURAL GAS PIPELINE PROJECT (CHINA–ASIA)

The West–East Natural Gas Pipeline is a set of natural gas pipelines that run from the western part of China to the east

The 4000-km (2500-mi)-long pipeline runs from Lunnan in Xinjiang to Shanghai. The pipeline passes through 66 cities in the 10 provinces in China. Natural gas transported by the pipeline is used for electricity production in the Yangtze River Delta area. The capacity of the pipeline is 12 bcm (420 billion cubic feet) of natural gas annually. The cost of the pipeline was US$5.7 billion. The capacity is planned to be upgraded to 17 bcm (600 billion cubic feet). For this purpose, 10 new gas compressor stations will be built and eight existing stations are to be upgraded.

The West–East Gas Pipeline is connected to the Shaan-Jing pipeline by three branch pipelines. The 886-km (551-mi)-long Ji-Ning branch between the Qingshan Distributing Station and the Anping Distributing Station became operational on December 30, 2005.

The pipeline is supplied from the Tarim Basin gas fields in Xinjiang province. The Changqing gas area in Shaanxi province is a secondary gas source. In the future, the planned Kazakhstan-China gas pipeline will be connected to the West-East Gas Pipeline.

Starting September 15, 2009, the pipeline was also supplied with coal bed methane from the Qinshui Basin in Shanxi.

Construction of the second West-East Gas Pipeline started on February 22, 2008. The pipeline with a total length of 9102 km (5656 mi), including 4843 km (3009 mi) of the main line and eight sublines, will run from Khorgas in northwestern Xinjiang to Guangzhou in Guangdong. Up to Gansu, it will be parallel and interconnected with the first west-east pipeline. The western part of the pipeline was commissioned by 2009, and the eastern part in June 2011.

The capacity of the second pipeline is 30 bcm (1.1 trillion cubic feet) of natural gas per year. It is mainly supplied by the Central Asia-China gas pipeline. The pipeline is expected to cost US$20 billion. The project is developed by China National Oil and Gas Exploration and Development Corp., a joint venture of China National Petroleum Corporation and PetroChina.

Construction of the third pipeline started in October 2012 and is to be completed by 2015. The third pipeline will run from Horgos in western Xinjiang to Fuzhou in Fujian. It will cross Xinjiang, Gansu, Ningxia, Shaanxi, Henan, Hubei, Hunan, Jiangxi, Fujian, and Guangdong provinces.

The total length of the third pipeline is 7378 km (4584 mi), including a 5220-km (3240-mi) mainline and eight branches. In addition, the project includes three gas storage units and a liquified natural gas plant. It will have a capacity of 30 bcm (1.1 trillion cubic feet) of natural gas per year with operating pressure of 10–12 MPa (1500–1700 psi). The pipeline will be supplied from Central Asia–China gas pipeline's Line C supplemented by supplies from the Tarim basin and coal bed methane in Xinjiang.

12. THE CASPIAN PIPELINE (RUSSIA–ASIA)

The Caspian Pipeline transports Caspian oil from Tengiz field to the Novorossiysk-2 Marine Terminal on Russia's Black Sea coast. It is also a major export route for oil from the Kashagan and Karachaganak fields.

The diameter of the 1510-km (940-mi)-long oil pipeline varies between 1016 mm (40.0 in) and 1067 mm (42.0 in). There are five pumping stations. The marine terminal includes two single-point moorings and the tank farm consists of four steel storage tanks of 100,000 cubic meters (3,500,000 cu ft) each. Pipeline flow started at 350,000 barrels per day (56,000 m^3/d) and has since increased to 700,000 barrels per day (110,000 m^3/d).

The Caspian Pipeline will allow maximum development of the Tengiz Field, which has potential reserves of 6 to 9 billion barrels of recoverable oil. The field produced over 600,000 barrels per day in 2011, which is expected to increase to about 1.4 billion by 2015 when it reaches peak production.

REFERENCES

[1] Wikipedia – The Free encyclopedia

CHAPTER TWO

Standards and Codes

1. CODES, STANDARDS, AND REGULATIONS

In the United States, Europe, and many parts of Asia, several organizations have been formed to develop and publish codes, standards, guides, and rules of engineering practice. For example, the American Society of Mechanical Engineers (ASME), American Society of Civil Engineers (ASCE), and Institute of Electronics and Electrical Engineers (IEEE) publish design, construction, and maintenance standards and guides related to the state of the art in the respective professions. These standards may be imposed by federal, state, or local laws and they are designated as codes. Similarly, the British Standards have been adopted by design and construction companies in the United Kingdom. The Deutsches Institut für Normung e.V. Standards are followed in Germany and France. However, the majority of the countries model their codes and standards based on US standards.

The following list of professional societies and organizations are responsible for US standards.

AA	Aluminum Association, Washington, DC
AASHTO	American Association of State Highway and Transportation Office, Washington, DC
ABMA	American Boiler Manufacturers Association, Arlington, VA
ACS	American Chemical Society, Washington, DC
ACI	American Concrete Institute, Detroit, MI
ACPA	American Concrete Pipe Association, Irving, TX
AGA	American Gas Association, Arlington, VA
AIChE	American Institute of Chemical Engineers, New York, NY
AIPE	American Institute of Plant Engineers, Cincinnati, OH
AISC	American Institute of Steel Construction, Chicago, IL
AISI	American Iron and Steel Institute, Washington, DC
ANSI	American National Standards Institute, New York, NY
ANS	American Nuclear Society, La Grange Park, IL
APL	American Petroleum Institute, Washington, DC
APFA	American Pipe Fitting Association, Springfield, VA

ASCE	American Society of Civil Engineers, Reston, VA
ASHRAE	American Society of Heating, Refrigeration and Air Conditioning Engineers, Atlanta, GA
ASME	American Society of Mechanical Engineers, New York, NY
ASNT	American Society of Non-Destructive Testing, Columbus, OH
ASPE	American Society of Plumbing Engineers, Westlake, CA
ASQC	American Society for Quality Control, Milwaukee, WI
ASTM	American Society of Testing and Materials
AWS	American Welding Society, Miami, FL
AWWA	American Water Works Association, Denver, CO
BOCA	Building Officials and Code Administration, International, Country Club Hills, IL
CABO	Council of American Building Officials, Falls Church, VA
CMA	Chemical Manufacturers Association, Washington, DC
CAGI	Compressed Air and Gas Institute, Cleveland, OH
CGA	Compressed Air Association, Arlington, VA
CISPI	Cast Iron Soil Pipe Institute, Chattanooga, TN
CSA	Construction Specifications Institute, Alexandria, VA
DIRA	Ductile Iron Research Association, Birmingham, AL
EEI	Edison Electric Institute, Washington, DC
EJMA	Expansion Joint Manufacturers Association, Tarrytown, NY
EMC	Equipment Maintenance Council, Lewisville, TX
EPRI	Electric Power Research Institute, Palo Alto, CA
EWI	Edison Welding Institute, Columbus, OH
FIA	Forging Industry Association, Cleveland, OH
HI	Hydraulic Institute, Parsippany, NJ
IAMPO	International Association of Mechanical and Plumbing Office, South Walnut, CA
ICBO	International Conference of Building Officials, Whittier, CA
ICRA	International Compressors Remanufacturers Association, Kansas City, MO
IEEE	Institute of Electronics and Electrical engineers, New York, NY
ISA	Instrument Society of America, Research Triangle, NC
MCA	Manufacturing Chemical Association, Washington, DC
MSS	Manufacturers Standard Society of Valves and Fittings Industry, Vienna, VA
NACE	National Association of Corrosion Engineers, Houston, TX
	National Board of Boiler and Pressure Vessel Inspectors, Columbus, OH
	National Certified Pipe Welding Bureau, Bethesda, MD
	National Corrugated Steel Pipe Association, Washington, DC
NCPI	National Clay Pipe Institute, Lake Geneva, WI
NEMA	National Electrical Manufacturers Association, Washington, DC
NFPA	National Fire Protection Association, Quincy, MA
NFSA	National Fire Sprinklers Association, Patterson, NY

NIST	National Institute of Standards and Technology, Gaithersburg, MD
NRC	Nuclear Regulatory Commission, Washington, DC
NTIAC	Non-Destructive Testing Information Analysis Center, Austin, TX
OSHA	Occupational Safety and Health Administration, Washington, DC
PEI	Petroleum Equipment Institute, Tulsa, OK
PFI	Pipe Fabricators Institute, Springdale, PA
PLCA	Pipe Line Contractors Association, Dallas, TX
PPFA	Plastic Pipe and Fittings Association, Glen Ellyn, IL
PMI	Plumbing Manufacturers Institute, Glen Ellyn, IL
PPI	Plastics Pipe Institute, Washington, DC
RETA	Refrigeration Engineers and Technician Association, Chicago, IL
RRF	Refrigeration Research Foundation, North Bethesda, MD
SBCCI	Southern Building Code Congress International, Washington, DC
SES	Standards Engineering Society, Dayton, OH
SFPE	Society of Fire Protection Engineers, Boston, MA
SME	Society of Manufacturing Engineers, Dearborn, MI
SPE	Society of Petroleum Engineers, Richardson, TX
SPE	Society of Plastics Engineers, Fairfield, CT
SSFI	Scaffolding, Shoring and Forming Institute, Cleveland, OH
SSPC	Steel Structures Painting Council, Pittsburg, PA
SMACNA	Sheet Metal and Air Conditioning Contractors National Association, Merrifield, VA
STI	Steel Tank Institute, Northbrook, IL
SWRI	Southwest Research Institute, San Antonio, TX
TEMA	Tubular Exchanger Manufacturers Association, Tarrytown, NY
TIMA	Thermal Insulation Manufacturers Association, Mt. Kisco, NY
TWI	The Welding Institute, Cambridge, UK
UL	Underwriters Laboratories, Northbrook, IL
UNI	Uni-bell PVC Pipe Association, Dallas, TX
VMAA	Valve Manufacturers Association of America, Washington, DC
	Vibration Institute, Willowbrook, IL
	Zinc Institute, New York, NY

In the United States, a series of documents that govern the design and construction of pressurized piping are the ASME B31 pressure piping code. "Pressure piping" refers to piping systems that operate at or above 15 psig. Piping systems that operate below atmospheric pressure down to vacuum are also addressed in certain ASME B31 standards.

The following is the list of ASME B31 standards.

ASME B31.1 Power Piping
Fossil-fueled power plant, nuclear power plant with a construction permit predating 1969 (B31.7 for 1969–1971 and ASME III post-1971).

ASME B31.12 Hydrogen Piping and Pipelines

International Piping Code

This code is applicable to piping in gaseous and liquid hydrogen service and to pipelines in gaseous hydrogen service. This code is applicable up to and including the joint connecting the piping to associated pressure vessels and equipment but not to the vessels and equipment themselves. It is applicable to the location and type of support elements but not to the structure to which the support elements are attached. This code is presented in the following parts:

1. General Requirements. This part contains definitions and requirements for materials, welding, brazing, heat treating, forming, testing, inspection, examination, operation, and maintenance.
2. Industrial Piping. This part includes requirements for components, design, fabrication, assembly, erection, inspection, examination, and testing of piping.
3. Pipelines. This part sets forth requirements for components, design, installation, and testing of hydrogen pipelines.

It is required that each part be used in conjunction with the General Requirements section but independent of the other parts. It is not intended that this edition of this code be applied retroactively to existing hydrogen systems.

ASME B31.2 Fuel Gas Piping (obsolete)

ASME B31.3 Process Piping

Hydrocarbons and others. Hydrocarbons include refining and petrochemicals. Others includes chemical process, making of chemical products, pulp and paper, pharmaceuticals, dye and colorings, food processing, laboratories, offshore platform separation of oil and gas, etc.

ASME B31.4 Liquid Petroleum Transportation Piping

Upstream liquid gathering lines and tank farms, downstream transport and distribution of hazardous liquids (refined products, liquid fuels, carbon dioxide).

ASME B31.5 Refrigeration Piping

Heating ventilation and air conditioning in industrial application.

ASME B31.6 Chemical Plant Piping (moved to B31.3)

ASME B31.7 Nuclear Power Plant Piping (moved to ASME III)

ASME B31.8 Gas Transmission and Distribution Piping

Upstream gathering lines, onshore and offshore, downstream transport pipelines, and distribution piping.

ASME B31.9 Building Service Piping

Low-pressure steam and water distribution.

ASME B31.10 Cryogenic Piping (moved to B31.3)
ASME B31.11 Slurry Transportation Piping
Mining, slurries, suspended solids transport, etc.

2. BOILER AND PRESSURE VESSEL CODE

The ASME Boiler and Pressure Vessel Code (B&PVC) is a set of rules for the design, fabrication, and inspection of boilers and pressure vessels. The mission of the B&PVC is to provide protection of life and property while assuring a long, useful service life to a pressure component designed and fabricated under this code. The B&PVC is written by volunteers who are nominated for seats on its various committees based on their expertise and their potential for making sound contributions to the writing, revising, interpreting, and administering of the document.

The ASME B&PVC was conceived in 1911 out of a need to protect the safety of the public. This need became apparent shortly after the conception of the steam engine in the late eighteenth century. In the nineteenth century, there were literally thousands of boiler explosions in the United States and Europe, some of which resulted in many deaths. The consequences of these failures were locally focused and, other than one or two, received minimal national or international attention. Undoubtedly, one of the most important failures that proved the need for Boiler Laws was the boiler explosion that occurred at the Grover Shoe Factory in Brockton, Massachusetts, on March 10, 1905. That incident resulted in 58 deaths and 117 injuries and completely leveled the factory. This catastrophe brought attention to the need to protect the public against such accidents with pressure-retaining equipment.

The first B&PVC (1914 edition) was published in 1915. Today there are 28 books, including 12 books dedicated to the Construction and Inspection of Nuclear Power Plant Components and two Code Case books. The 28 books are either standards that provide the rules for fabricating a component or they are support documents such as Materials (Section II, Parts A through D), Nondestructive Examination (Section V), and Welding (Section IX). Code cases provide rules that permit the use of materials and alternative methods of construction that are not covered by existing B&PVC rules.

Currently, all provinces of Canada and 49 of the 50 US states have adopted, by law, various sections of the B&PVC. Furthermore, the B&PVC is international. More than 25% of the companies are accredited by the

ASME Codes and Standards to manufacture pressure parts in accordance with various sections of the B&PVC are located outside of the United States and Canada.

3. FEDERAL AND STATE LAWS

Many pipelines in the United States are intrastate pipelines, meaning they are completely contained within state boundaries. Such pipelines are usually regulated by state regulators but more commonly use federal regulations such as Department of Transportation (US DOT) 192 for gas pipeline and DOT 195 for liquid pipeline. In any event the most stringent regulations generally conform.

Intrastate pipelines have to be approved by state agency such as Public Utilities Commission (e.g., California Public Utilities Commission). In addition, certain counties and local governments may require more stringent codes to be used in the design, construction, testing, and operation of pipeline. An example is certain parts of Los Angeles County, California, where population density is high. These areas are subject to the 67% rule in assessing the maximum allowable operating pressure based on hydrostatic testing of liquid pipeline. Compare this with the 72% rule for liquid pipeline under DOT 195. Thus, for example, if the hydrostatic test pressure is 1000 psig, the DOT regulation may require operating the pipeline on a continuous basis at not to exceed 720 psig where as a more stringent county requirement may be 670 psig.

Interstate pipelines on the other hand are those that cross state borders such as a typical pipeline that originates in Texas and then crosses into New Mexico and Arizona to finally end up in California for distribution. These pipelines are governed by Federal Energy Regulatory Commission (FERC) and follow the DOT Regulations and Codes for transmission pipelines.

For gas transmission pipeline, the pipeline operator should submit the so-called FERC filing (7c) to obtain a certificate of public necessity and convenience. Such FERC application can be sometimes long drawn out and may require a series of data requests from FERC looking for more information on design of pipeline, costs, and operating philosophy as well as calculations to determine the rate of return to the pipeline owners and bankers financing the project. The FERC is interested in making sure the public is not inconvenienced by exorbitant transportation tariffs and making sure the pipeline owners are not recouping the investment at a faster pace, thus violating the intent of federal regulations.

4. ASME COUNCIL FOR CODES AND STANDARDS

4.1 ASME B16 Standards

4.1.1 Metallic Gaskets for Pipe Flanges: Ring Joint Spiral Wound and Jacketed B16.20

This standard covers materials, dimensions, tolerances, and markings for metal ring-joint gaskets, spiral-wound metal gaskets, and metal jacketed gaskets and filler material. These gaskets are dimensionally suitable for use with flanges described in the reference flange standards ASME B16.5, ASME B16.47, and API-6A. This standard covers spiral-wound metal gaskets and metal jacketed gaskets for use with raised face and flat face flanges.

4.1.2 Metallic Gaskets for Pipe Flanges: Ring-Joint, Spiral-Wound, and Jacketed B16.20

Since 1922, ASME has been leading the effort to define safety in piping installations. This standard covers metal ring-joint gaskets, spiral-wound gaskets, metal-jacketed gaskets, and grooved metal gaskets with covering layers. It also covers spiral-wound metal gaskets and metal-jacketed gaskets for use with raised-face and flat-face flanges.

B16.20 offers comprehensive solutions applying to materials, dimensions, tolerances, and marking. It addresses gaskets that are dimensionally suitable for use with flanges described in reference flange standards ASME B16.5, ASME B16.47, API Specification 6A, and ISO 10423 as well as with other ASME standards, such as the Boiler and Pressure Vessel Code and the B31 Piping Codes. Notable revisions to the 2012 edition include a new chapter on Grooved Metal Gaskets with Covering Layers and updates to material tables.

Careful application of these B16 standards will help users comply with applicable regulations within their jurisdictions, while achieving the operational, cost, and safety benefits to be gained from the many industry best-practices detailed within these volumes.

4.1.3 Nonmetallic Flat Gaskets for Pipe Flanges B16.21

Upon request from industry and government, ASME has been defining piping safety since 1922.

This standard covers nonmetallic flat gaskets. It offers comprehensive solutions applying to materials, dimensions, tolerances, and marking.

B16.21 addresses gaskets that are dimensionally suitable for use with flanges described in reference flange standards ASME B16.5, ASME

B16.47, API Specification 6A, and ISO 10423 as well as with other ASME standards, such as the B&PVC and the B31 Piping Codes.

Careful application of these B16 standards will help users comply with applicable regulations within their jurisdictions, while achieving the operational, cost, and safety benefits to be gained from the many industry best-practices detailed within these volumes. These standards are intended for manufacturers, owners, employers, users, and others concerned with the specification, buying, maintenance, training, and safe use of nonmetallic gaskets with pressure equipment, plus all potential governing entities.

4.1.4 ASME/ANSI B16.5 Pipe Flanges and Flanged Fittings

The ASME B16.5 1996 Pipe Flanges and Flange Fittings standard covers pressure-temperature ratings, materials, dimensions, tolerances, marking, testing, and methods of designating openings for pipe flanges and flanged fittings.

The standard includes flanges with rating class designations 150, 300, 400, 600, 900, 1500, and 2500 in sizes NPS 1/2 through NPS 24, with requirements given in both metric and US units. The standard is limited to flanges and flanged fittings made from cast or forged materials, and blind flanges and certain reducing flanges made from cast, forged, or plate materials. Also included in this standard are requirements and recommendations regarding flange bolting, flange gaskets, and flange joints.

5. API STANDARDS AND RECOMMENDED PRACTICES

The **American Petroleum Institute**, commonly referred to as **API** (www.api.org) is the largest US trade association for the oil and natural gas industry. It claims to represent about 400 corporations involved in production, refinement, distribution, and many other aspects of the petroleum industry.

API was established on March 20, 1919:
- to afford a means of cooperation with the government in all matters of national concern
- to foster foreign and domestic trade in American petroleum products
- to promote in general the interests of the petroleum industry in all its branches
- to promote the mutual improvement of its members and the study of the arts and sciences connected with the oil and natural gas industry.

API maintains more than 500 documents that apply to many segments of the oil and gas industry, from drill bits to environmental protection. API standards advocate proven, sound engineering and operating practices and safe, interchangeable equipment, and materials.

API standards reference offshore production, drilling, structural pipe, pipeline, health and environmental issues, valves, and storage tanks, to name a few. API standards include manuals, standards, specifications, recommended practices, bulletins, guidelines, and technical reports.

API distributes thousands of copies of its publications each year. The publications, technical standards, and electronic and online products are designed, according to API, to help users improve the efficiency and cost-effectiveness of their operations, comply with legislative and regulatory requirements, and safeguard health, ensure safety, and protect the environment. Each publication is overseen by a committee of industry professionals, mostly member company engineers.

These technical standards tend to be uncontroversial. For example, API 610 is the specification for centrifugal pumps and API 675 is the specification for controlled volume positive displacement pumps and both packed-plunger and diaphragm types are included. Diaphragm pumps that use direct mechanical actuation are excluded. API 677 is the standard for gear units and API 682 governs mechanical seals.

API provides vessel codes and standards for the design and fabrication of pressure vessels that help safeguard the lives of people and environments all over the world.

API has entered petroleum industry nomenclature in a number of areas such as API gravity, a measure of the density of petroleum and API number, a unique identifier applied to each petroleum exploration or production well drilled in the United States.

6. MANUFACTURERS STANDARDIZATION SOCIETY

Officially founded in 1924, the Manufacturers Standardization Society (MSS) (www.mss-hq.org) of the valve and fittings industry is a nonprofit technical association organized for development and improvement of industry, national and international codes and standards for valves, fittings, flanges, and seals.

MSS provides its members the means to develop engineering standard practices for the use and benefit of the industry and users of its products. The society currently comprises 24 technical committees to write, revise, and reaffirm industry standards.

MSS also contributes to and monitors code and standard development activities worldwide. MSS, in cooperation with other standardizing bodies, appoints representatives to other standardization organization committees to share the views and objectives of the society.

Standards organizations in which MSS currently participates include, but are not limited to, the American Society of Mechanical Engineers (ASME), American National Standards Institute (ANSI), American Society for Testing and Materials (ASTM), American Petroleum Institute (API), American Water Works Association (AWWA), and National Fire Protection Association (NFPA).

6.1 MSS SP-44 Steel Pipeline Flanges

This standard practice covers pressure-temperature ratings, materials, dimensions, tolerances, marking, and testing. The welding neck type flanges shall be forged steel, and the blind flanges may be made of either forged steel or from steel plates.

Dimensional and tolerance requirements for sizes NPS 10 and smaller are provided by reference to ASME B16.5. When such flanges are made of materials meeting the requirements and meet all other stipulations of this standard, they shall be considered as complying therewith.

7. PIPE FABRICATION INSTITUTE STANDARDS

The Pipe Fabrication Institute (PFI) was formed in 1913. It is one of the oldest and most respected industry associations in the United States. It exists solely for the purpose of ensuring a level of quality in the pipe fabrication industry that is without compromise.

Over the years, PFI has fulfilled its charter by initiating engineering, studies, and research and proposing and maintaining suitable standards and technical bulletins and organizing meetings, technical exchanges, and presentations among members pertaining to materials and products, techniques of fabrication and operation, methods of examination and testing, and other topics.

8. AMERICAN INSTITUTE OF STEEL CONSTRUCTION

The American Institute of Steel Construction (AISC), is a not-for-profit technical institute and trade association established in 1921 to serve the structural steel design community and construction industry in the United States. AISC sets the standards to make structural steel the material of choice by being the leader in structural steel-related technical and market-building activities, including: specification and code development, research, education, technical assistance, quality certification, standardization, and market development.

For almost 90 years, AISC has conducted its numerous activities with a scrupulous sense of public responsibility. For this reason, and because of the high caliber of its staff, the Institute enjoys a close working relationship with architects, engineers, code officials, and educators who recognize its professional status in the fields of specification writing, structural research, design development, and performance standards.

AISC represents the total experience, judgment, and strength of the entire domestic industry of steel fabricators, distributors, and producers. The scope and success of its activities could not be achieved by any one member of the industry. The nation shares the rewards of these activities through better, safer, and more economical buildings, bridges, and other structures framed in structural steel.

9. AMERICAN CONCRETE INSTITUTE

The American Concrete Institute (ACI) is a nonprofit technical and educational society organized in 1904 and is one of the world's leading authorities on concrete technology. ACI is a forum for the discussion of all matters related to concrete and the development of solutions to problems. ACI conducts this forum through conventions and meetings; the *ACI Structural Journal*, the *ACI Materials Journal, Concrete International*, and technical publications; chapter activities; and technical committee work. As its mission states, its purpose is to "Provide knowledge and information for the best use of concrete."

This implies a willingness on the part of each member to contribute from his or her training and knowledge to the benefit of the public at large. By maintaining a high standard of professional and technical ability in its committee memberships and in the authorship of papers and publications, as well as in local chapter programs, ACI has contributed to a detailed knowledge of materials and their resulting structures.

ACI publishes reliable information on concrete and its applications, conducts educational seminars, provides a standard certification program for the industry, provides local forums for discussion through the chapter program, and encourages student involvement in the concrete field. Committee members involved with these activities meet at biannual conventions.

10. NATIONAL ASSOCIATION OF CORROSION ENGINEERS

NACE International is the world's leading professional organization for the corrosion control industry established in 1943. NACE was formerly known as the National Association of Corrosion Engineers. The founding engineers were originally part of a regional group formed in the 1930s when the study of cathodic protection was introduced. Since then, NACE International has become the global leader in developing corrosion prevention and control standards, certification and education. The members of NACE International still include engineers as well as numerous other professionals working in a range of areas related to corrosion control.

NACE International, The Corrosion Society, serves nearly 30,000 members in 116 countries and is recognized globally as the premier authority for corrosion control solutions. The organization offers technical training and certification programs, conferences, industry standards, reports, publications, technical journals, government relations activities and more. For more details check NACE Website: http://www.nace.org

NACE International is involved in every industry and area of corrosion prevention and control, from chemical processing and water systems, to transportation and infrastructure protection. NACE's main focus of activities includes cathodic protection, coatings for industry and material selection for specific chemical resistance.

11. FLUID CONTROL INSTITUTE STANDARDS

The Fluid Control Institute (FCI) is an association of manufacturers of equipment for fluid (liquid or gas) control and conditioning. The institute is organized into product-specific sections which address issues that are relevant to particular products and/or technologies.

FCI provides standards and other materials to assist purchasers and users in understanding and using fluid control and conditioning equipment. All FCI standards are voluntary, and most are submitted to ANSI for approval. FCI standards address issues relevant to particular products such as control valves, instruments, pipeline strainers, regulators, and solenoid valves.

12. HYDRAULICS INSTITUTE PUMP STANDARDS

The Hydraulic Institute (HI) publishes product standards for the North American pump industry. It is involved in the development of pump standards in North America and worldwide and the standards are developed within guidelines established by ANSI. Members work through a number of technical committees to develop draft standards. HI standards help define pump products, installation, operation, performance, testing, and pump life and quality. ANSI/HI standards are widely referenced in other standards such as those of API, AWWA, ASME B73 are accepted throughout North America and applied worldwide.

The ANSI/HI pump standards are specifically designed for use by pump users, consultants, engineering contractors, and manufacturers of pumps, pumping systems, and pump system integrators.

CHAPTER THREE

Physical Properties

1. PROPERTIES OF LIQUIDS AND GASES

In this chapter, we will discuss the various units of measurement employed in transmission pipeline hydraulics and the properties of liquids and gases that are significant and important properties of that affect hydraulic calculations. The importance of specific gravity, viscosity of pure liquids and gases and mixtures will be analyzed, and the concepts will be illustrated with sample problems. This chapter forms the foundation for all calculations involving pipeline pressure drops and horsepower requirements in subsequent chapters. In the tables, you will find the listing of properties of commonly used liquids and gases such as water and petroleum products and compressible gases such as methane, ethane as well as carbon dioxide that are pumped through transmission pipelines.

Common properties of petroleum fluids

Product	Viscosity cSt @ 60 °F	°API gravity	Specific gravity @ 60 °F	Reid vapor pressure
Regular gasoline				
Summer grade	0.70	62.0	0.7313	9.5
Interseasonal grade	0.70	63.0	0.7275	11.5
Winter grade	0.70	65.0	0.7201	13.5
Premium gasoline				
Summer grade	0.70	57.0	0.7467	9.5
Interseasonal grade	0.70	58.0	0.7165	11.5
Winter grade	0.70	66.0	0.7711	13.5
No. 1 fuel oil	2.57	42.0	0.8155	
No. 2 fuel oil	3.90	37.0	0.8392	
Kerosene	2.17	50.0	0.7796	
Jet fuel JP-4	1.40	52.0	0.7711	2.7
Jet fuel JP-5	2.17	44.5	0.8040	

API, American Petroleum Institute.

Specific gravity and American Petroleum Institute (API) gravity

Liquid	Specific gravity @ 60 °F	API gravity @ 60 °F
Propane	0.5118	N/A
Butane	0.5908	N/A
Gasoline	0.7272	63.0
Kerosene	0.7796	50.0
Diesel	0.8398	37.0
Light crude	0.8348	38.0
Heavy crude	0.8927	27.0
Very heavy crude	0.9218	22.0
Water	1.0000	10.0

2. UNITS OF MEASUREMENT

Before we discuss liquid properties, it would be appropriate to identify the different units of measurement used in pipeline hydraulics calculations.

Over the years, the English-speaking world adopted so-called "English units" of measurement, whereas most other European and Asian countries and South American countries adopted the "metric system of units."

The English system of units (referred to in the United States as customary US units) derives from the old foot-pound-second (FPS) and foot-slug-second (FSS) system that originated in England. The basic units are foot for length, slug for mass, and second for measurement of time. In the past, the FPS system used pound for mass. Because force, a derived unit, was also measured in pounds, there was evidently some confusion. To clarify the term pound-mass (lbm) and pound-force (lbf) were introduced. Numerically, the weight (which is a force resulting from gravity) of 1 lbm was equal to 1 lbf. However, the introduction of slug for unit of mass resulted in the adoption of pound exclusively for unit of force. Thus, in the FSS system which is now used in the United States, the unit of mass is slug. The relationship between a slug, lbf, and lbm will be explained later in this chapter.

In the metric system, originally known as centimeter-gram-second (CGS) system, the corresponding units for length, mass, and time were centimeter, gram, and second, respectively. In later years, modified metric units called meter-kilogram-second (MKS) system emerged. In MKS units, the meter was used for the measurement of length and kilogram for the measurement of mass. The measurement for time remained the second for all systems of units.

The scientific and engineering communities, during the past four decades have attempted to standardize on a universal system of units worldwide. Through the International Standards Organization, a policy for an International System of Units was formulated. These units are also known as "Systeme Internationale" (SI) units.

The conversion from the older system of units to SI units has advanced at different rates in different countries. Most countries of Western Europe and all of Eastern Europe, Russia, India, China, Japan, Australia, New Zealand, and South America have adopted the SI units completely. In North America, Canada and Mexico have adopted the SI units almost completely. However, engineers and scientists in these countries use both SI units and English units because of their business dealings with the United States. In the United States, SI units are used increasingly in colleges and the scientific community. However, the majority of work is still done using the English units referred to sometimes as customary US units.

The Metric Conversion Act of 1975 accelerated the adoption of the SI system of units in the United States. The American Society of Mechanical Engineers (ASME), American Society of Civil Engineers, and other professional societies and organizations have assisted in the process of conversion from English to SI units using the respective Institutions publications. For example, ASME through the ASME Metric Study Committee published a series of articles in its *Mechanical Engineering* magazine to help engineers master the SI system of units.

In the United States, the complete changeover to SI has not materialized fast enough. Therefore in this transition phase, engineering students, practicing engineers, technicians, and scientists must be familiar with the different systems of units such as English, metric CGS, metric MKS, and the SI units. In this book, we will use both English units (customary US) and the SI system of units.

Units of measurement are generally divided into three classes as follows:
Base units
Supplementary units
Derived units.

By definition, base units are dimensionally independent. These are units of length, mass, time, electric current, temperature, amount of substance, and luminous intensity.

Supplementary units are those used to measure plain angles and solid angles. Examples include radian and steradian.

Derived units are those that are formed by combination of base units, supplementary units, and other derived units. Examples of derived units are those of force, pressure, and energy.

2.1 Base Units

In the English (customary US) system of units, the following base units are used.

Length	Foot (ft)
Mass	Slug (slug)
Time	Second (s)
Electric current	Ampere (A)
Temperature	Degree Fahrenheit (°F)
Amount of substance	Mole (mol)
Luminous intensity	Candela (cd)

In SI units, the following base units are defined.

Length	Meter (m)
Mass	Kilogram (kg)
Time	Second (s)
Electric current	Ampere (A)
Temperature	Kelvin (K)
Amount of substance	Mole (mol)
Luminous intensity	Candela (cd)

2.2 Supplementary Units

Supplementary units in both English and SI system of units are as follows

Plain angle	Radian (rad)
Solid angle	Steradian (sr)

The radian is defined as the plain angle between two radii of a circle with an arc length equal to the radius. Thus, it represents the angle of a sector of a circle with the arc length the same as its radius.

The steradian is the solid angle having its apex at the center of a sphere such that the area of the surface of the sphere that it cuts out is equal to that of a square with sides equal to the radius of this sphere.

2.3 Derived Units

Derived units are generated from a combination of base units, supplementary units, and other derived units. Examples of derived units include those of area, volume, and so on.

In English units the following derived units are used:

Area	Square inches (in^2), square feet (ft^2)
Volume	Cubic inch (in^3), cubic feet (ft^3), gallons (gal), and barrels (bbl)
Speed/velocity	Feet per second (ft/s)
Acceleration	Feet per second per second (ft/s^2)
Density	Slugs per cubic foot (slugs/ft^3)
Specific weight	Pound per cubic foot (lb/ft^3)
Specific volume	Cubic foot per pound (ft^3/lb)
Dynamic viscosity	Pound second per square foot (lb-s/ft^2)
Kinematic viscosity	Square foot per second (ft^2/s)
Force	Pounds (lb)
Pressure	Pounds per square inch (lb/in^2)
Energy/work	Foot pound (ft lb)
Quantity of heat	British thermal units (Btu)
Power	Horsepower (HP)
Specific heat	Btu per pound per °F (Btu/lb/°F)
Thermal conductivity	Btu per hour per foot per °F (Btu/h/ft/°F)

In SI units the following derived units are used:

Area	Square meter (m^2)
Volume	Cubic meter (m^3)
Speed/velocity	Meter/second (m/s)
Acceleration	Meter per second per second (m/s^2)
Density	Kilogram per cubic meter (kg/m^3)
Specific volume	Cubic meter per kilogram (m^3/kg)
Dynamic viscosity	Pascal second (Pa s)
Kinematic viscosity	Square meters per second (m^2/s)
Force	Newton (N)
Pressure	Newton per square meter or Pascal (Pa)
Energy/work	Newton meter or Joule (J)
Quantity of heat	Joule (J)
Power	Joule per second or Watt (W)
Specific heat	Joule per kilogram per K (J/kg/K)
Thermal conductivity	Joule/second/meter/Kelvin (J/s/m/K)

Many other derived units are used in both English and SI units. A list of the more commonly used units in liquid pipeline hydraulics and their conversions are listed in the Appendix.

3. MASS, VOLUME, DENSITY, AND SPECIFIC WEIGHT

Several properties of liquids that affect liquid pipeline hydraulics will be discussed here. In steady-state hydraulics of liquid pipelines, the following properties are important.

3.1 Mass

Mass is defined as the quantity of matter. It is independent of temperature and pressure. Mass is measured in slugs in English units or kilograms (kg) in SI units. In the past, mass was used synonymously with weight. Strictly speaking, weight depends on acceleration because of gravity at a certain geographic location and therefore is considered to be a force. Numerically, mass and weight are interchangeable in the older FPS system of units. For example, a mass of 10 lbm is equivalent to a weight of 10 lbf. To avoid this confusion, in English units, the slug has been adopted for unit of mass. One slug is equal to 32.17 lb. Therefore, if a drum contains 55 gal of crude oil and weighs 410 lb, the mass of oil will be the same at any temperature and pressure. Hence the statement "conservation of mass."

3.2 Volume

Volume is defined as the space occupied by a given mass. In the case of the 55-gallon drum discussed previously, 410 lb of crude oil occupy the volume of the drum. Therefore the crude oil volume is 55 gal. Consider a solid block of ice measuring 12 in on each side. The volume of this block of ice is $12 \times 12 \times 12$ or 1728 in^3 or 1 ft^3. The volume of a certain petroleum product contained in a circular storage tank 100 ft in diameter and 50 ft high, may be calculated as follows, assuming the liquid depth is 40 ft:

$$\text{Liquid volumes} = (\pi/4) \times 100 \times 100 \times 40 = 314{,}160 \, ft^3$$

Liquids are practically incompressible, take the shape of their container, and have a free surface. Volume of a liquid varies with temperature and pressure. However for liquids, being practically incompressible, pressure has negligible effect on volume. Thus, if the liquid volume measured at 50 psi is 1000 gal, its volume at a 1000 psi will not be appreciably different, provided the liquid temperature remained constant. Temperature, however, has a more significant effect on volume. For example, the 55-gal volume of liquid in a drum at a temperature of 60 °F will increase to a slightly higher value (such as 56 gal) when the liquid temperature increases to 100 °F. The amount of increase in volume per unit temperature rise depends on the coefficient of expansion of the liquid. When measuring petroleum liquids, for the purpose of custody transfer, it is customary to correct volumes to a fixed temperature such as 60 °F. Volume correction factors from American Petroleum Institute (API) publications are commonly used in the petroleum industry.

In the petroleum industry, it is customary to measure volume in gallons or barrels. One barrel is equal to 42 US gallons. The Imperial gallon as used in the United Kingdom is a larger unit, approximately 20% larger than the US gallon. In SI units, volume is generally measured in cubic meters (m^3) or liters (L).

In a pipeline transporting crude oil or refined petroleum products, it is customary to talk about the "line fill volume" of the pipeline. The volume of liquid contained between two valves in a pipeline can be calculated simply by knowing the internal diameter of the pipe and the length of pipe between the two valves. By extension, the total volume or the line fill volume of the pipeline can be easily calculated.

As an example, if a 16-in pipeline, 0.250-in wall thickness is 5000 ft long from one valve to another, the line fill for this section of pipeline is.

$$\text{Line fill volume} = (\pi/4) \times (16 - 2 \times 0.250)^2 \times 5000$$
$$= 943{,}461.75 \text{ ft}^3 \text{ or } 168{,}038 \text{ bbl}$$

This calculation is based on conversion factors of:
1728 in^3 per ft^3
231 in^3 per gallon
and 42 gallons per barrel.

In a later chapter, we will discuss a simple formula for determining the line fill volume of a pipeline.

The volume flow rate in a pipeline is generally expressed in terms of cubic feet per second (ft^3/s), gallons per minute (gal/min), barrels per hour (bbl/h), and barrels per day (bbl/day) in customary English units. In the SI units, volume flow rate is referred to in cubic meters per hour (m^3/h) and liters per second (L/s).

It must be noted that because the volume of a liquid varies with temperature, the inlet flow rate and the outlet volume flow rate may be different in a long distance pipeline, even with no intermediate injections or deliveries. This is because the inlet flow rate may be measured at an inlet temperature of 70 °F to be 5000 bbl/h and the corresponding flow rate at the pipeline terminus, 100 miles away may be measured at an outlet temperature different than the inlet temperature. The temperature difference is due to heat loss or gain between the pipeline liquid and the surrounding soil or ambient conditions. Generally, significant variation in temperature is observed when pumping crude oils or other products that are heated at the pipeline inlet. In refined petroleum products and other pipelines that are not heated, temperature variations along the pipeline are insignificant. In any case if the volume measured at the pipeline inlet is corrected to a standard temperature such as 60 °F, the corresponding outlet volume can also be corrected to the same standard temperature. With temperature correction, it can be assumed that the same flow rate exists throughout the pipeline from inlet to outlet provided of course there are no intermediate injections or deliveries along the pipeline.

By the principle of conservation of mass, the mass flow rate at inlet will equal that at the pipeline outlet because the mass of liquid does not change with temperature or pressure.

3.3 Density

Density of a liquid is defined as the mass per unit volume. Customary units for density are slugs/ft^3 in the English units. The corresponding unit of density in SI units is kg/m^3. This is also referred to as mass density. The weight density is defined as the weight per unit volume. This term is more commonly called specific weight and will be discussed next.

Because mass does not change with temperature or pressure, but volume varies with temperature, we can conclude that density will vary with temperature. Density and volume are inversely related because density is defined as mass per unit volume. Therefore, with increase in temperature liquid volume increases while its density decreases. Similarly, with reduction in temperature, liquid volume decreases and its density increases.

3.4 Specific Weight

Specific weight of a liquid is defined as the weight per unit volume. It is measured in lb/ft^3 in English units and N/m^3 in SI units.

If a 55-gal drum of crude oil weighs 410 pounds (excluding weight of drum), the specific weight of crude oil is.

$$(410/55) \text{ or } 7.45 \text{ lb/gal.}$$

Similarly, consider the 5000-ft pipeline discussed in Section 3.2. The volume contained between the two valves was calculated to be 168,038 bbl. If we use specific weight calculated previously, we can estimate the weight of liquid contained in the pipeline as

$$7.45 \times 42 \times 168{,}038 = 52{,}579{,}090 \text{ lb} \quad \text{or} \quad 26{,}290 \text{ tons}$$

Similar to density, specific weight varies with temperature. Therefore, with increase in temperature specific weight will decrease. With reduction in temperature, liquid volume decreases and its specific weight increases.

Customary units for specific weight are lb/ft^3 and lb/gal in the English units. The corresponding unit of specific weight in SI units is N/m^3.

For example, water has a specific weight of 62.4 lb/ft^3 or 8.34 lb/gal at 60 °F. A typical gasoline has a specific weight of 46.2 lb/ft^3 or 6.17 lb/gal at 60 °F.

Although density and specific weight are dimensionally different, it is common to use the term *density* instead of specific weight and vice versa when calculating hydraulics of liquid pipelines. Thus you will find that the density of water and specific weight of water are both expressed as 62.4 lb/ft^3.

4. SPECIFIC GRAVITY AND API GRAVITY

Specific gravity of a liquid is the ratio of its density to the density of water at the same temperature and therefore has no units (dimensionless). It is a measure of how heavy a liquid is compared with water.

The term *relative density* is also used to compare the density of a liquid with another liquid such as water. In comparing the densities, it must be noted that both densities must be measured at the same temperature to be meaningful.

At 60 °F, a typical crude oil has a density of 7.45 lb/gal compared with a water density of 8.34 lb/gal. Therefore, the specific gravity of crude oil at 60 °F is

$$\text{Specific gravity} = 7.45/8.34 \quad \text{or} \quad 0.8933.$$

By definition, the specific gravity of water is 1.00 because the density of water compared with itself is the same. Specific gravity, like density, varies with temperature. As temperature increases, both density and specific gravity decrease. Similarly, decrease in temperature causes the density and specific gravity to increase in value. As with volume, pressure has very little effect on liquid specific gravity as long as pressures are within the range of most pipeline applications.

In the petroleum industry, it is customary to use units of °API for gravity. The API gravity is a scale of measurement using API = 10 on the low end for water at 60 °F. All liquids lighter than water will have API values higher than 10. Thus gasoline has an API gravity of 60, whereas a typical crude oil may be 35°API.

The API gravity of a liquid is a value determined in the laboratory comparing the density of the liquid versus the density of water at 60 °F. If the liquid is lighter than water its API gravity will be greater than 10.

The API gravity versus the specific gravity relationship is as follows

$$\text{Specific gravity } Sg = 141.5/(131.5 + API) \qquad (3.1)$$

or

$$API = 141.5/Sg - 131.5 \qquad (3.2)$$

Substituting an API value of 10 for water in Eqn (3.1), yields as expected the specific gravity of 1.00 for water. It is seen from this equation that the specific gravity of the liquid cannot be greater than 1.076 to result in a positive value of API.

Another scale of gravity for liquids heavier than water is known as the Baume scale. This scale is similar to the API scale with the exception of 140 and 130 being used in place of 141.5 and 131.5, respectively, in Eqns (3.1) and (3.2).

As another example, assume the specific gravity of gasoline at 60 °F is 0.736. Therefore, the API gravity of gasoline can be calculated from Eqn (3.2) as follows:

$$\text{API gravity} = 141.5/0.736 - 131.5 = 60.76°API$$

If diesel fuel is reported to have an API gravity of 35, the specific gravity can be calculated from Eqn (3.1) as follows:

$$\text{Specific gravity} = 141.5/(131.5 + 35) = 0.8498$$

API gravity is always referred to at 60 °F. Therefore in the Eqns (3.1) and (3.2), specific gravity must also be measured at 60 °F. Hence, it is meaningless to say that the API of a liquid is 35°API at 70 °F.

API gravity is measured in the laboratory using the method described in ASTM D1298 using a properly calibrated glass hydrometer. Also refer to API Manual of Petroleum Measurements for further discussion on API gravity.

Physical Properties

4.1 Specific Gravity Variation with Temperature

It was mentioned previously that the specific gravity of a liquid varies with temperature. It increases with decrease in temperature and vice versa.

For commonly encountered range of temperatures in liquid pipelines, the specific gravity of a liquid varies linearly with temperature. In other words, the specific gravity versus temperature can be expressed in the form of the following equation.

$$S_T = S_{60} - a(T - 60) \tag{3.3}$$

where

S_T – Specific gravity at temperature T
S_{60} – Specific gravity at 60 °F
T – Temperature, °F
a – A constant that depends on the liquid.

In Eqn (3.3), the specific gravity S_T at temperature T is related to the specific gravity at 60 °F by a straight line relationship. Because the terms S_{60} and a are unknown quantities, two sets of specific gravities at two different temperatures are needed to determine the specific gravity versus temperature relationship. If the specific gravity at 60 °F and the specific gravity at 70 °F are known, we can substitute these values in Eqn (3.3) to obtain the unknown constant a. Once the value of a is known, we can easily calculate the specific gravity of the liquid at any other temperature using Eqn (3.3). An example will illustrate how this is done.

Some handbooks such as *Hydraulic Institute Engineering Design* book and the *Crane Handbook* provide specific gravity versus temperature curves from which the specific gravity of most liquids can be calculated at any temperature.

Example Problem 3.1

The specific gravity of gasoline at 60 °F is 0.736. The specific gravity at 70 °F is 0.729. What is the specific gravity at 50 °F?

Solution

Using Eqn (3.3), we can write

$$0.729 = 0.736 - a(70 - 60)$$

Solving for a, we get

$$a = 0.0007$$

We can now calculate the specific gravity at 50 °F using Eqn (3.3) as

$$S_{50} = 0.736 - 0.0007(50 - 60) = 0.743$$

4.2 Specific Gravity of Blended Liquids

Suppose a crude oil of specific gravity 0.895 at 70 °F is blended with a lighter crude oil of specific gravity 0.815 at 70 °F, in equal volumes. What will be the specific gravity of the blended mixture? Common sense suggests that, because equal volumes are used, the resultant mixture should have a specific gravity of the average of the two liquids or

$$(0.895 + 0.815)/2 = 0.855$$

This is indeed the case, because specific gravity of a liquid is simply related to the mass and the volume of each liquid.

When two or more liquids are mixed homogenously, the resultant liquid specific gravity can be calculated using weighted average method. Thus, 10% of liquid A with specific gravity of 0.85 when blended with 90% of liquid B that has a specific gravity of 0.89 results in a blended liquid with specific gravity of

$$(0.1 \times 0.85) + (0.9 \times 0.89) = 0.886$$

It must be noted that when performing these calculations, both specific gravities must be measured at the same temperature.

Using the previous approach, the specific gravity of a mixture of two or more liquids can be calculated from the following equation:

$$S_b = [(Q_1 \times S_1) + (Q_2 \times S_2) + (Q_3 \times S_3) + ...]/[Q_1 + Q_2 + Q_3 + ...]$$
(3.4)

where

S_b – Specific gravity of the blended liquid
Q_1, Q_2, Q_3 etc. – Volume of each component
S_1, S_2, S_3 etc. – Specific gravity of each component.

This method of calculating the specific gravity of a mixture of two or more liquids cannot be directly applied when the gravities are expressed in °API values. If the component gravities of a mixture are given in °API we must first convert API values to specific gravities before applying Eqn (3.4).

Physical Properties

Example Problem 3.2
Three liquids A, B, and C are blended together in the ratio of 15%, 20%, and 65%, respectively. Calculate the specific gravity of the blended liquid if the individual liquids have the following specific gravities at 70 °F:

Solution
Specific gravity of liquid A: 0.815.
Specific gravity of liquid B: 0.850.
Specific gravity of liquid C: 0.895.
Using Eqn (3.4), we get the blended liquid specific gravity as

$$S_b = (15 \times 0.815 + 20 \times 0.850 + 65 \times 0.895)/100 = 0.874$$

5. VISCOSITY

Viscosity is a measure of sliding friction between successive layers of a liquid that flows in a pipeline. Imagine several layers of liquid that constitute a flow between two fixed parallel horizontal plates. A thin layer adjacent to the bottom plate will be at rest or zero velocity. Each subsequent layer above this will have a different velocity compared to the layer below. This variation in the velocity of the liquid layers results in a velocity gradient. If the velocity is V at the layer that is located a distance of y from the bottom plate, the velocity gradient is approximately

$$\text{Velocity gradient} = V/y \qquad (3.5)$$

If the variation of velocity with distance is not linear, using calculus we can write more accurately that

$$\text{Velocity gradient} = dV/dy \qquad (3.6)$$

where dV/dy represents the rate of change of velocity with distance or the velocity gradient.

Newton's law states that the shear stress between adjacent layers of a flowing liquid is proportional to the velocity gradient. The constant of proportionality is known as the absolute (or dynamic) viscosity of the liquid.

$$\text{Shear stress} = (\text{Viscosity})(\text{Velocity gradient})$$

Absolute viscosity of a liquid is measured in lb-s/ft² in English units and Pascal-s in SI units. Other commonly used units of absolute viscosity are poise and centipoise (cP).

The kinematic viscosity is defined as the absolute viscosity of a liquid divided by its density at the same temperature.

$$\nu = \mu/\rho \qquad (3.7)$$

where
ν = Kinematic viscosity
μ = Absolute viscosity
ρ = Density.

The units of kinematic viscosity are ft²/s in English units and m²/s in SI units. See Appendix A for conversion of units. Other commonly used units for kinematic viscosity are Stokes and centistokes (cSt). In the petroleum industry, two other units for kinematic viscosity are also used. These are Saybolt Seconds Universal (SSU) and Saybolt Seconds Furol (SSF). When expressed in these units, it represents the time taken for a fixed volume of a liquid to flow through an orifice of defined size. Both absolute and kinematic viscosities vary with temperature. As temperature increases, liquid viscosity decreases and vice versa. However, unlike specific gravity, viscosity versus temperature is not a linear relationship. We will discuss this in the next session.

Viscosity also varies somewhat with pressures. Significant variations in viscosity are found when pressures are several thousand psi. In most pipeline applications, viscosity of a liquid does not change appreciably with pressure.

For example, the viscosities of Alaskan North Slope crude oil may be reported as 200 SSU at 60 °F and 175 SSU at 70 °F. Viscosity in SSU and SSF may be converted to their equivalent in centistokes using the following equations:

Conversion from SSU to centistokes

$$\text{Centistokes} = 0.226(\text{SSU}) - 195/(\text{SSU}) \quad \text{for } 32 \leq \text{SSU} \leq 100 \qquad (3.8)$$

$$\text{Centistokes} = 0.220(\text{SSU}) - 135/(\text{SSU}) \quad \text{for SSU} > 100 \qquad (3.9)$$

Conversion from SSF to centistokes

$$\text{Centistokes} = 2.24(\text{SSF}) - 184/(\text{SSF}) \quad \text{for } 25 < \text{SSF} \leq 40 \qquad (3.10)$$

$$\text{Centistokes} = 2.16(\text{SSF}) - 60/(\text{SSF}) \quad \text{for SSU} > 40 \qquad (3.11)$$

Example Problem 3.3

Let us use these equations to convert viscosity of Alaskan North Slope crude oil from 200 SSU to its equivalent in centistokes.

Solution
Using Eqn (3.9)

$$\text{Centistokes} = 0.220 \times 200 - 135/200 = 43.33 \text{ cSt}$$

The reverse process of converting from viscosity in cSt to its equivalent in SSU using Eqns (3.8) and (3.9) is not quite so direct. Because Eqns (3.8) and (3.9) are valid for certain range of SSU values, we need to first determine which of the two equations to use. This is difficult because the equation to use depends on the SSU value, which itself is unknown. Therefore we will have to assume that the SSU value to be calculated falls in one of the two ranges shown and proceed to calculate by trial and error. We will have to solve a quadratic equation to determine the SSU value for a given viscosity in cSt. An example will illustrate this method.

Example Problem 3.4

Suppose we are given a liquid viscosity of 15 cSt and we are required to calculate the corresponding viscosity in SSU.

Solution
Let us assume that the calculated value in SSU is approximately $5 \times 15 = 75$ SSU. This is a good approximation because the SSU value generally is about 5 times the corresponding viscosity value in cSt. Because the assumed SSU value is 75, we need to use Eqn (3.8) for converting between cSt and SSU.

Substituting 15 cSt in Eqn (3.8) gives

$$15 = 0.226(\text{SSU}) - 195/(\text{SSU})$$

replacing SSU with variable x, this equation become, after transposition

$$15x = 0.226x^2 - 195$$

rearranging we get

$$0.226x^2 - 15x - 195 = 0$$

Solving for x, we get

$$x = \left[15 + (15 \times 15 + 4 \times 0.226 \times 195)^{1/2}\right] / (2 \times 0.226) = 77.5$$

Therefore, the viscosity is 77.5 SSU.

5.1 Viscosity Variation with Temperature

The viscosity of a liquid decreases as the liquid temperature increases and vice versa. For gases, the viscosity increases with temperature. Thus, if the viscosity of a liquid at 60 °F is 35 cSt, as the temperature increases to 100 °F, the viscosity could drop to a value of 15 cSt. The variation of liquid viscosity with temperature is not linear, unlike specific gravity variation with temperature discussed in a previous section. The viscosity temperature variation is found to be logarithmic in nature.

Mathematically, we can state the following:

$$\text{Log}_e(v) = A - B(T) \tag{3.12}$$

where
  v – Viscosity of liquid, cSt
  T – Absolute temperature, °R or °K

$$T = (t + 460)\,°R \quad \text{if temperature t is in °F} \tag{3.13}$$

$$T = (t + 273)\,°K \quad \text{if temperature t is in °C} \tag{3.14}$$

A and B are constants that depend on the specific liquid.

It can be seen from Eqn (3.12) that a graphic plot of $\text{Log}_e(v)$ against the temperature T will result in a straight line with a slope of B. Therefore, if we have two sets of viscosity versus temperature for a liquid we can determine the values of A and B by substituting the viscosity, temperature values in Eqn (3.12). Once A and B are known we can calculate the viscosity at any other temperature using Eqn (3.12). An example will illustrate this.

Example Problem 3.5

Suppose we are given the viscosities of a liquid at 60 °F and 100 °F as 43 cSt and 10 cSt. We will use Eqn (3.12) to calculate the values of A and B first.

$$\text{Log}_e(43) = A - B(60 + 460)$$

and

$$\text{Log}_e(10) = A - B(100 + 460)$$

Solution

Solving the above two equations for A and B results in

$$A = 22.7232 \quad B = 0.0365$$

Having found A and B, we can now calculate the viscosity of this liquid at any other temperature using Eqn (3.12). Let us calculate the viscosity at 80 °F

$$\text{Log}_e(v) = 22.7232 - 0.0365\,(80 + 460) = 3.0132$$

Viscosity at 80 °F = 20.35 cSt

In addition to Eqn (3.12), several researchers have put forth various equations that attempt to correlate viscosity variation of petroleum liquids with temperature. The most popular and accurate of the formulas is the one known as the ASTM method. In this method, also known as the ASTM D341 chart method, a special graph paper with logarithmic scales is used to plot the viscosity of a liquid at two known temperatures. Once the two points are plotted on the chart and a line drawn connecting them, the viscosity at any intermediate temperature can be interpolated. To some extent, values of viscosity may also be extrapolated from this chart. This is shown in Figure 3.1

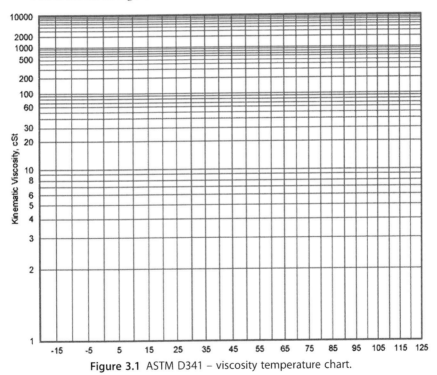

Figure 3.1 ASTM D341 – viscosity temperature chart.

In the following paragraphs, we will discuss how to calculate the viscosity variations with temperature, using the ASTM method, without using the special logarithmic graph paper.

$$\text{Log Log}(Z) = A - B \text{ Log}(T) \qquad (3.15)$$

where

Log is the logarithm to base 10
Z — depends on viscosity of the liquid v
v — Viscosity of liquid, cSt
T — Absolute temperature, °R or °K
A and B are constants that depend on the specific liquid.
The variable Z is defined as follows:

$$Z = (v + 0.7 + C - D) \qquad (3.16)$$

where C and D are

$$C = \exp[-1.14883 - 2.65868(v)] \qquad (3.17)$$

$$D = \exp[-0.0038138 - 12.5645(v)] \qquad (3.18)$$

C, D, and Z are all functions of the kinematic viscosity v.

Given two sets of temperature viscosity values (T_1, v_1) and (T_2, v_2), we can calculate the corresponding values of C, D, and Z from Eqns (3.16)–(3.18).

We can then come up with two equations using the pairs of (T_1, Z_1) and (T_2, Z_2) values by substituting these values into Eqn (3.15) as shown below:

$$\text{Log Log}(Z_1) = A - B \text{ Log}(T_1) \qquad (3.19)$$

$$\text{Log Log}(Z_2) = A - B \text{ Log}(T_2) \qquad (3.20)$$

From these equations, the two unknown constants A and B can be easily calculated, because T_1, Z_1 and T_2, Z_2 values are known.

The following example will illustrate this approach for viscosity temperature variation.

Example Problem 3.6

A certain liquid has a temperature versus viscosity relationship as given here:

Temperature, °F	60	180
Viscosity, cSt	750	25

1. Calculate the constants A and B that define the viscosity versus temperature correlation for this liquid using Eqn (3.15).
2. What is the estimated viscosity of this liquid at 85 °F?

Solution
1. At the first temperature 60 °F
C, D, and Z are calculated using Eqns (3.16)–(3.18)

$$C_1 = \exp[-1.14883 - 2.65868 \times 750] = 0$$

$$D_1 = \exp[-0.0038138 - 12.5645 \times 750] = 0$$

$$Z_1 = (750 + 0.7) = 750.7$$

Similarly, at the second temperature of 180 °F, the corresponding values of C, D, and Z are calculated to be

$$C_2 = \exp[-1.14883 - 2.65868 \times 25] = 0$$

$$D_2 = \exp[-0.0038138 - 12.5645 \times 25] = 0$$

$$Z_2 = (25 + 0.7) = 25.7$$

Substituting in Eqn (3.19), we get

$$\text{Log Log}(750.7) = A - B \text{ Log}(60 + 460)$$

or

$$0.4587 = A - 2.716B$$

$$\text{Log Log}(25.7) = A - B \text{ Log}(180 + 460)$$

or

$$0.1492 = A - 2.8062B$$

Solving for A and B, we get

$$A = 9.778$$

$$B = 3.4313$$

2. At temperature of 85 °F using Eqn (3.15), we get

$$\text{Log Log}(Z) = A - B \text{ Log}(85 + 460)$$

$$\text{Log Log}(Z) = 9.778 - 3.4313 \times 2.7364 = 0.3886$$

$$Z = 279.78$$

Therefore

$$\text{Viscosity at 85 °F} = 279.78 - 0.7 = 279.08 \text{ cSt}$$

5.2 Viscosity of Blended Products

Suppose a crude oil of viscosity 10 cSt at 60 °F is blended with a lighter crude oil of viscosity 30 cSt at 60 °F, in equal volumes. What will be the viscosity of the blended mixture? We cannot average the viscosities as we did with specific gravities blending earlier. This is due to the nonlinear nature of viscosity with mass and volumes of liquids.

When blending two or more liquids, the specific gravity of the blended product can be calculated directly, by using the weighted average approach as demonstrated in an earlier section. However, the viscosity of a blend of two or more liquids cannot be calculated by simply using the ratio of each component. Thus if 20% of liquid A of viscosity 10 cSt is blended with 80% of liquid B with a viscosity of 30 cSt, the blended viscosity is *not* the following.

$$0.2 \times 10 + 0.8 \times 30 = 26 \text{ cSt}$$

In fact, the actual blended viscosity would be 23.99 cSt, as will be demonstrated in the following section.

The viscosity of a blend of two or more products can be estimated using the following equation:

$$\sqrt{V_b} = \frac{[Q_1 + Q_2 + Q_3 + \ldots]}{(Q_1/\sqrt{V_1}) + (Q_2/\sqrt{V_2}) + (Q_3/\sqrt{V_3}) + \ldots} \qquad (3.21)$$

where

V_b – Viscosity of blend, SSU

Q_1, Q_2, Q_3 etc. – Volumes of each component

V_1, V_2, V_3 etc. – Viscosity of each component, SSU.

Because Eqn (3.19) requires the component viscosities to be in SSU, we cannot use this equation to calculate the blended viscosity when viscosity is less than 32 SSU (1.0 cSt).

Another method of calculating the viscosity of blended products has been in use in the pipeline industry for over four decades. This method is referred to as the Blending Index method. In this method, a Blending Index is calculated for each liquid based on its viscosity. Next the Blending Index of the mixture is calculated from the individual blending indices by using the weighted average of the composition of the mixture. Finally, the viscosity of the blended mixture is calculated using the Blending Index of the mixture. The equations used are described here:

$$H = 40.073 - 46.414 \, Log_{10} \, Log_{10} (V + B) \quad (3.22)$$

$$B = 0.931 \, (1.72)^V \quad \text{for } 0.2 < V < 1.5 \quad (3.23)$$

$$B = 0.6 \text{ for } V \geq 1.5 \quad (3.24)$$

$$Hm = [H1(pct1) + H2(pct2) + H3(pct3) + \ldots]/100 \quad (3.25)$$

where
H, H1, H2… – Blending index of liquids
Hm – Blending index of mixture
B – Constant in Blending Index equation
V – Viscosity in centistokes
pct1, pct2,… – Percentage of liquids 1, 2,… in blended mixture.

Example Problem 3.7
Calculate the blended viscosity obtained by mixing 20% of liquid A with a viscosity of 10 cSt and 80% of liquid B with a viscosity of 30 cSt at 70 °F.

Solution
First, convert the given viscosities to SSU to use Eqn (3.21).
Viscosity of liquid A is calculated using Eqns (3.8) and (3.9).

$$10 = 0.226 \, (V_A) - \frac{195}{V_A}$$

Rearranging we get

$$0.226 \, V_A^2 - 10 \, V_A - 195 = 0$$

Solving the quadratic equation for V_A, we get

$$V_A = 58.90 \text{ SSU}$$

Similarly, viscosity of liquid B is.

$$V_B = 140.72 \text{ SSU}$$

From Eqn (3.21), the blended viscosity is

$$\sqrt{V_{blnd}} = \frac{20 + 80}{(20/\sqrt{58.9}) + (80/\sqrt{140.72})} = 10.6953$$

Therefore the viscosity of the blend is

$$V_{blnd} = 114.39 \text{ SSU}$$

or

> Viscosity of blend = 23.99 cSt after converting from SSU to cSt

A graphical method is also available to calculate the blended viscosities of two petroleum products using ASTM D341-77. This method involves using a logarithmic chart with viscosity scales on the left and right sides of the paper. The horizontal axis is for selecting the percentage of each product as shown in Figure 3.2. This chart is also available in handbooks such as *Crane Handbook* and the *Hydraulic Institute Engineering Data Book*. The viscosities of both products must be plotted at the same temperature.

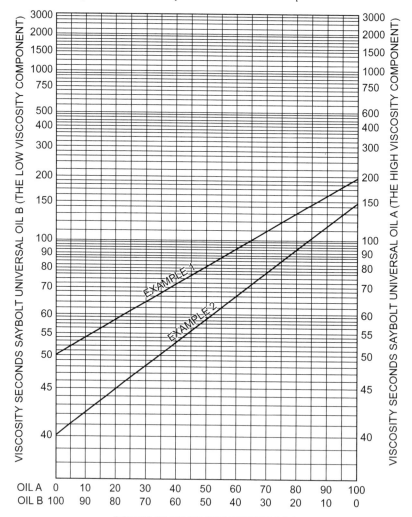

Figure 3.2 Viscosity blending chart.

Using this method, the blended viscosity of two products at a time is calculated and the process repeated for multiple products. Thus if three products are blended in the ratios of 10%, 20%, and 70%, we would first calculate the blend using the first two liquids considering 10 parts of liquid A mixed with 20 parts of liquid B. This means that the blend would be calculated on the basis of one-third of liquid A and two-thirds of liquid B. Next this blended liquid will be mixed with liquid C in the proportion of 30% and 70%, respectively.

6. VAPOR PRESSURE

Vapor pressure of a liquid is defined as the pressure at a given temperature at which the liquid and vapor exist in equilibrium. The normal boiling point of a liquid can thus be defined as the temperature at which the vapor pressure equals the atmospheric pressure. In the laboratory, the vapor pressure is measured at a fixed temperature of 100 °F and is then reported as the Reid vapor pressure. The vapor pressure of a liquid increases with temperature. Charts are available to determine the actual vapor pressure of a liquid at any temperature once its Reid vapor pressure is known. Refer to the *Crane Handbook* for vapor pressure charts.

The importance of vapor pressure will be evident when we discuss the operation of centrifugal pumps on pipelines. To prevent cavitation of pumps, the liquid vapor pressure at the flowing temperature must be taken into account in the calculation of net positive suction head available at the pump suction. Centrifugal pumps are discussed in Chapter 9.

7. BULK MODULUS

The bulk modulus of a liquid is a measure of the compressibility of the liquid. It is defined as the pressure required to produce a unit change in its volume. Mathematically, bulk modulus is expressed as

$$\text{Bulk modulus } K = V dP/dV \tag{3.26}$$

where dV is the change in volume corresponding to a change in pressure of dP.

The units of bulk modulus, K, are psi or kPa. For most liquids, the bulk modulus is approximately in the range of 250,000 to 300,000 psi. The fairly high number demonstrates the incompressibility of liquids.

Let us demonstrate the incompressibility of liquids by performing a calculation using bulk modulus. Assume the bulk modulus of a petroleum product is 250,000 psi. To calculate the pressure required to change the volume of a given quantity of liquid by 1%, we would proceed as follows.

From Eqn (3.26), with some rearrangement.

Bulk modulus = change in pressure/(change in volume/volume)

Therefore

$$250,000 = \text{change in pressure}/(0.01)$$

Therefore

$$\text{change in pressure} = 2500 \text{ psi}$$

It can be seen that a fairly large pressure is required to produce a very small (1%) change in the liquid volume. Hence we say that liquids are fairly incompressible.

Bulk modulus is used in line pack calculations and transient flow analysis. There are two bulk modulus values used in practice: isothermal and adiabatic. The bulk modulus of a liquid depends on temperature, pressure, and specific gravity. The following empirical equations, also known as ARCO formulas, may be used to calculate the bulk modulus.

7.1 Adiabatic Bulk Modulus

$$Ka = A + B(P) - C(T)^{1/2} - D(API) - E(API)^2 + F(T)(API) \quad (3.27)$$

where
  $A = 1.286 \times 10^6$
  $B = 13.55$
  $C = 4.122 \times 10^4$
  $D = 4.53 \times 10^3$
  $E = 10.59$
  $F = 3.228$
  P – Pressure in psig
  T – Temperature in °R
  API – API gravity of liquid.

7.2 Isothermal Bulk Modulus

$$Ki = A + B(P) - C(T)^{1/2} + D(T)^{3/2} - E(API)^{3/2} \quad (3.28)$$

where
 A = 2.619×10^6
 B = 9.203
 C = 1.417×10^5
 D = 73.05
 E = 341.0
 P – Pressure in psig
 T – Temperature in °R
 API – API gravity of liquid.

For a typical crude oil of 35°API gravity at 1000 psig pressure and 80 °F temperature, the bulk modulus calculated from Eqns (3.27) and (3.28) are:

$$\text{Adiabatic bulk modulus} = 231{,}426 \text{ psi}$$

$$\text{Isothermal bulk modulus} = 181{,}616 \text{ psi}$$

The bulk modulus of water at 70 °F is 320,000 psi. Refer to Appendix A for physical properties of various liquids used in transmission pipelines.

8. FUNDAMENTAL CONCEPTS OF FLUID FLOW

In this section, we will discuss some fundamental concepts of fluid flow that will set the stage for the succeeding chapter. The basic principles of continuity and energy equations are introduced first.

8.1 Continuity

One of the fundamental concepts that must be satisfied in any type of pipe flow is the principle of continuity of flow. This principle states that the total amount of fluid passing through any section of a pipe is fixed. This may also be thought of as the principle of conservation of mass. Basically, it means that liquid is neither created nor destroyed as it flows through a pipeline. Because mass is the product of the volume and density, we can write the following equation for continuity.

$$M = \text{Vol} \times \rho = \text{Constant} \quad (3.29)$$

where

M – Mass flow rate at any point in the pipeline, slugs/s
Vol – Volume flow rate at any point in the pipeline, ft^3/s
ρ – Density of liquid at any point in the pipeline, slugs/ft^3.

Because the volume flow rate at any point in a pipeline is the product of the area of cross-section of the pipe and the average liquid velocity, we can rewrite Eqn (3.29) as follows:

$$M = A \times V \times \rho = \text{Constant} \tag{3.30}$$

where

M – Mass flow rate at any point in the pipeline, slugs/s
A – Area of cross-section of pipe, ft^2
V – Average liquid velocity, ft/s
ρ – Density of liquid at any point in the pipeline, slugs/ft^3.

Because liquids are generally considered to be incompressible and therefore density does not change appreciably, the continuity equation reduces to

$$AV = \text{Constant} \tag{3.31}$$

8.2 Energy Equation

The basic principle of conservation of energy applied to liquid hydraulics is embodied in the Bernoulli's equation. This equation simply states that the total energy of the fluid contained in the pipeline at any point is a constant. Obviously, this is an extension of the principle of conservation of energy, which states that energy is neither created nor destroyed, but transformed from one form to another.

Consider a pipeline shown in Figure 3.3 that depicts flow from point A to point B with elevation of point A being Z_A and elevation at B being Z_B above some chosen datum. The pressure in the liquid at point A is P_A and that at B is P_B. Assuming a general case, where the pipe diameter at A may be different from that at B, we will designate the velocities at A and B to be V_A and V_B respectively. Consider a particle of the liquid of weight W at point A in the pipeline. This liquid particle at A may be considered to possess a total energy E that consists of three components:

$$\text{Energy due to position, or potential energy} = W\,Z_A$$

$$\text{Energy due to pressure, or pressure energy} = WP_A/\gamma$$

$$\text{Energy due to velocity, or kinetic energy} = W(V_A/2g)^2$$

where γ is the specific weight of liquid.

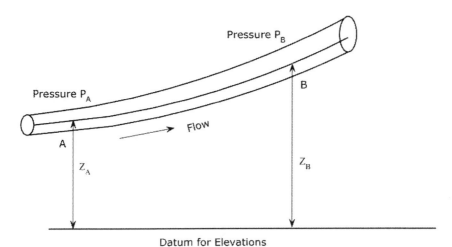

Figure 3.3 Energy of a liquid in pipe flow.

We can thus state that

$$E = WZ_A + WP_A/\gamma + WV_A^2/2g \qquad (3.32)$$

Dividing by W throughout, we get the total energy per unit weight of liquid as

$$H_A = Z_A + P_A/\gamma + V_A^2/2g \qquad (3.33)$$

where H_A is the total energy per unit weight at point A.

Considering the same liquid particle as it arrives at point B, the total energy per unit weight at B is.

$$H_B = Z_B + P_B/\gamma + V_B^2/2g \qquad (3.34)$$

Because conservation of energy

$$H_A = H_B$$

Therefore

$$Z_A + P_A/\gamma + V_A^2/2g = Z_B + P_B/\gamma + V_B^2/2g \qquad (3.35)$$

Equation (3.35) is one form of the Bernoulli's equation for fluid flow.

In real-world pipeline transportation, there is energy loss between point A and point B because friction in pipe. We include the energy loss because friction by modifying Eqn (3.35) as follows

$$Z_A + P_A/\gamma + V_A^2/2g = Z_B + P_B/\gamma + V_B^2/2g + \sum h_L \qquad (3.36)$$

where Σh_L represents all the head losses between points A and B, resulting from friction.

In Bernoulli's equation, we must also include any energy added to the liquid, such as when there is a pump between points A and B. Thus the left hand side of the equation will have a positive term added to it that will represent the energy generated by a pump.

Equation (3.36) will be modified as follows to include a pump at point A that will add a certain amount of pump head to the liquid.

$$Z_A + P_A/\gamma + V_A^2/2g + H_P = Z_B + P_B/\gamma + V_B^2/2g + \sum h_L \qquad (3.37)$$

where

H_P represents the pump head added to the liquid at point A.

In the next chapter, we will further explore the concepts of pressure, velocity, flow rates, and energy lost because of pipe friction.

9. GAS PROPERTIES

Next we will discuss the properties of gases that affect how gas flow through a pipeline, what pressures are required, and how the properties change with the gas temperature and pressure. We will start with ideal or perfect gases that obey the ideal gas equation and then examine how real gases differ from ideal gases. The concept of compressibility or gas deviation factor will be introduced and methods of calculating the compressibility factor using several popular correlations and empirical methods explored. The properties of a mixture of gases will be analyzed that will serve as a starting point for natural gas mixtures that are encountered in practice.

10. MASS

Mass is the quantity of matter in a substance. It is sometimes used interchangeably with weight. Strictly speaking, mass is a scalar quantity, whereas weight is a force and therefore a vector quantity. Mass is independent of the geographic location whereas weight depends upon the acceleration because gravity and therefore varies with geographic location. Mass is measured in slugs in the US customary system (USCS) of units and kilograms (kg) in SI units.

However, for most purposes we say that a 10-lb mass has a weight of 10 lb. The pound (lb) is a more convenient unit for mass and to distinguish between mass and weight the terms lbm and lbf are sometimes used. A slug is approximately equal to 32.2 lb.

If some gas is contained in a volume and the temperature and pressure change, the mass will remain constant unless some gas is taken out or added to the container. This is known as conservation of mass.

11. VOLUME

Volume of a gas is the space a given mass of gas occupies at a particular temperature and pressure. Because gas is compressible, it will expand to fill available space. Therefore, the gas volume will vary with temperature and pressure. Hence, a certain volume of a given mass of gas at some temperature and pressure will decrease in volume as the pressure is increased and vice versa. Suppose a quantity of gas is contained in a volume of 100 ft^3 at a temperature of 80 °F and a pressure of 200 psi. If the temperature is increased to 100 °F, keeping the volume constant, the pressure will change according to Charles's law. Charles's law states that for constant volume the pressure of a fixed mass of gas will vary directly with the temperature. Thus if temperature increases by 20%, the pressure will also rise by 20%. Similarly, if pressure is maintained constant, by Charles's law, the volume will increase in direct proportion with temperature. Charles's law, Boyle's law, and other gas laws will be discussed in detail later in this chapter.

Volume of gas is measured in ft^3 in USCS units and m^3 in SI units. Other units include thousand ft^3 (Mft^3), million ft^3 (MM ft^3) in USCS units, and km^3 and mm^3 in SI units. When referred to standard conditions of temperature and pressure (60 °F and 14.7 psia in USCS units), the volume is stated as standard volume and therefore measured in standard ft^3 (SCF) or million standard ft^3. Volume flow rate of gas is measured per unit time and may be expressed as ft^3/min, ft^3/h, ft^3/day, SCFD, million standard ft^3, and so on in USCS units. In SI units, gas flow rate is expressed in m^3/h or mm^3/day. For years, in the USCS units the practice has been to use M to represent a 1000 and MM to refer to a million. This goes back to the Roman days of numerals. In SI units, a more logical step is followed. For 1000, the letter k (for kilo) is used and the letter M (for mega) is used for a million. Therefore,

500 MSCFD in USCS units refers to 500 thousand standard cubic feet per day whereas 15 mm³/day means 15 million cubic meters per day in SI units.

12. DENSITY AND SPECIFIC WEIGHT

Density of a gas represents the amount of gas that can be packed in a given volume. Therefore, it is measured in terms of mass per unit volume. If 5 lb of gas is contained in 100 ft^3 of volume, at some temperature and pressure, we say that the gas density is $5/100 = 0.05$ lb/ft^3.

Strictly speaking, in USCS units density must be expressed as slug/ft^3 because mass is customarily referred to in slug.

$$\text{Thus, Density, } \rho = m/V \tag{3.38}$$

where
ρ = Density of gas
m = Mass of gas
V = Volume of gas.

Density is expressed in slug/ft^3 or lb/ft^3 in USCS units and kg/m^3 in SI units.

A companion term called specific weight is also used when referring to the density of gas. Specific weight (γ) is measured in lb/ft^3 in USCS units and is therefore contrasted with density when the latter is measured in slug/ft^3. In SI units, the specific weight is expressed in Newton per m^3 (N/m^3).

The reciprocal of the specific weight is known as the specific volume. By definition therefore, specific volume represents the volume occupied by a unit weight of gas. It is measured in ft^3/lb in USCS units and m^3/N in SI units.

13. SPECIFIC GRAVITY

Specific gravity, sometimes called gravity is a measure of how heavy the gas is compared with air at a particular temperature. It may also be called relative density, expressed as the ratio of the gas density to density of air. Being a ratio, it is dimensionless.

$$\text{Gas gravity} = \text{density of gas}/\text{density of air}$$

As an example, natural gas has a specific gravity of 0.60 (air = 1.00) at 60 °F. This means that the gas is 60% as heavy as air.

If we know the molecular weight of a particular gas, we can calculate its gravity by dividing the molecular weight by the molecular weight of air.

$$\text{Specific Gravity} = M_g/M_{air}$$

or

$$G = M_g/28.9625 \tag{3.39}$$

where
  G – Specific gravity of gas
  M_g – Molecular weight of gas
  M_{air} – Molecular weight of air.

Rounding off the molecular weight of air to 29, we can state that the gas gravity is simply $M_g/29$. With natural gas, consisting of a mixture of several gases (methane, ethane, etc.) the molecular weight M_g in Eqn (3.39) is referred to as the apparent molecular weight of the gas mixture.

When the molecular weight and the percentage of the individual components of a natural gas mixture are known, we can calculate the molecular weight of the natural gas mixture by using a weighted average method. Thus a natural gas mixture consisting of 90% methane, 8% ethane, and 2% propane has a specific gravity of

$$G = (0.9 \times M_1 + 0.08 \times M_2 + 0.02 \times M_3)/29$$

where M_1, M_2, and M_3 are the molecular weights of methane, ethane, and propane, respectively, and 29 represents the molecular weight of air. Table 3.1 lists the molecular weights and other properties of several hydrocarbon gases.

Refer to Table 3.1 for a typical list of properties of hydrocarbon gases transported through transmission pipelines.

14. VISCOSITY

The viscosity of a fluid represents its resistance to flow. The higher the viscosity, the more difficult it is to flow. Lower viscosity fluids flow easily in pipes and cause less pressure drop. Liquids have much larger values of viscosity compared with gases. For example, water has a viscosity of 1.0 cP, whereas viscosity of natural gas is approximately 0.0008 cP. Even though the gas viscosity is a small number, it has an important function in assessing the type of flow in pipelines. The Reynolds number is a dimensionless parameter that is used to classify flow rate in pipelines. It depends on the

Table 3.1 Properties of hydrocarbon gases

Gas	Formula	Molecular weight	Vapor pressure psia at 100 °F	Critical constants Pressure psia	Critical constants Temp. °F	Critical constants Volume ft³/lb	Ideal gas Spgr (air = 1.00)	Ideal gas 14.696 psia, 60 °F ft³/lb gas	Specific heat, Btu/lb/°F 14.696 psia, 60 °F Ideal gas
Methane	CH_4	16.0430	5000	666.0	−116.66	0.0988	0.5539	23.654	0.52676
Ethane	C_2H_6	30.0700	800	707.0	90.07	0.0783	1.0382	12.620	0.40789
Propane	C_3H_8	44.0970	188.65	617.0	205.93	0.0727	1.5226	8.6059	0.38847
Isobutane	C_4H_{10}	58.1230	72.581	527.9	274.4	0.0714	2.0068	6.5291	0.38669
n-butane	C_4H_{10}	58.1230	51.706	548.8	305.52	0.0703	2.0068	6.5291	0.39500
Iso-pentane	C_5H_{12}	72.1500	20.443	490.4	368.96	0.0684	2.4912	5.2596	0.38448
n-pentane	C_5H_{12}	72.1500	15.575	488.1	385.7	0.0695	2.4912	5.2596	0.38831
Neo-pentane	C_5H_{12}	72.1500	36.72	464.0	321.01	0.0673	2.4912	5.2596	0.39038
n-hexane	C_6H_{14}	86.1770	4.9596	436.9	453.8	0.0688	2.9755	4.4035	0.38631
2-Methylpentane	C_6H_{14}	86.1770	6.769	436.6	435.76	0.0682	2.9755	4.4035	0.38526
3-Methylpentane	C_6H_{14}	86.1770	6.103	452.5	448.2	0.0682	2.9755	4.4035	0.37902
Neo hexane	C_6H_{14}	86.1770	9.859	446.7	419.92	0.0667	2.9755	4.4035	0.38231
2,3-dimethylbutane	C_6H_{14}	86.1770	7.406	454.0	440.08	0.0665	2.9755	4.4035	0.37762
n-heptane	C_7H_{16}	100.2040	1.621	396.8	512.8	0.0682	3.4598	3.7872	0.38449
2-Methylhexane	C_7H_{16}	100.2040	2.273	396.0	494.44	0.0673	3.4598	3.7872	0.38170
3-Methylhexane	C_7H_{16}	100.2040	2.13	407.6	503.62	0.0646	3.4598	3.7872	0.37882
3-Ethylpentane	C_7H_{16}	100.2040	2.012	419.2	513.16	0.0665	3.4598	3.7872	0.38646
2,2-Dimethylpentane	C_7H_{16}	100.2040	3.494	401.8	476.98	0.0665	3.4598	3.7872	0.38651

2,4-Dimethylpentane	C_7H_{16}	100.2040	3.294	397.4	475.72	0.0667	3.4598	3.7872	0.39627
3,3-Dimethylpentane	C_7H_{16}	100.2040	2.775	427.9	505.6	0.0662	3.4598	3.7872	0.38306
Triptane	C_7H_{16}	100.2040	3.376	427.9	496.24	0.0636	3.4598	3.7872	0.37724
n-octane	C_8H_{18}	114.2310	0.5371	360.7	564.15	0.0673	3.9441	3.322	0.38334
Di isobutyl	C_8H_{18}	114.2310	1.1020	361.1	530.26	0.0676	3.9441	3.322	0.37571
Isooctane	C_8H_{18}	114.2310	1.7090	372.7	519.28	0.0657	3.9441	3.322	0.38222
n-nonane	C_9H_{20}	128.2580	0.17155	330.7	610.72	0.0693	4.4284	2.9588	0.38248
n-decane	$C_{10}H_{22}$	142.2850	0.06088	304.6	652.1	0.0702	4.9127	2.6671	0.38181
Cyclopentane	C_5H_{10}	70.1340	9.917	653.8	461.1	0.0594	2.4215	5.411	0.27122
Methylcyclopentane	C_6H_{12}	84.1610	4.491	548.8	499.28	0.0607	2.9059	4.509	0.30027
Cyclohexane	C_6H_{12}	84.1610	3.267	590.7	536.6	0.0586	2.9059	4.509	0.29012
Methylcyclohexane	C_7H_{14}	98.1880	1.609	503.4	570.2	0.0600	3.3902	3.8649	0.31902
Ethylene	C_2H_4	28.0540	1400	731.0	48.54	0.0746	0.9686	13.527	0.35789
Propylene	C_3H_6	42.0810	232.8	676.6	198.31	0.0717	1.4529	9.0179	0.35683
Butylene	C_4H_8	56.1080	62.55	586.4	296.18	0.0683	1.9373	6.7636	0.35535
Cis–2–butene	C_4H_8	56.1080	45.97	615.4	324.31	0.0667	1.9373	6.7636	0.33275
Trans–2–butene	C_4H_8	56.1080	49.88	574.9	311.8	0.0679	1.9373	6.7636	0.35574
Isobutene	C_4H_8	56.1080	64.95	580.2	292.49	0.0681	1.9373	6.7636	0.36636
1-Pentene	C_5H_{10}	70.1340	19.12	509.5	376.86	0.0674	2.4215	5.411	0.35944
1,2-Butadiene	C_4H_6	54.0920	36.53	656.0	354	0.0700	1.8677	7.0156	0.34347
1,3-Butadiene	C_4H_6	54.0920	59.46	620.3	306	0.0653	1.8677	7.0156	0.34223
Isoprene	C_5H_8	68.1190	16.68	582.0	403	0.0660	2.3520	5.571	0.35072
Acetylene	C_2H_2	26.0380		890.4	95.29	0.0693	0.8990	14.574	0.39754
Benzene	C_6H_6	78.1140	3.225	710.4	552.15	0.0531	2.6971	4.8581	0.24295
Toluene	C_7H_8	92.1410	1.033	595.5	605.5	0.0549	3.1814	4.1184	0.26005
Ethyl-benzene	C_8H_{10}	106.1670	0.3716	523	651.22	0.0564	3.6657	3.5744	0.27768
o-xylene	C_8H_{10}	106.1670	0.2643	541.6	674.85	0.0557	3.6657	3.5744	0.28964

(Continued)

Table 3.1 Properties of hydrocarbon gases—Cont'd

Gas	Formula	Molecular weight	Vapor pressure psia at 100 °F	Critical constants Pressure psia	Critical constants Temp. °F	Critical constants Volume ft³/lb	Spgr (air = 1.00)	Ideal gas 14.696 psia, 60 °F	Ideal gas ft³/lb — gas	Specific heat, Btu/lb/°F 14.696 psia, 60 °F Ideal gas
m-xylene	C_8H_{10}	106.1670	0.3265	512.9	650.95	0.0567	3.6657	3.6657	3.5744	0.27427
p-xylene	C_8H_{10}	106.1670	0.3424	509.2	649.47	0.0572	3.6657	3.6657	3.5744	0.27470
Styrene	C_8H_8	104.1520	0.2582	587.8	703	0.0534	3.5961	3.5961	3.6435	0.26682
Isopropylbenzene	C_9H_{12}	120.1940	0.188	465.4	676.2	0.0569	4.1500	4.1500	3.1573	0.30704
Methyl alcohol	CH_4O	32.0420	4.631	1174	463.01	0.0590	1.1063	1.1063	11.843	0.32429
Ethyl alcohol	C_2H_6O	46.0690	2.313	891.7	465.31	0.0581	1.5906	1.5906	8.2372	0.33074
Carbon monoxide	CO	28.0100		506.8	−220.51	0.0527	0.9671	0.9671	13.548	0.24847
Carbon dioxide	CO_2	44.0100		1071	87.73	0.0342	1.5196	1.5196	8.6229	0.19909
Hydrogen sulfide	H_2S	34.0820	394.59	1306	212.4	0.0461	1.1768	1.1768	11.134	0.23838
Sulfur dioxide	SO_2	64.0650	85.46	1143	315.7	0.0305	2.2120	2.2120	5.9235	0.14802
Ammonia	NH_3	17.0305	211.9	1647	270.2	0.0681	0.5880	0.5880	22.283	0.49678
Air	$N_2 + O_2$	28.9625		546.9	−221.29	0.0517	1.0000	1.0000	13.103	0.2398
Hydrogen	H_2	2.0159		187.5	−400.3	0.5101	0.06960	0.06960	188.25	3.4066
Oxygen	O_2	31.9988		731.4	−181.4	0.0367	1.1048	1.1048	11.859	0.21897
Nitrogen	N_2	28.0134		493	−232.48	0.0510	0.9672	0.9672	13.546	0.24833
Chlorine	Cl_2	70.9054	157.3	1157	290.69	0.0280	2.4482	2.4482	5.3519	0.11375
Water	H_2O	18.0153	0.95	3200.1	705.1	0.04975	0.62202	0.62202	21.065	0.44469
Helium	He	4.0026		32.99	−450.31	0.2300	0.1382	0.1382	94.814	1.24040
Hydrogen chloride	HCl	36.4606	906.71	1205	124.75	0.0356	1.2589	1.2589	10.408	0.19086

gas viscosity, flow rate, pipe diameter, temperature, and pressure. The absolute viscosity also call the dynamic viscosity is expressed in lb/ft-s in USCS units and poise (P) in SI units. Another related term is the kinematic viscosity. The latter is simply the absolute viscosity divided by the density. The two viscosities are related by the following.

$$\text{Kinematic viscosity } \nu = \mu/\rho \tag{3.40}$$

where
ν – Kinematic viscosity
μ – Dynamic viscosity
ρ – density.

Kinematic viscosity is expressed in ft^2/s in USCS units and Stokes (St) in SI units. Other units of viscosity are cP and cSt. The viscosity of a gas depends on its temperature and pressure. Unlike liquids, the viscosity of a gas.

See Table 3.2 for viscosities of common gases in pipelines; Figure 3.4 shows the variation of viscosity with temperature for a gas.

Table 3.2 Viscosities of common gases

Gas	Viscosity (cP)
Methane	0.0107
Ethane	0.0089
Propane	0.0075
i-Butane	0.0071
n-Butane	0.0073
i-Pentane	0.0066
n-Pentane	0.0066
Hexane	0.0063
Heptane	0.0059
Octane	0.0050
Nonane	0.0048
Decane	0.0045
Ethylene	0.0098
Carbon monoxide	0.0184
Carbon dioxide	0.0147
Hydrogen sulfide	0.0122
Air	0.0178
Nitrogen	0.0173
Helium	0.0193

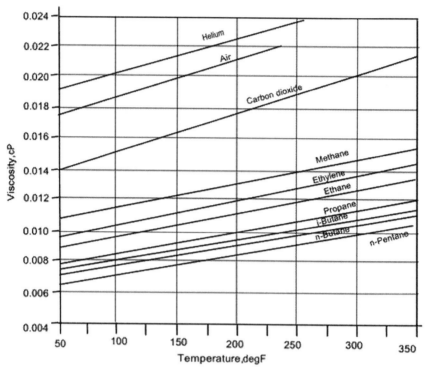

Figure 3.4 Variation of gas viscosity with temperature.

Because natural gas is a mixture of pure gases such as methane or ethane, the following formula is used to calculate the viscosity from the component gases.

$$\mu = \frac{\Sigma'(\mu_i y_i \sqrt{M_i})}{\Sigma(y_i \sqrt{M_i})} \quad (3.41)$$

where
 μ – Dynamic viscosity of gas mixture
 μ_i – Dynamic viscosity of gas component i
 y_i – Mole percent of gas component i
 M_i – Molecular weight of gas component i.

15. IDEAL GASES

An ideal gas is defined as a gas in which the volume of the gas molecules is negligible compared to the volume occupied by the gas. Also, the attraction or repulsion between the individual gas molecules and the container are negligible. Further, for an ideal gas, the molecules are considered to be perfectly elastic and there is no internal energy loss resulting from collision between the molecules. Such ideal gases are said to obey several classical equations such as the Boyle's law, Charles's law and the ideal gas equation or the perfect gas equation. We will first discuss the behavior of ideal gases and then follow it up with the behavior of real gases.

If M represents the molecular weight of a gas and the mass of a certain quantity of gas is m, the number of moles is given by

$$n = m/M \qquad (3.42)$$

where n is the number that represents the number of moles in the given mass.

As an example, the molecular weight of methane is 16.043. Therefore, 50 lb of methane will contain approximately 3 mol.

The ideal gas law, sometimes referred to as the perfect gas equation simply states that the pressure, volume, and temperature of the gas are related to the number of moles by the following equation.

$$PV = nRT \qquad (3.43)$$

where

P – Absolute pressure, psia
V – Gas volume, ft^3
n – Number of lb moles as defined in Equation (3.42)
R – Universal gas constant
T – Absolute temperature of gas, °R (°F + 460).

The universal gas constant R has a value of 10.732 psia ft^3/lb mole °R in USCS units. We can combine Eqn (3.42) with Eqn (3.43) and express the ideal gas equation as follows

$$PV = mRT/M \qquad (3.44)$$

where all symbols have been defined previously. It has been found that the ideal gas equation is correct only at low pressures close to the atmospheric pressure. Because gas pipelines generally operate at pressures higher than atmospheric pressures, we must modify Eqn (3.44) to take into account the effect of compressibility. The latter is accounted for by using a term called the compressibility factor or gas deviation factor. We will discuss the compressibility factor later in this chapter.

In the perfect gas Eqn (3.44), the pressures and temperatures must be in absolute units. Absolute pressure is defined as the gauge pressure (as measured by a gauge) plus the local atmospheric pressure. Therefore

$$P_{abs} = P_{gauge} + P_{atm} \qquad (3.45)$$

Thus if the gas pressure is 20 psig and the atmospheric pressure is 14.7 psia, we get the absolute pressure of the gas as

$$P_{abs} = 20 + 14.7 = 34.7 \text{ psia}$$

Absolute pressure is expressed as psia, whereas the gauge pressure is referred to as psig. The adder to the gauge pressure, which is the local atmospheric pressure, is also called the base pressure. In SI units, 500 kPa gauge pressure is equal to 601 kPa absolute pressure if the base pressure is 101 kPa.

The absolute temperature is measured above a certain datum. In USCS units, the absolute scale of temperatures is designated as degree Rankin (°R) and is equal to the sum of the temperature in °F and the constant 460. In SI units, the absolute temperature scale is referred to as degree Kelvin (K). Absolute temperature in K is equal to °C + 273.

Therefore,

Absolute temperature, °R = Temp °F + 460.
Absolute temperature, K = Temp °C + 460.

It is customary to drop the degree symbol for absolute temperature in Kelvin.

Ideal gases also obey Boyle's law and Charles's law. Boyle's law is used to relate the pressure and volume of a given quantity of gas when the temperature is kept constant. Constant temperature is also called isothermal condition. Boyle's law is as follows

$$P_1/P_2 = V_2/V_1$$

or

$$P_1 V_1 = P_2 V_2 \qquad (3.46)$$

where P_1 and V_1 are the pressure and volume at condition 1 and P_2 and V_2 are the corresponding value at some other condition 2 where the temperature is not changed.

Charles's law states that for constant pressure, the gas volume is directly proportional to the gas temperature. Similarly, if volume is kept constant, the pressure varies directly as the temperature. Therefore we can state the following.

Physical Properties

$$V_1/V_2 = T_1/T_2 \text{ at constant pressure} \quad (3.47)$$

$$P_1/P_2 = T_1/T_2 \text{ at constant volume} \quad (3.48)$$

Example Problem 3.8

A certain mass of gas has a volume of 1000 ft^3 at 60 psig. If temperature is constant and the pressure increases to 120 psig, what is the final volume of the gas? The atmospheric pressure is 14.7 psi.

Solution

Boyle's law can be applied because the temperature is constant. Using Eqn (3.46), we can write.

$$V_2 = P_1 V_1 / P_2$$

or

$$V_2 = (60 + 14.7) \times 1000/(120 + 14.7) = 554.57 \text{ ft}^3$$

Example Problem 3.9

At 75 psig and 70 °F, a gas has a volume of 1000 ft^3. If the volume is kept constant and the gas temperature increases to 120 °F, what is the final pressure of the gas? For constant pressure at 75 psig, if the temperature increases to 120 °F, what is the final volume? Use 14.7 psi for the base pressure.

Solution

Because the volume is constant in the first part of the problem, Charles's law applies.

$$(75 + 14.7)/(P_2) = (70 + 460)/(120 + 460)$$

Solving for P_2 we get

$$P_2 = 98.16 \text{ psia or } 88.46 \text{ psig}$$

For the second part, the pressure is constant and Charles's law can be applied.

$$V_1/V_2 = T_1/T_2$$

$$1000/V_2 = (70 + 460)/(120 + 460)$$

Solving for V_2 we get

$$V_2 = 1094.34 \text{ ft}^3$$

Example Problem 3.10
An ideal gas occupies a tank volume of 250 ft³ at a pressure of 80 psig and temperature of 110 °F.
1. What is the gas volume at standard conditions of 14.73 psia and 60 °F? Assume atmospheric pressure is 14.6 psia.
2. If the gas is cooled to 90 °F, what is the gas pressure?

Solution
1. Using the ideal gas Eqn (3.43), we can state that

$$P_1 V_1 / T_1 = P_2 V_2 / T_2$$

$$P_1 = 80 + 14.6 = 94.6 \text{ psia}$$

$$V_1 = 250 \text{ ft}^3$$

$$T_1 = 110 + 460 = 570 \text{ °R}$$

$$P_2 = 14.73$$

V_2 is to be calculated
and

$$T_2 = 60 + 460 = 520 \text{ °R}$$

$$94.6 \times 250/570 = 14.73 \times V_2/520$$

$$V_2 = 1{,}464.73 \text{ ft}^3$$

2. When the gas is cooled to 90 °F, the final conditions are:

$$T_2 = 90 + 460 = 550 \text{ °R}$$

$$V_2 = 250 \text{ ft}^3$$

P_2 is to be calculated.
The initial conditions are:

$$P_1 = 80 + 14.6 = 94.6 \text{ psia}$$

$$V_1 = 250 \text{ ft}^3$$

$$T_1 = 110 + 460 = 570 \text{ °R}$$

It can be seen that the volume of gas is constant and the temperature reduces from 110 °F to 90 °F. Therefore using Charles's law, we can calculate as follows

$$P_1/P_2 = T_1/T_2$$

$$94.6/P_2 = 570/550$$

$$P_2 = 94.6 \times 550/570 = 91.28 \text{ psia or the } 91.28 - 14.6 = 76.68 \text{ psig}$$

Physical Properties

16. REAL GASES

The ideal gas equation is applicable only when the pressure of the gas is very low or near atmospheric pressure. When gas pressures and temperatures are higher, the ideal gas equation will not give accurate results. The calculation errors may be as high as 500%. An equation of state is generally used for calculating the properties of gases at a higher temperatures and pressures.

Real gases behave according to a modified version of the ideal gas law discussed earlier. The modifying factor is known as the compressibility factor Z. This is also called the gas deviation factor. Z is a dimensionless number less than 1.0 and varies with temperature, pressure, and physical properties of the gas.

The real gas equation can be written as follows:

$$PV = ZnRT \tag{3.49}$$

where

P – Absolute pressure, psia

V – Gas volume, ft^3

Z – Gas deviation factor or compressibility factor, dimensionless

T – Absolute temperature of gas, °R

n – Number of lb moles as defined in Eqn (3.42)

R – Universal gas constant, 10.732 psia ft^3/lb mole °R.

The calculation of the compressibility factor will be discussed in the next few pages.

17. NATURAL GAS MIXTURES

The critical temperature of a pure gas is the temperature above which it cannot be liquefied regardless of the pressure. The critical pressure of a pure substance is defined as the pressure above which liquid and gas cannot coexist, regardless of the temperature. With multicomponent mixtures, these properties are referred to as pseudo critical temperature and pseudo critical pressure. If the composition of the gas mixture is known, we can calculate the pseudo critical pressure and the pseudo critical temperature of the gas mixture knowing the critical pressure and temperature of the pure components.

The reduced temperature is simply the temperature of the gas divided by its critical temperature. Similarly, the reduced pressure is simply the pressure of the gas divided by its critical pressure, both temperature and pressure being in absolute units. Similar to pseudo critical temperature and pseudo critical pressure, for a gas mixture we can calculate the pseudo reduced temperature and the pseudo reduced pressure.

Example Problem 3.11
Calculate the pseudo critical temperature and the pseudo critical pressure of a natural gas mixture consisting of 85% of methane, 10% ethane, and 5% propane.

The C_1, C_2, and C_3 components of the gas mixture have the following critical properties.

Components	Critical temperature, °R	Critical pressure, psia
C_1	343	666
C_2	550	707
C_3	666	617

Some numbers have been rounded off for simplicity.

Solution
From the given mole fractions of components, we use Kay's rule to calculate the average pseudo critical temperature and pressure of gas.

$$T_{pc} = \sum yT_c \tag{3.50}$$

$$P_{pc} = \sum yP_c \tag{3.51}$$

where T_c and P_c are the critical temperature and pressure of the pure component (C_1, C_2, etc.) and y represents the mole fraction of the component. The calculated values T_{pc} and P_{pc} are the average pseudo critical temperature and pressure of the gas mixture.

Using the given mole fractions, the pseudo critical properties are

$$T_{pc} = (0.85 \times 343) + (0.10 \times 550) \\ + (0.05 \times 666) = 379.85 \, °R$$

and

$$P_{pc} = (0.85 \times 666) + (0.10 \times 707) \\ + (0.05 \times 617) = 667.65 \, \text{psia}$$

Example Problem 3.12
The temperature of the gas in the previous example is 80 °F and the average pressure is 1000 psig. What are the pseudo reduced temperature and pseudo reduced pressure? Base pressure is 14.7 psia.

Solutions
The pseudo reduced temperature $T_{pr} = (80 + 460)/379.85 = 1.4216$ °R

The pseudo reduced pressure $P_{pr} = (1000 + 14.7)/667.65 = 1.5198$ psia

18. PSEUDO CRITICAL PROPERTIES FROM GRAVITY

If the gas composition data are not available, we can calculate an approximate value of the pseudo critical temperature and pseudo critical pressure of the gas from the gas gravity as follows

$$T_{pc} = 170.491 + 307.344\,G \qquad (3.52)$$

$$P_{pc} = 709.604 - 58.718\,G \qquad (3.53)$$

where G is the gas gravity (air = 1.00) and T_{pc} is the pseudo critical temperature and P_{pc} is the pseudo critical pressure of the gas.

Example Problem 3.13
Calculate the gas gravity of a natural gas mixture consisting of 85% of methane, 10% ethane, and 5% propane. Using the gas gravity, calculate the pseudo critical temperature and pseudo critical pressure for this natural gas.

Solution
Using Kay's rule for molecular weight of gas mixture and Eqn (3.39)

Gas gravity G $= [(0.85 \times 16.04) + (0.10 \times 30.07) + (0.05 \times 44.10)]$
$\times /29.0$

$$G = 0.6499$$

Using Eqns (3.51) and (3.52) we get for the pseudo critical properties.

$$T_{pc} = 170.491 + 307.344 \times (0.6499) = 370.22\ °R$$

$$P_{pc} = 709.604 - 58.718 \times (0.6499) = 671.44\ \text{psia}$$

Comparing these calculated values with the more accurate solution in the previous example, we see that the T_{pc} is off by 2.5% and P_{pc} is off by 0.6%. These discrepancies are acceptable for most engineering calculations dealing with natural gas pipeline transportation.

19. ADJUSTMENT FOR SOUR GAS AND NONHYDROCARBON COMPONENTS

The Standing-Katz chart for compressibility factor calculation (discussed next) can be used only if there are small amounts of nonhydrocarbon components, up to 50% by volume. Adjustments must be made for sour gases containing carbon dioxide and hydrogen sulfide. The adjustments are made to the pseudo critical temperature and pseudo critical pressure as follows.

First an adjustment factor ε is calculated based on the amounts of carbon dioxide and hydrogen sulfide present in the sour gas, as follows

$$\varepsilon = 120\left(A^{0.9} - A^{1.6}\right) + 15\left(B^{0.5} - B^{4.0}\right) \qquad (3.54)$$

where
A – Sum of the mole fractions of CO_2 and H_2S
B – Mole fraction of H_2S

The adjustment factor ε is in °R. We can then apply this adjustment to the pseudo critical temperature to get the adjusted pseudo critical temperature T'_{pc} as follows

$$T'_{pc} = T_{pc} - \varepsilon \qquad (3.55)$$

Similarly, the adjusted pseudo critical pressure, P'_{pc} is

$$P'_{pc} = \frac{P_{pc} \times T'_{pc}}{T_{pc} + B(1-B)\varepsilon} \qquad (3.56)$$

20. COMPRESSIBILITY FACTOR

The concept of compressibility factor or gas deviation factor was introduced earlier in this chapter. It is a measure of how close a real gas is to an ideal gas. The compressibility factor, Z is a dimensionless number close to 1.00. It is independent of the quantity of gas. It depends on the gravity of gas, its temperature, and pressure. For example a sample of natural gas may have a Z value of 0.8595 at 1000 psia and 70 °F. Charts are available that show the variation of Z with temperature and pressure. A related term called the supercompressibility factor, F_{pv}, is defined as follows:

$$F_{pv} = 1/Z^{1/2} \qquad (3.57)$$

or

$$Z = 1/(F_{pv})^2 \qquad (3.58)$$

Several methods are available to calculate the value of Z at a temperature T and pressure P. One approach requires knowledge of the critical temperature and critical pressure of the gas mixture. The reduced temperature and reduced pressure are the calculated from the critical temperatures and pressures as follows

$$\text{Reduced temperature} = T/T_c \qquad (3.59)$$

$$\text{Reduced pressure} = P/P_c \qquad (3.60)$$

Temperatures in above equation are in absolute units as are the pressures.

The value of compressibility factor Z is calculated using one of the following methods.
1. Standing-Katz method
2. Hall-Yarborough method
3. Dranchuk, Purvis, and Robinson method
4. AGA method
5. CNGA method

20.1 Standing-Katz

This method uses a chart based on binary mixtures and saturated hydrocarbon vapor data. This approach is reliable for sweet natural gas compositions. Corrections must be applied for hydrogen sulfide and carbon dioxide content of natural gas, using the adjustment factor ε discussed earlier. See Figure 3.5 for the compressibility factor chart.

20.2 Hall-Yarborough Method

This method was developed using the equation of state proposed by Starling-Carnahan and requires knowledge of the pseudo critical temperature and pseudo critical pressure of the gas. At a given temperature T and pressure P, we first calculate the pseudo reduced temperature and pseudo reduced pressure. Next, a parameter, y, known as reduced density is calculated from the following equation.

$$-0.06125 P_{pr} t e^{-1.2(1-t)^2} + \frac{y + y^2 + y^3 - y^4}{(1-y)^3} - Ay^2 + By^{(2.18+2.82t)} = 0 \qquad (3.61)$$

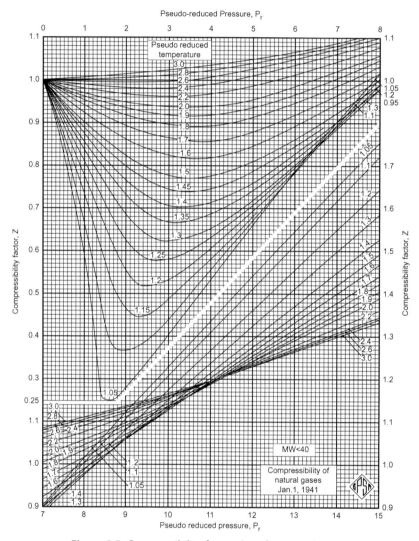

Figure 3.5 Compressibility factor chart for natural gases.

where A and B are defined as follows

$$A = \left(14.76t - 9.76t^2 + 4.58t^3\right)$$

$$B = \left(90.7t - 242.2t^2 + 42.4t^3\right)$$

and

P_{pr} – Pseudo reduced pressure
T_{pr} – Pseudo reduced temperature

t − 1/T_{pr}
y − Reduced density, dimensionless.

The calculation of y is not straightforward and requires a trial and error approach. Once y is calculated, the compressibility factor Z is found from the following equation.

$$Z = \frac{-0.06125 P_{pr} t e^{-1.2(1-t)^2}}{y} \qquad (3.62)$$

20.3 Dranchuk, Purvis, and Robinson Method

In this method, the Benedict-Webb-Rubin equation of state is used to correlate the Standing-Katz Z factor chart. Eight coefficients (e.g., A_1, A_2) are used in this equation as indicated here.

$$Z = 1 + \left(A_1 + \frac{A_2}{T_{pr}} + \frac{A_3}{T_{pr}^3}\right)\rho_r + \left(A_4 + \frac{A_5}{T_{pr}}\right)\rho_r^2 + \frac{A_5 A_6 \rho_r^5}{T_{pr}}$$
$$+ \frac{A_7 \rho_r^3}{T_{pr}^3\left(1 + A_8 \rho_r^2\right)\exp\left(-A_8 \rho_r^2\right)} \qquad (3.63)$$

where ρ_r and the constants A_1 through A_8 are given as follows

$$\rho_r = 0.27\, P_{pr}/(Z T_{pr}) \qquad (3.64)$$

$A_1 = 0.31506237$ $A_2 = -1.04670990$
$A_3 = -0.57832729$ $A_4 = 0.53530771$
$A_5 = -0.61232032$ $A_6 = -0.10488813$
$A_7 = 0.68157001$ $A_8 = 0.68446549$

20.4 AGA Method

The AGA method of calculating the compressibility factor Z involves a complicated mathematical approach using the gas properties. A computer program is necessary to calculate the Z factor.

It may be stated as follows

$$Z = \text{Function (gas properties, pressures, temperature)} \qquad (3.65)$$

The AGA method for calculating Z is outlined in AGA-IGT, Report No. 10. This correlation is valid for gas temperatures ranging from 30 °F to 120 °F and for gas pressures up to 1380 psig. The calculated values are fairly accurate and within 0.03% of the chart method in this range of temperatures and pressures. With higher temperatures and pressures, the difference between the AGA method and the chart method may be as high as 0.07%.

For details of other methods of compressibility calculations refer to American Gas Association publication Report No. 8, Second Edition, November 1992.

20.5 CNGA Method

This is referred to variously as the California or Canadian Natural Gas Association (CNGA) method and is one of the easiest equations for calculating the compressibility factor from given gas gravity, temperature, and pressure. Using this method, the compressibility factor Z is calculated from the following formula.

$$Z = 1 \Big/ \left[1 + \left(P_{avg} 344400 (10)^{(1.785G)} \right) \Big/ T_f^{3.825} \right]^2 \quad (3.66)$$

This formula is valid for the average gas pressure $P_{avg} > 100$ psia. When $P_{avg} <= 100$, we can assume that $Z = 1.00$.

where
P_{avg} – Average gas pressure, psia
T_f – Average gas temperature, °R
G – Gas gravity (air = 1.00).

In the case of a gas flowing through a pipeline, because the pressure varies along the pipeline, the compressibility factor Z must be calculated based on an average pressure at a particular location on the pipeline. If two locations have pressures of P_1 and P_2, we could use a simple average pressure of $(P_1 + P_2)/2$. However, a more accurate value of the average pressure is calculated using the following equation.

$$P_{avg} = \frac{2}{3} \left(P_1 + P_2 - \frac{P_1 \times P_2}{P_1 + P_2} \right) \quad (3.67)$$

Example Problem 3.14
Using the Standing-Katz chart and the calculated values of T_{pc} and P_{pc}, calculate the compressibility factor for the gas in the previous example at 80 °F and 100 psig.

Solutions
From previous example we get:
  The pseudo reduced temperature $T_{pr} = 1.4216$ °R
  The pseudo reduced pressure $P_{pr} = 1.5198$ psia
  Using the Standing-Katz chart Figure 3.5, we read the value of Z as

$$Z = 0.83$$

Example Problem 3.15
A natural gas sample has the following molecular composition.

Component	y
C_1	0.780
C_2	0.005
C_3	0.002
N_2	0.013
CO_2	0.016
H_2S	0.184

where y represents the mole fraction.
1. Calculate the molecular weight of the gas, its gravity, and the pseudo critical temperature and pseudo critical pressure.
2. Determine the compressibility factor of this gas at 100 °F temperature and 1000 psia pressure.

Solution
From the properties of hydrocarbon components, we created the following spreadsheet showing the molecular weight M, critical temperature T_c, and critical pressure P_c for each of the component gases and calculated the molecular weight of the mixture and the pseudo critical temperature and pseudo critical pressure using Kay's rule discussed earlier under Eqns (3.49) and (3.50).

Component	y	M	yM	T_c	P_c	yT_c	yP_c
C_1	0.780	16.04	12.5112	343	666	267.54	519.48
C_2	0.005	30.07	0.1504	550	707	2.75	3.54
C_3	0.002	44.10	0.0882	666	617	1.33	1.23
N_2	0.013	28.01	0.3641	227	493	2.95	6.41
CO_2	0.016	44.01	0.7042	548	1071	8.77	17.14
H_2S	0.184	34.08	6.2707	672	1306	123.65	240.30
Total	1.000		20.0888			406.99	788.10

Therefore, molecular weight of the natural gas sample is
$$Mw = \sum yM = 20.09$$
and the gas gravity is
$$G = Mw/29.0 = 20.09/29.0 = 0.6928$$
Also:
Pseudo critical temperature = $\Sigma \, yTc = 406.99$ °R
Pseudo critical pressure = $\Sigma \, yP_c = 788.1$ psia

Because this is a sour gas that contains more than 5% nonhydrocarbons, we must adjust the pseudo critical temperature and pseudo critical pressure.

The temperature adjustment factor ε is calculated from Eqn (3.54) as follows

$$A = (0.016 + 0.184) = 0.20 \text{ and } B = 0.184$$

Therefore

$$\varepsilon = 120\left[(0.2)^{0.9} - (0.2)^{1.6}\right] + 15\left[(0.184)^{0.5} - (0.184)^{4.0}\right] = 25.47 \,°R$$

Therefore, the adjusted pseudo critical temperature and pseudo critical pressure are

$$T'_{pc} = 406.99 - 25.47 = 381.52 \,°R$$

$$P'_{pc} = \frac{788.1 \times 381.52}{406.99 + 0.184 \times (1 - 0.184) \times 25.47} = 731.90 \text{ psia}$$

We can now calculate the compressibility factor Z at 100 °F and 1000 psia pressure using the pseudo reduced temperature and pseudo reduced pressure as follows

$$\text{Pseudo reduced temperature} = (100 + 460)/381.52 = 1.468$$

$$\text{Pseudo reduced pressure} = 1000/731.9 = 1.366$$

Using the Standing-Katz chart for the pseudo reduced temperature and pseudo reduced pressure, we get

$$Z = 0.855$$

Example Problem 3.16
The gas gravity of a sample of natural gas is 0.65. Calculate the compressibility factor of this gas at 1000 psia pressure and a temperature of 80 °F using the CNGA method. Use a base temperature of 60 °F.

Solution

$$\text{Gas temperature } T_f = 80 + 460 = 540 \,°R$$

Using Eqn (3.66), with slight simplification, the Z factor is given by

$$\frac{1}{\sqrt{Z}} = 1 + \frac{1000 \times 344{,}400 \times (10)^{1.785 \times 0.65}}{540^{3.825}} = 1.1762$$

Solving for Z, we get

$$Z = 0.7228$$

Physical Properties

21. HEATING VALUE

The heating value of a gas represents the thermal energy available per unit volume of the gas. For natural gas, the heating value ranges from 900 to 1000 Btu/ft^3. Two heating values are used in practice: lower heating value and higher heating value. The gross heating value of a gas mixture is calculated from the heating value of the component gases using the following equation.

$$H_m = \sum (yH) \qquad (3.68)$$

where y represents percentage of each component gas with heating value H.

Calculating properties of gas mixtures.

The specific gravity and viscosity of gas mixtures may be calculated from that of the component gases as follows.

Specific gravity of a mixture of gases is calculated from the percentage composition of each component gas and its molecular weight. If the gas mixture consists of three components with molecular weights, M_1, M_2, and M_3, and the respective percentages are pct_1, pct_2, and pct_3. The apparent molecular weight of the mixture.

$$M_m = (pct_1 M_1 + pct_2 M_2 + pct_3 M_3)/100$$

or

$$M_m = \sum (yM)/100 \qquad (3.69)$$

where y represents percentage of each component gas with molecular weight M.

The specific gravity G_m of the gas mixture (relative to air $= 1.00$) is

$$G_m = M_m/28.9625 \qquad (3.70)$$

Example Problem 3.17

A typical natural gas mixture consists of 85% methane, 10% ethane, and 5% butane. Assuming the molecular weights of the three component gases to be 16.043, 30.070, and 44.097, respectively, calculate the specific gravity of this natural gas mixture. Use 28.9625 for the molecular weight of air.

Solution

Applying the percentages to each component in the mixture, we get the molecular weight of the mixture as

$$(0.85 \times 16.043) + (0.10 \times 30.070) + (0.05 \times 44.097) = 18.8484$$

The specific gravity of gas = molecular weight of gas/molecular weight of air.

$$G = 18.8484/28.9625 = 0.6508$$

The viscosity of a mixture of gases at a specified pressure and temperature can be calculated if the viscosities of the component gases in the mixture are known. The following formula can be used to calculate the viscosity of a mixture of gases.

$$\mu = \frac{\Sigma'\left(\mu_i y_i \sqrt{M_i}\right)}{\Sigma\left(y_i \sqrt{M_i}\right)} \tag{3.71}$$

where y_i represents the mole fraction of each component gas with molecular weight M_i and μ_i is the viscosity of the component. The viscosity of the mixture is μ_m. Viscosities of common gases at atmospheric conditions are shown in Figure 3.4.

Example Problem 3.18
The viscosities of components C_1, C_2, C_3, and C_4 of a natural gas mixture and their percentages are as follows

Component	y
C_1	0.8500
C_2	0.0900
C_3	0.0400
nC_4	0.0200
Total	1.000

Determine the viscosity of the gas mixture.

Solution

Component	y	M	$M^{1/2}$	$yM^{1/2}$	μ	$\mu y M^{1/2}$
C_1	0.8500	16.04	4.00	3.4042	0.0130	0.0443
C_2	0.0900	30.07	5.48	0.4935	0.0112	0.0055
C_3	0.0400	44.10	6.64	0.2656	0.0098	0.0026
nC_4	0.0200	58.12	7.62	0.1525	0.0091	0.0014
Total	1.000			4.3159		0.0538

The viscosity of the gas mixture is calculated as follows

The viscosity of the gas mixture = $0.0538/4.3158 = 0.0125$

Physical Properties

22. SUMMARY

In this chapter, we discussed the more important properties of liquids that determine the nature of liquid flow in pipelines. The specific gravity and viscosity of liquids were explained along with how to calculate these properties in liquid mixtures and at various temperatures. We also introduced the basic concepts of liquid flow consisting of the continuity equation and the energy equation embodied in Bernoulli's equation.

23. PROBLEMS

1. Calculate the specific weight and specific gravity of a liquid that weighs 312 lb, contained in volume of 5.9 ft^3. Assume water weighs 62.4 lb/ft^3.
2. The specific gravity of a liquid at 60 °F and 100 °F are reported to be 0.895 and 0.815, respectively. Determine the specific gravity of the liquid at 85 °F. Assume a linear relationship between gravity and temperature.
3. The gravity of a petroleum product is 59° API. Calculate the corresponding specific gravity at 60 °F.
4. The viscosity of a liquid at 70 °F is 45 cSt. Express this viscosity in SSU. If the specific gravity at 70 °F is 0.885, determine the absolute or dynamic viscosity.
5. The viscosities of a crude oil at 60 °F and 100 °F are 40 cSt and 15 cSt, respectively. Using the ASTM correlation method, calculate the viscosity of this product at 80 °F.
6. Two liquids are blended to form a homogeneous mixture. The first liquid A has a specific gravity of 0.815 at 70 °F and a viscosity of 15 cSt at 70 °F. At the same temperature, liquid B has a specific gravity of 0.85 and viscosity of 25 cSt. If 20% of liquid A is blended with 80% of liquid B, calculate the specific gravity and viscosity of the blended product.
7. Using the viscosity blending chart, calculate the blended viscosity for two liquids as follows:

Product	Percentage	Viscosity (SSU)
Liquid A	15	50
Liquid B	85	200

8. If liquid A with a viscosity of 40 SSU is blended with liquid B of viscosity 150 SSU. What percentage of each component would be required to obtain a blended viscosity of 46 SSU?
9. In Figure 3.3, consider the pipe to have a 20-in diameter. The liquid is water with specific gravity = 1.00. Point A is at elevation 100 ft and B is at elevation 200 ft. The pressure at A is 500 psi and that at B is 400 psi. Specific weight of water is 62.34 lb/ft^3. Write down Bernoulli's equation for energy conservation between points A and B.

CHAPTER FOUR

Pipeline Stress Design

In this chapter, we discuss the strength capabilities of a transmission pipeline that is subject to internal pressure. We will also cover pipeline hydrotesting and various safety factors along with design code implications for gas and liquid pipelines. To transport a liquid or a gas through a pipeline, pumping pressure is required at the origin of the pipeline as well as at intermediate pump and compressor stations along the length of the pipeline. These internal pressures subject the pipe material to circumferential (or hoop stress), axial, and radial stresses. We must therefore select proper pipe material and adequate wall thickness to withstand the pressure during the normal course of operation of the pipeline. Also in buried pipelines the pipe is subject to external loads and these cause additional stresses that have to be considered in selecting the required pipe wall thickness. We will discuss different materials used to construct the pipelines, the standards and codes that apply, and the method of calculating the internal pressure that a given pipe can withstand based on pipe material and wall thickness.

For a specific internal pressure, the minimum wall thickness required will be calculated depending on the pipe diameter and yield strength of the pipe material, based on design and construction codes. This approach uses the Barlow's equation for internal pressure. Next, we establish the hydrostatic test pressure the pipeline will be subject to, such that the pipeline can be operated safely at the internal design pressure specified.

1. ALLOWABLE OPERATING PRESSURE AND HYDROSTATIC TEST PRESSURE

To transport a liquid or gas through a transmission pipeline, the fluid must be under sufficient pressure to compensate for the pressure loss because of friction and the pressure required for any elevation changes. Longer pipelines operating at high flow rates cause higher frictional pressure drop and consequently higher internal pressure to transport the product from the beginning of the pipeline to its terminus.

In gravity flow systems, liquids flow from a storage tank located at a higher elevation down to a terminus at lower elevation without additional pump pressure. Even if no external pumping pressure is required for a gravity flow system, the pipeline will still be subject to internal pressure because of the static elevation difference between the two ends of the pipeline. The allowable internal operating pressure in a pipeline is defined as the maximum safe continuous pressure that the pipeline can be operated at without causing rupture. This is frequently referred to as the maximum allowable operating pressure (MAOP). At this internal pressure, the pipe material is subjected to stresses that are safely below the yield strength of the pipe material. As indicated previously, the pipe subjected to internal pressure results in stresses in the pipe material in three different directions as follows:

1. Circumferential (or hoop) stress
2. Longitudinal (or axial) stress
3. Radial stress

The last item is significant only in thick wall pipes. BecauseBecause most transmission pipelines are considered to be thin-walled, the radial stress component is neglected. Thus the two important stresses in a transmission pipeline are hoops stress S_h and axial stress S_a as shown in Figure 4.1.

It will be shown shortly that S_a equals $S_h/2$. Therefore, the hoop stress is the controlling stress that determines the amount of internal pressure the pipeline can withstand. In liquid transmission pipelines the hoop stress may be allowed to reach 72% of the yield strength of the pipe material.

If pipe material has 70,000 psi yield strength, the maximum hoop stress that the pipeline can be subject to because of internal pressure is

$$0.72 \times 70,000 = 50,400 \text{ psi}.$$

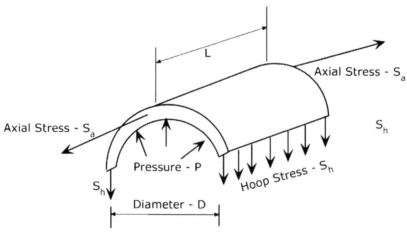

Figure 4.1 Pipe stress.

To ensure that the pipeline can be safely operated at a particular MAOP, the pipeline must be tested at a higher internal pressure prior to be put into operation. Usually this is done using water and the process is called hydrostatic testing. The pipeline is divided into test sections and is filled with water. Each test section is subjected to the required hydrostatic test pressure and maintained for a specified period such as 8 h and the pipeline inspected for leaks.

The hydrostatic test pressure is the pressure (higher than MAOP) that the pipeline is tested to. Generally, the pipeline is tested for a specified period of time, such as 4 h (for above-ground piping) or 8 h (buried pipeline) as required by the pipeline design code or by city, county, state, or federal government regulations. In the United States, the Department of Transportation (DOT) Code of Federal Regulations, Part 192 applies to gas pipelines. For liquid transmission pipelines the corresponding code is Part 195. Generally, the hydrostatic test pressure must be a minimum of 125% of the MAOP. Thus, if the MAOP is 1440 psig, the pipeline will be hydrostatically tested at a minimum pressure of $1.25 \times 1440 = 1800$ psig. Because the MAOP is based on the hoop stress equal to 72% of the yield strength of the pipeline, during hydrotesting S_h will reach a value of $1.25 \times 72\% = 90\%$ of pipe yield strength. To summarize, under normal operating conditions the MAOP results in hoop stress S_h equals 72% specified minimum yield strength (SMYS) for liquid pipelines. During hydrotesting, S_h equals 90% SMYS, where SMYS represents the specified minimum yield strength of pipe material.

Transmission pipelines are constructed using pipe materials conforming to API 5LX standards. Commonly used pipe materials are designated as API 5LX 42, 46, 52, 60, 65, 70, and 80.

API 5LX-42 pipe has an SMYS equal to 42,000 psi, whereas API 5LX-80 has an SMYS of 80,000 psi.

Calculation of internal design pressure in a pipeline is based on Barlow's equation for thin-walled cylindrical pipes and will be discussed in the next section.

2. BARLOW'S EQUATION FOR INTERNAL PRESSURE

When a thin-walled cylindrical pipe is subject to internal pressure, the hoop stress in the pipe material is calculated using Barlow's equation as follows:

$$S_h = \frac{PD}{2t} \tag{4.1}$$

where

S_h - hoop (or circumferential) stress, psi
P - internal pressure, psi
D - pipe outside diameter, in
t - pipe wall thickness, in

In SI units, the same formula applies, with pressure and stress in kPa and Dia and wall thickness in mm.

In addition to the hoop stress, the axial stress that occurs in the longitudinal direction designated as S_a is calculated as follows:

$$S_a = \frac{PD}{4t} \qquad (4.2)$$

In SI units, the same equation applies with appropriate units as above.

The hoop stress S_h equals double the value of the axial stress S_a. Therefore, to determine the minimum wall thickness required for a pipe of diameter D subject to an internal pressure P, we use Eqn (4.1) based on the hoop stress. Suppose the internal design pressure required is 1200 psig. If the pipe outside diameter is 20 in and the allowable hoop stress is 50,400 psi (corresponding to 72% of API 5LX-70 pipe), the wall thickness required will be calculated from Eqn (4.1) as follows:

$$50,400 = 1200 \times 20/(2t)$$

$$t = 1200 \times 20/(2 \times 50,400) = 0.2381 \text{ in.}$$

Thus a minimum wall thickness of $t = 0.2381$ in is required to withstand the internal pressure of 1200 psi without exceeding a hoop stress of 50,400 psi in the 20-in outside diameter pipeline. In this calculation, we used the hoop stress value of 50,400 psi. For liquid pipelines and gas pipelines in a class 1 location, 72% SMYS is the maximum allowable hoop stress. If the pipe material were API 5LX-80, the allowable hoop stress is

$$S_h = 0.72 \times 80,000 = 57,600 \text{ psi}$$

Therefore, the minimum wall thickness required with X-80 pipe is, using Eqn (4.2).

$$t = 1200 \times 20/(2 \times 57,600) = 0.2083 \text{ in.}$$

Therefore, by changing the pipe material from X-70 to X-80 pipe, the required wall thickness for the 20-in outside diameter pipe to withstand an internal pressure of 1200 psi is reduced from 0.2381 to 0.2083 in. This represents a percentage reduction of

$$0.2381 - 0.2083/0.2381 \times 100 = 12.52\%$$

If the pipe length is 100 miles, this will mean a reduction in total pipe requirement of

$$(12.52\%) \times 10.68 \times (20 - 0.2381) \times 0.2381 \times 5280 \times 100/2000 \text{ tons}$$

or a reduced cost of $1660.99 \times 1500) = $2.491 million based on a cost of $1500 per ton.

The derivation of Barlow's Eqn (4.1) is as follows:

Consider one half of a length of pipe L as shown in Figure 4.1. The internal pressure P causes a bursting force on one half of the pipe equal to pressure multiplied by the projected area as follows:

$$\text{Bursting force} = P \times D \times L$$

This bursting force is exactly balanced by the hoop stress S_h acting along both edges of the half pipe section. Therefore

$$(S_h \times t \times L) \times 2 = P \times D \times L$$

Solving for S_h we get

$$S_h = \frac{PD}{2t}$$

Equation (4.1) for axial stress S_a is derived as follows:

The axial stress S_a acts on an area of cross-section of pipe represented by $\pi D t$. This is balanced by the internal pressure P acting on the internal cross-sectional area of pipe $\pi D^2/4$. Equating the two, we get

$$S_a \times \pi D t = P \times \pi D^2/4$$

Solving for S_a, we get

$$S_a = \frac{PD}{4t}$$

In the design of transmission pipeline the internal design pressure is calculated from Barlow's equation with some modifications as follows. We introduce three factors E, F, and T called seam joint factor, design factor, and temperature de-ration factor, respectively. In US customary system units, the internal design pressure in a pipe is calculated using the following equation:

$$P = \frac{2tSEFT}{D} \qquad (4.3)$$

where

P - internal pipe design pressure, psig
D - nominal pipe outside diameter, in
t - nominal pipe wall thickness, in
S - SMYS of pipe material, psig
E - seam joint factor, 1.0 for seamless and submerged arc welded (SAW) pipes. See table in Appendix.
F - design factor, usually 0.72 for liquid pipelines, except that a design factor of 0.60 is used for pipe, including risers, on a platform located off shore or on a platform in inland navigable waters, and 0.54 is used for pipe that has been subjected to cold expansion to meet the SMYS and subsequently heated, other than by welding or stress relieving as a part of the welding, to a temperature higher than 900 °F (482 °C) for any period or over 600 °F (316 °C) for more than 1 h.
T - temperature de-ration factor equals 1.00 for temperatures below 250 °F (121 °C). See Table 4.1 for details.

For gas transmission pipelines, the design factor F ranges from 0.72 for cross-country gas pipelines to as low as 0.4 in class 4 locations. Class locations for gas pipelines depend on the population density in the vicinity of the pipeline and are listed in Table 4.1.

This form of Barlow's equation may be found in Parts 192 and 195 of Code of Federal Regulations, Title 49, and ASME standards B31.4 and B31.8 for liquid and gas pipelines, respectively.

In SI units, Barlow's equation can be written as:

$$P = \frac{2tSEFT}{D} \qquad (4.4)$$

Table 4.1 Class Location Factor

Class Location	Design Factor, F
1	0.72
2	0.60
3	0.50
4	0.40

Pipeline Stress Design

where

    P - pipe internal design pressure, kPa
    D - nominal pipe outside diameter, mm
    t - nominal pipe wall thickness, mm
    S - SMYS of pipe material, kPa
    E - seam joint factor, 1.0 for seamless and SAW pipes. See table in Appendix.
    F - design factor, usually 0.72 for liquid pipelines, except that a design factor of 0.60 is used for pipe, including risers, on a platform located off shore or on a platform in inland navigable waters, and 0.54 is used for pipe that has been subjected to cold expansion to meet the SMYS and subsequently heated, other than by welding or stress relieving as a part of the welding, to a temperature higher than 900 °F (482 °C) for any period or over 600 °F (316 °C) for more than 1 h.
    T - temperature de-ration factor equals 1.00 for temperatures below 250 °F (121 °C).

3. GAS TRANSMISSION PIPELINE: CLASS LOCATION

The following definitions of class 1 through class 4 are taken from DOT 49 Code of Federal Regulations, Part 192. The class location unit is defined as an area that extends 220 yards on either side of the centerline of a 1-mile section of pipe as indicated in Figure 4.2.

3.1 Class 1

Offshore gas pipelines are class 1 locations. For onshore pipelines, any class location unit that has 10 or fewer buildings intended for human occupancy is termed class 1.

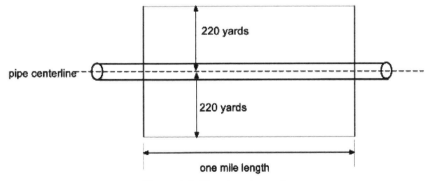

Figure 4.2 Class location unit.

Table 4.2 Temperature De-ration Factor T

Temperature		Derating Factor, T
°F	°C	
250 or less	121 or less	1.000
300	149	0.967
350	177	0.033
400	204	0.900
450	232	0.867

3.2 Class 2

This is any class location unit that has more than 10 but fewer than 46 buildings intended for human occupancy.

3.3 Class 3

This is any class location unit that has 46 or more buildings intended for human occupancy or an area where the pipeline is within 100 yards of a building or a playground, recreation area, outdoor theater, or other place of public assembly that is occupied by 20 or more people on at least 5 days a week for 10 weeks in any 12-month period. The days and weeks needed not be consecutive.

3.4 Class 4

This is any class location unit where buildings with four or more storey above ground exist.

The temperature de-ration factor T is equal to 1.00 up to gas temperature 250 °F as indicated in Table 4.2.

Problem 4.1

A gas pipeline is constructed of API 5LX-65 steel, NPS 16 0.250-in wall thickness. Calculate the MAOP of this pipeline for class 1 through class 4 locations. Use a temperature de-ration factor of 1.00.

Solution

Using Eqn (4.4), the MAOP is given by

$$P = \frac{2 \times 0.250 \times 65000 \times 1.0 \times 0.72 \times 1.0}{16} = 1462.5 \text{ psig for class 1}$$

Similarly, for

$$\text{Class 2, } \text{MAOP} = 1462.5 \times \frac{0.6}{0.72} = 1218.8 \text{ psig}$$

$$\text{Class 3, MAOP} = 1462.5 \times \frac{0.5}{0.72} = 1015.62 \text{ psig}$$

$$\text{Class 4, MAOP} = 1462.5 \times \frac{0.4}{0.72} = 812.5 \text{ psig}$$

In summary, Barlow's equation for internal pressure is based on calculation of the hoop stress (circumferential) in the pipe material. Within the stressed pipe material, there are two stresses on a pipe element called hoop stress and axial or longitudinal stress. It can be shown that the controlling stress is the hoop stress being twice the axial stress as depicted in Figure 4.3.

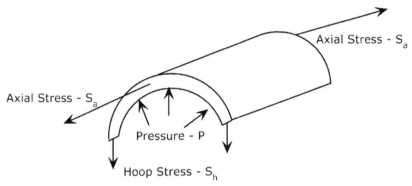

Figure 4.3 Hoop stress and axial stress in pipe.

The strength of pipe material designated as SMYS in Eqns (4.1) and (4.2) depend on pipe material and grade. In the United States, steel pipeline material used in the oil and gas industry is manufactured in accordance with American Petroleum Institute (API) standards 5L and 5LX. For example, grades 5LX-42, 5LX-52, 5LX-60, 5LX-65, 5LX-70, and 5LX-80 are used commonly in pipeline applications. The numbers after 5LX indicate the SMYS values in thousands of psi. Thus, 5LX-52 pipe has a minimum yield strength of 52,000 psi. The lowest grade of pipe material used is 5L grade B, which has an SMYS of 35,000 psi. In addition, seamless steel pipe designated as ASTM A106 and grade B pipe are also used for liquid pipeline systems. These have an SMYS value of 35,000 psi.

It is obvious from Barlow's Eqn (4.4), that for a given pipe diameter, pipe material, and seam joint factor, the allowable internal pressure P is directly proportional to the pipe wall thickness. For example, 16-in diameter pipe with a wall thickness of 0.250 in made of 5LX-52 pipe has an allowable internal design pressure of 1170 psi calculated as follows:

$$P = (2 \times 0.250 \times 52,000 \times 1.0 \times 0.72)/16 = 1170 \text{ psig}$$

Therefore if the wall thickness is increased to 0.375 in the allowable internal design pressure increases to

$$(0.375/0.250) \times 1170 = 1755 \text{ psig}$$

On the other hand, if the pipe material is changed to 5LX-70, keeping the wall thickness at 0.250 in, the new internal pressure is

$$(70,000/52,000)1170 = 1575 \text{ psig}$$

Note that we used the Barlow's equation to calculate the allowable internal pressure-based on the pipe material being stressed to 72% of SMYS. In some situations, more stringent city or government regulations may require that the pipe be operated at a lower pressure. Thus, instead of using a 72% factor in Eqn (4.4), we may be required to use a more conservative factor (lower number) in place of $F = 0.72$. As an example, in certain areas of the City of Los Angeles, liquid pipelines are only allowed to operate at a 66% factor instead of the 72% factor. Therefore, in the earlier example, the 16 in/0.250 in/X52 pipeline can only be operated at

$$1170(66/72) = 1073 \text{ psig.}$$

As mentioned previously, to operate a pipeline at 1170 psig, it must be hydrostatically tested at 25% higher pressure. Because 1170 psig internal pressure is based on the pipe material being stressed to 72% of SMYS, the hydrostatic test pressure will cause the hoop stress to reach

$$1.25(72) = 90\% \text{ of SMYS.}$$

Generally, the hydrostatic test pressure is specified as a range of pressures, such as 90% SMYS to 95% SMYS. This is called the hydrotest pressure envelope. Therefore, in the present example, the hydrotest pressure range is

$$1.25(1170) = 1463 \text{ psig} - \text{lower limit } (90\% \text{ SMYS})$$

$$(95/90)1463 = 1544 \text{ psig} - \text{higher limit } (95\% \text{ SMYS})$$

To summarize, a pipeline with an MAOP of 1170 psig needs to be hydrotested at a pressure range of 1463–1544 psig. According to the design code, the test pressure will be held for a minimum 4-h period for aboveground pipelines and 8 h for buried pipelines.

In calculating the allowable internal pressure in older pipelines, consideration must be given to wall thickness reduction because of corrosion over the life of the pipeline. A pipeline that was installed 25 years ago with 0.250-in wall thickness may have reduced in wall thickness to 0.200 in or less because of corrosion. Therefore, the allowable internal pressure will have to be reduced in the ratio of the wall thickness compared with the original design pressure.

4. LINE FILL VOLUME AND BATCHES

Frequently we need to know how much liquid is contained in a pipeline between two points along its length, such as between valves or pump stations.

For a circular pipe, we can calculate the volume of a given length of pipe by multiplying the internal cross-sectional area by the pipe length. If the pipe inside diameter is D in and the length is L ft, the volume of this length of pipe is

$$V = 0.7854 \left(D^2/144\right)L \quad (4.5)$$

where
  V - volume, ft^3
Simplifying:

$$V = 5.4542 \times 10^{-3}D^2L \quad (4.6)$$

We will now restate this equation in terms of conventional pipeline units, such as the volume in barrels (bbl) in a mile of pipe.

The quantity of liquid contained in a mile of pipe also called the line fill volume is calculated as follows:

$$V_L = 5.129(D)^2 \quad (4.7)$$

where
  V_L - line fill volume of pipe, bbl/mile
  D - pipe inside diameter, in

In SI units, we can express the line fill volume per km of pipe as follows:

$$V_L = 7.855 \times 10^{-4}D^2 \quad (4.8)$$

where
  V_L - line fill volume, m^3/km
  D - pipe inside diameter, mm

Using Eqn (4.7), a pipeline 100 miles long, 16-in diameter, and 0.250-in wall thickness, has a line fill volume of

$$5.129(15.5)^2(100) = 123,224 \text{ bbl}$$

Many crude oil and refined product pipelines operate in a batched mode. Multiple products are simultaneously pumped through the pipeline as batches. For example, 50,000 bbl of product C will enter the pipeline

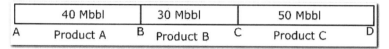

Figure 4.4 Batched pipeline.

followed by 30,000 bbl of product B and 40,000 bbl of product A. If the line fill volume of the pipeline is 120,000 bbl, an instantaneous snap shot condition of a batched pipeline is as shown in Figure 4.4.

Problem 4.2
A 50-mile pipeline consists of a 20 mile of 16-in diameter, 0.375-in wall thickness pipe followed by 30 miles of 14-in diameter, 0.250-in wall thickness pipe. Calculate the total volume contained in the 50-mile long pipeline.

Solution
Using Eqn (4.7) we get

For the 16-in pipeline

$$\text{Volume per mile} = 5.129\ (15.25)^2 = 1192.81\ \text{bbl/mi}$$

For the 14-in pipeline

$$\text{Volume per mile} = 5.129\ (13.5)^2 = 934.76\ \text{bbl/mi}$$

Total line fill volume is

$$20 \times 1192.81 + 30 \times 934.76 = 51899\ \text{bbl}.$$

Problem 4.3
A pipeline 100 km long is 600 mm outside diameter and 25 mm wall thickness. If batches of three liquids A (3000 m³), B (5000 m³), and C occupy the pipe, at a particular instant, calculate the interface locations of the batches, considering the origin of the pipeline to be at 0.0 km.

Solution
Using Eqn (4.8), we get the line fill volume per km to be

$$V_L = 7.855 \times 10^{-4}(600 - 25)^2 = 216.65\ \text{m}^3/\text{km}$$

The first batch A will start at 0.0 km and will end at a distance of

$$3000/216.65 = 16.86\ \text{km}$$

The second batch B starts at 16.86 km and ends at

$$16.86 + (5000/177.9754) = 44.95\ \text{km}$$

The third batch C starts at 44.95 km and ends at 100 km.
The total volume in pipe is

$$177.9754 \times 100 = 17,798 \ m^3$$

Thus the volume of the third batch C is

$$17798 - 300 - 5000 = 9,798 \ m^3$$

It can thus be seen that line fill volume calculation is important when dealing with batched pipelines. We need to know the boundaries of each liquid batch, so that the correct liquid properties can be used to calculate pressure drops for each batch.

The total pressure drop in a batched pipeline would be calculated by adding up the individual pressure drop for each batch. Because intermixing of the batches is not desirable, batched pipelines must run in turbulent flow. In laminar flow, there will be extensive mixing of the batches, which defeats the purpose of keeping each product separate so that at the end of the pipeline each product may be diverted into a separate tank. Some intermixing will occur at the product interfaces and this contaminated liquid is generally pumped into a slop tank at the end of the pipeline and may be blended with a less critical product. The amount of contamination that occurs at the batch interface depends on the physical properties of the batched products, batch length, and Reynolds number. Several correlations have been developed for determining the contamination volumes and will be discussed in a subsequent chapter on batching.

5. GAS PIPELINES

5.1 Pipe Wall Thickness

In Chapter 3, we calculated the pressure needed to transport a given volume of gas through a pipeline. The internal pressure in a pipe causes the pipe wall to be stressed and if allowed to reach the yield strength of the pipe material it could cause permanent deformation of the pipe and ultimate failure. Obviously, the pipe should have sufficient strength to handle the internal pressure safely. In addition to the internal pressure resulting from gas flowing through the pipe, the pipe may also be subjected to external pressure.

External pressure can result from the weight of the soil above the pipe in a buried pipeline and also by the loads transmitted from vehicular traffic in areas where the pipeline is located below roads, highways, and railroads. The deeper the pipe is buried, the higher will be the soil load on the pipe.

However, the pressure transmitted to the pipe from vehicles above ground will diminish with the depth of the pipe below the ground surface. Thus, the external pressure from vehicular loads on a buried pipeline that is 6 ft below ground will be less than that on a pipeline that is at a depth of 4 ft. In most cases involving buried pipelines, transporting gas and other compressible fluids, the effect of the internal pressure is more than that of external loads. Therefore, the necessary minimum wall thickness will be dictated by the internal pressure in a gas pipeline.

The minimum wall thickness required to withstand the internal pressure in a gas pipeline will depend on the pressure, pipe diameter, and pipe material. The larger the pressure or diameter larger would be the wall thickness required. Higher strength steel pipe will require less wall thickness to withstand the given pressure compared to low strength materials. The commonly used formula to determine the wall thickness for internal pressure is known as Barlow's equation. This equation has been modified to take into account design factors and type of pipe joints (seamless, welded, etc.) and incorporated into design codes such as DOT Code of Federal Regulations Part 192 and ASME B31.8 Standards. Refer to Chapter 2 for a full list of design codes and standards used in the design, construction, and operation of gas pipelines.

6. BARLOW'S EQUATION

When a circular pipe is subject to internal pressure, the pipe material at any point will have two stress components at right angles to each other. The larger of the two stresses is known as the hoop stress and acts along the circumferential direction. Hence it is also called the circumferential stress. The other stress is the longitudinal stress, also known as the axial stress, which acts in a direction parallel to the pipe axis. Figure 4.1 shows a cross-section of a pipe subject to internal pressure. An element of the pipe wall material is shown with the two stresses S_h and S_a in perpendicular directions. Both stresses will increase as the internal pressure is increased. As will be shown shortly, the hoop stress S_h is the larger of the two stresses and hence will govern the minimum wall thickness required for a given internal pressure.

In its basic form, Barlow's equation relates the hoop stress in the pipe wall to the internal pressure, pipe diameter, and wall thickness as follows:

$$S_h = \frac{PD}{2t} \qquad (4.9)$$

where
S_h = hoop or circumferential stress in pipe material, psi
P = internal pressure, psi
D = pipe outside diameter, in
t = pipe wall thickness, in

Similar to Eqn (4.9), the axial (or longitudinal) stress, S_a is given by the following equation:

$$S_a = \frac{PD}{4t} \qquad (4.10)$$

Note that in these equations, the pipe diameter used is the outside diameter, not the inside diameter as used in previous chapter.

For example, consider an NPS 20 pipe, 0.500 in wall thickness, which is subject to an internal gas pressure of 1200 psig. The pipe wall material will be stressed in the circumferential direction by the hoop stress given by Eqn (4.9) as follows:

$$S_h = \frac{1200 \times 20}{2 \times 0.500} = 24,000 \text{ psig}$$

And in accordance with Eqn (4.10) the axial stress in the pipe wall is

$$S_a = \frac{1200 \times 20}{4 \times 0.500} = 12,000 \text{ psig}$$

Barlow's equation is valid only for thin-walled cylindrical pipes. Most pipelines transporting gases and liquids generally fall in this category. There are instances in which pipes carrying gases and petroleum liquids, subject to high external loads, such as deep submarine pipelines that may be classified as thick walled pipes. The governing equations for such thick wall pipes are different and more complex. We will introduce these formulas for information only.

7. THICK WALL PIPES

Consider a thick-walled pipe with an outside diameter D_o and inside diameter of D_i subject to and internal pressure of P. The largest stress in the pipe wall will be found to occur in the circumferential direction near the inner surface of the pipe. This stress can be calculated from the following equation.

$$S_{max} = \frac{P(D_o^2 + D_i^2)}{(D_o^2 - D_i^2)} \qquad (4.11)$$

The pipe wall thickness is

$$t = \frac{D_o - D_i}{2} \tag{4.12}$$

Rewriting Eqn (4.3) in terms of outside diameter and wall thickness, we get

$$S_{max} = P\left[\frac{D_o^2 + (D_o - 2t)^2}{D_o^2 - (D_o - 2t)^2}\right]$$

Simplifying further:

$$S_{max} = \frac{PD_o}{2t}\left[\frac{1 - \left(\frac{t}{D_o}\right) + 2\left(\frac{t}{D_o}\right)^2}{1 + \left(\frac{t}{D_o}\right)}\right] \tag{4.13}$$

In the limiting case, a thin wall pipe is one in which the wall thickness is very small compared with the diameter D_o. In this case (t/D) is small compared with 1 and therefore can be neglected in Eqn (4.13). Therefore, the approximation for thin-walled pipes from Eqn (4.13) becomes:

$$S_{max} = \frac{PD_o}{2t}$$

which is the same as Barlow's Eqn (4.1) for hoop stress.

Problem 4.4

A gas pipeline is subject to an internal pressure of 1400 psig. It is constructed of steel pipe 36 in outside diameter and 0.75 in wall thickness. Calculate the maximum hoop stress in the pipeline considering both the thin walled approach and the thick wall equation. What is the error in assuming that the pipe is thin walled?

Solution

Pipe inside diameter = 36−2 × 0.75 = 34.5 in

From Eqn (4.9) for thin-walled pipe, Barlow's equation gives the maximum hoop stress as

$$S_h = \frac{1400 \times 36}{2 \times 0.75} = 33,600 \text{ psig}$$

Considering thick wall pipe formula Eqn (4.11)

$$S_{max} = \frac{1400(36^2 + 34.5^2)}{(36^2 - 34.5^2)} = 32,915 \text{ psig}$$

Therefore, by assuming thin walled pipe, the hoop stress is overestimated by approximately

$$\frac{33,600 - 32,915}{32,915} = 0.0208 \quad \text{or} \quad 2.08\%.$$

8. DERIVATION OF BARLOW'S EQUATION

Because Barlow's equation is the basic equation for pipes under internal pressure, it is appropriate to understand how the formula is derived, which is the subject of this section.

Consider a circular pipe of length L, outside diameter D, and wall thickness t as shown in Figure 4.1. We consider the cross-section of one half portion of this pipe. The pipe is subject to an internal pressure of P psig. Within the pipe material, the hoop stress S_h and the axial stress S_a act at right angles to each other as shown.

Considering the one half section of the pipe, for balancing the forces in the direction of the hoop stress S_h we can say, that S_h acting on the two rectangular areas $L \times t$ balances the internal pressure on the projected area $D \times L$.

Therefore

$$P \times D \times L = S_h \times L \times t \times 2 \tag{4.14}$$

Solving for S_h we get the derivation of Eqn (4.1) as

$$S_h = \frac{PD}{2t}$$

Now we will look at the balancing of longitudinal forces. The internal pressure P acting on the cross-sectional area of pipe $\frac{\pi}{4}D^2$ produces the bursting force. This is balanced by the axial resisting force S_a acting on the area πDt.

Therefore

$$\frac{\pi}{4}D^2 = S_a \times \pi Dt \tag{4.15}$$

Solving for S_a we get the derivation of Eqn (4.15) as

$$S_a = \frac{PD}{4t}$$

It can be seen from the preceding equations that the hoop stress is twice the axial stress, and therefore is the governing stress. Consider a pipe of a 20-in outside diameter and 0.500 in wall thickness subject to an internal pressure of 1000 psig. From Barlow's Eqns (4.9) and (4.10) we calculate the hoop stress and axial stress as follows:

$$S_h = \frac{1000 \times 20}{2 \times 0.500} = 20,000 \text{ psig}$$

$$S_a = \frac{1000 \times 20}{4 \times 0.500} = 10,000 \text{ psig}$$

Therefore we are able to determine the stress levels in the pipe material for a given internal pressure, pipe diameter, and wall thickness. If the calculated values are within the stress limits of the pipe material, we can conclude that the NPS 20 pipe with 0.500 in wall thickness is adequate for the internal pressure of 1000 psig. The yield stress of the pipe material represents the stress at which the pipe material yields and undergoes permanent deformation. Therefore, we must ensure that the stress calculation above do not come dangerously close to the yield stress.

Frequently, we have to solve the reverse problem of determining the wall thickness of a pipeline for a given pressure. For example, suppose the pipe is constructed of steel with a yield strength of 52,000 psi and we are required to determine what wall thickness is needed for NPS 20 pipe to withstand 1400 psig internal pressure. If we are allowed to stress the pipe material to more than 60% of the yield stress, we can easily calculate the minimum wall thickness required using Eqn (4.9) as follows:

$$0.6 \times 52000 = \frac{1400 \times 20}{2t}$$

Here we have equated the hoop stress per Barlow's equation to 60% yield strength of the pipe material.

Solving for pipe wall thickness, we get

$$t = 0.4487 \text{ in.}$$

Suppose we used the nearest standard wall thickness of 0.500 in. The actual hoop stress can then be calculated from Barlow's equation as

$$S_h = \frac{1400 \times 20}{2 \times 0.5} = 28,000 \text{ psi}$$

Therefore, the pipe will be stressed to $\frac{28000}{52000} = 0.54$ or 54% of yield stress, which is less than the 60% we started off with.

Incidentally, the actual axial or longitudinal stress in the preceding example will be one half the hoop stress or 14,000 psi.

Therefore, in this basic example we used the Barlow's equation to calculate the pipe wall thickness required for an NPS 20 pipe to withstand an internal pressure of 1400 psig, without stressing the pipe material beyond 60% of its yield strength.

In the foregoing, we arbitrarily picked 60% of the yield stress of pipe material to calculate the pipe wall thickness. We did not use 100% of the yield stress because in this case the pipe material would yield at the given pressure which obviously cannot be allowed. In design we generally use a design factor that is a number less than 1.00 that represents the fraction of the yield stress of the pipe material that the pipe may be stressed to. Gas pipelines are designed with various design factors ranging from 0.4 to 0.72. This simply means that the pipe hoop stress is allowed to be between 40% and 72% of the yield strength of pipe material. The actual percentage will depend on various factors and will be discussed shortly. The yield stress used in calculation of pipe wall thickness is called the SMYS of pipe material. Thus in the preceding example, we calculated the pipe wall thickness based on a design factor of 0.6 or allowed the pipe stress to go up to 60% of SMYS.

9. PIPE MATERIAL AND GRADE

Steel pipe used in gas pipeline systems generally conform to API 5L and 5LX specifications. These are manufactured in grades ranging from X42 to X80 with SMYS as shown in Table 4.3.

Sometimes API 5L grade B pipe with 35,000 psi SMYS is also used in certain installations.

Table 4.3 Pipe Material Yield Strength

Pipe Material API 5LX Grade	Specified Minimum Yield Strength, psi
X42	42,000
X46	46,000
X52	52,000
X56	56,000
X60	60,000
X65	65,000
X70	70,000
X80	80,000
X90	90,000

10. INTERNAL DESIGN PRESSURE EQUATION

We indicated earlier in this chapter that Barlow's equation in a modified form is used in designing gas pipelines. The following form of the Barlow's equation is used in design codes for petroleum transportation systems, for calculating the allowable internal pressure in a pipeline based on given diameter, wall thickness, and pipe material.

$$P = \frac{2tSEFT}{D} \quad (4.16)$$

where

P = internal pipe design pressure, psig
D = pipe outside diameter, in
t = pipe wall thickness, in
S = SMYS of pipe material, psig
E = seam joint factor, 1.0 for seamless and SAW pipes
F = design factor, usually 0.72 for cross-country gas pipelines, but may be as low as 0.4 depending on class location and type of construction
T = temperature deration factor = 1.00 for temperatures below 250 °F

In the foregoing, we use the outside diameter of the pipe and not the inside diameter, as used in pressure drop calculations.

The seam joint factor E varies with the type of pipe material and welding employed, which is given in Table 4.4 for most commonly used pipe and joint types.

The internal design pressure calculated from Eqn (4.16) is known as the MAOP of the pipeline. This term has been shortened to maximum operating pressure (MOP) in recent years. Throughout this book, we will use MOP and MAOP interchangeably. The design factor F has values ranging

Table 4.4 Pipe Seam Joint Factor

Specification	Pipe Class	Seam Joint Factor, E
ASTM A53	Seamless	1
	Electric resistance welded	1
	Furnace lap welded	0.8
	Furnace butt welded	0.6
ASTM A106	Seamless	1
ASTM A134	Electric fusion arc welded	0.8
ASTM A135	Electric resistance welded	1
ASTM A139	Electric fusion welded	0.8
ASTM A211	Spiral welded pipe	0.8
ASTM A333	Seamless	1
ASTM A333	Welded	1
ASTM A381	Double submerged Arc welded	1
ASTM A671	Electric fusion welded	1
ASTM A672	Electric fusion welded	1
ASTM A691	Electric fusion welded	1
API 5L	Seamless	1
	Electric resistance welded	1
	Electric flash welded	1
	Submerged arc welded	1
	Furnace lap welded	0.8
	Furnace butt welded	0.6
API 5LX	Seamless	1
	Electric resistance welded	1
	Electric flash welded	1
	Submerged arc welded	1
API 5LS	Electric resistance welded	1
	Submerged arc welded	1

from 0.4 to 0.72 as mentioned earlier. Table 4.1 lists the values of the design factor based on class locations. The class locations in turn depend on the population density in the vicinity of the pipeline.

11. MAINLINE VALVES

Mainline valves are installed in gas pipelines so that portions of the pipeline may be isolated for hydrostatic testing and maintenance. Valves are also necessary to separate sections of pipe and minimize gas loss that may occur because of pipe rupture from construction damage. Design codes specify the spacing of these valves based on class location, which in turn depends on the population density around the pipeline. The following lists the maximum spacing between mainline valves in gas transmission piping. These are taken from ASME B31.8 code.

Class Location	Valve Spacing
1	20 miles
2	15 miles
3	10 miles
4	5 miles

It can be seen from the preceding that the valve spacing is shorter as the pipeline traverses high population areas. This is necessary as a safety feature, to protect the inhabitants in the vicinity of the pipeline by restricting the amount of gas that may escape because of pipeline rupture. These mainline valves must be full opening, through-conduit type valves such that scraper pigs and inspection tools may pass through these valves without any obstruction. Therefore, ball valves and gate valves are used of the welded construction rather than flanged type. Buried valves have extended stems with elevated valve operators located above ground with lubrication and bleed lines brought up above ground for easy access and maintenance.

12. HYDROSTATIC TEST PRESSURE

When a pipeline is designed to operate at a certain MOP, it must be tested to ensure that it is structurally sound and can withstand safely the internal pressure before being put into service. Generally, gas pipelines are hydrotested with water by filling the test section of the pipe with water and pumping the pressure up to a value higher than the MAOP and holding it at this test pressure for a period of 4–8 h. The magnitude of the test pressure is specified by design code and it is usually 125% of the operating pressure. Thus a pipeline designed to operate continuously 1000 psig will be hydrotested to a minimum pressure of 1250 psig.

Consider a pipeline NPS 24 with 0.375 in wall thickness constructed of API 5L X65 pipe. Using a temperature de-ration factor of 1.00, we calculate the MOP of this pipeline from Eqn (4.8) for class 1 location as follows:

$$P = \frac{2 \times 0.375 \times 65000 \times 1.0 \times 0.72 \times 1.0}{24} = 1462.5 \text{ psig}$$

Because the pipe fittings and valves will be ANSI 600, we will establish a MOP of 1440 psig for this pipeline.

Therefore, the hydrotest pressure will be:

$$1.25 \times 1440 = 1800 \text{ psig.}$$

If the pipeline is designed to be below ground, the test pressure is held constant for 8 h and it is thoroughly checked for leaks. Above-ground pipelines are tested for 4 h. If the design factor used in the MOP calculation is 0.72 (class 1), the hoop stress is allowed to reach 72% of the SMYS of pipe material. Testing this pipe at 125% of MOP will result in the hoop stress reaching a value of 1.25 × 0.72 = 0.90 or 90% of SMYS.

Thus, by hydrotesting the pipe at 1.25 times the operating pressure, we are stressing the pipe material to 90% of the yield strength.

Generally, the hydrotest pressure is given such that the hoop stress has as a range of values such as 90–95% of SMYS. Therefore, in the preceding example the minimum and maximum hydrotest pressures will be as follows:

$$\text{Minimum hydrotest pressure} = 1.25 \times 1440 = 1800 \text{ psig}$$

$$\text{Minimum hydrotest pressure} = 1800 \times (95/90) = 1900 \text{ psig}$$

It can be seen from Eqn (4.1) that the 1800 psig internal pressure will cause a hoop stress of

$$S_h = \frac{1800 \times 24}{2 \times 0.375} = 57,600 \text{ psi}$$

Dividing this hoop stress by the SMYS, we get the lower limit of the hydrotest pressure as

$$\frac{57,600}{65,000} = 0.89 = 89\% \text{ of SMYS}$$

Similarly, by proportion the maximum hydrotest pressure of 1900 psig will cause a hoop stress of

$$S_h = \frac{1900 \times 57,600}{1800} = \frac{60,800}{65,000} = 0.94 = 94\% \text{ of SMYS}$$

Therefore, in this example the hydrotest envelope of 1800–1900 psig is equivalent to stressing the pipe in the range of 89–94% of SMYS.

In the preceding analysis, we have not taken into consideration the pipeline elevation profile in calculating the hydrotest pressures. Generally a long pipeline is divided into test sections and the hydrotest pressures are established for each section taking into account the elevations along the pipeline profiles. The reason for subdividing the pipeline into sections for hydrotest will be evident for the following example.

Consider for example a pipeline 50 miles long with an elevation profile as shown in Figure 4.5.

The elevation of the starting point, Norwalk is 300 ft, whereas the pipeline terminus Lakewood is at an elevation of 1200 ft. If the entire 50-miles length

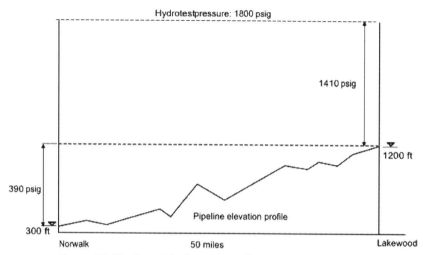

Figure 4.5 Pipeline with elevation profile: impact on hydrotest.

of the pipeline were filled with water for hydrotesting, the static pressure difference between the two ends resulting from elevation will be as follows:

Pressure difference $= (1200 - 300) \times 0.433 = 389.7$ psig.

The factor 0.433 is the conversion factor from ft of water to pressure in psig.

It can be seen that the pipe section at the low elevation point at Norwalk will be at a higher pressure than the pipeline at the high elevation end at Lakewood by almost 390 psig. Therefore, if we pump the water in the line to the required hydrotest pressure of 1800 psig at Norwalk, the corresponding water pressure at Lakewood will be

$$1800 - 390 = 1410 \text{ psig}$$

Conversely, if we pump the water in the line to the required hydrotest pressure of 1800 psig at Lakewood, the corresponding water pressure at Norwalk will be

$$1800 - 390 = 2190 \text{ psig}$$

The pressure of 2190 psig at Norwalk will result in a hoop stress of

$$S_h = \frac{2190 \times 24}{2 \times 0.375} = 70,080 \text{ psi}$$

This is equivalent to

$$\frac{70,080}{65,000} = 1.08 = 108\% \text{ of SMYS}$$

Obviously, we have exceeded the yield strength of the pipe material and hence this is not acceptable.

On the other hand, with 1800 psig test pressure at Norwalk, the corresponding test pressure at Lakewood was calculated to be 1410 psig. Even though the pipe section at the low end at Norwalk has the requisite test pressure (125% MOP), the pipe section at the higher elevation at Lakewood will see only

$$\frac{1410}{1800} \times 125 = 98\% \text{ MOP}.$$

This will not be an acceptable hydrotest because we have not been able to test the entire pipeline at the correct hydrotest pressure.

This is not acceptable because the hydrotest pressures in a pipeline must at least be 125% of the MAOP. The solution to this dilemma is to break the length of 500 miles into several sections such that each section can be tested separately at the required test pressure. These test sections will have smaller elevation difference between the ends of the test sections. Therefore, each section will be hydrotested to pressures close to the required minimum pressure. Figure 4.6 shows such a pipeline subdivided into sections suitable for hydrotesting. Using the hydrotest envelope of 90–95% of SMYS, we will be able to adjust the test pressures for each section such that even with some elevation difference between the ends of each test section, the hydrotest pressures may be close to the required pressures. This will not be possible if we have one single test section with significant elevation difference between the two ends as illustrated in Figure 4.6.

Tables 4.5–4.13 lists the internal design pressure and hydrostatic test pressure for various pipe diameters and pipe material ranging from X42 to X90.

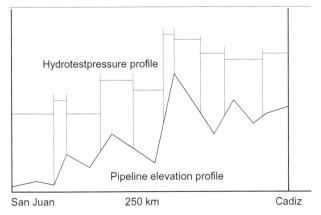

Figure 4.6 Hydrotesting by subdividing pipeline.

Table 4.5 Pipeline Internal Design Pressures and Test Pressures for API 5L X42 Pipe Material API 5L X42 SMYS 42,000 psig

Diameter in	Wall Thickness in	Weight lb/ft	Internal Design Pressure, Psig				Hydrostatic Test Pressure, psig		
			Class 1	Class 2	Class 3	Class 4	90% SMYS	95% SMYS	100% SMYS
4.5	0.237	10.79	3185	2654	2212	1770	3982	4203	4424
	0.337	14.98	4529	3774	3145	2516	5662	5976	6291
	0.437	18.96	5873	4894	4079	3263	7342	7749	8157
	0.531	22.51	7137	5947	4956	3965	8921	9416	9912
6.625	0.250	17.02	2282	1902	1585	1268	2853	3011	3170
	0.280	18.97	2556	2130	1775	1420	3195	3373	3550
	0.432	28.57	3944	3286	2739	2191	4930	5204	5477
	0.562	36.39	5131	4275	3563	2850	6413	6769	7126
8.625	0.250	22.36	1753	1461	1217	974	2191	2313	2435
	0.277	24.70	1942	1619	1349	1079	2428	2563	2698
	0.322	28.55	2258	1882	1568	1254	2822	2979	3136
	0.406	35.64	2847	2372	1977	1582	3559	3756	3954
10.75	0.250	28.04	1407	1172	977	781	1758	1856	1953
	0.307	34.24	1727	1439	1199	960	2159	2279	2399
	0.365	40.48	2054	1711	1426	1141	2567	2709	2852
	0.500	54.74	2813	2344	1953	1563	3516	3712	3907
12.75	0.250	33.38	1186	988	824	659	1482	1565	1647
	0.330	43.77	1565	1304	1087	870	1957	2065	2174
	0.375	49.56	1779	1482	1235	988	2224	2347	2471
	0.406	53.52	1926	1605	1337	1070	2407	2541	2675
	0.500	65.42	2372	1976	1647	1318	2965	3129	3294

Pipeline Stress Design

14.00	0.250	36.71	1080	900	750	600	1350	1425	1500
	0.312	45.61	1348	1123	936	749	1685	1778	1872
	0.375	54.57	1620	1350	1125	900	2025	2138	2250
	0.437	63.30	1888	1573	1311	1049	2360	2491	2622
	0.500	72.09	2160	1800	1500	1200	2700	2850	3000
16.00	0.250	42.05	945	788	656	525	1181	1247	1313
	0.312	52.27	1179	983	819	655	1474	1556	1638
	0.375	62.58	1418	1181	984	788	1772	1870	1969
	0.437	72.64	1652	1377	1147	918	2065	2180	2294
	0.500	82.77	1890	1575	1313	1050	2363	2494	2625
18.00	0.250	47.39	840	700	583	467	1050	1108	1167
	0.312	58.94	1048	874	728	582	1310	1383	1456
	0.375	70.59	1260	1050	875	700	1575	1663	1750
	0.437	81.97	1468	1224	1020	816	1835	1937	2039
	0.500	93.45	1680	1400	1167	933	2100	2217	2333
20.00	0.312	65.60	943	786	655	524	1179	1245	1310
	0.375	78.60	1134	945	788	630	1418	1496	1575
	0.437	91.30	1321	1101	918	734	1652	1744	1835
	0.500	104.13	1512	1260	1050	840	1890	1995	2100
	0.562	116.67	1699	1416	1180	944	2124	2242	2360
22.00	0.375	86.61	1031	859	716	573	1289	1360	1432
	0.500	114.81	1375	1145	955	764	1718	1814	1909
	0.625	142.68	1718	1432	1193	955	2148	2267	2386
	0.750	170.21	2062	1718	1432	1145	2577	2720	2864
24.00	0.375	94.62	945	788	656	525	1181	1247	1313
	0.437	109.97	1101	918	765	612	1377	1453	1530
	0.500	125.49	1260	1050	875	700	1575	1663	1750

(*Continued*)

Table 4.5 Pipeline Internal Design Pressures and Test Pressures for API 5L X42—cont'd
Pipe Material API 5L X42   SMYS 42,000 psig

Diameter in	Wall Thickness in	Weight lb/ft	Internal Design Pressure, Psig				Hydrostatic Test Pressure, psig			
			Class 1	Class 2	Class 3	Class 4	90% SMYS	95% SMYS	100% SMYS
26.00	0.562	140.68	1416	1180	984	787	1770	1869	1967
	0.625	156.03	1575	1313	1094	875	1969	2078	2188
	0.750	186.23	1890	1575	1313	1050	2363	2494	2625
	0.375	102.63	872	727	606	485	1090	1151	1212
	0.500	136.17	1163	969	808	646	1454	1535	1615
	0.625	169.38	1454	1212	1010	808	1817	1918	2019
	0.750	202.25	1745	1454	1212	969	2181	2302	2423
28.00	0.375	110.64	810	675	563	450	1013	1069	1125
	0.500	146.85	1080	900	750	600	1350	1425	1500
	0.625	182.73	1350	1125	938	750	1688	1781	1875
	0.750	218.27	1620	1350	1125	900	2025	2138	2250
30.00	0.375	118.65	756	630	525	420	945	998	1050
	0.500	157.53	1008	840	700	560	1260	1330	1400
	0.625	196.08	1260	1050	875	700	1575	1663	1750
	0.750	234.29	1512	1260	1050	840	1890	1995	2100

Pipeline Stress Design

32.00	0.375	126.66	709	591	492	394	886	935	984
	0.500	168.21	945	788	656	525	1181	1247	1313
	0.625	209.43	1181	984	820	656	1477	1559	1641
	0.750	250.31	1418	1181	984	788	1772	1870	1969
34.00	0.375	134.67	667	556	463	371	834	880	926
	0.500	178.89	889	741	618	494	1112	1174	1235
	0.625	222.78	1112	926	772	618	1390	1467	1544
	0.750	266.33	1334	1112	926	741	1668	1760	1853
36.00	0.375	142.68	630	525	438	350	788	831	875
	0.500	189.57	840	700	583	467	1050	1108	1167
	0.625	236.13	1050	875	729	583	1313	1385	1458
	0.750	282.35	1260	1050	875	700	1575	1663	1750
42.00	0.375	166.71	540	450	375	300	675	713	750
	0.500	221.61	720	600	500	400	900	950	1000
	0.625	276.18	900	750	625	500	1125	1188	1250
	0.750	330.41	1080	900	750	600	1350	1425	1500
	1.000	437.88	1440	1200	1000	800	1800	1900	2000

Table 4.6 Pipeline Internal Design Pressures and Test Pressures for API 5L X46
Pipe Material API 5L X46   SMYS 46,000 psig

Diameter	Wall Thickness	Weight	Internal Design Pressure, psig				Hydrostatic Test Pressure, psig		
in	in	lb/ft	Class 1	Class 2	Class 3	Class 4	90% SMYS	95% SMYS	100% SMYS
4.5	0.237	10.79	3489	2907	2423	1938	4361	4603	4845
	0.337	14.98	4961	4134	3445	2756	6201	6545	6890
	0.437	18.96	6433	5361	4467	3574	8041	8488	8934
	0.531	22.51	7816	6514	5428	4342	9770	10,313	10,856
6.625	0.250	17.02	2500	2083	1736	1389	3125	3298	3472
	0.280	18.97	2800	2333	1944	1555	3499	3694	3888
	0.432	28.57	4319	3599	3000	2400	5399	5699	5999
	0.562	36.39	5619	4683	3902	3122	7024	7414	7804
8.625	0.250	22.36	1920	1600	1333	1067	2400	2533	2667
	0.277	24.70	2127	1773	1477	1182	2659	2807	2955
	0.322	28.55	2473	2061	1717	1374	3091	3263	3435
	0.406	35.64	3118	2598	2165	1732	3898	4114	4331
10.75	0.250	28.04	1540	1284	1070	856	1926	2033	2140
	0.307	34.24	1892	1576	1314	1051	2365	2496	2627
	0.365	40.48	2249	1874	1562	1249	2811	2968	3124
	0.500	54.74	3081	2567	2140	1712	3851	4065	4279
12.75	0.250	33.38	1299	1082	902	722	1624	1714	1804
	0.330	43.77	1714	1429	1191	952	2143	2262	2381
	0.375	49.56	1948	1624	1353	1082	2435	2571	2706
	0.406	53.52	2109	1758	1465	1172	2637	2783	2930
	0.500	65.42	2598	2165	1804	1443	3247	3427	3608
14.00	0.250	36.71	1183	986	821	657	1479	1561	1643
	0.312	45.61	1476	1230	1025	820	1845	1948	2050
	0.375	54.57	1774	1479	1232	986	2218	2341	2464

16.00	0.437	63.30	2068	1723	1436	1149	2585	2728	2872
	0.500	72.09	2366	1971	1643	1314	2957	3121	3286
	0.250	42.05	1035	863	719	575	1294	1366	1438
	0.312	52.27	1292	1076	897	718	1615	1704	1794
	0.375	62.58	1553	1294	1078	863	1941	2048	2156
	0.437	72.64	1809	1508	1256	1005	2261	2387	2513
	0.500	82.77	2070	1725	1438	1150	2588	2731	2875
18.00	0.250	47.39	920	767	639	511	1150	1214	1278
	0.312	58.94	1148	957	797	638	1435	1515	1595
	0.375	70.59	1380	1150	958	767	1725	1821	1917
	0.437	81.97	1608	1340	1117	893	2010	2122	2234
	0.500	93.45	1840	1533	1278	1022	2300	2428	2556
20.00	0.312	65.60	1033	861	718	574	1292	1363	1435
	0.375	78.60	1242	1035	863	690	1553	1639	1725
	0.437	91.30	1447	1206	1005	804	1809	1910	2010
	0.500	104.13	1656	1380	1150	920	2070	2185	2300
	0.562	116.67	1861	1551	1293	1034	2327	2456	2585
22.00	0.375	86.61	1129	941	784	627	1411	1490	1568
	0.500	114.81	1505	1255	1045	836	1882	1986	2091
	0.625	142.68	1882	1568	1307	1045	2352	2483	2614
	0.750	170.21	2258	1882	1568	1255	2823	2980	3136
24.00	0.375	94.62	1035	863	719	575	1294	1366	1438
	0.437	109.97	1206	1005	838	670	1508	1591	1675
	0.500	125.49	1380	1150	958	767	1725	1821	1917
	0.562	140.68	1551	1293	1077	862	1939	2047	2154
	0.625	156.03	1725	1438	1198	958	2156	2276	2396
	0.750	186.23	2070	1725	1438	1150	2588	2731	2875
26.00	0.375	102.63	955	796	663	531	1194	1261	1327
	0.500	136.17	1274	1062	885	708	1592	1681	1769
	0.625	169.38	1592	1327	1106	885	1990	2101	2212

(*Continued*)

Table 4.6 Pipeline Internal Design Pressures and Test Pressures for API 5L X46—cont'd Pipe Material API 5L X46

SMYS 46,000 psig

Diameter in	Wall Thickness in	Weight lb/ft	Internal Design Pressure, psig					Hydrostatic Test Pressure, psig		
			Class 1	Class 2	Class 3	Class 4		90% SMYS	95% SMYS	100% SMYS
28.00	0.750	202.25	1911	1592	1327	1062		2388	2521	2654
	0.375	110.64	887	739	616	493		1109	1171	1232
	0.500	146.85	1183	986	821	657		1479	1561	1643
	0.625	182.73	1479	1232	1027	821		1848	1951	2054
	0.750	218.27	1774	1479	1232	986		2218	2341	2464
30.00	0.375	118.65	828	690	575	460		1035	1093	1150
	0.500	157.53	1104	920	767	613		1380	1457	1533
	0.625	196.08	1380	1150	958	767		1725	1821	1917
	0.750	234.29	1656	1380	1150	920		2070	2185	2300
32.00	0.375	126.66	776	647	539	431		970	1024	1078
	0.500	168.21	1035	863	719	575		1294	1366	1438
	0.625	209.43	1294	1078	898	719		1617	1707	1797
	0.750	250.31	1553	1294	1078	863		1941	2048	2156
34.00	0.375	134.67	731	609	507	406		913	964	1015
	0.500	178.89	974	812	676	541		1218	1285	1353
	0.625	222.78	1218	1015	846	676		1522	1607	1691
	0.750	266.33	1461	1218	1015	812		1826	1928	2029
36.00	0.375	142.68	690	575	479	383		863	910	958
	0.500	189.57	920	767	639	511		1150	1214	1278
	0.625	236.13	1150	958	799	639		1438	1517	1597
	0.750	282.35	1380	1150	958	767		1725	1821	1917
42.00	0.375	166.71	591	493	411	329		739	780	821
	0.500	221.61	789	657	548	438		986	1040	1095
	0.625	276.18	986	821	685	548		1232	1301	1369
	0.750	330.41	1183	986	821	657		1479	1561	1643
	1.000	437.88	1577	1314	1095	876		1971	2081	2190

Pipeline Stress Design

Table 4.7 Pipeline Internal Design Pressures and Test Pressures for API 5L X52 Pipe Material API 5L X52   SMYS 52,000 psig

Diameter	Wall Thickness	Weight	Internal Design Pressure, psig				Hydrostatic Test Pressure, psig			
in	in	lb/ft	Class 1	Class 2	Class 3	Class 4	90% SMYS	95% SMYS	100% SMYS
4.5	0.237	10.79	3944	3286	2739	2191	4930	5203	5477
	0.337	14.98	5608	4673	3894	3115	7010	7399	7788
	0.437	18.96	7272	6060	5050	4040	9090	9595	10,100
	0.531	22.51	8836	7363	6136	4909	11,045	11,658	12,272
6.625	0.250	17.02	2826	2355	1962	1570	3532	3728	3925
	0.280	18.97	3165	2637	2198	1758	3956	4176	4395
	0.432	28.57	4883	4069	3391	2713	6103	6443	6782
	0.562	36.39	6352	5293	4411	3529	7940	8381	8822
8.625	0.250	22.36	2170	1809	1507	1206	2713	2864	3014
	0.277	24.70	2405	2004	1670	1336	3006	3173	3340
	0.322	28.55	2796	2330	1941	1553	3494	3689	3883
	0.406	35.64	3525	2937	2448	1958	4406	4651	4896
10.75	0.250	28.04	1741	1451	1209	967	2177	2298	2419
	0.307	34.24	2138	1782	1485	1188	2673	2822	2970
	0.365	40.48	2542	2119	1766	1412	3178	3355	3531
	0.500	54.74	3483	2902	2419	1935	4353	4595	4837
12.75	0.250	33.38	1468	1224	1020	816	1835	1937	2039
	0.330	43.77	1938	1615	1346	1077	2423	2557	2692
	0.375	49.56	2202	1835	1529	1224	2753	2906	3059
	0.406	53.52	2384	1987	1656	1325	2981	3146	3312
	0.500	65.42	2936	2447	2039	1631	3671	3875	4078

(Continued)

Table 4.7 Pipeline Internal Design Pressures and Test Pressures for API 5L X52—cont'd
Pipe Material API 5L X52   SMYS 52,000 psig

Diameter	Wall Thickness	Weight	Internal Design Pressure, psig				Hydrostatic Test Pressure, psig		
in	in	lb/ft	Class 1	Class 2	Class 3	Class 4	90% SMYS	95% SMYS	100% SMYS
14.00	0.250	36.71	1337	1114	929	743	1671	1764	1857
	0.312	45.61	1669	1391	1159	927	2086	2202	2318
	0.375	54.57	2006	1671	1393	1114	2507	2646	2786
	0.437	63.30	2337	1948	1623	1299	2922	3084	3246
	0.500	72.09	2674	2229	1857	1486	3343	3529	3714
16.00	0.250	42.05	1170	975	813	650	1463	1544	1625
	0.312	52.27	1460	1217	1014	811	1825	1927	2028
	0.375	62.58	1755	1463	1219	975	2194	2316	2438
	0.437	72.64	2045	1704	1420	1136	2556	2698	2841
	0.500	82.77	2340	1950	1625	1300	2925	3088	3250
18.00	0.250	47.39	1040	867	722	578	1300	1372	1444
	0.312	58.94	1298	1082	901	721	1622	1713	1803
	0.375	70.59	1560	1300	1083	867	1950	2058	2167
	0.437	81.97	1818	1515	1262	1010	2272	2399	2525
	0.500	93.45	2080	1733	1444	1156	2600	2744	2889
20.00	0.312	65.60	1168	973	811	649	1460	1541	1622
	0.375	78.60	1404	1170	975	780	1755	1853	1950
	0.437	91.30	1636	1363	1136	909	2045	2159	2272
	0.500	104.13	1872	1560	1300	1040	2340	2470	2600
	0.562	116.67	2104	1753	1461	1169	2630	2776	2922
22.00	0.375	86.61	1276	1064	886	709	1595	1684	1773
	0.500	114.81	1702	1418	1182	945	2127	2245	2364
	0.625	142.68	2127	1773	1477	1182	2659	2807	2955
	0.750	170.21	2553	2127	1773	1418	3191	3368	3545

24.00	0.375	94.62	1170	975	813	650	1463	1544	1625
	0.437	109.97	1363	1136	947	757	1704	1799	1894
	0.500	125.49	1560	1300	1083	867	1950	2058	2167
	0.562	140.68	1753	1461	1218	974	2192	2314	2435
	0.625	156.03	1950	1625	1354	1083	2438	2573	2708
	0.750	186.23	2340	1950	1625	1300	2925	3088	3250
26.00	0.375	102.63	1080	900	750	600	1350	1425	1500
	0.500	136.17	1440	1200	1000	800	1800	1900	2000
	0.625	169.38	1800	1500	1250	1000	2250	2375	2500
	0.750	202.25	2160	1800	1500	1200	2700	2850	3000
28.00	0.375	110.64	1003	836	696	557	1254	1323	1393
	0.500	146.85	1337	1114	929	743	1671	1764	1857
	0.625	182.73	1671	1393	1161	929	2089	2205	2321
	0.750	218.27	2006	1671	1393	1114	2507	2646	2786
30.00	0.375	118.65	936	780	650	520	1170	1235	1300
	0.500	157.53	1248	1040	867	693	1560	1647	1733
	0.625	196.08	1560	1300	1083	867	1950	2058	2167
	0.750	234.29	1872	1560	1300	1040	2340	2470	2600
32.00	0.375	126.66	878	731	609	488	1097	1158	1219
	0.500	168.21	1170	975	813	650	1463	1544	1625
	0.625	209.43	1463	1219	1016	813	1828	1930	2031
	0.750	250.31	1755	1463	1219	975	2194	2316	2438
34.00	0.375	134.67	826	688	574	459	1032	1090	1147
	0.500	178.89	1101	918	765	612	1376	1453	1529
	0.625	222.78	1376	1147	956	765	1721	1816	1912
	0.750	266.33	1652	1376	1147	918	2065	2179	2294

(*Continued*)

Table 4.7 Pipeline Internal Design Pressures and Test Pressures for API 5L X52—cont'd
Pipe Material API 5L X52   SMYS 52,000 psig

Diameter	Wall Thickness	Weight	Internal Design Pressure, psig				Hydrostatic Test Pressure, psig		
in	in	lb/ft	Class 1	Class 2	Class 3	Class 4	90% SMYS	95% SMYS	100% SMYS
36.00	0.375	142.68	780	650	542	433	975	1029	1083
	0.500	189.57	1040	867	722	578	1300	1372	1444
	0.625	236.13	1300	1083	903	722	1625	1715	1806
	0.750	282.35	1560	1300	1083	867	1950	2058	2167
42.00	0.375	166.71	669	557	464	371	836	882	929
	0.500	221.61	891	743	619	495	1114	1176	1238
	0.625	276.18	1114	929	774	619	1393	1470	1548
	0.750	330.41	1337	1114	929	743	1671	1764	1857
	1.000	437.88	1783	1486	1238	990	2229	2352	2476

Pipeline Stress Design

Table 4.8 Pipeline Internal Design Pressures and Test Pressures for API 5L X56
Pipe Material API 5L X56   SMYS 56,000 psig

Diameter	Wall Thickness	Weight	Internal Design Pressure, psig				Hydrostatic Test Pressure, psig		
in	in	lb/ft	Class 1	Class 2	Class 3	Class 4	90% SMYS	95% SMYS	100% SMYS
4.5	0.237	10.79	4247	3539	2949	2359	5309	5604	5899
	0.337	14.98	6039	5033	4194	3355	7549	7968	8388
	0.437	18.96	7831	6526	5438	4351	9789	10,333	10,876
	0.531	22.51	9516	7930	6608	5286	11,894	12,555	13,216
6.625	0.250	17.02	3043	2536	2113	1691	3804	4015	4226
	0.280	18.97	3408	2840	2367	1893	4260	4497	4734
	0.432	28.57	5258	4382	3652	2921	6573	6938	7303
	0.562	36.39	6841	5701	4750	3800	8551	9026	9501
8.625	0.250	22.36	2337	1948	1623	1299	2922	3084	3246
	0.277	24.70	2590	2158	1798	1439	3237	3417	3597
	0.322	28.55	3011	2509	2091	1673	3763	3972	4181
	0.406	35.64	3796	3163	2636	2109	4745	5009	5272
10.75	0.250	28.04	1875	1563	1302	1042	2344	2474	2605
	0.307	34.24	2303	1919	1599	1279	2879	3039	3199
	0.365	40.48	2738	2282	1901	1521	3423	3613	3803
	0.500	54.74	3751	3126	2605	2084	4688	4949	5209
12.75	0.250	33.38	1581	1318	1098	878	1976	2086	2196
	0.330	43.77	2087	1739	1449	1160	2609	2754	2899
	0.375	49.56	2372	1976	1647	1318	2965	3129	3294
	0.406	53.52	2568	2140	1783	1427	3210	3388	3566
	0.500	65.42	3162	2635	2196	1757	3953	4173	4392

(Continued)

Table 4.8 Pipeline Internal Design Pressures and Test Pressures for API 5L X56—cont'd
Pipe Material API 5L X56 SMYS 56,000 psig

Diameter	Wall Thickness	Weight	Internal Design Pressure, psig				Hydrostatic Test Pressure, psig		
in	in	lb/ft	Class 1	Class 2	Class 3	Class 4	90% SMYS	95% SMYS	100% SMYS
14.00	0.250	36.71	1440	1200	1000	800	1800	1900	2000
	0.312	45.61	1797	1498	1248	998	2246	2371	2496
	0.375	54.57	2160	1800	1500	1200	2700	2850	3000
	0.437	63.30	2517	2098	1748	1398	3146	3321	3496
	0.500	72.09	2880	2400	2000	1600	3600	3800	4000
16.00	0.250	42.05	1260	1050	875	700	1575	1663	1750
	0.312	52.27	1572	1310	1092	874	1966	2075	2184
	0.375	62.58	1890	1575	1313	1050	2363	2494	2625
	0.437	72.64	2202	1835	1530	1224	2753	2906	3059
	0.500	82.77	2520	2100	1750	1400	3150	3325	3500
18.00	0.250	47.39	1120	933	778	622	1400	1478	1556
	0.312	58.94	1398	1165	971	777	1747	1844	1941
	0.375	70.59	1680	1400	1167	933	2100	2217	2333
	0.437	81.97	1958	1631	1360	1088	2447	2583	2719
	0.500	93.45	2240	1867	1556	1244	2800	2956	3111
20.00	0.312	65.60	1258	1048	874	699	1572	1660	1747
	0.375	78.60	1512	1260	1050	840	1890	1995	2100
	0.437	91.30	1762	1468	1224	979	2202	2325	2447
	0.500	104.13	2016	1680	1400	1120	2520	2660	2800
	0.562	116.67	2266	1888	1574	1259	2832	2990	3147
22.00	0.375	86.61	1375	1145	955	764	1718	1814	1909
	0.500	114.81	1833	1527	1273	1018	2291	2418	2545
	0.625	142.68	2291	1909	1591	1273	2864	3023	3182
	0.750	170.21	2749	2291	1909	1527	3436	3627	3818

OD	t	wt							
24.00	0.375	94.62	1260	1050	875	700	1575	1663	1750
	0.437	109.97	1468	1224	1020	816	1835	1937	2039
	0.500	125.49	1680	1400	1167	933	2100	2217	2333
	0.562	140.68	1888	1574	1311	1049	2360	2492	2623
	0.625	156.03	2100	1750	1458	1167	2625	2771	2917
	0.750	186.23	2520	2100	1750	1400	3150	3325	3500
26.00	0.375	102.63	1163	969	808	646	1454	1535	1615
	0.500	136.17	1551	1292	1077	862	1938	2046	2154
	0.625	169.38	1938	1615	1346	1077	2423	2558	2692
	0.750	202.25	2326	1938	1615	1292	2908	3069	3231
28.00	0.375	110.64	1080	900	750	600	1350	1425	1500
	0.500	146.85	1440	1200	1000	800	1800	1900	2000
	0.625	182.73	1800	1500	1250	1000	2250	2375	2500
	0.750	218.27	2160	1800	1500	1200	2700	2850	3000
30.00	0.375	118.65	1008	840	700	560	1260	1330	1400
	0.500	157.53	1344	1120	933	747	1680	1773	1867
	0.625	196.08	1680	1400	1167	933	2100	2217	2333
	0.750	234.29	2016	1680	1400	1120	2520	2660	2800
32.00	0.375	126.66	945	788	656	525	1181	1247	1313
	0.500	168.21	1260	1050	875	700	1575	1663	1750
	0.625	209.43	1575	1313	1094	875	1969	2078	2188
	0.750	250.31	1890	1575	1313	1050	2363	2494	2625
34.00	0.375	134.67	889	741	618	494	1112	1174	1235
	0.500	178.89	1186	988	824	659	1482	1565	1647
	0.625	222.78	1482	1235	1029	824	1853	1956	2059
	0.750	266.33	1779	1482	1235	988	2224	2347	2471

(Continued)

Table 4.8 Pipeline Internal Design Pressures and Test Pressures for API 5L X56—cont'd
Pipe Material API 5L X56   SMYS 56,000 psig

Diameter in	Wall Thickness in	Weight lb/ft	Internal Design Pressure, psig				Hydrostatic Test Pressure, psig		
			Class 1	Class 2	Class 3	Class 4	90% SMYS	95% SMYS	100% SMYS
36.00	0.375	142.68	840	700	583	467	1050	1108	1167
	0.500	189.57	1120	933	778	622	1400	1478	1556
	0.625	236.13	1400	1167	972	778	1750	1847	1944
	0.750	282.35	1680	1400	1167	933	2100	2217	2333
42.00	0.375	166.71	720	600	500	400	900	950	1000
	0.500	221.61	960	800	667	533	1200	1267	1333
	0.625	276.18	1200	1000	833	667	1500	1583	1667
	0.750	330.41	1440	1200	1000	800	1800	1900	2000
	1.000	437.88	1920	1600	1333	1067	2400	2533	2667

Table 4.9 Pipeline Internal Design Pressures and Test Pressures for API 5L X60
Pipe Material API 5L X60   SMYS 60,000 psig

Diameter in	Wall Thickness in	Weight lb/ft	Internal Design Pressure, psig				Hydrostatic Test Pressure, psig		
			Class 1	Class 2	Class 3	Class 4	90% SMYS	95% SMYS	100% SMYS
4.5	0.237	10.79	4550	3792	3160	2528	5688	6004	6320
	0.337	14.98	6470	5392	4493	3595	8088	8537	8987
	0.437	18.96	8390	6992	5827	4661	10,488	11,071	11,653
	0.531	22.51	10,195	8496	7080	5664	12,744	13,452	14,160
6.625	0.250	17.02	3260	2717	2264	1811	4075	4302	4528
	0.280	18.97	3652	3043	2536	2029	4565	4818	5072
	0.432	28.57	5634	4695	3912	3130	7042	7434	7825
	0.562	36.39	7329	6108	5090	4072	9162	9671	10,180
8.625	0.250	22.36	2504	2087	1739	1391	3130	3304	3478
	0.277	24.70	2775	2312	1927	1542	3469	3661	3854
	0.322	28.55	3226	2688	2240	1792	4032	4256	4480
	0.406	35.64	4067	3389	2824	2259	5084	5366	5649
10.75	0.250	28.04	2009	1674	1395	1116	2512	2651	2791
	0.307	34.24	2467	2056	1713	1371	3084	3256	3427
	0.365	40.48	2934	2445	2037	1630	3667	3871	4074
	0.500	54.74	4019	3349	2791	2233	5023	5302	5581
12.75	0.250	33.38	1694	1412	1176	941	2118	2235	2353
	0.330	43.77	2236	1864	1553	1242	2795	2951	3106
	0.375	49.56	2541	2118	1765	1412	3176	3353	3529
	0.406	53.52	2751	2293	1911	1528	3439	3630	3821
	0.500	65.42	3388	2824	2353	1882	4235	4471	4706

(Continued)

Table 4.9 Pipeline Internal Design Pressures and Test Pressures for API 5L X60—cont'd
Pipe Material API 5L X60   SMYS 60,000 psig

Diameter	Wall Thickness	Weight	Internal Design Pressure, psig				Hydrostatic Test Pressure, psig		
in	in	lb/ft	Class 1	Class 2	Class 3	Class 4	90% SMYS	95% SMYS	100% SMYS
14.00	0.250	36.71	1543	1286	1071	857	1929	2036	2143
	0.312	45.61	1925	1605	1337	1070	2407	2541	2674
	0.375	54.57	2314	1929	1607	1286	2893	3054	3214
	0.437	63.30	2697	2247	1873	1498	3371	3558	3746
	0.500	72.09	3086	2571	2143	1714	3857	4071	4286
16.00	0.250	42.05	1350	1125	938	750	1688	1781	1875
	0.312	52.27	1685	1404	1170	936	2106	2223	2340
	0.375	62.58	2025	1688	1406	1125	2531	2672	2813
	0.437	72.64	2360	1967	1639	1311	2950	3114	3278
	0.500	82.77	2700	2250	1875	1500	3375	3563	3750
18.00	0.250	47.39	1200	1000	833	667	1500	1583	1667
	0.312	58.94	1498	1248	1040	832	1872	1976	2080
	0.375	70.59	1800	1500	1250	1000	2250	2375	2500
	0.437	81.97	2098	1748	1457	1165	2622	2768	2913
	0.500	93.45	2400	2000	1667	1333	3000	3167	3333
20.00	0.312	65.60	1348	1123	936	749	1685	1778	1872
	0.375	78.60	1620	1350	1125	900	2025	2138	2250
	0.437	91.30	1888	1573	1311	1049	2360	2491	2622
	0.500	104.13	2160	1800	1500	1200	2700	2850	3000
	0.562	116.67	2428	2023	1686	1349	3035	3203	3372
22.00	0.375	86.61	1473	1227	1023	818	1841	1943	2045
	0.500	114.81	1964	1636	1364	1091	2455	2591	2727
	0.625	142.68	2455	2045	1705	1364	3068	3239	3409
	0.750	170.21	2945	2455	2045	1636	3682	3886	4091

Size	Thk	Wt						
24.00	0.375	94.62	1350	1125	938	750	1688	1781	1875
	0.437	109.97	1573	1311	1093	874	1967	2076	2185
	0.500	125.49	1800	1500	1250	1000	2250	2375	2500
	0.562	140.68	2023	1686	1405	1124	2529	2670	2810
	0.625	156.03	2250	1875	1563	1250	2813	2969	3125
	0.750	186.23	2700	2250	1875	1500	3375	3563	3750
26.00	0.375	102.63	1246	1038	865	692	1558	1644	1731
	0.500	136.17	1662	1385	1154	923	2077	2192	2308
	0.625	169.38	2077	1731	1442	1154	2596	2740	2885
	0.750	202.25	2492	2077	1731	1385	3115	3288	3462
28.00	0.375	110.64	1157	964	804	643	1446	1527	1607
	0.500	146.85	1543	1286	1071	857	1929	2036	2143
	0.625	182.73	1929	1607	1339	1071	2411	2545	2679
	0.750	218.27	2314	1929	1607	1286	2893	3054	3214
30.00	0.375	118.65	1080	900	750	600	1350	1425	1500
	0.500	157.53	1440	1200	1000	800	1800	1900	2000
	0.625	196.08	1800	1500	1250	1000	2250	2375	2500
	0.750	234.29	2160	1800	1500	1200	2700	2850	3000
32.00	0.375	126.66	1013	844	703	563	1266	1336	1406
	0.500	168.21	1350	1125	938	750	1688	1781	1875
	0.625	209.43	1688	1406	1172	938	2109	2227	2344
	0.750	250.31	2025	1688	1406	1125	2531	2672	2813
34.00	0.375	134.67	953	794	662	529	1191	1257	1324
	0.500	178.89	1271	1059	882	706	1588	1676	1765
	0.625	222.78	1588	1324	1103	882	1985	2096	2206
	0.750	266.33	1906	1588	1324	1059	2382	2515	2647

(*Continued*)

Table 4.9 Pipeline Internal Design Pressures and Test Pressures for API 5L X60—cont'd Pipe Material API 5L X60   SMYS 60,000 psig

Diameter	Wall Thickness	Weight	Internal Design Pressure, psig				Hydrostatic Test Pressure, psig		
in	in	lb/ft	Class 1	Class 2	Class 3	Class 4	90% SMYS	95% SMYS	100% SMYS
36.00	0.375	142.68	900	750	625	500	1125	1188	1250
	0.500	189.57	1200	1000	833	667	1500	1583	1667
	0.625	236.13	1500	1250	1042	833	1875	1979	2083
	0.750	282.35	1800	1500	1250	1000	2250	2375	2500
42.00	0.375	166.71	771	643	536	429	964	1018	1071
	0.500	221.61	1029	857	714	571	1286	1357	1429
	0.625	276.18	1286	1071	893	714	1607	1696	1786
	0.750	330.41	1543	1286	1071	857	1929	2036	2143
	1.000	437.88	2057	1714	1429	1143	2571	2714	2857

Pipeline Stress Design 127

Table 4.10 Pipeline Internal Design Pressures and Test Pressures for API 5L X65
Pipe Material API 5L X65 SMYS 65,000 psig

Diameter	Wall Thickness	Weight	Internal Design Pressure, psig				Hydrostatic Test Pressure, psig		
in	in	lb/ft	Class 1	Class 2	Class 3	Class 4	90% SMYS	95% SMYS	100% SMYS
4.5	0.237	10.79	4930	4108	3423	2739	6162	6504	6847
	0.337	14.98	7010	5841	4868	3894	8762	9249	9736
	0.437	18.96	9090	7575	6312	5050	11,362	11,993	12,624
	0.531	22.51	11,045	9204	7670	6136	13,806	14,573	15,340
6.625	0.250	17.02	3532	2943	2453	1962	4415	4660	4906
	0.280	18.97	3956	3297	2747	2198	4945	5220	5494
	0.432	28.57	6103	5086	4238	3391	7629	8053	8477
	0.562	36.39	7940	6617	5514	4411	9925	10,477	11,028
8.625	0.250	22.36	2713	2261	1884	1507	3391	3580	3768
	0.277	24.70	3006	2505	2088	1670	3758	3966	4175
	0.322	28.55	3494	2912	2427	1941	4368	4611	4853
	0.406	35.64	4406	3672	3060	2448	5507	5813	6119
10.75	0.250	28.04	2177	1814	1512	1209	2721	2872	3023
	0.307	34.24	2673	2228	1856	1485	3341	3527	3713
	0.365	40.48	3178	2648	2207	1766	3973	4193	4414
	0.500	54.74	4353	3628	3023	2419	5442	5744	6047
12.75	0.250	33.38	1835	1529	1275	1020	2294	2422	2549
	0.330	43.77	2423	2019	1682	1346	3028	3196	3365
	0.375	49.56	2753	2294	1912	1529	3441	3632	3824
	0.406	53.52	2981	2484	2070	1656	3726	3933	4140
	0.500	65.42	3671	3059	2549	2039	4588	4843	5098

(*Continued*)

Table 4.10 Pipeline Internal Design Pressures and Test Pressures for API 5L X65—cont'd
Pipe Material API 5L X65 SMYS 65,000 psig

Diameter	Wall Thickness	Weight	Internal Design Pressure, psig				Hydrostatic Test Pressure, psig		
in	in	lb/ft	Class 1	Class 2	Class 3	Class 4	90% SMYS	95% SMYS	100% SMYS
14.00	0.250	36.71	1671	1393	1161	929	2089	2205	2321
	0.312	45.61	2086	1738	1449	1159	2607	2752	2897
	0.375	54.57	2507	2089	1741	1393	3134	3308	3482
	0.437	63.30	2922	2435	2029	1623	3652	3855	4058
	0.500	72.09	3343	2786	2321	1857	4179	4411	4643
16.00	0.250	42.05	1463	1219	1016	813	1828	1930	2031
	0.312	52.27	1825	1521	1268	1014	2282	2408	2535
	0.375	62.58	2194	1828	1523	1219	2742	2895	3047
	0.437	72.64	2556	2130	1775	1420	3196	3373	3551
	0.500	82.77	2925	2438	2031	1625	3656	3859	4063
18.00	0.250	47.39	1300	1083	903	722	1625	1715	1806
	0.312	58.94	1622	1352	1127	901	2028	2141	2253
	0.375	70.59	1950	1625	1354	1083	2438	2573	2708
	0.437	81.97	2272	1894	1578	1262	2841	2998	3156
	0.500	93.45	2600	2167	1806	1444	3250	3431	3611
20.00	0.312	65.60	1460	1217	1014	811	1825	1927	2028
	0.375	78.60	1755	1463	1219	975	2194	2316	2438
	0.437	91.30	2045	1704	1420	1136	2556	2698	2841
	0.500	104.13	2340	1950	1625	1300	2925	3088	3250
	0.562	116.67	2630	2192	1827	1461	3288	3470	3653
22.00	0.375	86.61	1595	1330	1108	886	1994	2105	2216
	0.500	114.81	2127	1773	1477	1182	2659	2807	2955
	0.625	142.68	2659	2216	1847	1477	3324	3509	3693
	0.750	170.21	3191	2659	2216	1773	3989	4210	4432

Size	Wall								
24.00	0.375	94.62	1463	1219	1016	813	1828	1930	2031
	0.437	109.97	1704	1420	1184	947	2130	2249	2367
	0.500	125.49	1950	1625	1354	1083	2438	2573	2708
	0.562	140.68	2192	1827	1522	1218	2740	2892	3044
	0.625	156.03	2438	2031	1693	1354	3047	3216	3385
	0.750	186.23	2925	2438	2031	1625	3656	3859	4063
26.00	0.375	102.63	1350	1125	938	750	1688	1781	1875
	0.500	136.17	1800	1500	1250	1000	2250	2375	2500
	0.625	169.38	2250	1875	1563	1250	2813	2969	3125
	0.750	202.25	2700	2250	1875	1500	3375	3563	3750
28.00	0.375	110.64	1254	1045	871	696	1567	1654	1741
	0.500	146.85	1671	1393	1161	929	2089	2205	2321
	0.625	182.73	2089	1741	1451	1161	2612	2757	2902
	0.750	218.27	2507	2089	1741	1393	3134	3308	3482
30.00	0.375	118.65	1170	975	813	650	1463	1544	1625
	0.500	157.53	1560	1300	1083	867	1950	2058	2167
	0.625	196.08	1950	1625	1354	1083	2438	2573	2708
	0.750	234.29	2340	1950	1625	1300	2925	3088	3250
32.00	0.375	126.66	1097	914	762	609	1371	1447	1523
	0.500	168.21	1463	1219	1016	813	1828	1930	2031
	0.625	209.43	1828	1523	1270	1016	2285	2412	2539
	0.750	250.31	2194	1828	1523	1219	2742	2895	3047
34.00	0.375	134.67	1032	860	717	574	1290	1362	1434
	0.500	178.89	1376	1147	956	765	1721	1816	1912
	0.625	222.78	1721	1434	1195	956	2151	2270	2390
	0.750	266.33	2065	1721	1434	1147	2581	2724	2868

(Continued)

Table 4.10 Pipeline Internal Design Pressures and Test Pressures for API 5L X65—cont'd
Pipe Material API 5L X65   SMYS 65,000 psig

Diameter	Wall Thickness	Weight	Internal Design Pressure, psig				Hydrostatic Test Pressure, psig		
in	in	lb/ft	Class 1	Class 2	Class 3	Class 4	90% SMYS	95% SMYS	100% SMYS
36.00	0.375	142.68	975	813	677	542	1219	1286	1354
	0.500	189.57	1300	1083	903	722	1625	1715	1806
	0.625	236.13	1625	1354	1128	903	2031	2144	2257
	0.750	282.35	1950	1625	1354	1083	2438	2573	2708
42.00	0.375	166.71	836	696	580	464	1045	1103	1161
	0.500	221.61	1114	929	774	619	1393	1470	1548
	0.625	276.18	1393	1161	967	774	1741	1838	1935
	0.750	330.41	1671	1393	1161	929	2089	2205	2321
	1.000	437.88	2229	1857	1548	1238	2786	2940	3095

Pipeline Stress Design

Table 4.11 Pipeline Internal Design Pressures and Test Pressures for API 5L X70 Pipe Material API 5L X70   SMYS 70,000 psig

Diameter	Wall Thickness	Weight	Internal Design Pressure, psig				Hydrostatic Test Pressure, psig		
in	in	lb/ft	Class 1	Class 2	Class 3	Class 4	90% SMYS	95% SMYS	100% SMYS
4.5	0.237	10.79	5309	4424	3687	2949	6636	7005	7373
	0.337	14.98	7549	6291	5242	4194	9436	9960	10,484
	0.437	18.96	9789	8157	6798	5438	12,236	12,916	13,596
	0.531	22.51	11,894	9912	8260	6608	14,868	15,694	16,520
6.625	0.250	17.02	3804	3170	2642	2113	4755	5019	5283
	0.280	18.97	4260	3550	2958	2367	5325	5621	5917
	0.432	28.57	6573	5477	4565	3652	8216	8673	9129
	0.562	36.39	8551	7126	5938	4750	10,689	11,282	11,876
8.625	0.250	22.36	2922	2435	2029	1623	3652	3855	4058
	0.277	24.70	3237	2698	2248	1798	4047	4271	4496
	0.322	28.55	3763	3136	2613	2091	4704	4965	5227
	0.406	35.64	4745	3954	3295	2636	5931	6261	6590
10.75	0.250	28.04	2344	1953	1628	1302	2930	3093	3256
	0.307	34.24	2879	2399	1999	1599	3598	3798	3998
	0.365	40.48	3423	2852	2377	1901	4278	4516	4753
	0.500	54.74	4688	3907	3256	2605	5860	6186	6512
12.75	0.250	33.38	1976	1647	1373	1098	2471	2608	2745
	0.330	43.77	2609	2174	1812	1449	3261	3442	3624
	0.375	49.56	2965	2471	2059	1647	3706	3912	4118
	0.406	53.52	3210	2675	2229	1783	4012	4235	4458
	0.500	65.42	3953	3294	2745	2196	4941	5216	5490

(*Continued*)

Table 4.11 Pipeline Internal Design Pressures and Test Pressures for API 5L X70—cont'd
Pipe Material API 5L X70   SMYS 70,000 psig

Diameter	Wall Thickness	Weight	Internal Design Pressure, psig				Hydrostatic Test Pressure, psig		
in	in	lb/ft	Class 1	Class 2	Class 3	Class 4	90% SMYS	95% SMYS	100% SMYS
14.00	0.250	36.71	1800	1500	1250	1000	2250	2375	2500
	0.312	45.61	2246	1872	1560	1248	2808	2964	3120
	0.375	54.57	2700	2250	1875	1500	3375	3563	3750
	0.437	63.30	3146	2622	2185	1748	3933	4152	4370
	0.500	72.09	3600	3000	2500	2000	4500	4750	5000
16.00	0.250	42.05	1575	1313	1094	875	1969	2078	2188
	0.312	52.27	1966	1638	1365	1092	2457	2594	2730
	0.375	62.58	2363	1969	1641	1313	2953	3117	3281
	0.437	72.64	2753	2294	1912	1530	3441	3633	3824
	0.500	82.77	3150	2625	2188	1750	3938	4156	4375
18.00	0.250	47.39	1400	1167	972	778	1750	1847	1944
	0.312	58.94	1747	1456	1213	971	2184	2305	2427
	0.375	70.59	2100	1750	1458	1167	2625	2771	2917
	0.437	81.97	2447	2039	1699	1360	3059	3229	3399
	0.500	93.45	2800	2333	1944	1556	3500	3694	3889
20.00	0.312	65.60	1572	1310	1092	874	1966	2075	2184
	0.375	78.60	1890	1575	1313	1050	2363	2494	2625
	0.437	91.30	2202	1835	1530	1224	2753	2906	3059
	0.500	104.13	2520	2100	1750	1400	3150	3325	3500
	0.562	116.67	2832	2360	1967	1574	3541	3737	3934
22.00	0.375	86.61	1718	1432	1193	955	2148	2267	2386
	0.500	114.81	2291	1909	1591	1273	2864	3023	3182
	0.625	142.68	2864	2386	1989	1591	3580	3778	3977
	0.750	170.21	3436	2864	2386	1909	4295	4534	4773

Pipeline Stress Design

24.00	0.375	94.62	1575	1313	1094	875	1969	2078	2188
	0.437	109.97	1835	1530	1275	1020	2294	2422	2549
	0.500	125.49	2100	1750	1458	1167	2625	2771	2917
	0.562	140.68	2360	1967	1639	1311	2951	3114	3278
	0.625	156.03	2625	2188	1823	1458	3281	3464	3646
	0.750	186.23	3150	2625	2188	1750	3938	4156	4375
26.00	0.375	102.63	1454	1212	1010	808	1817	1918	2019
	0.500	136.17	1938	1615	1346	1077	2423	2558	2692
	0.625	169.38	2423	2019	1683	1346	3029	3197	3365
	0.750	202.25	2908	2423	2019	1615	3635	3837	4038
28.00	0.375	110.64	1350	1125	938	750	1688	1781	1875
	0.500	146.85	1800	1500	1250	1000	2250	2375	2500
	0.625	182.73	2250	1875	1563	1250	2813	2969	3125
	0.750	218.27	2700	2250	1875	1500	3375	3563	3750
30.00	0.375	118.65	1260	1050	875	700	1575	1663	1750
	0.500	157.53	1680	1400	1167	933	2100	2217	2333
	0.625	196.08	2100	1750	1458	1167	2625	2771	2917
	0.750	234.29	2520	2100	1750	1400	3150	3325	3500
32.00	0.375	126.66	1181	984	820	656	1477	1559	1641
	0.500	168.21	1575	1313	1094	875	1969	2078	2188
	0.625	209.43	1969	1641	1367	1094	2461	2598	2734
	0.750	250.31	2363	1969	1641	1313	2953	3117	3281
34.00	0.375	134.67	1112	926	772	618	1390	1467	1544
	0.500	178.89	1482	1235	1029	824	1853	1956	2059
	0.625	222.78	1853	1544	1287	1029	2316	2445	2574
	0.750	266.33	2224	1853	1544	1235	2779	2934	3088

(Continued)

Table 4.11 Pipeline Internal Design Pressures and Test Pressures for API 5L X70—cont'd Pipe Material API 5L X70   SMYS 70,000 psig

Diameter	Wall Thickness	Weight	Internal Design Pressure, psig				Hydrostatic Test Pressure, psig		
in	in	lb/ft	Class 1	Class 2	Class 3	Class 4	90% SMYS	95% SMYS	100% SMYS
36.00	0.375	142.68	1050	875	729	583	1313	1385	1458
	0.500	189.57	1400	1167	972	778	1750	1847	1944
	0.625	236.13	1750	1458	1215	972	2188	2309	2431
	0.750	282.35	2100	1750	1458	1167	2625	2771	2917
42.00	0.375	166.71	900	750	625	500	1125	1188	1250
	0.500	221.61	1200	1000	833	667	1500	1583	1667
	0.625	276.18	1500	1250	1042	833	1875	1979	2083
	0.750	330.41	1800	1500	1250	1000	2250	2375	2500
	1.000	437.88	2400	2000	1667	1333	3000	3167	3333

Pipeline Stress Design 135

Table 4.12 Pipeline Internal Design Pressures and Test Pressures for API 5L X80
Pipe Material API 5L X80   SMYS 80,000 psig

Diameter	Wall Thickness	Weight	Internal Design Pressure, psig				Hydrostatic Test Pressure, psig		
in	in	lb/ft	Class 1	Class 2	Class 3	Class 4	90% SMYS	95% SMYS	100% SMYS
4.5	0.237	10.79	6067	5056	4213	3371	7584	8005	8427
	0.337	14.98	8627	7189	5991	4793	10,784	11,383	11,982
	0.437	18.96	11,187	9323	7769	6215	13,984	14,761	15,538
	0.531	22.51	13,594	11,328	9440	7552	16,992	17,936	18,880
6.625	0.250	17.02	4347	3623	3019	2415	5434	5736	6038
	0.280	18.97	4869	4057	3381	2705	6086	6424	6762
	0.432	28.57	7512	6260	5217	4173	9390	9912	10,433
	0.562	36.39	9772	8144	6786	5429	12,216	12,894	13,573
8.625	0.250	22.36	3339	2783	2319	1855	4174	4406	4638
	0.277	24.70	3700	3083	2569	2055	4625	4882	5139
	0.322	28.55	4301	3584	2987	2389	5376	5675	5973
	0.406	35.64	5423	4519	3766	3013	6778	7155	7532
10.75	0.250	28.04	2679	2233	1860	1488	3349	3535	3721
	0.307	34.24	3290	2742	2285	1828	4112	4341	4569
	0.365	40.48	3911	3260	2716	2173	4889	5161	5433
	0.500	54.74	5358	4465	3721	2977	6698	7070	7442
12.75	0.250	33.38	2259	1882	1569	1255	2824	2980	3137
	0.330	43.77	2982	2485	2071	1656	3727	3934	4141
	0.375	49.56	3388	2824	2353	1882	4235	4471	4706
	0.406	53.52	3668	3057	2547	2038	4585	4840	5095
	0.500	65.42	4518	3765	3137	2510	5647	5961	6275

(*Continued*)

Table 4.12 Pipeline Internal Design Pressures and Test Pressures for API 5L X80—cont'd
Pipe Material API 5L X80   SMYS 80,000 psig

Diameter	Wall Thickness	Weight	Internal Design Pressure, psig				Hydrostatic Test Pressure, psig			
in	in	lb/ft	Class 1	Class 2	Class 3	Class 4	90% SMYS	95% SMYS	100% SMYS	
14.00	0.250	36.71	2057	1714	1429	1143	2571	2714	2857	
	0.312	45.61	2567	2139	1783	1426	3209	3387	3566	
	0.375	54.57	3086	2571	2143	1714	3857	4071	4286	
	0.437	63.30	3596	2997	2497	1998	4495	4745	4994	
	0.500	72.09	4114	3429	2857	2286	5143	5429	5714	
16.00	0.250	42.05	1800	1500	1250	1000	2250	2375	2500	
	0.312	52.27	2246	1872	1560	1248	2808	2964	3120	
	0.375	62.58	2700	2250	1875	1500	3375	3563	3750	
	0.437	72.64	3146	2622	2185	1748	3933	4152	4370	
	0.500	82.77	3600	3000	2500	2000	4500	4750	5000	
18.00	0.250	47.39	1600	1333	1111	889	2000	2111	2222	
	0.312	58.94	1997	1664	1387	1109	2496	2635	2773	
	0.375	70.59	2400	2000	1667	1333	3000	3167	3333	
	0.437	81.97	2797	2331	1942	1554	3496	3690	3884	
	0.500	93.45	3200	2667	2222	1778	4000	4222	4444	
20.00	0.312	65.60	1797	1498	1248	998	2246	2371	2496	
	0.375	78.60	2160	1800	1500	1200	2700	2850	3000	
	0.437	91.30	2517	2098	1748	1398	3146	3321	3496	
	0.500	104.13	2880	2400	2000	1600	3600	3800	4000	
	0.562	116.67	3237	2698	2248	1798	4046	4271	4496	
22.00	0.375	86.61	1964	1636	1364	1091	2455	2591	2727	
	0.500	114.81	2618	2182	1818	1455	3273	3455	3636	
	0.625	142.68	3273	2727	2273	1818	4091	4318	4545	
	0.750	170.21	3927	3273	2727	2182	4909	5182	5455	

24.00	0.375	94.62	1800	1500	1250	1000	2250	2375	2500
	0.437	109.97	2098	1748	1457	1165	2622	2768	2913
	0.500	125.49	2400	2000	1667	1333	3000	3167	3333
	0.562	140.68	2698	2248	1873	1499	3372	3559	3747
	0.625	156.03	3000	2500	2083	1667	3750	3958	4167
	0.750	186.23	3600	3000	2500	2000	4500	4750	5000
26.00	0.375	102.63	1662	1385	1154	923	2077	2192	2308
	0.500	136.17	2215	1846	1538	1231	2769	2923	3077
	0.625	169.38	2769	2308	1923	1538	3462	3654	3846
	0.750	202.25	3323	2769	2308	1846	4154	4385	4615
28.00	0.375	110.64	1543	1286	1071	857	1929	2036	2143
	0.500	146.85	2057	1714	1429	1143	2571	2714	2857
	0.625	182.73	2571	2143	1786	1429	3214	3393	3571
	0.750	218.27	3086	2571	2143	1714	3857	4071	4286
30.00	0.375	118.65	1440	1200	1000	800	1800	1900	2000
	0.500	157.53	1920	1600	1333	1067	2400	2533	2667
	0.625	196.08	2400	2000	1667	1333	3000	3167	3333
	0.750	234.29	2880	2400	2000	1600	3600	3800	4000
32.00	0.375	126.66	1350	1125	938	750	1688	1781	1875
	0.500	168.21	1800	1500	1250	1000	2250	2375	2500
	0.625	209.43	2250	1875	1563	1250	2813	2969	3125
	0.750	250.31	2700	2250	1875	1500	3375	3563	3750
34.00	0.375	134.67	1271	1059	882	706	1588	1676	1765
	0.500	178.89	1694	1412	1176	941	2118	2235	2353
	0.625	222.78	2118	1765	1471	1176	2647	2794	2941
	0.750	266.33	2541	2118	1765	1412	3176	3353	3529

(*Continued*)

Table 4.12 Pipeline Internal Design Pressures and Test Pressures for API 5L X80—cont'd
Pipe Material API 5L X80   SMYS 80,000 psig

Diameter	Wall Thickness	Weight	Internal Design Pressure, psig				Hydrostatic Test Pressure, psig		
in	in	lb/ft	Class 1	Class 2	Class 3	Class 4	90% SMYS	95% SMYS	100% SMYS
36.00	0.375	142.68	1200	1000	833	667	1500	1583	1667
	0.500	189.57	1600	1333	1111	889	2000	2111	2222
	0.625	236.13	2000	1667	1389	1111	2500	2639	2778
	0.750	282.35	2400	2000	1667	1333	3000	3167	3333
42.00	0.375	166.71	1029	857	714	571	1286	1357	1429
	0.500	221.61	1371	1143	952	762	1714	1810	1905
	0.625	276.18	1714	1429	1190	952	2143	2262	2381
	0.750	330.41	2057	1714	1429	1143	2571	2714	2857
	1.000	437.88	2743	2286	1905	1524	3429	3619	3810

Table 4.13 Pipeline Internal Design Pressures and Test Pressures for API 5L X90
Pipe Material API 5L X90

SMYS 90,000 psig

Diameter in	Wall Thickness in	Weight lb/ft	Internal Design Pressure, psig				Hydrostatic Test Pressure, psig		
			Class 1	Class 2	Class 3	Class 4	90% SMYS	95% SMYS	100% SMYS
4.5	0.237	10.79	6826	5688	4740	3792	8532	9006	9480
	0.337	14.98	9706	8088	6740	5392	12,132	12,806	13,480
	0.437	18.96	12,586	10,488	8740	6992	15,732	16,606	17,480
	0.531	22.51	15,293	12,744	10,620	8496	19,116	20,178	21,240
6.625	0.250	17.02	4891	4075	3396	2717	6113	6453	6792
	0.280	18.97	5477	4565	3804	3043	6847	7227	7608
	0.432	28.57	8451	7042	5869	4695	10,564	11,150	11,737
	0.562	36.39	10,994	9162	7635	6108	13,742	14,506	15,269
8.625	0.250	22.36	3757	3130	2609	2087	4696	4957	5217
	0.277	24.70	4162	3469	2890	2312	5203	5492	5781
	0.322	28.55	4838	4032	3360	2688	6048	6384	6720
	0.406	35.64	6101	5084	4237	3389	7626	8049	8473
10.75	0.250	28.04	3014	2512	2093	1674	3767	3977	4186
	0.307	34.24	3701	3084	2570	2056	4626	4883	5140
	0.365	40.48	4400	3667	3056	2445	5500	5806	6112
	0.500	54.74	6028	5023	4186	3349	7535	7953	8372
12.75	0.250	33.38	2541	2118	1765	1412	3176	3353	3529
	0.330	43.77	3354	2795	2329	1864	4193	4426	4659
	0.375	49.56	3812	3176	2647	2118	4765	5029	5294
	0.406	53.52	4127	3439	2866	2293	5159	5445	5732
	0.500	65.42	5082	4235	3529	2824	6353	6706	7059

(*Continued*)

Table 4.13 Pipeline Internal Design Pressures and Test Pressures for API 5L X90—cont'd
Pipe Material API 5L X90 SMYS 90,000 psig

Diameter in	Wall Thickness in	Weight lb/ft	Internal Design Pressure, psig				Hydrostatic Test Pressure, psig		
			Class 1	Class 2	Class 3	Class 4	90% SMYS	95% SMYS	100% SMYS
14.00	0.250	36.71	2314	1929	1607	1286	2893	3054	3214
	0.312	45.61	2888	2407	2006	1605	3610	3811	4011
	0.375	54.57	3471	2893	2411	1929	4339	4580	4821
	0.437	63.30	4045	3371	2809	2247	5057	5338	5619
	0.500	72.09	4629	3857	3214	2571	5786	6107	6429
16.00	0.250	42.05	2025	1688	1406	1125	2531	2672	2813
	0.312	52.27	2527	2106	1755	1404	3159	3335	3510
	0.375	62.58	3038	2531	2109	1688	3797	4008	4219
	0.437	72.64	3540	2950	2458	1967	4425	4670	4916
	0.500	82.77	4050	3375	2813	2250	5063	5344	5625
18.00	0.250	47.39	1800	1500	1250	1000	2250	2375	2500
	0.312	58.94	2246	1872	1560	1248	2808	2964	3120
	0.375	70.59	2700	2250	1875	1500	3375	3563	3750
	0.437	81.97	3146	2622	2185	1748	3933	4152	4370
	0.500	93.45	3600	3000	2500	2000	4500	4750	5000
20.00	0.312	65.60	2022	1685	1404	1123	2527	2668	2808
	0.375	78.60	2430	2025	1688	1350	3038	3206	3375
	0.437	91.30	2832	2360	1967	1573	3540	3736	3933
	0.500	104.13	3240	2700	2250	1800	4050	4275	4500
	0.562	116.67	3642	3035	2529	2023	4552	4805	5058
22.00	0.375	86.61	2209	1841	1534	1227	2761	2915	3068
	0.500	114.81	2945	2455	2045	1636	3682	3886	4091
	0.625	142.68	3682	3068	2557	2045	4602	4858	5114
	0.750	170.21	4418	3682	3068	2455	5523	5830	6136

24.00	0.375	94.62	2025	1688	1406	1125	2531	2672	2813
	0.437	109.97	2360	1967	1639	1311	2950	3114	3278
	0.500	125.49	2700	2250	1875	1500	3375	3563	3750
	0.562	140.68	3035	2529	2108	1686	3794	4004	4215
	0.625	156.03	3375	2813	2344	1875	4219	4453	4688
	0.750	186.23	4050	3375	2813	2250	5063	5344	5625
26.00	0.375	102.63	1869	1558	1298	1038	2337	2466	2596
	0.500	136.17	2492	2077	1731	1385	3115	3288	3462
	0.625	169.38	3115	2596	2163	1731	3894	4111	4327
	0.750	202.25	3738	3115	2596	2077	4673	4933	5192
28.00	0.375	110.64	1736	1446	1205	964	2170	2290	2411
	0.500	146.85	2314	1929	1607	1286	2893	3054	3214
	0.625	182.73	2893	2411	2009	1607	3616	3817	4018
	0.750	218.27	3471	2893	2411	1929	4339	4580	4821
30.00	0.375	118.65	1620	1350	1125	900	2025	2138	2250
	0.500	157.53	2160	1800	1500	1200	2700	2850	3000
	0.625	196.08	2700	2250	1875	1500	3375	3563	3750
	0.750	234.29	3240	2700	2250	1800	4050	4275	4500
32.00	0.375	126.66	1519	1266	1055	844	1898	2004	2109
	0.500	168.21	2025	1688	1406	1125	2531	2672	2813
	0.625	209.43	2531	2109	1758	1406	3164	3340	3516
	0.750	250.31	3038	2531	2109	1688	3797	4008	4219
34.00	0.375	134.67	1429	1191	993	794	1787	1886	1985
	0.500	178.89	1906	1588	1324	1059	2382	2515	2647
	0.625	222.78	2382	1985	1654	1324	2978	3143	3309
	0.750	266.33	2859	2382	1985	1588	3574	3772	3971

(*Continued*)

Table 4.13 Pipeline Internal Design Pressures and Test Pressures for API 5L X90—cont'd
Pipe Material API 5L X90   SMYS 90,000 psig

Diameter	Wall Thickness	Weight	Internal Design Pressure, psig				Hydrostatic Test Pressure, psig		
in	in	lb/ft	Class 1	Class 2	Class 3	Class 4	90% SMYS	95% SMYS	100% SMYS
36.00	0.375	142.68	1350	1125	938	750	1688	1781	1875
	0.500	189.57	1800	1500	1250	1000	2250	2375	2500
	0.625	236.13	2250	1875	1563	1250	2813	2969	3125
	0.750	282.35	2700	2250	1875	1500	3375	3563	3750
42.00	0.375	166.71	1157	964	804	643	1446	1527	1607
	0.500	221.61	1543	1286	1071	857	1929	2036	2143
	0.625	276.18	1929	1607	1339	1071	2411	2545	2679
	0.750	330.41	2314	1929	1607	1286	2893	3054	3214
	1.000	437.88	3086	2571	2143	1714	3857	4071	4286

Problem 4.5

A gas pipeline, NPS 20, 0.500-in wall thickness, is constructed of API 5L X52 pipe.
(a) Calculate the design pressures for class 1 through class 4 locations.
(b) What is the range of hydrotest pressures for each of these class locations?
Assume joint factor = 1.00. Temperature deration factor = 1.00.

Solution

Using Eqn (4.8) the internal design pressure is

$$P = \frac{2 \times 0.500 \times 52000 \times 1.00 \times 1.0 \times F}{20} = 2600\ F$$

where

F = design factor = 0.72 for class 1

Therefore, the design pressures for class 1 through class 4 are as follows:

$$\text{Class 1} = 2600 \times 0.72 = 1872\ \text{psig}$$

$$\text{Class 2} = 2600 \times 0.60 = 1560\ \text{psig}$$

$$\text{Class 3} = 2600 \times 0.50 = 1300\ \text{psig}$$

$$\text{Class 4} = 2600 \times 0.40 = 1040\ \text{psig}$$

The range of hydrotest pressures is such that the hoop stress will be between 90% and 95% of SMYS.

For class 1, the range of hydrotest pressures is

$$1.25 \times 1872\ \text{to}\ 1.3194 \times 1872 = 2340\ \text{psig to}\ 2470\ \text{psig}$$

where 1.3194 is equal to the factor $1.25 \times 95/90$ representing the upper limit of the hydrotest envelope.

For class 2, the range of hydrotest pressures is

$$1.25 \times 1560\ \text{to}\ 1.3194 \times 15650 = 1950\ \text{psig to}\ 2058\ \text{psig}$$

For class 3, the range of hydrotest pressures is

$$1.25 \times 1300\ \text{to}\ 1.3194 \times 1300 = 1625\ \text{psig to}\ 1715\ \text{psig}$$

For class 4, the range of hydrotest pressures is

$$1.25 \times 1040\ \text{to}\ 1.3194 \times 1040 = 1300\ \text{psig to}\ 1372\ \text{psig}$$

13. BLOWDOWN CALCULATIONS

Blowdown valves and piping system are installed around the mainline valve in a gas transmission piping system to evacuate gas from sections of pipeline in the event of an emergency or maintenance purposes. The objective of the blowdown assembly is to remove gas from the pipeline once the pipe section is isolated by closing the mainline block valves, in a reasonable period of time. The pipe size required to blowdown a section of pipe will depend on the gas gravity, pipe diameter, length of pipe section, the pressure in the pipeline, and the blowdown time. AGA recommends the following equation to estimate the blowdown time.

$$T = \frac{0.0588 P_1^{\frac{1}{3}} G^{\frac{1}{2}} D^2 L F_c}{d^2} \quad \text{(USCS units)} \quad (4.17)$$

where,
T = blowdown time, min
P_1 = initial pressure, psia
G = gas gravity (air = 1.00)
D = pipe inside diameter, in.
L = length of pipe section, mile
d = inside diameter of blowdown pipe, in
F_c = choke factor (as follows)
Choke factor list.
Ideal nozzle = 1.0
Through gate = 1.6
Regular gate = 1.8
Regular lube plug = 2.0
Venture lube plug = 3.2
In SI units:

$$T = \frac{0.0886 P_1^{\frac{1}{3}} G^{\frac{1}{2}} D^2 L F_c}{d^2} \quad \text{(SI units)} \quad (4.18)$$

where,
P_1 = initial pressure, kg/cm^2
D = pipe inside diameter, mm
L = length of pipe section, km
d = pipe inside diameter of blowdown, mm
Other symbols are as defined before.

Problem 4.6
Calculate the blowdown time required for an NPS 14, 0.250-in wall thickness blowdown assembly on an NPS 36 pipe, 0.500-in wall thickness considering 10-mile pipe section starting at a pressure of 1000 psia. The gas gravity is 0.6 and choke factor = 1.8.

Solution
Pipe inside diameter = 36−2 × 0.500 = 35.0 in
Blowdown pipe inside diameter = 14.0−2 × 0.250 = 13.5 in
Using Eqn (4.9), we get:

$$T = \frac{0.0588 \times (1000)^{\frac{1}{3}}(0.6)^{\frac{1}{2}}(35)^2 \times 5 \times 1.8}{13.5^2} = 28 \text{ min approximately}$$

14. DETERMINING PIPE TONNAGE

Frequently, in pipeline design we are interested in knowing the amount of pipe used so that we can determine the total cost of pipe. A convenient formula for calculating the weight per unit length of pipe, used by pipe vendors is given in Eqn (4.19).

In US customary system units, pipe weight in lb/ft is calculated for a given diameter and wall thickness as follows:

$$w = 10.68 \times t \times (D-t) \quad \text{(USCS units)} \quad (4.19)$$

where
w = pipe weight, lb/ft
D = pipe outside diameter, in
t = pipe wall thickness, in

The constant 10.68 in Eqn (4.19) includes the density of steel and therefore the equation is only applicable to steel pipe. For other pipe material we can ratio the densities to obtain the pipe weight for non steel pipe.

In SI units, the pipe weight in kg/m is found from

$$w = 0.0246 \times t \times (D-t) \quad \text{(SI units)} \quad (4.20)$$

where
w = pipe weight, kg/m
D = pipe outside diameter, mm
t = pipe wall thickness, mm

Problem 4.7
Calculate the total amount of pipe in a 10-mile pipeline, NPS 20, 0.500-in wall thickness. If pipe costs $700 per ton, determine the total pipeline cost.

Solution
Using Eqn (4.11), the weight per foot of pipe is:
$$w = 10.68 \times 0.500 \times (20 - 0.500) = 101.46 \text{ lb/ft}$$

Therefore, total pipe tonnage in 10 miles of pipe is:
$$\text{Tonnage} = 101.46 \times 5280 \times 10/2000 = 2679 \text{ tons}$$

$$\text{Total pipeline cost} = 2679 \times 700 = \$1,875,300$$

Problem 4.8
A 60-km pipeline consists of 20 km of DN 500, 12-mm wall thickness pipe connected to a 40-km length of DN 400, 10-mm wall thickness pipe. What is the total metric tons of pipe?

Solution
Using Eqn (4.12), the weight per meter of DN 500 pipe is
$$w = 0.0246 \times 12 \times (500 - 12) = 144.06 \text{ kg/m}$$

And the weight per meter of DN 400 pipe is
$$w = 0.0246 \times 10 \times (400 - 10) = 95.94 \text{ kg/m}$$

Therefore, the total pipe weight for 20 km of DN 500 pipe and 40 km of DN 400 pipe is
$$\text{Weight} = (20 \times 144.06) + (40 \times 95.94) = 6719 \text{ tons}$$

$$\text{Total metric tons} = 6719$$

Problem 4.9
Calculate the MOP for NPS 16 pipeline, 0.250-in wall thickness constructed of API 5LX-52 steel. What minimum wall thickness is required for an internal working pressure of 1440 psi? Use class 2 construction with design factor $F = 0.60$ and for an operating temperature below 250 °F.

Solution
Using Eqn (4.8), the internal design pressure is

$$P = \frac{2 \times 0.250 \times 52000 \times 0.60 \times 1.0 \times 1.0}{16} = 975 \text{ pisg}$$

For an internal working pressure of 1440 psi, the wall thickness required is

$$1440 = \frac{2 \times t \times 52000 \times 0.6 \times 1.0}{16}$$

Solving for t, we get:
Wall thickness $t = 0.369$ in
The nearest standard pipe wall thickness is 0.375 in

Problem 4.10
A natural gas pipeline, 600 km long, is constructed of DN 800 pipe and has a required operating pressure of 9 MPa. Compare the cost of using X-60 or X-70 steel pipe. The material cost of the two grades of pipe is as follows:

Pipe Grade	Material Cost ($/ton)
X-60	800
X-70	900

Use class 1 design factor and temperature deration factor of 1.00.

Solution
We will first determine the wall thickness of pipe required to withstand the operating pressure of 9 MPa.

Using Eqn (4.8), the pipe wall thickness required for X-60 pipe (60,000 psi = 414 MPa)

$$t = \frac{9 \times 800}{2 \times 414 \times 1.0 \times 0.72 \times 1.0}$$
$$= 12.08 \text{ mm}. \quad \text{Use 13 mm wall thickness}$$

Similarly, the pipe wall thickness required for X-70 pipe (70,000 psi = 483 MPa)

$$t = \frac{9 \times 800}{2 \times 483 \times 1.0 \times 0.72 \times 1.0} = 10.35 \text{ mm}.$$
Use 11 mm wall thickness

Pipe weight in kg/m will be calculated using Eqn (4.12).

For X-60 pipe, weight per meter = $0.0246 \times 13 \times (800-13)$ = 251.68 kg/m.
Therefore total cost of a 600-km pipeline at $800 per ton of X-60 pipe is:
Total cost = $600 \times 251.68 \times 800$ = $120.81 million.
Similarly, pipe weight in kg/m for X-70 pipe is
For X-70 pipe, weight per meter = $0.0246 \times 11 \times (800-11)$ = 213.50 kg/m.
Therefore total cost of 600 km pipeline at $900 per ton of X-70 pipe is:
Total cost = $600 \times 213.50 \times 900$ = $115.29 million.
Therefore the X-70 pipe will cost less than the X-60 pipe. The difference in cost is:
$120.81 − $115.29 = $5.52 million.

15. SUMMARY

In this chapter, we discussed how to calculate the pipe wall thickness required to with stand an internal pressure in a gas pipeline using Barlow's equation. The influence of the population density in the vicinity of the pipeline on the required pipe wall thickness by reducing the allowable hoop stress in the high-population areas were explained by way of class locations. We explored the range of pressures required to hydrotest pipeline sections to ensure safe operation of the pipeline. The effect of pipeline elevations on determining a testing plan by sectioning the pipeline was covered. The need for isolating portions of the pipeline by properly spaced mainline valves and the method of calculating the time required for evacuating gas from the pipeline sections were also discussed. Finally, a simple method of calculating the pipe tonnage was explained.

We discussed how allowable internal pressure in a pipeline is calculated depending on pipe size and material. We showed that for pipe under internal pressure the hoop stress in the pipe material will be a controlling factor. The importance of design factor in selecting pipe wall thickness was illustrated using an example. Based on Barlow's equation, the internal design pressure calculation as recommended by ASME standard B31.4 and US Code of Federal Regulation, Part 195, of the DOT was illustrated. The need for hydrostatic testing pipelines for safe operation was discussed. The line fill volume calculation was introduced and its importance in batched pipelines was shown using an example.

CHAPTER FIVE

Fluid Flow in Pipes

In this chapter, we introduce the concept of pressure in a liquid and how it is measured. We will also discuss what is meant by the velocity or speed of the fluid in a pipe, the different types of flow, and we also introduce the concept of Reynolds number and its importance in determining the type of flow. Depending upon the flow regime, such as laminar, critical, or turbulent methods will be discussed in terms of how to calculate the pressure drop from friction. Several popular formulas such as Colebrook–White and Hazen–Williams equations will be presented and compared. We will also discuss the noniterative or straightforward formulas such as the Swamee–Jain equation, the Churchill's equation, and some other less popular formulas. Next we will also cover minor pressure losses in piping because of fittings and valves and those resulting from pipe enlargements and contractions. The importance of pipe internal roughness will be discussed as well. We will also explore drag reduction as a means of reducing energy loss in pipe flow.

In addition, in the gas section of this chapter, we will discuss the various methods of calculating the pressure drop from friction in a gas pipeline. Commonly used formulas will be reviewed and illustrated using examples. The impact of internal conditions of the pipe on the pipe capacity will also be explored.

1. LIQUID PRESSURE

Hydrostatics is the study of hydraulics that deals with liquid pressures and forces resulting from the weight of the liquid at rest. The force per unit area at a certain point within a liquid is called the pressure p. This pressure at a certain depth h, below the free surface of the liquid consists of equal pressures in all directions. This is known as Pascal's law. Consider an imaginary flat surface within the liquid located at a depth h below the liquid surface as shown in Figure 5.1. The pressure on this surface must act normal to the surface at all points along the surface because liquids at rest cannot transmit shear. The variation of pressure with the depth of the liquid is calculated by considering forces acting on a thin vertical cylinder of height Δh and a cross sectional area Δa as shown in Figure 5.1.

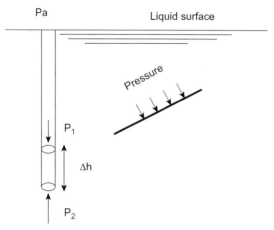

Figure 5.1 Pressure in a liquid.

Because the liquid is at rest, the cylindrical volume is in equilibrium from the forces acting upon it. By the principles of statics, the algebraic sum of all forces acting on this cylinder in the vertical and horizontal directions must equal zero. The vertical forces on the cylinder consists of the weight of the cylinder and the forces resulting from liquid pressure P_1 at the top and P_2 at the bottom as shown in Figure 5.1. Because the specific weight of the liquid γ does not change with pressure, we can write the following equation for the summation of forces in the vertical direction.

$$P_2 \Delta a = \gamma\, \Delta h\, \Delta a + P_1 \Delta a$$

where the product term $\gamma\, \Delta h\, \Delta a$ represents the weight of the cylindrical element.

Simplifying above we get

$$P_2 = \gamma \Delta h + P_1 \qquad (5.1)$$

If we now imagine that the cylinder is extended to the liquid surface, P_1 becomes the pressure at the liquid surface (atmospheric pressure P_a) and Δh becomes h, the depth of the point in the liquid where the pressure is P_2. Replacing P_2 with P, the pressure in the liquid at depth h, Eqn (5.1) becomes.

$$P = \gamma h + P_a \qquad (5.2)$$

From Eqn (5.2), we conclude that the pressure in a liquid at a depth h increases with the depth. If the term P_a (atmospheric pressure) is neglected, we can state that the gauge pressure (based on zero atmospheric pressure) at a depth h is simply Δh.

Therefore, the gauge pressure is

$$P = \gamma h \tag{5.3}$$

Dividing both sides by γ and transposing we can write

$$h = P/\gamma \tag{5.4}$$

In Eqn (5.4), the term h represents the "pressure head" corresponding to the pressure P. It represents the depth in feet of liquid of specific weight γ to produce the pressure P. Values of absolute pressure $(P + P_a)$ are always positive, whereas the gauge pressure P may be positive or negative depending on whether the pressure is greater or less than the atmospheric pressure. Negative gauge pressure means that a partial vacuum exists in the liquid.

From this discussion, it is clear that the absolute pressure within a liquid consists of the head pressure because of the depth of liquid and the atmospheric pressure at the liquid surface. The atmospheric pressure at a geographic location varies with the elevation above sea level. Because the density of the atmospheric air varies with the altitude, a straight line relationship does not exist between the altitude and the atmospheric pressure (unlike the linear relationship between liquid pressure and depth). For most purposes, we can assume that the atmospheric pressure at sea level is approximately 14.7 psi. In SI units, the atmospheric pressure is approximately 101 kPa.

The instrument used to measure the atmospheric pressure at a given location is called a "barometer." A typical barometer is shown in Figure 5.2.

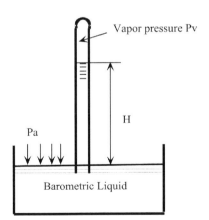

Figure 5.2 Barometer for measuring pressure.

In such an instrument, the tube is filled with a heavy liquid (usually mercury) and quickly inverting the tube and positioning it in the container full of the liquid as shown in Figure 5.2.

If the tube is sufficiently long, the level of liquid will fall slightly to cause a vapor space at the top of the tube just above the liquid surface. Equilibrium will be reached when the liquid vaporizes in the vapor space and creates a pressure P_v. Because of the density of mercury (approximately 13 times as heavy as water) and its vapor pressure being low, it is an ideal liquid for a barometer. If a liquid such as water were used, we would need a fairly long tube to measure the atmospheric pressure as we shall see shortly.

From Figure 5.2, the atmospheric pressure Pa exerted at the surface of the liquid is equal to the sum of the vapor pressure P_v and the pressure generated by the column of the barometric liquid of height H_b.

$$P_a = P_v + \gamma H_b \tag{5.5}$$

where:

P_a – atmospheric pressure
P_v – vapor pressure of barometric liquid
γ – specific weight of barometric liquid
H_b – barometric reading

In the previous equation, if pressures are in psi and liquid specific weight is in lb/ft^3, the pressures must be multiplied by 144 to obtain the barometric reading in ft of liquid. Equation (5.5) is valid for barometers with any liquid. Because the vapor pressure of mercury is negligible, we can rewrite Eqn (5.5) for a mercury barometer as follows:

$$P_a = \gamma H_b \tag{5.6}$$

Let us compare using water and mercury as barometric liquids to measure the atmospheric pressure.

Example 5.1

Assume the vapor pressure of water at 70 F is 0.3632 psi and its specific weight is 62.3 lb/ft^3. Mercury used in a manometer has a specific gravity of 13.54 and negligible vapor pressure. If the sea level atmospheric pressure is 14.7 psi, determine the barometric heights for water and mercury.

Solution

From Eqn (5.5) for water:

$$14.7 = 0.3632 + (62.3/144)H_b$$

The barometric height for water is

$$H_b = (14.7 - 0.3632) \times 144/62.3 = 33.14 \text{ ft}$$

Similarly, for mercury, neglecting the vapor pressure, using Eqn (5.6), we get

$$14.7 \times 144 = (13.54 \times 62.3)H_b$$

The barometric height for mercury is

$$H_b = (14.7 \times 144)/(13.54 \times 62.3) = 2.51 \text{ ft}$$

It can be seen from the previous equation that the mercury barometer requires a much shorter tube than a water barometer.

The pressure in a liquid is measured in lb/in^2 (psi) in the English units or kilopascals (kPa) in SI units. Because pressure is measured using a gauge and is relative to the atmospheric pressure at the specific location, it is also reported as psig (psi gauge). The absolute pressure in a liquid is the sum of the gauge pressure and the atmospheric pressure at the location. Thus:

Absolute pressure in psia = gauge pressure in psig + atmospheric pressure.

For example, if the pressure gauge reading is 800 psig, the absolute pressure in the liquid is

$$P_{abs} = 800 + 14.7 = 814.7 \text{ psia}$$

This is based on the assumption that atmospheric pressure at the location is 14.7 psia.

Pressure in a liquid may also be referred to in terms of feet (or meters in SI units) of liquid head. By dividing the pressure in lb/ft^2 by the liquid specific weight in lb/ft^3, we get the pressure head in feet of liquid. When expressed this way, the head represents the height of the liquid column required to match the given pressure in psig. For example, if the pressure in a liquid is 1000 psig, the head of liquid corresponding to this pressure is calculated as follows:

$$\text{Head} = 2.31(\text{psig})/\text{Spgr} \quad \text{ft (English units)} \tag{5.7}$$

$$\text{Head} = 0.102(\text{kPa})/\text{Spgr} \quad \text{m (SI units)}. \tag{5.8}$$

where

Spgr — liquid specific gravity

The factor 2.31 in the above equation comes from the ratio

$$\frac{144 \text{ in}^2/\text{ft}^2}{62.34 \text{ lb/ft}^3}$$

where 62.34 lb/ft^3 is the specific weight of water.

Therefore, if the liquid specific gravity is 0.85, the equivalent liquid head is

$$\text{Head} = (1{,}000)(2.31)/0.85 = 2{,}717.65 \text{ ft}$$

This means that the liquid pressure of 1000 psig is equivalent to the pressure exerted at the bottom of a liquid column, of specific gravity 0.85, 2717.65 ft in height. If such a column of liquid had a cross-sectional area of 1 square in, the weight of the column will be

$$2{,}717.65(1/144)(62.34)(0.85) = 1{,}000 \text{ lb}$$

where 62.34 lb/ft^3 is the density of water.

These weights act on area of 1 square in. The calculated pressure is therefore 1000 psig.

We can analyze head pressure resulting from a column of liquid in another way:

Consider a cylindrical column of liquid, of height H ft and area of cross-section A in^2. If the top surface of the liquid column is open to the atmosphere, we can calculate the pressure exerted by this column of liquid at its base as

$$\text{Pressure} = \frac{\text{Weight of liquid column}}{\text{Area of cross section}}$$

or

$$\text{Pressure} = \frac{\text{Volume} \times \text{Specific weight}}{\text{Area of cross section}} = (AH\gamma)/(144 \times A)$$

or

$$\text{Pressure} = H\gamma/144 \tag{5.9}$$

where:

γ — specific weight of liquid, lb/ft^3

The factor 144 is used to convert from in^2 to ft^2.

If we use 62.34 lb/ft^3 for specific weight of water, the pressure for a water column from Eqn (5.9) is

$$\text{Pressure} = H \times 62.34/144 = H/2.31$$

2. LIQUID: VELOCITY

Velocity of flow in a pipeline is the average velocity based on the pipe diameter and liquid flow rate. It may be calculated as follows:

$$\text{Velocity} = \text{Flow rate}/\text{area of flow}$$

Depending on the type of flow (laminar, turbulent, etc.), the liquid velocity in a pipeline at a particular pipe cross-section will vary along the pipe radius. The liquid molecules at the pipe wall are at rest and therefore have zero velocity. As we approach the center line of the pipe, the liquid molecules are increasingly free and therefore have increasing velocity. The variation of velocity for laminar flow and turbulent flow are as shown in Figure 5.3. In laminar flow, the variation of velocity at a pipe cross-section is parabolic. In turbulent flow, there is an approximate trapezoidal shape to the velocity profile. Laminar flow is also known as viscous flow or streamline flow.

The average or bulk velocity of a liquid flowing in a pipe can be easily calculated as a function of the actual flow rate and the inside dimensions of the pipe.

To calculate the average velocity of liquid flow in barrels (bbl)/day, bbl/h and gal/min, the following equations may be used.

$$V = 0.0119(\text{bbl/day})/D^2 \tag{5.10}$$

$$V = 0.2859(\text{bbl/hr})/D^2 \tag{5.11}$$

$$V = 0.4085(\text{gal/min})/D^2 \tag{5.12}$$

where:
V – velocity, ft/s
D – inside diameter, in
In SI units, the velocity is calculated as follows:

$$V = 353.6777(\text{m}^3/\text{hr})/D^2 \tag{5.13}$$

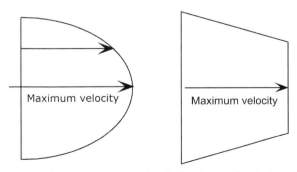

Figure 5.3 Velocity variation in pipe for laminar and turbulent flow.

where:

V – velocity, m/s
D – inside diameter, mm

For example, liquid flowing through a 16-in pipeline (wall thickness 0.250 in) at the rate of 100,000 bbl/day, has an average velocity of:

$$0.0119(100,000)/(15.5)^2 = 4.95 \text{ ft/s}$$

This represents the average velocity at a particular cross-section of pipe. The velocity at the centerline will be higher than this, depending on whether the flow is turbulent or laminar.

3. LIQUID: REYNOLDS NUMBER

Flow in a liquid pipeline may be smooth, laminar flow also known as viscous flow. In this type of flow, the liquid flows in layers or laminations without causing eddies or turbulence. If the pipe is transparent and we inject a dye into the flowing stream, it would flow smoothly in a straight line confirming smooth or laminar flow. As the liquid flow rate is increased, the velocity increases and the flow will change from laminar flow to turbulent flow with eddies and disturbances. This can be seen clearly when a dye is injected into the flowing stream.

An important dimensionless parameter called the Reynolds number is used in classifying the type of flow in pipelines.

Reynolds number of flow, R, is calculated as follows:

$$R = VD\rho/\mu \qquad (5.14)$$

where:

V – average velocity, ft/s
D – pipe internal diameter, ft
ρ – liquid density, slugs/ft^3
μ – absolute viscosity, lb-s/ft^2
R – Reynolds number is a dimensionless value

Because of, the kinematic viscosity $\nu = \mu/\rho$, the Reynolds number can also be expressed as

$$R = VD/\nu \qquad (5.15)$$

where:

ν – kinematic viscosity, ft^2/s

Care should be taken to ensure that proper units are used in Eqns (5.14) and (5.15) such that R is dimensionless.

Flow-through pipes are classified into three main flow regimes.
1. Laminar flow – R < 2000
2. Critical flow – R > 2000 and R < 4000
3. Turbulent flow – R > 4000

Depending upon the Reynolds number, flow-through pipes will fall in one of these three flow regimes. Let us first examine the concepts of the Reynolds number. Sometimes an R value of 2100 is used as the limit of laminar flow.

Using customary units in the pipeline industry, the Reynolds number can be calculated using the following formula:

$$R = 92.24 \, Q/(vD) \qquad (5.16)$$

where:
Q – flow rate, bbl/day
D – internal diameter, in
v – kinematic viscosity, cSt

Equation (5.16) is simply a modified form of Eqn (5.15) after performing conversions to commonly used pipeline units. R is still a dimensionless value.

Another version of the Reynolds number in English units is as follows:

$$R = 3{,}160 \, Q/(vD) \qquad (5.17)$$

where:
Q – flow rate, gal/min
D – internal diameter, in
v – kinematic viscosity, cSt

A similar equation for the Reynolds number in SI units is

$$R = 353{,}678 \, Q/(vD) \qquad (5.18)$$

where
Q – flow rate, m^3/h
D – internal diameter, mm
v – kinematic viscosity, cSt

As indicated previously, if the Reynolds number is less than 2000, the flow is considered to be laminar. This is also known as viscous flow. This means that the various layers of liquid flow without turbulence in the form of laminations. We will now illustrate the various flow regimes using an example.

Consider a 16-in pipeline, 0.250-in wall thickness transporting a liquid of viscosity 250 cSt. At a flow rate of 50,000 bbl/day, the Reynolds number is, using Eqn (5.16),

$$R = 92.24(50,000)/(250 \times 15.5) = 1,190$$

Because R is less than 2000, this flow is laminar. If the flow rate is tripled to 150,000 bbl/day, the Reynolds number becomes 3570 and the flow will be in the critical region. At flow rates above 168,040 bbl/day, the Reynolds number exceeds 4000 and the flow will be in the turbulent region. Thus, for this 16-in pipeline and given liquid viscosity of 250 cSt, flow will be fully turbulent at flow rates above 168,040 bbl/day.

As the flow rate and velocity increase, the flow regime changes. With a change in flow regime, the energy lost from pipe friction increases. At laminar flow, there is less frictional energy lost compared to turbulent flow.

4. FLOW REGIMES

In summary, the three flow regimes may be distinguished as follows:
Laminar – R < 2000.
Critical – R > 2000 and R < 4000.
Turbulent – R > 4000.

As liquid flows through a pipeline, energy is lost because of friction between the pipe surface and the liquid and because of the interaction between liquid molecules. This energy lost is at the expense of liquid pressure. Hence we refer to the frictional energy lost as pressure drop because of friction.

The pressure drop because of friction in a pipeline depends on the flow rate, pipe diameter, pipe roughness, liquid specific gravity, and viscosity. In addition, frictional pressure drop depends on the Reynolds number (and hence the flow regime). Our objective would be to calculate the pressure drop given these pipe and liquid properties and the flow regime.

In classical mechanics, the pressure drop in a given length of pipe, expressed in feet of liquid head (h), can be calculated using the Darcy–Weisbach equation as follows:

$$h = f(L/D)(V^2/2g) \qquad (5.19)$$

where the pressure drop results from friction friction, h is expressed in feet of liquid head, and the other symbols are defined here.

f – Darcy friction factor, dimensionless, usually a number 0.008 to 0.10
L – pipe length, ft
D – pipe internal diameter, ft
V – average liquid velocity, ft/s
g – acceleration because of gravity, 32.2 ft/s² in English units.

Some textbooks refer to a friction factor called Fannings friction factor. This is numerically equal to one-fourth the Darcy friction factor. Thus a value of f = 0.02 translates to a Fanning friction factor = 0.02/4 = 0.05.

Henceforth, we will always use the Darcy friction factor f. In laminar flow, the Darcy friction factor f depends only on the Reynolds number. In comparison, in turbulent flow, f depends on pipe diameter, internal pipe roughness, and Reynolds number, as we will explain shortly.

Example 5.2

Consider a pipeline transporting 5000 bbl/h of gasoline (Spgr = 0.736). Calculate the pressure drop in a 3000 ft length of 24-in pipe (wall thickness 0.500 in) using the Darcy–Weisbach equation.

Assume the friction factor is 0.02.

Solution

Using Eqn (5.10):

$$\text{Average liquid velocity} = 0.0119(5{,}000 \times 24)/(24.0)^2 = 2.48 \text{ ft/s}$$

Using the Darcy–Weisbach Eqn (5.19):

$$\text{Pressure drop} = 0.02(5{,}000)(12/15.5)(4.76^2/64.4) = 27.24 \text{ ft of head}$$

Converting to pressure in psi, using Eqn (5.7):

$$\text{Pressure drop} = 27.24(0.736)/2.31 = 8.68 \text{ psi}.$$

In these calculations, the friction factor f was assumed to be 0.02. However, the actual friction factor for a particular flow depends on various factors as explained previously. In the next section, we will see how the friction factor is calculated for the various flow regimes.

5. FRICTION FACTOR

For laminar flow, with Reynolds number R < 2000, the Darcy friction factor f is calculated from the simple relationship:

$$f = 64/R \tag{5.20}$$

It can be seen from Eqn (5.20) that for laminar flow, the friction factor depends only on the Reynolds number and is independent of the internal condition of the pipe. Thus, regardless of whether the pipe is smooth or rough, the friction factor for laminar flow is a number that varies inversely as the Reynolds number.

Therefore, if the Reynolds number $R = 1800$, the friction factor becomes

$$f = 64/1{,}800 = 0.0356$$

It might appear that because f for laminar flow decreases with Reynolds number, using the Darcy–Weisbach equation the pressure drop will decrease with increase in flow rate. This is not true. Because pressure drop is proportional to the velocity V squared (Eqn (5.19)), the influence of V is greater than that of f. Therefore, pressure drop will increase with flow rate in the laminar region.

To illustrate, consider the Reynolds number example in Section 5.3 discussed earlier. If the flow rate is increased from 50,000 bbl/day to 80,000 bbl/day, the Reynolds number R will increase from 1190 to 1904 (still laminar). The velocity will increase from V_1 to V_2 as follows:

$$V_1 = 0.0119(50{,}000)/(15.5)^2 = 2.48 \text{ ft/s}$$

$$V_2 = 0.0119(80{,}000)/(15.5)^2 = 3.96 \text{ ft/s}$$

Friction factors at 50,000 bbl/day and 80,000 bbl/day flow rate are

$$f_1 = 64/1{,}190 = 0.0538$$

$$f_2 = 64/1{,}904 = 0.0336$$

Considering 5000-ft length of pipe, the head loss because of friction using Darcy–Weisbach Eqn (5.19):

$$h_{L1} = 0.0538 \times (5{,}000 \times 12/15.5) \times (2.48^2/64.4) = 19.89 \text{ ft}$$

$$h_{L2} = 0.0336 \times (5{,}000 \times 12/15.5) \times (3.96^2/64.4) = 31.67 \text{ ft}$$

Therefore, it is clear in laminar flow even though the friction factor decreases with flow increase, the pressure drop still increases with increase in flow rate.

For turbulent flow, when the Reynolds number $R > 4000$ the friction factor f depends not only on R, but also on the internal roughness of the pipe. As the pipe roughness increases, so does the friction factor. Therefore, smooth pipes have less friction factor compared to rough pipes. More correctly, friction factor depends on the relative roughness (e/D) rather than the absolute pipe roughness e.

Various correlations exist for calculating friction factor f. These are based on experiments conducted by scientists and engineers over the past 60 years or more. A good all-purpose equation for the friction factor f in the turbulent region is called the Colebrook–White equation as follows:

$$1/\sqrt{f} = -2\log_{10}\left[(e/3.7D) + 2.51/(R\sqrt{f})\right] \qquad (5.21)$$

and applies only for turbulent flow R > 4000
where
f – Darcy friction factor, dimensionless
D – pipe internal diameter, in
e – absolute pipe roughness, in
R – Reynolds number of flow, dimensionless

In SI units, the previous equation for f remains the same, as long as the absolute roughness e and the pipe diameter D are both expressed in mm. All other terms in the equation are dimensionless.

It can be seen from Eqn (5.21), that the calculation of f is not easy because it appears on both sides of the equation. A trial-and-error approach needs to be used. We assume a starting value of f (say 0.02) and substitute it in the right hand side of Eqn (5.21). This will yield a second approximation for f, which can then be used to recalculate a better value of f by successive iteration. Generally, three to four iterations will yield a satisfactory result for f correct to within 0.001.

During the past two or three decades, several formulas for friction factor for turbulent flow have been put forth by various researchers. All of these equations attempt to simplify calculation of the friction factor compared to the Colebrook–White equation discussed previously. Two such equations that are explicit equations in f, afford easy solution of friction factor compared with the implicit Eqn (5.21) that requires a trial-and-error solution. These are called the Churchill equation and Swamee–Jain equation and will be discussed later on in this chapter.

In the critical zone, where the Reynolds number is between 2000 and 4000, there is no generally accepted formula for determining the friction factor. This is because the flow is unstable in this region and therefore the friction factor is indeterminate. Most users calculate the value of f based upon turbulent flow.

To make matters more complicated, the turbulent flow region ($R > 4000$) actually consists of three separate regions:
Turbulent flow in smooth pipes
Turbulent flow in fully rough pipes
Transition flow between smooth and rough pipes
For turbulent flow in smooth pipes, pipe roughness has a negligible effect on the friction factor. Therefore, friction factor in this region depends only on the Reynolds number as follows:

$$1/\sqrt{f} = -2 \log_{10}[2.51/(R/\sqrt{f})] \qquad (5.22)$$

For turbulent flow in fully rough pipes, the friction factor f appears to be less dependent on Reynolds number as the latter increases in magnitude. It depends only on the pipe roughness and diameter. It can be calculated from the following equation:

$$1/\sqrt{f} = -2 \log_{10}[(e/3.7D)] \qquad (5.23)$$

For the transition region between turbulent flow in smooth pipes and turbulent flow in fully rough pipes, the friction factor f is calculated using the Colebrook–White equation given previously:

$$1/\sqrt{f} = -2 \log_{10}[(e/3.7D) + 2.51/(R/\sqrt{f})] \qquad (5.24)$$

As mentioned previously, in SI units, the previous equation for f remains the same, if e and D are both in mm.

The friction factor equations discussed previously are also plotted on the Moody diagram as shown in Figure 5.4. Relative roughness is defined as e/D. It is simply the result of dividing the absolute pipe roughness by the pipe inside diameter. The relative roughness term is a dimensionless parameter.

The Moody diagram represents the complete friction factor map for laminar and all turbulent regions of pipe flows. It is used commonly in estimating friction factor in pipe flow. If the Moody diagram is not available, we must use trial and error solution of Eqn (5.24) to calculate the friction factor.

To use the Moody diagram for determining the friction factor f we first calculate the Reynolds number R for the flow. Next, we find the location on the horizontal axis of Reynolds number for the value of R and draw a vertical line that intersects with the appropriate e/D curve. From this point of intersection on the e/D curve, we go horizontally

Fluid Flow in Pipes

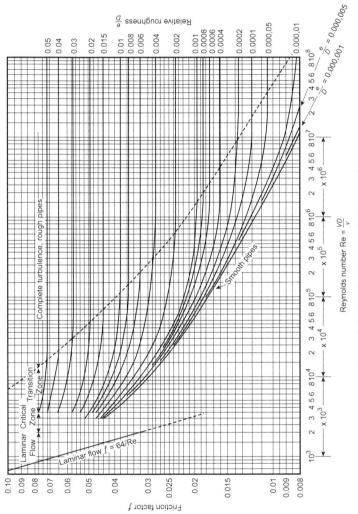

Figure 5.4 Moody diagram for friction factor.

to the left and read the value of the friction factor f on the vertical axis on the left.

Before leaving the discussion of friction factor, we must mention an additional term called the Fanning friction factor. Some publications use this friction factor instead of the Darcy friction factor.

The Fanning friction factor is defined as follows:

$$f_f = f_d/4 \tag{5.25}$$

where

f_f – Fanning friction factor
f_d – Darcy friction factor

Unless otherwise specified, we will use the Darcy friction factor f throughout this book.

Example 5.3

Water flows through a 20-in pipe at 5700 gal/min. Calculate the friction factor using Colebrook–White equation. Assume 0.375-in. pipe wall thickness and an absolute roughness of 0.002 in. Use specific gravity of 1.00 and viscosity of 1.0 cSt. What is the head loss resulting from friction in 2500 ft of pipe?

Solution

First we calculate the Reynolds number from Eqn (5.17) as follows:

$$R = 3{,}160 \times 5{,}700/(19.25 \times 1.0) = 935{,}688$$

The flow is fully turbulent and the friction factor f is calculated using Eqn (5.21) as follows:

$$1/\sqrt{f} = -2 \log_{10}\left[(0.002/(3.7 \times 19.25)) + 2.51/(935{,}688\sqrt{f})\right]$$

This implicit equation for f must be solved by trial and error.

First assume a trial value of $f = 0.02$. Substituting in equation above, we get a successive approximations for f as follows:

$$f = 0.0133,\ 0.0136\ \text{and}\ 0.0136$$

Therefore the solution is $f = 0.0136$.

Using Eqn (5.12),

$$\text{velocity} = 0.4085(5700)/19.25^2 = 6.28\ \text{ft/s}$$

Using Eqn (5.19), head loss from friction is

$$h = 0.0136 \times (2{,}500 \times 12/19.25) \times 6.28^2/64.4 = 12.98\ \text{ft}$$

6. PRESSURE DROP FROM FRICTION

In the previous section, we introduced the Darcy–Weisbach equation as follows:

$$h = f(L/D)(V^2/2g) \quad (5.26)$$

where the pressure drop h is expressed in feet of liquid head and the other symbols are defined below.

f – Darcy friction factor, dimensionless
L – pipe length, ft
D – pipe internal diameter, ft
V – average liquid velocity, ft/sec
g – acceleration from gravity, 32.2 ft/s^2 in English units.

A more practical equation, using customary pipeline units, is given below for calculating the pressure drop in pipelines.

Pressure drop from friction per unit length of pipe, in English units is

$$P_m = 0.0605 \, fQ^2 (Sg/D^5) \quad (5.27)$$

and in terms of transmission factor F

$$P_m = 0.2421 (Q/F)^2 (Sg/D^5) \quad (5.28)$$

where

P_m – pressure drop from friction, lb/in^2 per mile (psi/mi) of pipe length
Q – liquid flow rate, bbl/day
f – Darcy friction factor, dimensionless
F – transmission factor, dimensionless
Sg – liquid specific gravity
D – pipe internal diameter, in

The transmission factor F is directly proportional to the volume that can be transmitted through the pipeline and therefore has an inverse relationship with the friction factor f. The transmission factor F is calculated from the following equation.

$$F = 2/\sqrt{f} \quad (5.29)$$

Because friction factor f ranges from 0.008 to 0., it can be seen from Eqn (5.29) that the transmission factor F ranges from approximately 6 to 22.

The Colebrook–White Eqn (5.21) can be rewritten in terms of the transmission factor F as follows:

$$F = -4 \log[(e/3.7D) + 1.255(F/R)] \quad (5.30)$$

for turbulent flow R > 4000

Similar to the calculation of the friction factor f using Eqn (5.21), the calculation of transmission factor F from Eqn (5.30) will also be a trial-and-error approach. We assume a starting value of F (say 10.0) and substitute it in the right hand side of Eqn (5.30). This will yield a second approximation for F, which can then be used to recalculate a better value, by successive iteration. Generally, three to four iterations will yield a satisfactory result for F.

In SI units, the Darcy equation (in pipeline units) for the pressure drop in terms of the friction factor is represented as follows:

$$P_{km} = 6.2475 \times 10^{10} \, fQ^2 (Sg/D^5) \quad (5.31)$$

and the corresponding equation in terms of transmission factor F is written as follows:

$$P_{km} = 24.99 \times 10^{10} \, (Q/F)^2 (Sg/D^5) \quad (5.32)$$

where

P_{km} – pressure drop because of friction in kPa/km
Q – liquid flow rate, m^3/h
f – Darcy friction factor, dimensionless
F – transmission factor, dimensionless
Sg – liquid specific gravity
D – pipe internal diameter, mm

In SI units, the transmission factor F is calculated using Eqn (5.30) as follows:

$$F = -4 \log[(e/3.7D) + 1.255(F/R)] \quad (5.33)$$

for turbulent flow R > 4000
where

D – pipe internal diameter, mm
e – absolute pipe roughness, mm
R – Reynolds number of flow, dimensionless

Example 5.4

Consider a 100-mi pipeline, 16-in diameter, 0.250-in wall thickness, transporting a liquid (specific gravity of 0.815 and viscosity of 15 cSt at 70 °F) at a flow rate of 90,000 bbl/day. Calculate the friction factor and pressure drop per unit length of pipeline using Colebrook–White equation. Assume 0.002-in pipe roughness.

Solution

The Reynolds number is calculated first.

$$R = \frac{92.24 \times 90,000}{15.5 \times 15} = 35,706$$

Using Colebrook–White Eqn (5.30), the transmission factor is

$$F = -4 \log[(0.002/(3.7 \times 15.5)) + 1.255F/35,706]$$

Solving the previous equation by trial and error yields

$$F = 13.21$$

To calculate the friction factor f, we use Eqn (5.29) after some transposition and simplification as follows:
Friction factor $f = 4/F^2 = 4/(13.21)^2 = 0.0229$.
The pressure drop per mile is calculated using Eqn (5.28).

$$Pm = 0.2421(90,000/13.21)^2 (0.815/15.5^5) = 10.24 \text{ psi/mi}$$

The total pressure drop in 100-mile length is then:
Total pressure drop $= 100 \times 10.24 = 1024$ psi.

7. COLEBROOK–WHITE EQUATION

In 1956, the US Bureau of Mines conducted experiments and recommended a modified version of the Colebrook–White equation. The modified Colebrook–White equation yields a more conservative transmission factor F. The pressure drop calculated using the modified Colebrook–White equation is slightly higher than that calculated using the original Colebrook–White equation. This modified Colebrook–White equation, in terms of transmission factor F is defined as follows:

$$F = -4 \log[(e/3.7D) + 1.4125(F/R)] \quad (5.34)$$

In SI units, the transmission factor equation above remains the same with e and D expressed in mm, other terms being dimensionless.

Comparing Eqn (5.34) with Eqn (5.30) or (5.33), it can be seen that the only change is in the substitution of the constant 1.4125 in place of 1.255 in the original Colebrook–White equation. Some companies use the modified Colebrook–White equation stated in Eqn (5.34).

An explicit form of an equation to calculate the friction factor was proposed by Swamee and Jain. This equation does not require trial-and-error solution like the Colebrook–White equation. It correlates very closely to the Moody diagram values. Refer to the Appendix for a version of the Swamee–Jain Equation for friction factor.

8. HAZEN–WILLIAMS EQUATION

Hazen–Williams equation is commonly used in the design of water distribution lines and in calculation of frictional pressure drop in refined petroleum products such as gasoline, diesel etc. This method involves the use of Hazen–Williams C-factor instead of pipe roughness or liquid viscosity.

The pressure drop calculation using Hazen–Williams equation takes into account flow rate, pipe diameter, and specific gravity as follows:

$$h = 4.73 \, L(Q/C)^{1.852}/D^{4.87} \tag{5.35}$$

where
- h – head loss because of friction, ft
- L – length of pipe, ft
- D – internal diameter of pipe, ft
- Q – flow rate, ft^3/s
- C – Hazen–Williams coefficient or C-Factor, dimensionless

Typical values of Hazen–Williams C-factor are given in Appendix A.

In customary pipeline units, the Hazen–Williams equation can be rewritten as follows.

In English units:

$$Q = 0.1482(C) \, (D)^{2.63} (Pm/Sg)^{0.54} \tag{5.36}$$

where
- Q – flow rate, bbl/day
- D – pipe internal diameter, in
- P_m – frictional pressure drop, psi/mi
- Sg – liquid specific gravity
- C – Hazen–Williams C-factor

Another form of Hazen–Williams equation, when the flow rate is in gal/min and head loss is measured in feet of liquid per 1000 feet of pipe is as follows:

$$GPM = 6.7547 \times 10^{-3}(C) \, (D)^{2.63}(H_L)^{0.54} \tag{5.37}$$

where
- GPM – flow rate, gal/min
- H_L – friction loss, ft of liquid per 1000 ft of pipe
- Other symbols remain the same.

Fluid Flow in Pipes

In SI units, the Hazen–Williams formula is as follows:

$$Q = 9.0379 \times 10^{-8} \, (C) \, (D)^{2.63} (P_{km}/Sg)^{0.54} \quad (5.38)$$

where
  Q – flow rate, m³/h
  D – pipe internal diameter, mm
  P_{km} – frictional pressure drop, kPa/km
  Sg – liquid specific gravity
  C – Hazen–Williams C-factor

Historically, many empirical formulas have been used to calculate frictional pressure drop in pipelines. Hazen–Williams formula has been used widely in the analysis of pipeline networks and water distribution systems, because of its simple form and ease of use. A review of the Hazen–Williams formula shows that the pressure drop from friction depends on the liquid specific gravity, pipe diameter, and the Hazen–Williams coefficient or C factor.

Unlike the Colebrook–White equation in which the friction factor is calculated based on pipe roughness, pipe diameter, and the Reynolds number, which further depends on liquid specific gravity and viscosity, the Hazen–Williams C factor appears to not take into account liquid viscosity or pipe roughness. It could be argued that the C factor is in fact a measure of the pipe internal roughness. However, there does not seem to be any indication of how the C factor varies from laminar flow to turbulent flow.

We could compare the Darcy–Weisbach equation with the Hazen–Williams equation and infer that the C factor is a function of Darcy friction factor and Reynolds number. Based on this comparison, it can be concluded that the C factor is indeed an index of relative roughness of the pipe. It must be remembered that the Hazen–Williams equation, although convenient from the standpoint of its explicit nature, must be regarded as an empirical equation and that it is difficult to apply to all fluids under all conditions. Nevertheless, in real-world pipelines, with sufficient field data, we could determine specific C factors for specific pipelines and fluids pumped.

Example 5.5
A 3-in (internal diameter) smooth pipeline is used to pump 100 gal/min of water. Using the Hazen–Williams formula, calculate the head loss in 3000 ft of this pipe. Assume C factor = 140.

Solution

Using Eqn (5.37) substituting given values, we get

$$100 = 6.7547 \times 10^{-3} \times 140(3.0)^{2.63}(H_L)^{0.54}$$

Solving for the head loss we get

$$H_L = 26.6 \text{ ft per } 1000 \text{ ft}$$

Therefore head loss for 3000 ft = $26.6 \times 3 = 79.8$ ft of water.

9. SHELL-MIT EQUATION

The Shell-MIT equation, sometimes called the MIT equation, is used in calculation of pressure drop in heavy crude oil and heated liquid pipelines. Using this method, a modified Reynolds number Rm is calculated first from the Reynolds number as follows:

$$R = 92.24(Q)/(Dv) \qquad (5.39)$$

$$Rm = R/(7742) \qquad (5.40)$$

where

R – Reynolds number, dimensionless
Rm – modified Reynolds number, dimensionless
Q – flow rate, bbl/day
D – internal diameter, in
v – kinematic viscosity, cSt.

Next, depending on the flow (laminar or turbulent), the friction factor is calculated from one of the following equations.

$$f = 0.00207/Rm - \text{Laminar flow} \qquad (5.41)$$

$$f = 0.0018 + 0.00662(1/Rm)^{0.355} - \text{Turbulent flow} \qquad (5.42)$$

Note that this friction factor f in these equations is not the same as the Darcy friction factor f discussed earlier. In fact, the friction factor f in these equations is more like the Fanning friction factor discussed previously.

Finally, the pressure drop from friction is calculated using the following equation.

$$P_m = 0.241\left(f \, SgQ^2\right)/D^5 \qquad (5.43)$$

where

P_m – frictional pressure drop, psi/mi
f – friction factor, dimensionless
Sg – liquid specific gravity
Q – flow rate, bbl/day
D – pipe internal diameter, in

In SI units, the MIT equation is expresses as follows:

$$P_m = 6.2191 \times 10^{10} (f\, SgQ^2)/D^5 \quad (5.44)$$

where

P_m – frictional pressure drop, kPa/km
f – friction factor, dimensionless
Sg – liquid specific gravity
Q – flow rate, m³/h
D – pipe internal diameter, mm

Comparing Eqn (5.43) with Eqns (5.27) and (5.28) and recognizing the relationship between transmission factor F and Darcy friction factor f, using Eqn (5.29) it is evident that the friction factor f in Eqn (5.43) is not the same as the Darcy friction factor. It appears to be one-fourth of the Darcy friction factor.

Example 5.6

A 500-mm outside diameter, 10-mm wall thickness steel pipeline is used to transport heavy crude oil at a flow rate of 800 m³/h at 100 °C. Using the MIT equation, calculate the friction loss per kilometer of pipe assuming internal pipe roughness of 0.05 mm. The heavy crude oil has a specific gravity of 0.89 at 100 °C and a viscosity of 120 cSt at 100 °C.

Solution

First calculate the Reynolds number.

Reynolds number = 353,678 × 800/(120 × 480) = 4912
The flow is therefore turbulent.
Modified Reynolds number = 4912/7742 = 0.6345
Friction factor = 0.0018 + 0.00662 (1/0.6345)$^{0.355}$ = 0.0074
Pressure drop from Eqn (5.44) is

$$P_m = 6.2191 \times 10^{10} (0.0074 \times 0.89 \times 800 \times 800)/480^5$$
$$= 10.29 \text{ kPa/km}$$

10. MILLER EQUATION

The Miller equation also known as the Benjamin Miller formula is used in hydraulics studies involving crude oil pipelines. This equation does not consider pipe roughness and is an empirical formula for calculating the flow rate from a given pressure drop. The equation can also be rearranged to calculate the pressure drop from a given flow rate. One of the popular versions of this equation is as follows:

$$Q = 4.06(M)\left(D^5 P_m / S_g\right)^{0.5} \tag{5.45}$$

where M is defined as follows:

$$M = \log_{10}\left(D^3 S_g P_m / cp^2\right) + 4.35 \tag{5.46}$$

and

Q – flow rate, bbl/day
D – pipe internal diameter, in
P_m – pressure drop, psi/mi
S_g – liquid specific gravity
cp – liquid viscosity, centipoise

In SI units, the Miller equation is as follows:

$$Q = 3.996 \times 10^{-6}(M)\left(D^5 P_m / S_g\right)^{0.5} \tag{5.47}$$

and M is defined as follows:

$$M = \log_{10}\left(D^3 S_g P_m / cp^2\right) - 0.4965 \tag{5.48}$$

where

Q – flow rate, m³/h
D – pipe internal diameter, mm
P_m – frictional pressure drop, kPa/km
S_g – liquid specific gravity
cp – liquid viscosity, centipoise

It can be seen from this version of Miller equation that calculating the pressure drop P_m from the flow rate Q is not straightforward. This is because the parameter M depends on the pressure drop P_m. Therefore, if we solve for P_m in terms of Q and other parameters from Eqn (5.45), we get:

$$P_m = (Q/4.06M)^2 \left(S_g / D^5\right) \tag{5.49}$$

where M is calculated from Eqn (5.46).

Fluid Flow in Pipes

To calculate P_m from a given value of flow rate Q, we use a trial-and-error approach. First, we assume a value of P_m to get a starting value of M from Eqn (5.46). This value of M is then substituted in Eqn (5.49) to determine a second approximation for P_m. This value of P_m will be used to generate a better value of M form Eqn (5.46), which is then used to recalculate P_m. Once the successive values of P_m are within an allowable tolerance, such as 0.01 psi/mile, the iteration can be terminated and the value of pressure drop P_m is calculated.

Example 5.7
Using the Miller equation, determine the pressure drop in a 14-in, 0.250-in wall thickness, and crude oil pipeline at a flow rate of 3000 gal/min. The crude oil properties are: Specific gravity = 0.825 at 60 °F and viscosity = 15 cSt at 60 °F.

Solution
Liquid viscosity in centipoise = $0.825 \times 15 = 12.375$ cP.

First the parameter M is calculated from Eqn (5.46) using an initial value of $P_m = 10.0$.

$$M = \log_{10}\left(13.5^3 \times 0.825 \times 10.0/12.375^2\right) + 4.35$$
$$= 6.4724$$

Using this value of M in Eqn (5.49), we get:

$$P_m = [3000 \times 34.2857/(4.06 \times 6.4724)]^2 \left(0.825/13.5^5\right)$$
$$= 28.19 \text{ psi/mi}$$

We were quite far off in our initial estimate of P_m.
Using this value of Pm, a new value of M is calculated as:

$$M = 6.9225$$

Substituting this value of M in Eqn (5.49), we get:

$$P_m = 24.64$$

By successive iteration, we get the final value for Pm = 25.02 psi/mi.

11. T.R. AUDE EQUATION

Another pressure drop equation used in the pipeline industry that is popular among companies that transport refined petroleum products is the T.R. Aude equation, sometimes referred to simply as the Aude equation. This equation is named after the engineer that conducted experiments on pipelines in the 1950s.

It must be noted that the Aude equation is based on field data collected from 6 to 8-in refined products pipelines. Therefore, use caution when applying this formula to larger pipelines. The Aude equation is used in pressure drop calculations for 6–12-in pipelines. This method requires the use of the Aude K factor, representing pipeline efficiency. One version of this formula is described below:

$$P_m = \left[Q(z^{0.104})(Sg^{0.448})/(0.871(K)(D^{2.656}))\right]^{1.812} \quad (5.50)$$

where

P_m – pressure drop because of friction, psi/mi
Q – flow rate, bbl/h
D – pipe internal diameter, in
Sg – liquid specific gravity
z – liquid viscosity, centipoise
K – T.R. Aude K-factor, usually 0.90 to 0.95

In SI units, the Aude equation is as follows:

$$P_m = 8.888 \times 10^8 \left[Q(z^{0.104})(Sg^{0.448})/(K(D^{2.656}))\right]^{1.812} \quad (5.51)$$

where

P_m – frictional pressure drop, kPa/km
Sg – liquid specific gravity
Q – flow rate, m³/h
D – pipe internal diameter, mm
z – liquid viscosity, centipoise
K – T.R. Aude K factor, usually 0.90 to 0.95

Because the Aude formula for pressure drop given here does not contain pipe roughness, it can be deduced that the K factor somehow must take into account the internal condition of the pipe. As with the Hazen–Williams C factor discussed earlier, the Aude K factor is also an experience based factor and must be determined by field measurement and calibration of an existing pipeline. If field data are not available, engineers usually approximate using a value such as $K = 0.90$–0.95. A higher value of K will result in lower pressure drop for a given flow rate or higher flow rate for a given pressure drop.

Example 5.8
Crude oil pipeline NPS 20 with 0.375 wall thickness flows at 100 MBD. Use $K = 0.92$ to calculate the pressure drop in the pipeline for $z = 25$ cP and $Sg = 0.895$.

Solution

$$P_m = \left[Q(z^{0.104})(Sg^{0.448})/(0.871(K)(D^{2.656}))\right]^{1.812}$$

pressure drop $= = \left[100{,}000/24(25^{0.104})\right.$
$\left.\times (0.895^{0.448})/(0.871(0.92)(19.25^{2.656}))\right]^{1.812}$

or

$$P_m = 5.97 \text{ psi/mi}$$

12. MINOR LOSSES

In most long-distance pipelines, such as trunk lines, the pressure drop from friction in the straight lengths of pipe form the significant portion of the total frictional pressure drop. Valves and fittings contribute very little to the total pressure drop in the entire pipeline. Hence, in such cases, pressure losses through valves, fittings, and other restrictions are generally classified as *minor losses*. Minor losses include energy losses resulting from rapid changes in the direction or magnitude of liquid velocity in the pipeline. Thus pipe enlargements, contractions, bends, and restrictions such as check valves and gate valves are included in minor losses.

In short pipelines, such as terminal and plant piping, the pressure loss from valves, fittings, etc., may be a substantial portion of the total pressure drop. In such cases, the term minor losses is a misnomer.

Therefore, in long pipelines, the pressure loss through bends, elbows, valves, fittings, etc., are classified as "minor" losses and in most instances may be neglected without significant error. However, in shorter pipelines, these losses must be included for correct engineering calculations. Experiments with water at high Reynolds numbers have shown that the minor losses varied approximately as the square of the velocity. This leads to the conclusion that minor losses can be represented by a function of the liquid velocity head or kinetic energy ($V^2/2g$).

Accordingly, the pressure drop through valves and fittings is generally expressed in terms of the liquid kinetic energy $V^2/2g$ multiplied by a head loss coefficient K. Comparing this with the Darcy–Weisbach equation for head loss in a pipe, we can see the following analogy. For straight pipe, the head loss, h, is $V^2/2g$ multiplied by the factor (fL/D). Thus, the head loss coefficient for straight pipe is fL/D where L and D are the pipe length and diameter both expressed in feet.

Therefore, the pressure drop in a valve or fitting is calculated as follows:

$$h = KV^2/2g \tag{5.52}$$

where
- h – head loss because of valve or fitting, ft
- K – head loss coefficient for the valve or fitting, dimensionless
- V – velocity of liquid through valve or fitting, ft/s
- g – acceleration from gravity, 32.2 ft/s^2 in English units

The head loss coefficient K is, for a given flow geometry, considered practically constant at high Reynolds number. K increases with pipe roughness and with lower Reynolds numbers. In general the value of K is determined mainly by the flow geometry or by the shape of the pressure loss device.

It can be seen from Eqn (5.45) that K is analogous to the term (fL/D) for straight length of pipe. Values of K are available for various types of valves and fittings in standard handbooks, such as Crane Handbook and Cameron Hydraulic Data. A table of K values, commonly used for valves and fittings, is included in Appendix A.

12.1 Gradual Enlargement

Consider liquid flowing through a pipe of diameter D_1. If at a certain point, the diameter enlarges to D_2, the energy loss that occurs because of the enlargement can be calculated as follows:

$$h = K(V_1 - V_2)^2/2g \tag{5.53}$$

where V_1 and V_2 are the velocity of the liquid in the smaller diameter and the larger diameter, respectively. The value of K depends upon the diameter ratio D_1/D_2 and the different cone angle because of the enlargement. A gradual enlargement is shown in Figure 5.5

For a sudden enlargement, K = 1.0 and the corresponding head loss is

$$h = (V_1 - V_2)^2/2g \tag{5.54}$$

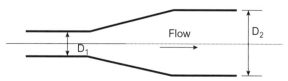

Figure 5.5 Gradual enlargement.

Example 5.9

Calculate the head loss resulting from a gradual enlargement in a pipe that flows 100 gal/min of water from a 2-in diameter to a 3-in diameter with an included angle of 30°. Both pipe sizes are internal diameters.

Solution

The liquid velocities in the two pipe sizes are as follows:

$$V_1 = 0.4085 \times 100/2^2 = 10.21 \text{ ft/s}$$

$$V_1 = 0.4085 \times 100/3^2 = 4.54 \text{ ft/s}$$

$$\text{Diameter ratio} = 3/2 = 1.5$$

From charts, for diameter ratio = 1.5 and cone angle = 30° the value of K is

$$K = 0.38$$

Therefore head loss because of gradual enlargement is

$$h = 0.38 \times (10.21 - 4.54)^2/64.4 = 0.19 \text{ ft}$$

If the expansion was a sudden enlargement from 2 to 3 in, the head loss would be

$$h = (10.21 - 4.54)^2/64.4 = 0.50 \text{ ft}$$

12.2 Abrupt Contraction

For flow through an abrupt contraction, the flow from the larger pipe (diameter D_1 and velocity V_1) to a smaller pipe (diameter D_2 and velocity V_2) results in the formation of a vena contracta or throat, immediately after the diameter change. At the vena contracta, the flow area reduces to A_c with increased velocity of V_c. Subsequently, the flow velocity decreases to V_2 in the smaller pipe. From velocity V_1, the liquid first accelerates to velocity V_c at the vena contracta and subsequently decelerates to V_2. This is shown in Figure 5.6.

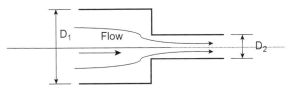

Figure 5.6 Abrupt contraction.

Table 5.1 Head loss coefficient K for abrupt contraction

A_2/A_1	C_c	K
0.0	0.617	0.50
0.1	0.624	0.46
0.2	0.632	0.41
0.3	0.643	0.36
0.4	0.659	0.30
0.5	0.681	0.24
0.6	0.712	0.18
0.7	0.755	0.12
0.8	0.813	0.06
0.9	0.892	0.02
1.0	1.000	0.00

The energy loss because of the sudden contraction depends upon the ratio of the pipe diameters D_2 and D_1 and the ratio A_c/A_2. The value of the head loss coefficient K can be found using the Table 5.1, where $C_c = A_c/A_2$. The ratio A_2/A_1 can be calculated from the ratio of the diameters D_2/D_1.

A pipe connected to a large storage tank represents a type of abrupt contraction. If the storage tank is a large body of liquid, we can state that this is a limiting case of the abrupt contraction. For such a square-edged pipe entrance from a large tank, $A_2/A_1 = 0$. From Table 5.1 for this case $K = 0.5$ for turbulent flow.

Another type of pipe entrance from a large tank is called a reentrant pipe entrance. If the pipe is thin-walled and the opening within the tank is located more than one pipe diameter upstream from the tank wall, the K value will be close to 0.8.

If the edges of the pipe entrance in a tank are rounded or bell shaped, the head loss coefficient is considerably smaller. An approximate value for K for a bell-mouth entrance is 0.1.

12.3 Head Loss and L/D Ratio for Pipes and Fittings

We have discussed how minor losses can be accounted for using the head loss coefficient K in conjunction with the liquid velocity head. The Appendix A lists K values for common valves and fittings.

Referring to the Table in Appendix A for K values, we see that for a 16-in gate valve:

$$K = 0.10$$

Therefore, compared with a 16-in straight pipe, we can write from Darcy–Weisbach equation:

$$\frac{fL}{D} = 0.10$$

or

$$\frac{L}{D} = 0.10f$$

If we assume f = 0.0125, we get:

$$L/D = 8$$

This means that compared with a straight pipe with a 16-in diameter, a 16-in gate valve has an (L/D) ratio of 8, which causes the same friction loss. The L/D ratio represents the equivalent length of straight pipe in terms of its diameter that will equal the pressure loss in the valve or fitting. In Appendix A, the (L/D) ratio for various valves and fittings are given.

Using the (L/D) ratio, we can replace a 16-in. gate valve with 8 × 16-in = 128-in of straight 16-in pipe. This length of pipe will have the same friction loss as the 16-in gate valve. Thus we can use the K values or (L/D) ratios to calculate the friction loss in valves and fittings.

13. INTERNALLY COATED PIPES AND DRAG REDUCTION

In turbulent flow, we saw that the pressure drop from friction depends on the pipe roughness. Therefore, if the internal pipe surface can be smoothened, the frictional pressure drop can be reduced. Internally coating a pipeline with an epoxy will considerably reduce the pipe roughness, compared to uncoated pipe.

For example, if the uncoated pipe has an absolute roughness of a 0.002-in coated pipe can reduce roughness to a value as low as 0.0002 in. The friction factor f may therefore reduce from 0.02 to 0.01 depending on the flow rate, Reynolds number, etc. Because pressure drop is directly proportional to the friction factor in accordance with the Darcy–Weisbach equation, the total pressure drop in the internally coated pipeline in this example would be 50% of that in the uncoated pipeline.

Another method of reducing frictional pressure drop in a pipeline is by using drag reduction. Drag reduction is the process of reducing the pressure drop from friction in a pipeline by continuously injecting a very small quantity (in parts per million or ppm) of a high molecular weight hydrocarbon, called the drag reduction agent (DRA) into the flowing liquid stream. The DRA is effective, only in pipe segments between two pump stations. It degrades in performance as it flows through the pipeline for long distances. It also completely breaks up or suffers shear degradation as it passes through pump stations, meters, and other restrictions. DRA works only in turbulent flow and with low viscosity liquids. Thus, it works well with refined petroleum products (gasoline, diesel, etc.) and light crude oils. It is ineffective in heavy crude oil pipelines, particularly in laminar flow. Currently, in the United States, two leading vendors of DRA products are Baker Petrolite and Conoco-Phillips.

To determine the amount of drag reduction using DRA, we proceed as follows.

If the pressure drops from friction with and without DRA are known, we can calculate the percentage drag reduction.

$$\text{Percentage Drag Reduction} = 100(DP_0 - DP_1)/DP_0 \quad (5.55)$$

where

DP_0 – friction drop in pipe segment without DRA, psi
DP_1 – friction drop in pipe segment with DRA, psi

The pressure drops are also referred to as untreated versus treated pressure drops. It is fairly easy to calculate the value of untreated pressure drop, using the pipe size, liquid properties, etc. The pressure drop with DRA is obtained using DRA vendor information. In most cases involving DRA, we are interested in calculating how much DRA we need to use to reduce the pipeline friction drop, and hence the pumping horsepower required. It must be noted that DRA may not be effective at the higher flow rate, if existing pump and driver limitations preclude operating at higher flow rates because of pump driver horsepower limitations.

Consider a situation where a pipeline is limited in throughput because of maximum allowable operating pressures (MAOP). Let us assume the friction drop in this MAOP limited pipeline is 800 psi at 100,000 bbl/day. We are interested in increasing pipeline flow rate to 120,000 bbl/day using DRA and we would proceed as follows:

Flow improvement desired = $(120,000 - 100,000)/100,000 = 20\%$

If we calculate the actual pressure drop in the pipeline at the increased flow rate of 120,000 bbl/day (ignoring the MAOP violation) and assume we get the following pressure drop:

Frictional pressure drop at 120,000 bbl/day = 1,150 psi

and

Frictional pressure drop at 100,000 bbl/day = 800 psi

The percentage drag reduction is then calculated from Eqn (5.55) as:

Percentage Drag Reduction = $100(1150 - 800)/1150 = 30.43\%$

In these calculations, we have tried to maintain the same frictional drop (800 psi) using DRA at the higher flow rate as the initial pressure-limited case. Knowing the drag reduction percent required, we can get the DRA vendor to tell us how much DRA will be required to achieve the 30.43% drag reduction, at the flow rate of 120,000 bbl/day. If the answer is 15 ppm of brand X DRA, we can calculate the daily DRA requirement as follows:

Quantity of DRA required = $(15/10^6)(120,000)(42) = 75.6$ gal/day

If DRA costs $10 per gallon, this equates to a daily DRA cost of $756. In this example, a 20% flow improvement requires a drag reduction of 30.43% and 15 ppm of DRA, costing $756 per day. Of course, these are simply rough numbers used to illustrate the DRA calculations methods. The quantity of DRA required will depend on the pipe size, liquid viscosity, flow rate, and Reynolds number, in addition to the percentage drag reduction required. Most DRA vendors will confirm that drag reduction is effective only in turbulent flow (R > 4000) and that it does not work with heavy (high viscosity) crude oil and other liquids.

Also, drag reduction cannot be increased indefinitely by injecting more DRA. There is a theoretical limit to the drag reduction attainable. For a certain range of flow rates, the percentage drag reduction will increase as DRA ppm is increased. At some point, depending on the pumped liquid, flow characteristics, etc., the drag reduction levels off. No further increase in drag reduction is possible by increasing the DRA ppm. We would have reached the point of diminishing returns in this case.

In a later chapter on feasibility studies and pipeline economics, we will explore the subject of DRA further.

14. FLUID FLOW IN GAS PIPELINES

14.1 Bernoulli's Equation

As gas flows through a pipeline, the total energy of the gas at various points consists of energy from pressure, energy from velocity, and energy

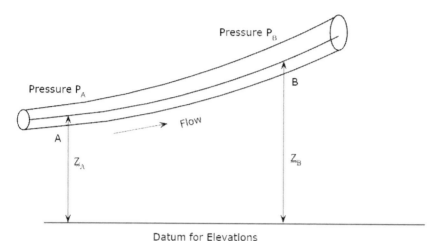

Figure 5.7 Energy of flow of a fluid.

from position or elevation above an established datum. Bernoulli's equation simply connects these components of the energy of the flowing fluid to form an energy conservation equation. Bernoulli's equation is stated as follows considering two points, A and B, as shown in Figure 5.7.

Energy from position, or potential energy $= WZ_A$
Energy from pressure, or pressure energy $= WP_A/\gamma$
Energy from velocity, or kinetic energy $= W(V_A/2g)^2$
where γ is the specific weight of liquid.

We can thus state that:

$$E = WZ_A + WP_A/\gamma + WV_A^2/2g \qquad (5.56)$$

Dividing by W throughout, we get the total energy per unit weight of liquid as

$$H_A = Z_A + P_A/\gamma + V_A^2/2g \qquad (5.57)$$

where H_A is the total energy per unit weight at point A.

Considering the same liquid particle as it arrives at point B, the total energy per unit weight at B is

$$H_B = Z_B + P_B/\gamma + V_B^2/2g \qquad (5.58)$$

Because of conservation of energy:

$$H_A = H_B$$

Therefore:

$$Z_A + P_A/\gamma + V_A^2/2g = Z_B + P_B/\gamma + V_B^2/2g \qquad (5.59)$$

Equation (5.59) is one form of the Bernoulli's equation for fluid flow. Also a more general form is as follows:

$$Z_A + P_A/\gamma + V_A^2/2g = Z_B + P_B/\gamma + V_B^2/2g + H_p + h_r \qquad (5.60)$$

where H_P is the equivalent head added to the fluid by a compressor at A and h_f represents the total frictional pressure loss between points A and B. Starting with the basic energy Eqn (5.59), applying gas laws and after some simplification various formulas were developed over the years to predict the performance of a pipeline transporting gas. These formulas are intended to show the relationship between the gas properties such as gravity and compressibility factor with the flow rate, pipe diameter, and length and the pressures along the pipeline

Thus for a given pipe size and length, we can predict the flow rate possible through a pipeline based upon an inlet pressure and an outlet pressure of a pipe segment. Simplifications are sometimes introduced such as uniform gas temperature and no heat transfer between the gas and the surrounding soil in a buried pipeline to adopt these equations for manual calculations. With the advent of microcomputers, we are able to introduce heat transfer effects and therefore more accurately model gas pipelines taking into consideration gas flow temperatures, soil temperatures and thermal conductivities of pipe material, insulation, and soil. In this chapter, we will concentrate on steady-state isothermal flow of gas in pipelines. The Appendices include an output report from a commercial gas pipeline simulation model GASMOD developed by SYSTEK Technologies, Inc. (www.systek.us) that takes into account heat transfer. For most practical purposes, the assumption of isothermal flow is good enough because in long transmission lines the gas temperature reaches constant values anyway.

15. FLOW EQUATIONS

Several equations are available that relate the gas flow rate with gas properties, pipe diameter and length, upstream and downstream pressures. These equations are listed as follows.
1. General flow equation
2. Colebrook–White equation
3. Modified Colebrook–White equation

4. American Gas Association (AGA) equation
5. Weymouth equation
6. Panhandle A equation
7. Panhandle B equation
8. IGT Equation

We will discuss each of these equations, their limitations, and applicability to compressible fluids, such as natural gas, using illustrated examples. A comparison of these equations will also be discussed using an example pipeline.

16. GENERAL FLOW EQUATION

The general flow equation, also called the fundamental flow equation for the steady-state isothermal flow in a gas pipeline, is the basic equation for relating the pressure drop with flow rate. The most common form of this equation in US customary system (USCS) of units is given in terms of the pipe diameter, gas properties, pressures, temperatures, and flow rate as follows. Refer to Figure 5.8 for explanation of symbols used.

$$Q = 77.54 \left(\frac{T_b}{P_b}\right) \left(\frac{P_1^2 - P_2^2}{GT_f LZf}\right)^{0.5} D^{2.5} \quad \text{(USCS units)} \tag{5.61}$$

where

Q – gas flow rate, measured at standard conditions, ft³/day (standard cubic feet per day [SCFD])
f – friction factor, dimensionless
P_b – base pressure, psia
T_b – base temperature, °R (460 + °F)
P_1 – upstream pressure, psia
P_2 – downstream pressure, psia
G – gas gravity (air = 1.00)

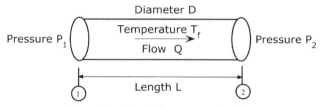

Figure 5.8 Steady flow in gas pipeline.

T_f – average gas flowing temperature, °R (460 + °F)
L – pipe segment length, miles
Z – gas compressibility factor at the flowing temperature, dimensionless
D – pipe inside diameter, in

It must be noted that for the pipe segment from Section 1 to Section 2, the gas temperature T_f is assumed to be constant (isothermal flow).

In SI units, the general flow equation is stated as follows:

$$Q = 1.1494 \times 10^{-3} \left(\frac{T_b}{P_b}\right) \left(\frac{P_1^2 - P_2^2}{GT_f LZf}\right)^{0.5} D^{2.5} \quad \text{(SI units)} \quad (5.62)$$

where
Q – gas flow rate, measured at standard conditions, m³/day
f – friction factor, dimensionless
P_b – base pressure, kPa
T_b – base temperature, K (273 + °C)
P_1 – upstream pressure, kPa
P_2 – downstream pressure, kPa
G – gas gravity (air = 1.00)
T_f – average gas flowing temperature, K (273 + °C)
L – pipe segment length, km
Z – gas compressibility factor at the flowing temperature, dimensionless
D – pipe inside diameter, mm

Because of the nature of Eqn (5.62), the pressures may also be in MPa or Bar as long as the same consistent unit is used.

Eqn (5.61) relates the capacity (flow rate or throughput) of a pipe segment of length L based on an upstream pressure of P_1 and a downstream pressure of P_2 as shown in Figure 5.8. It is assumed that there is no elevation difference between the upstream and downstream points; therefore, the pipe segment is horizontal.

Upon examining the general flow Eqn (5.61), we see that for a pipe segment of length L and diameter D, the gas flow rate Q (at standard conditions) depends on several factors. Q depends on gas properties represented by the gravity G and the compressibility factor Z. If the gas gravity is increased (heavier gas), the flow rate will decrease. Similarly, as the compressibility factor Z increases, flow rate will decrease. Also as the gas flowing temperature T_f increases, throughput will decrease. Thus hotter the gas, lower will be the flow rate. Therefore, to increase flow rate, it helps to keep the gas temperature low. The impacts of pipe length and inside diameter are also clear. As the pipe segment length increases for given pressure P_1 and P_2, the flow rate will decrease. On the other hand, larger the diameter, larger will be the flow rate. The term $P_1^2 - P_2^2$ represents the driving force that causes the flow rate from the upstream end to the downstream end. As the downstream pressure P_2 is reduced, keeping the upstream pressure P_1 constant, the flow rate will increase. It is obvious that when there is no flow rate, P_1 is equal to P_2. It is because of friction between the gas and pipe walls that the pressure drop ($P_1 - P_2$) occurs from the upstream point 1 to downstream point 2. The friction factor f depends on the internal condition of the pipe as well as the type of flow (laminar or turbulent) and will be discussed in detail beginning in Section 22.

Sometimes the general flow equation is represented in terms of the transmission factor F instead of the friction factor f. This form of the equation is as follows:

$$Q = 38.77 F \left(\frac{T_b}{P_b}\right)\left(\frac{P_1^2 - P_2^2}{GT_f LZf}\right)^{0.5} D^{2.5} \quad \text{(USCS units)} \quad (5.63)$$

where the transmission factor F and friction factor f are related by

$$F = \frac{2}{\sqrt{f}} \quad (5.64)$$

And in SI units:

$$Q = 5.747 \times 10^{-4} F \left(\frac{T_b}{P_b}\right)\left(\frac{P_1^2 - P_2^2}{GT_f LZf}\right)^{0.5} D^{2.5} \quad \text{(SI units)} \quad (5.65)$$

We will discuss several aspects of the general flow equation before moving on to the other formulas for pressure drop calculation.

$$Q = 38.77 F \left(\frac{T_b}{P_b}\right)\left(\frac{P_1^2 - P_2^2}{GT_f LZf}\right)^{0.5} D^{2.5} \quad \text{(USCS units)} \quad (5.66)$$

where the transmission factor F and friction factor f are related by

$$F = \frac{2}{\sqrt{f}} \quad (5.67)$$

and in SI units:

$$Q = 5.747 \times 10^{-4} F \left(\frac{T_b}{P_b}\right) \left(\frac{P_1^2 - P_2^2}{GT_f LZf}\right)^{0.5} D^{2.5} \qquad (5.68)$$

We will discuss several aspects of the general flow equation before moving on to the other formulas for pressure drop calculation.

17. EFFECT OF PIPE ELEVATIONS

When elevation difference between the ends of a pipe segment is included, the general flow equation is modified as follows:

$$Q = 38.77 F \left(\frac{T_b}{P_b}\right) \left(\frac{P_1^2 - e^s P_2^2}{GT_f LZf}\right)^{0.5} D^{2.5} \qquad (5.69)$$

and in SI units

$$Q = 5.747 \times 10^{-4} F \left(\frac{T_b}{P_b}\right) \left(\frac{P_1^2 - e^s P_2^2}{GT_f LZf}\right)^{0.5} D^{2.5} \qquad (5.70)$$

where

$$L_e = \frac{L(e^s - 1)}{s} \qquad (5.71)$$

The equivalent length L_e and the term e^s take into account the elevation difference between the upstream and downstream ends of the pipe segment. The parameter s depends upon the gas gravity, gas compressibility factor, the flowing temperature, and the elevation difference. It is defined as follows in USCS units.

$$s = 0.0375 G \left(\frac{H_2 - H_1}{T_f Z}\right) \qquad (5.72)$$

where
  s – elevation adjustment parameter, dimensionless
  H_1 – upstream elevation, ft
  H_2 – downstream elevation, ft
  Other symbols are as defined earlier.

In SI units, the elevation adjustment parameter s is defined as follows:

$$s = 0.0684 G \left(\frac{H_2 - H_1}{T_f Z}\right) \qquad (5.73)$$

where

H_1 – upstream elevation, m
H_2 – downstream elevation, m
Other symbols are as defined earlier.

In the calculation of L_e in Eqn (5.64), we have assumed that there is a single slope between the upstream point 1 and the downstream point 2 in Figure 5.8. If, however, the pipe segment of length L has a series of slopes, then we introduce a parameter j as follows for each individual pipe subsegment that composes the pipe length from point 1 to point 2.

$$j = \frac{e^s - 1}{s} \quad (5.74)$$

The parameter j is calculated for each slope of each pipe sub segment of length L_1, L_2, etc., that make up the total length L. The equivalent length term Le in Eqn (5.62) and Eqn (5.63) is calculated by summing the individual slopes as defined here:

$$L_e = j_1 L_1 + j_2 L_2 \, e^{s_1} + j_3 L_3 e^{s_2} + \ldots \ldots \quad (5.75)$$

The terms j_1, j_2, etc., for each rise or fall in the elevations of individual pipe subsegments are calculated for the parameters s_1, s_2, etc., for each segment in accordance with Eqn (5.74).

In the subsequent sections of this chapter, we will discuss how the friction factor and transmission factor are calculated using various equations such as Colebrook–White and AGA. The general flow equation is the most commonly used equation to calculate the flow rate and pressure in a gas pipeline. To apply it correctly, we must use the correct friction factor or transmission factor. The Colebrook equation, AGA equation, and other empirical equations are used to calculate the friction factor to be used in the general flow equation. Several other equations, such as Panhandle A, Panhandle B, and Weymouth, calculate the flow rate for a given pressure without using a friction factor or transmission factor. However, an equivalent friction factor (or transmission factor) can be calculated using these methods as well.

18. AVERAGE PIPE SEGMENT PRESSURE

In the general flow equation, the compressibility factor Z is used. This must be calculated at the gas flowing temperature and average pressure in the pipe segment. Therefore, it is important to first calculate the average pressure in a pipe segment described in Figure 5.8.

Consider a pipe segment with the upstream pressure P_1 and downstream pressure P_2, as in Figure 5.x. An average pressure for this segment must be used to calculate the approximation we may use an arithmetic average of $(P_1 + P_2)/2$. However, it has been found that a more accurate value of the average gas pressure in a pipe segment is compressibility factor of gas at the average gas temperature Tf. First:

$$P_{avg} = \frac{2}{3}\left(P_1 + P_2 - \frac{P_1 P_2}{P_1 + P_2}\right) \tag{5.76}$$

Another form of the average pressure in a pipe segment is

$$P_{avg} = \frac{2}{3}\left(\frac{P_1^3 - P_2^3}{P_1^3 + P_2^3}\right) \tag{5.77}$$

The pressures used in the general flow equation are all in absolute units. Therefore, gauge pressure units should be converted to absolute pressure by adding the base pressure.

For example, the upstream and downstream pressures are 1000 psia and 900 psia, respectively. From Eqn (5.62) the average pressure is

$$P_{avg} = \frac{2}{3}\left(1000 + 900 - \frac{1000 \times 900}{1900}\right) = 950.88 \text{ psia}$$

Compare this to the arithmetic average of

$$P_{avg} = \frac{1}{2}(1000 + 900) = 950 \text{ psia}$$

19. VELOCITY OF GAS IN A PIPELINE

The velocity of gas flow in a pipeline represents the speed at which the gas molecules move from one point to another. Unlike a liquid pipeline, because of compressibility, the gas velocity depends upon the pressure and hence will vary along the pipeline even if the pipe diameter is constant. The highest velocity will be at the downstream end where the pressure is the least. Correspondingly, the least velocity will be at the upstream end, where the pressure is higher.

Consider a pipe transporting gas from point A to point B as shown in Figure 5.8. Under steady-state flow, at A the mass flow rate of gas is designated as M and will be the same as the mass flow rate at point B if between A and B there is no injection or delivery of gas. The mass being the product of volume and density we can write the following relationship for point A:

$$M = Q\rho \tag{5.78}$$

The volume rate Q can be expressed in terms of the flow velocity u and cross-sectional area A as follows:

$$Q = uA \tag{5.79}$$

Therefore, combining Eqns (5.78) and (5.79) and applying the conservation of mass to points A and B, we get:

$$M_1 = u_1 A_1 \rho_1 = M_2 = u_2 A_2 \rho_2 \tag{5.80}$$

where subscript 1 and 2 refer to points A and B, respectively.

If the pipe is of uniform cross-section between A and B, then $A_1 = A_2 = A$.

Therefore, the area term in Eqn (5.80) may be dropped and the velocity at A and B are related by the following equation.

$$u_1 \rho_1 = u_2 \rho_2 \tag{5.81}$$

Because the flow of gas in a pipe may result in variation of temperature from point A to point B, the gas density will also vary with temperature and pressure. If the density and velocity at one point are known the corresponding velocity at the other point may be calculated using Eqns (5.88) and (5.89).

If inlet conditions are represented by point A and the volume flow rate Q at standard conditions of 60 °F and 14.7 psia are known, we can calculate the velocity at any point along the pipeline at which the pressure and temperature of the gas are P and T, respectively.

The velocity of gas at section 1 is related to the flow rate Q_1 at Section 1 and pipe cross-sectional area A as follows:

$$Q_1 = u_1 A$$

The mass flow rate M at Sections 1 and 2 are the same for steady state flow.

Therefore

$$M_1 = Q_1 \rho_1 = Q_2 \rho_2 = Q_b \rho_b \tag{5.82}$$

where Q_b is the gas flow rate at standard conditions and ρ_b is the corresponding gas density.

Therefore, simplifying Eqn (5.82):

$$Q_1 = Q_b \left(\frac{\rho_b}{\rho_1} \right) \tag{5.83}$$

Applying the gas law equation, we get:

$$\frac{P_1}{\rho_1} = Z_1 R T_1 \quad \text{or} \quad \rho_1 = \frac{P_1}{Z_1 R T_1} \tag{5.84}$$

where P_1 and T_1 are the pressure and temperature at pipe Section 1. Similarly, at standard conditions:

$$\rho_b = \frac{P_b}{Z_b R T_b} \tag{5.85}$$

From Eqns (5.83), (5.84), and (5.85) we get:

$$Q_1 = Q_b \left(\frac{P_b}{T_b}\right)\left(\frac{T_1}{P_1}\right)\left(\frac{Z_1}{Z_b}\right) \tag{5.86}$$

Because $Z_b = 1.00$ approximately, we can simplify this to:

$$Q_1 = Q_b \left(\frac{P_b}{T_b}\right)\left(\frac{T_1}{P_1}\right) Z_1 \tag{5.87}$$

Therefore, the gas velocity at Section 1 is

$$u_1 = \frac{Q_b P_1}{A}\left(\frac{P_b}{T_b}\right)\left(\frac{T_1}{P_1}\right) = \frac{4 \times 144}{\pi D^2} Q_b Z_1 \left(\frac{P_b}{T_b}\right)\left(\frac{T_1}{P_1}\right)$$

Or

$$u_1 = 0.002122 \left(\frac{Q_b}{D^2}\right)\left(\frac{P_b}{T_b}\right)\left(\frac{Z_1 T_1}{P_1}\right) \tag{5.88}$$

where

u_1 - upstream gas velocity, ft/s
Q_b - gas flow rate, measured at standard conditions, ft³/day (SCFD)
D - pipe inside diameter, in
P_b - base pressure, psia
T_b - base temperature, °R (460 + °F)
P_1 - upstream pressure, psia.
T_1 - upstream gas temperature, °R (460 + °F)
Z_1 - gas compressibility factor at upstream conditions, dimensionless

Similarly, the gas velocity at Section 2 is given by

$$u_2 = 0.002122 \left(\frac{Q_b}{D^2}\right)\left(\frac{P_b}{T_b}\right)\left(\frac{Z_2 T_2}{P_2}\right) \tag{5.89}$$

In general, the gas velocity at any point in a pipeline is given by

$$u = 0.002122 \left(\frac{Q_b}{D^2}\right)\left(\frac{P_b}{T_b}\right)\left(\frac{ZT}{P}\right) \quad (5.90)$$

In SI units the gas velocity at any point in a gas pipeline is given by

$$u = 14.7349 \left(\frac{Q_b}{D^2}\right)\left(\frac{P_b}{T_b}\right)\left(\frac{ZT}{P}\right) \quad (5.91)$$

where
- u – gas velocity, m/s
- Q_b – gas flow rate, measured at standard conditions, m^3/day
- D – pipe inside diameter, mm
- P_b – base pressure, kPa
- T_b – base temperature, K (273 + °C)
- P – pressure, kPa
- T – average gas flowing temperature, K (273 + °C)
- Z – gas compressibility factor at the flowing temperature, dimensionless

Because the right hand side of the Eqn (5.91) contains ratios of pressures, any consistent unit may be used such as kPa, MPa, or Bar.

20. EROSIONAL VELOCITY

We have seen from the preceding section that the gas velocity is directly related to the flow rate. As flow rate increases, so does the gas velocity. As the velocity increases, vibration and noise are evident. In addition, higher velocities will cause erosion of the pipe interior over a long period. The upper limit of the gas velocity is usually calculated approximately from the following equation:

$$u_{max} = \frac{100}{\sqrt{\rho}} \quad (5.92)$$

where
- u_{max} – maximum or erosional velocity, ft/s
- ρ – gas density at flowing temperature, lb/ft^3

Because the gas density ρ may be expressed in terms of pressure and temperature, using the gas law equation, the maximum velocity Eqn (5.92) may be rewritten as:

$$u_{max} = 100\sqrt{\frac{ZRT}{29GP}} \quad (5.93)$$

where
    Z – compressibility factor of gas, dimensionless
    R – gas constant = 10.73 ft³ psia/lb-mole R
    T – gas temperature, °R
    G – gas gravity (air = 1.00)
    P – gas pressure, psia

Example 5.10 (USCS)
A gas pipeline NPS 20 with 0.500-in wall thickness transports natural gas (specific gravity = 0.6) at a flow rate of 250 million SCFD (MMSCFD) at an inlet temperature of 60 °F. Assuming isothermal flow, calculate the velocity of gas at inlet and outlet of the pipe if the inlet pressure is 1000 psig and the outlet pressure is 850 psig. The base pressure and base temperature are 14.7 psia and 60 °F, respectively. Assume compressibility factor $Z = 1.00$. What is the erosional velocity for this pipeline based on above data and a compressibility factor $Z = 0.90$?

Solution
If we assume compressibility factor $Z = 1.00$.

The velocity of gas at the inlet pressure of 1000 psig is

$$u_1 = 0.002122 \left(\frac{250 \times 10^6}{19.0^2}\right)\left(\frac{14.7}{60+460}\right)\left(\frac{60+460}{1014.7}\right) = 21.29 \text{ ft/s}$$

And the gas velocity at the outlet is by proportions.

$$u_2 = 21.29 \times \frac{1014.7}{864.7} = 24.98 \text{ ft/s}$$

The erosional velocity is found for $Z = 0.90$.

$$u_{max} = 100\sqrt{\frac{0.9 \times 10.73 \times 520}{29 \times 0.6 \times 1014.7}} = 53.33 \text{ ft/s}$$

Usually an acceptable operational velocity is 50% of the above.

Example 5.11
A gas pipeline DN 500 with 12-mm wall thickness transports natural gas (specific gravity = 0.6) at a flow rate of 7.5 Mm³/day at an inlet temperature of 15 °C. Assuming isothermal flow, calculate the velocity of gas at the inlet and outlet of the pipe if the inlet pressure is 7 MPa and the outlet pressure is 6 MPa. The base pressure and base temperature are 0.1 MPa and 15 °C.

Assume compressibility factor Z = 0.95.

Solution

Inside diameter of pipe D = 500 − (2 × 12) = 476 mm.
Flow rate at standard conditions $Q_b = 7.5 \times 10^6$ m³/day.
The velocity of gas at the inlet pressure of 7 MPa is

$$u_1 = 14.7349 \left(\frac{7.5 \times 10^6}{476^2}\right) \left(\frac{0.1}{15 + 273}\right) \left(\frac{0.95 \times 288}{7.0}\right) = 6.62 \text{ m/s}$$

And the gas velocity at the outlet is by proportions.

$$u_2 = 6.62 \times \frac{7.0}{6.0} = 7.72 \text{ m/s}$$

In the preceding example (5.11), we have assumed the value of compressibility factor Z as constant. A more accurate solution will be to calculate the value of Z using one of the methods such as California Natural Gas Association (CNGA) or the Standing–Katz method.

For example, if we used the CNGA equation, the compressibility factor in example 5.11 will be

$$Z_1 = \frac{1}{\left[1 + \frac{1000 \times 344400 \times (10)^{1.785 \times 0.6}}{520^{3.825}}\right]}$$

= 0.8578 at inlet pressure of 1000 psig.

And

$$Z_2 = \frac{1}{\left[1 + \frac{850 \times 344400 \times (10)^{1.785 \times 0.6}}{520^{3.825}}\right]}$$

= 0.8765 at outlet pressure of 850 psig.

The inlet and outlet gas velocities then will be modified as follows:
Inlet velocity $u_1 = 0.8578 \times 21.29 = 18.26$ ft/s.
Outlet velocity $u_2 = 0.8765 \times 24.98 = 21.90$ ft/s.

21. REYNOLDS NUMBER OF FLOW

An important parameter in flow of fluids in a pipe is the nondimensional term Reynolds number. The Reynolds number is used to characterize the type of flow in a pipe, such as laminar, turbulent, or critical flow. It is also used to calculate the friction factor in pipe flow. We will first outline the calculation of the Reynolds number based upon the properties of the gas and pipe diameter and then discuss the range of Reynolds number for the various types of flow and how to calculate the friction factor. Reynolds number is a function of the gas flow rate, pipe inside diameter, and the gas density and viscosity and is calculated from the following equation.

$$R = \frac{uD\rho}{\mu} \qquad (5.94)$$

where
    R – Reynolds number, dimensionless
    u – average velocity of gas in pipe, ft/s
    D – inside diameter of pipe, ft
    ρ – gas density, lb/ft^3
    μ – gas viscosity, lb/ft-s

The previous equation for Reynolds number is in the USCS units. The corresponding equation for Reynolds number in SI units is as follows:

$$R = \frac{uD\rho}{\mu} \qquad (5.95)$$

where
    R – Reynolds number, dimensionless
    u – average velocity of gas in pipe, m/s
    D – inside diameter of pipe, m
    ρ – gas density, kg/m^3
    μ – gas viscosity, kg/m-s

In gas pipeline hydraulics, using customary units, a more suitable equation for Reynolds number is as follows:

$$R = 0.0004778 \left(\frac{P_b}{T_b}\right)\left(\frac{GQ}{\mu D}\right) \qquad (5.96)$$

where
    P_b – base pressure, psia
    T_b – base temperature, °R (460 + °F)
    G – specific gravity of gas (Air = 1.0)
    Q – gas flow rate, standard ft^3/day (SCFD)
    D – pipe inside diameter, in.
    μ – viscosity of gas, lb/ft-s

In SI units, the Reynolds number is

$$R = 0.5134 \left(\frac{P_b}{T_b}\right)\left(\frac{GQ}{\mu D}\right) \qquad (5.97)$$

where

P_b – base pressure, kPa
T_b – base temperature, °K (273 + °C)
G – specific gravity of gas (air = 1.0)
Q – gas flow rate, m³/day (standard conditions)
D – pipe inside diameter, mm
μ – viscosity of gas, poise

Laminar flow occurs in a pipeline when Reynolds number is below a value of approximately 2000. Turbulent flow occurs when the Reynolds number is greater than 4000. For Reynolds numbers between 2000 and 4000, the flow is undefined and is referred to as critical flow.

Thus

For laminar flow, $R \le 2000$.
For turbulent flow, $R > 4000$.
For critical flow, $R > 2000$ and $Re \le 4000$.

Most natural gas pipelines operate in the turbulent flow region. Therefore, the Reynolds number is greater than 4000. Turbulent flow is further divided into three regions known as smooth pipe flow, fully rough pipe flow, and transition flow. We will discuss these flow regions in more detail in the subsequent sections of this chapter.

Example 5.12

A natural gas pipeline, NPS 20 with 0.500-in wall thickness transports 100 MMSCFD. The specific gravity of gas is 0.6 and viscosity is 0.000,008 lb/ft-s. Calculate the value of Reynolds number of flow. Assume the base temperature and base pressure are 60 °F and 14.7 psia, respectively.

Solution

Pipe inside diameter $= 20 - 2 \times 0.5 = 19.0$ in
The base temperature $= 60 + 460 = 520$ °R
We get Reynolds number as:

$$R = 0.0004778 \left(\frac{14.7}{520}\right) \left(\frac{0.6 \times 100 \times 10^6}{0.000008 \times 19}\right) = 5{,}331{,}726$$

Because R is greater than 4000, the flow is in the turbulent region.

Example 5.13 (SI Units)

A natural gas pipeline, DN 500 with 12-mm wall thickness transports 3 Mm³/day. The specific gravity of gas is 0.6 and viscosity is 0.00012 poise.

Calculate the value of Reynolds number. Assume the base temperature and base pressure are 15 °C and 101 kPa, respectively.

Solution

Pipe inside diameter = 500 − 2 × 12 = 476 mm.
  The base temperature = 15 + 273 = 288 K.
  The Reynolds number is

$$R = 0.5134 \left(\frac{101}{15+273}\right)\left(\frac{0.6 \times 3 \times 10^6}{0.00012 \times 476}\right) = 5{,}673{,}735$$

Because R is greater than 4000, the flow is in the turbulent region.

22. FRICTION FACTOR

To calculate the pressure drop in a pipeline at a given flow rate, we must first understand the concept of friction factor. The term friction factor is a dimensionless parameter that depends upon the Reynolds number of flow. In engineering literature, we find two different friction factors mentioned. The Darcy friction factor is more common and will be used throughout this book. Another friction factor known as the Fanning friction factor is preferred by some engineers. The Fanning friction factor is numerically equal to one-fourth the Darcy friction factor as below.

$$f_f = \frac{f_d}{4} \qquad (5.98)$$

where
  f_f - Fanning friction factor
  f_d - Darcy friction factor

To avoid confusion, in subsequent discussions the Darcy friction factor is used and will be represented by the symbol f. For laminar flow, the friction factor is inversely proportional to the Reynolds number, as indicated below.

$$f = \frac{64}{R} \qquad (5.99)$$

For turbulent flow, the friction factor is a function of Reynolds number, pipe inside diameter, and internal roughness of the pipe. Many empirical relationships for calculating f have been put forth by researchers. The more popular correlations include Colebrook–White and the AGA equations.

Before we discuss the equations for calculating the friction factor in turbulent flow, it is appropriate to analyze the turbulent flow regime.

Turbulent flow in pipes (Re > 4000) is subdivided into three separate regions as follows:
1. Turbulent flow in smooth pipes
2. Turbulent flow in fully rough pipes
3. Transition flow between smooth pipes and rough pipes.

For turbulent flow in smooth pipes, the friction factor f depends only on the Reynolds number. For fully rough pipes, f depends more on the pipe internal roughness and less on the Reynolds number. In the transition zone between smooth pipe flow and flow in fully rough pipes, f depends on the pipe roughness, pipe inside diameter, and the Reynolds number. The various flow regimes are depicted in the Moody diagram, shown in Figure 5.9.

The Moody diagram is a graphic plot of the variation of the friction factor with the Reynolds number for various values of relative pipe roughness. The latter term is simply a dimensionless parameter obtained by dividing the absolute (or internal) pipe roughness by the pipe inside diameter as follows:

$$\text{Relative roughness} = e/D \qquad (5.100)$$

where
 e – absolute or internal roughness of pipe, in
 D – pipe inside diameter, in

The term absolute pipe roughness or internal pipe roughness means the same.

Generally, the internal pipe roughness is expressed in micro inches (one-millionth of an inch). For example an internal roughness of 0.0006 in is referred to as 600 micro inches or 600 micro in. If the pipe inside diameter is 15.5 in, the relative roughness is, in this case:

$$\text{Relative roughness} = \frac{0.0006}{15.5} = 0.0000387 = 3.87 \times 10^{-5}$$

For example, from the Moody diagram, Figre 5.9 for Re = 10 million and e/D = 0.0001, we find that f = 0.012.

23. COLEBROOK–WHITE EQUATION

The Colebrook–White equation, sometimes referred to simply as the Colebrook equation is a relationship between the friction factor and the Reynolds number, pipe roughness, and inside diameter of pipe.

Fluid Flow in Pipes

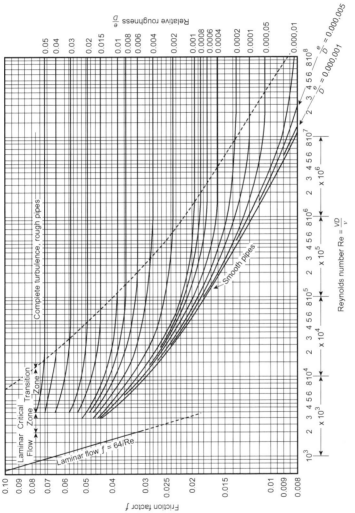

Figure 5.9 Moody diagram.

The following form of the Colebrook equation is used to calculate the friction factor in gas pipelines in turbulent flow.

$$1/\sqrt{f} = -2\log_{10}[(e/3.7D) + 2.51/(R\sqrt{f})] \quad \text{for } R > 4,000 \quad (5.101)$$

where
  f – friction factor
  D – pipe inside diameter, in
  e – absolute pipe roughness, in
  R – Reynolds number of flow for R > 4000

Because R and f are dimensionless, as long as consistent units are used for both e and D, the Colebrook equation is the same regardless of the units employed. Therefore, in SI units Equation above is used with e and D expressed in mm.

It can be seen from Eqn (5.101), that to calculate the friction factor f we must use a trial-and-error approach. It is an implicit equation in f because f appears on both sides of the equation. We first assume a value of f (such as 0.01) and substitute it in the right hand side of the equation. This will yield a second approximation for f which can then be used to calculate a better value of f and so on. Generally three to four iterations are sufficient to converge on a reasonably good value of the friction factor.

It can be seen from the Colebrook equation, for turbulent flow in smooth pipes the first term within the square brackets is negligible compared to the second term because pipe roughness e is very small.

Therefore for smooth pipe flow, the friction factor equation reduces to

$$\frac{1}{\sqrt{f}} = -2\log_{10}\left(\frac{2.51}{R\sqrt{f}}\right) \quad \text{for turbulent flow in smooth pipes} \quad (5.102)$$

Similarly, for turbulent flow in fully rough pipes, with R being a large number, f depends mostly on the roughness e and therefore, the friction factor equation reduces to

$$1/\sqrt{f} = -2\log_{10}[(e/3.7D)] \quad \text{for turbulent flow in fully rough pipes} \quad (5.103)$$

Table 5.2 lists typical values of pipe internal roughness used to calculate the friction factor.

Table 5.2 Pipe internal roughness

Pipe material	Roughness (in)	Roughness (mm)
Riveted steel	0.0354–0.354	0.9–9.0
Commercial steel/welded steel	0.0018	0.045
Cast iron	0.0102	0.26
Galvanized iron	0.0059	0.15
Asphalted cast iron	0.0047	0.12
Wrought iron	0.0018	0.045
Polyvinyl chloride, drawn tubing, glass	0.000059	0.0015
Concrete	0.0118–0.118	0.3–3.0

As an example, if $R = 100$ million or larger and $e/D = 0.0002$, the friction factor from Eqn (5.103) is

$$\frac{1}{\sqrt{f}} = -2 \log_{10}\left(\frac{0.0002}{3.7}\right)$$

Or $f = 0.0137$, which correlates well with the friction factor obtained from the Moody diagram Figure 5.9.

Example 5.14 (USCS)

A natural gas pipeline, NPS 20 with 0.500-in wall thickness transports 200 MMSCFD. The specific gravity of gas is 0.6 and viscosity is 0.000,008 lb/ft-s. Calculate the friction factor using the Colebrook equation. Assume absolute pipe roughness = 600 micro in.

The base temperature and base pressure are 60 °F and 14.7 psia, respectively.

Solution

Pipe inside diameter $= 20 - 2 \times 0.5 = 19.0$ in
Absolute pipe roughness $= 600$ micro in $= 0.0006$ in
First we calculate the Reynolds number.

$$R = 0.0004778 \left(\frac{14.7}{60 + 460}\right)\left(\frac{0.6 \times 200 \times 10^6}{0.000008 \times 19}\right) = 10{,}663{,}452$$

Using Eqn (5.101):

$$\frac{1}{\sqrt{f}} = -2 \log_{10}\left(\frac{0.0006}{3.7 \times 19} + \frac{2.51}{10{,}663{,}452\sqrt{f}}\right)$$

This equation will be solved by successive iteration.

Assume f = 0.01 initially and substituting above we get a better approximation as:
f = 0.0101. Repeating the iteration, we get the final value as:
f = 0.0101.
Therefore, the friction factor is 0.0101.

Example 5.15 (SI Units)
A natural gas pipeline, DN 500 with 12-mm wall thickness transports 6 Mm^3/day. The specific gravity of gas is 0.6 and viscosity is 0.00012 poise. Calculate the friction factor using the Colebrook equation. Assume absolute pipe roughness = 0.03 mm and the base temperature and base pressure are 15 °C and 101 kPa, respectively.

Solution
Pipe inside diameter = 500 − 2 × 12 = 476 mm.
First we calculate the Reynolds number.

$$Re = 0.5134 \left(\frac{101}{15 + 273}\right)\left(\frac{0.6 \times 6 \times 10^6}{0.00012 \times 476}\right) = 11{,}347{,}470$$

The friction factor is

$$\frac{1}{\sqrt{f}} = -2 \log_{10}\left(\frac{0.030}{3.7 \times 476} + \frac{2.51}{11{,}347{,}470\sqrt{f}}\right)$$

This equation will be solved by successive iteration.
Assume f = 0.01 initially and substituting above we get a better approximation as:
f = 0.0112. Repeating the iteration, we get the final value as:
f = 0.0112.
Therefore, the friction factor is 0.0112.

24. TRANSMISSION FACTOR

The transmission factor F is considered the opposite of the friction factor f. Whereas the friction factor indicates how difficult it is to move a certain quantity of gas through a pipeline, the transmission factor is a direct measure of how much gas can be transported through the pipeline. As the friction factor increases, the transmission factor decreases and therefore, the gas flow rate also decreases. Conversely, the higher the transmission factor, the lower the friction factor and therefore higher will be the flow rate.

The transmission factor F is related to the friction factor f as follows:

$$F = \frac{2}{\sqrt{f}} \tag{5.104}$$

Therefore:

$$f = \frac{4}{F^2} \tag{5.105}$$

where
 f – friction factor
 F – transmission factor

It must be noted that the friction factor f in above equation is the Darcy friction factor. Because some engineers prefer to use the Fanning friction factor, the relationship between the transmission factor F and the Fanning friction factor is given below for reference.

$$F = \frac{1}{\sqrt{f_f}} \tag{5.106}$$

where f_f is the Fanning friction factor.

For example, if the Darcy friction factor is 0.025, the transmission factor is, using Eqn (5.104):

$$F = \frac{2}{\sqrt{0.025}} = 12.65$$

The Fanning friction factor in this case will be $0.025/4 = 0.00625$. Therefore the transmission factor is

$F = \frac{1}{\sqrt{0.00625}} = 12.65$, which is the same as calculated using the Darcy friction factor.

Thus it must be noted that there is only one transmission factor, whereas there are two different friction factors.

Having defined a transmission factor, we can rewrite the Colebrook Eqn (5.101) in terms of the transmission factor as follows:

$$F = -4 \log_{10}\left(\frac{e}{3.7D} + \frac{1.255F}{R}\right) \tag{5.107}$$

Because R and F are dimensionless, as long as consistent units are used for both e and D, the transmission factor equation is the same regardless of the units employed. Therefore, in SI units e and D expressed in mm.

Similar to the calculation of the friction factor f, the transmission factor F is calculated in an iterative approach. This will be illustrated using an example.

Example 5.16 (USCS)
For a gas pipeline, flowing 100 MMSCFD gas of specific gravity 0.6 and viscosity of 0.000008 lb/ft-s, calculate the friction factor and transmission factor considering an NPS 20 pipeline, 0.500-in wall thickness and an internal roughness of 600 micro inches. Assume the base temperature and base pressure are 60 °F and 14.7 psia, respectively. If the flow rate increases by 50%, what is the impact on the friction factor and transmission factor?

Solution
The base temperature = 60 + 460 = 520 °R
Pipe inside diameter = 20 − 2 × 0.500 = 19.0 in
the Reynolds number is

$$R = 0.0004778 \left(\frac{14.7}{520}\right)\left(\frac{0.6 \times 100 \times 10^6}{0.000008 \times 19}\right) = 5{,}331{,}726$$

$$\text{The relative roughness} = \frac{600 \times 10^{-6}}{19} = 0.0000316$$

Next calculate the friction factor.

$$\frac{1}{\sqrt{f}} = -2 \log_{10}\left(\frac{0.0000316}{3.7} + \frac{2.51}{5{,}331{,}726\sqrt{f}}\right)$$

Solving by successive iteration, we get.

$$f = 0.0105$$

Therefore, the transmission factor, F is as follows:

$$F = \frac{2}{\sqrt{0.0105}} = 19.53$$

The friction factor calculated previously is the Darcy friction factor. The corresponding Fanning friction factor will be one-fourth the calculated value.

When flow rate is increased by 50%, the Reynolds number becomes by proportion:

$$R = 1.5 \times 5{,}331{,}726 = 7{,}997{,}589$$

The new friction factor is

$$\frac{1}{\sqrt{f}} = -2 \log_{10}\left(\frac{0.0000316}{3.7} + \frac{2.51}{7{,}997{,}589\sqrt{f}}\right)$$

Fluid Flow in Pipes

Solving for f by successive iteration, we get:
$$f = 0.0103$$

Corresponding transmission factor is
$$F = \frac{2}{\sqrt{0.0103}} = 19.74$$

Compared with the previous values of 0.0105 for friction factor and 19.53 for transmission factor, we see the following changes.

$$\text{Decrease in friction factor} = \frac{0.0105 - 0.0103}{0.0105} = 0.019 \text{ or } 1.9 \text{ percent}$$

$$\text{Increase in transmission factor} = \frac{19.74 - 19.53}{19.53} = 0.0108 \text{ or } 1.08 \text{ percent}$$

Thus increasing the flow rate by 50% reduces the friction factor by 1.9% and increases transmission factor by 1.08%.

Example 5.17 (SI Units)

For a gas pipeline, flowing 3 Mm^3/day gas of specific gravity 0.6 and viscosity of 0.000119 poise, calculate the friction factor and transmission factor considering a DN 400 pipeline, 10-mm wall thickness, and an internal roughness of 0.02 mm. The base temperature and base pressure are 15 °C and 101 kPa, respectively. If the flow rate is doubled, what is the impact on the friction factor and transmission factor?

Solution

The base temperature = 15 + 273 = 288 K.
Pipe inside diameter = 400 − 2 × 10 = 380 mm.
We calculate the Reynolds number as:

$$R = 0.5134 \left(\frac{101}{288}\right) \left(\frac{0.6 \times 3 \times 10^6}{0.000119 \times 380}\right) = 7{,}166{,}823$$

$$\text{The relative roughness} = \frac{0.02}{380} = 0.0000526$$

The friction factor is

$$\frac{1}{\sqrt{f}} = -2 \log_{10}\left(\frac{0.0000526}{3.7} + \frac{2.51}{7{,}166{,}823\sqrt{f}}\right)$$

Solving by iteration, we get:

$$f = 0.0111$$

Therefore, the transmission factor, F is

$$F = \frac{2}{\sqrt{0.0111}} = 18.98$$

The friction factor calculated here is the Darcy friction factor. The corresponding Fanning friction factor will be one-fourth the calculated value. When the flow rate is doubled, the Reynolds number becomes

$$R = 2 \times 7{,}166{,}823 = 14{,}333{,}646$$

The new value of friction factor is

$$\frac{1}{\sqrt{f}} = -2\log_{10}\left(\frac{0.0000526}{3.7} + \frac{2.51}{14{,}333{,}646\sqrt{f}}\right)$$

Solving for f by successive iteration, we get:

$$f = 0.0109$$

and the transmission factor is

$$F = \frac{2}{\sqrt{0.0109}} = 19.16$$

Therefore doubling the flow rate increased the transmission factor and decreased the friction factor as follows:

$$\text{Decrease in friction factor} = \frac{0.0111 - 0.0109}{0.0111} = 0.018 \text{ or } 1.8 \text{ percent}$$

$$\text{Increase in transmission factor} = \frac{19.16 - 18.98}{18.98} = 0.0095 \text{ or } 0.95 \text{ percent}$$

25. MODIFIED COLEBROOK–WHITE EQUATION

The Colebrook–White equation discussed in the preceding section has been in use for many years in both liquid flow and gas flow. The US Bureau of Mines published a report in 1956 that introduced a modified form of the Colebrook–White equation. The modification results in a higher friction factor and hence a smaller value of the transmission factor. Because of this, a conservative value of flow rate is obtained because of

Fluid Flow in Pipes

the higher friction and pressure drop. The modified version of the Colebrook–White equation for turbulent flow is as follows:

$$1/\sqrt{f} = -2\log[(e/3.7D) + 2.825/(R\sqrt{f})] \quad \text{for Turbulent flow} \tag{5.108}$$

Rewriting Eqn (5.108) in terms of the transmission factor, we get the following version of the modified Colebrook–White equation.

$$F = -4\log[(e/3.7D) + 1.4125\,F/R] \quad \text{for Turbulent flow} \tag{5.109}$$

Upon comparing Eqn (5.101) with Eqn (5.108), the difference between the Colebrook equation and the modified Colebrook equation lies in the second constant term within the square brackets. The constant 2.51 in the first case is replaced with the constant 2.825 in the second case. Similarly, in the transmission factor equations, the modified equation has 1.4125 instead of 1.255 in the unmodified equation.

Because Re, f, and F are dimensionless, as long as consistent units are used for both e and D, the modified Colebrook equation is the same regardless of the units employed. Therefore, in SI units the e and D expressed in mm.

Many commercial hydraulic simulation programs list both Colebrook–White equations. Some use only the original Colebrook-White equation.

Example 5.18 (USCS)
A gas pipeline flows 100 MMSCFD gas of specific gravity 0.6 and viscosity of 0.000008 lb/ft-s; calculate, using the modified Colebrook–White equation, the friction factor, and transmission factor assuming NPS 20 pipeline, 0.500-in wall thickness, and an internal roughness of 600 micro in. The base temperature and base pressure are 60 °F and 14.7 psia, respectively. How do these numbers compare with those calculated, using the original Colebrook equation?

Solution
The base temperature = 60 + 460 = 520 °R
 Pipe inside diameter = 20 − 2 × 0.500 = 19.0 in
 We calculate the Reynolds number as:

$$R = 0.0004778 \left(\frac{14.7}{520}\right)\left(\frac{0.6 \times 100 \times 10^6}{0.000008 \times 19}\right) = 5{,}331{,}726$$

The relative roughness is

$$\frac{e}{D} = \frac{600 \times 10^6}{19} = 3.16 \times 10^{-5}$$

The friction factor using the modified Colebrook equation is

$$\frac{1}{\sqrt{f}} = -2 \log_{10}\left(\frac{0.0000316}{3.7} + \frac{2.825}{5,331,726\sqrt{f}}\right)$$

Solving by successive iteration, we get:

$$f = 0.0106$$

Therefore, the transmission factor, F is found as follows:

$$F = \frac{2}{\sqrt{0.0106}} = 19.43$$

By comparing these results with the friction factor and the transmission factor calculated using the unmodified Colebrook equation, it can be seen that the modified friction factor is approximately 0.95% higher than that calculated using the original Colebrook–White equation, whereas the transmission factor is approximately 0.51% lower than that calculated using the original Colebrook–White equation.

Example 5.19 (USCS)

A gas pipeline NPS 20 with 0.500-in wall thickness, flows 200 MMSCFD gas of specific gravity 0.6, and viscosity of 0.000008 lb/ft-s. Using the modified Colebrook–White equation calculate the pressure drop in a 50-mile segment of pipe based on an upstream pressure of 1000 psig. Assume an internal pipe roughness of 600 micro in and the base temperature and base pressure are 60 °F and 14.73 psia, respectively. Neglect elevation effects and use 60 °F for gas flowing temperature and compressibility factor $Z = 0.88$.

Solution

Inside diameter of pipe $= 20 - 2 \times 0.5 = 19.0$ in
The base temperature $= 60 + 460 = 520$ °R
Gas flow temperature $= 60 + 460 = 520$ °R

$$R = 0.0004778 \left(\frac{14.73}{520}\right)\left(\frac{0.6 \times 200 \times 10^6}{0.000008 \times 19}\right) = 10,685,214$$

The transmission factor F is calculated as follows:

$$F = -4 \log_{10}\left(\frac{600 \times 10^{-6}}{3.7 \times 19} + \frac{1.4125F}{10,685,214}\right)$$

Solving for F by successive iteration.

$$F = 19.81$$

Next, using general flow equation we calculate the downstream pressure P_2 as follows:

$$200 \times 10^6 = 38.77 \times 19.81 \left(\frac{60+460}{14.73}\right) \left[\frac{1014.73^2 - P_2^2}{0.6 \times 520 \times 50 \times 0.88}\right]^{0.5}$$
$$\times 19^{2.5}$$

Solving for P_2, we get:

$$P_2 = 853.23 \text{ psia} = 838.5 \text{ psig}$$

Therefore, the pressure drop $= 1014.73 - 853.23 = 161.5$ psi.

26. AGA EQUATION

In 1964 and 1965, the AGA published a report on how to calculate the transmission factor for gas pipelines to be used in the general flow equation. This is sometimes referred to as the AGA NB-13 method. Using the method outlined in this report, the transmission factor F is calculated using two different equations. First F is calculated for the rough pipe law (referred to as fully turbulent zone). Next F is calculated based on the smooth pipe law (referred to as the partially turbulent zone). Finally, the smaller of the two values of the transmission factor is used in the general flow equation for pressure drop calculation. Even though the AGA method uses the transmission factor F instead of the friction factor f, we can still calculate the friction factor using the relationship shown in Eqn (5.104).

For the fully turbulent zone, AGA recommends using the following formula for F based on relative roughness e/D and independent of the Reynolds number.

$$F = 4 \log(3.7D/e) \quad (5.110)$$

For the partially turbulent zone, F is calculated from the following equations using the Reynolds number and a parameter known as pipe drag factor.

$$F = 4 D_f \log[R/(1.4125 F_t)] \quad (5.111)$$

$$F_t = 4 \log(R/F_t) - 0.6 \quad (5.112)$$

where

F_t – the smooth pipe transmission factor, also known as the Von Karman's smooth pipe transmission factor

D_f – pipe drag factor that depends on the bend index of the pipe.

Table 5.3 Bend index and drag factor

	Bend index		
	Extremely low 5°–10°	Average 60°–80°	Extremely high 200°–300°
Bare steel	0.975–0.973	0.960–0.956	0.930–0.900
Plastic lined	0.979–0.976	0.964–0.960	0.936–0.910
Pig burnished	0.982–0.980	0.968–0.965	0.944–0.920
Sand-blasted	0.985–0.983	0.976–0.970	0.951–0.930

Note: The drag factors above are based on 40-ft joints of pipelines and mainline valves at 10-mile spacing.

Equation (5.112) is also known as the Von Karman rough pipe flow equation.

For the partially turbulent zone, F is calculated from the following equations using the Reynolds number, a parameter D_f known as the pipe drag factor, and the Von Karman smooth pipe transmission factor F_t.

$$F = 4D_f \log_{10}\left(\frac{Re}{1.4125F_t}\right) \quad (5.113)$$

and

$$F_t = 4 \log_{10}\left(\frac{Re}{F_t}\right) - 0.6 \quad (5.114)$$

where

F_t – Von Karman smooth pipe transmission factor
D_f – pipe drag factor that depends on the bend index of the pipe.

The pipe drag factor, D_f, is a parameter that takes into account the number of bends, fittings, etc. Its value ranges from 0.90 to 0.99. The bend index is the sum of all the angles and bends in the pipe segment, divided by the total length of the pipe section under consideration.

The value of D_f is generally chosen from Table 5.3.

For further discussion on bend index and drag factor, the reader is referred to AGA NB-13 Committee Report.

Example 5.20 (USCS)

Using the AGA method, calculate the transmission factor and friction factor for gas flow in an NPS 20 pipeline with 0.500-in wall thickness. The flow rate is 200 MMSCFD and gas gravity = 0.6 and viscosity = 0.000008 lb/ft-sec. The absolute pipe roughness is 700 micro in. Assume a bend index of 60°, base pressure = 14.73 psia, and base temperature = 60 °F.

Fluid Flow in Pipes

Solution
Inside diameter of pipe = $20 - 2 \times 0.5 = 19.0$ in
The base temperature = $60 + 460 = 520\ °R$
We will first calculate the Reynolds number.

$$R = \frac{0.0004778 \times 200 \times 10^6 \times 0.6 \times 14.73}{19 \times 0.000008 \times 520} = 10{,}685{,}214$$

Next calculate the two transmission factors:
The fully turbulent transmission factor is

$$F = 4 \log_{10}\left(\frac{3.7 \times 19}{0.0007}\right) = 20.01$$

For the smooth pipe zone, we get the Von Karman transmission factor as

$$F_t = 4 \log_{10}\left(\frac{10{,}685{,}214}{F_t}\right) - 0.6$$

Solving this equation by trial and error we get $F_t = 22.13$.
From Table 5.3 for a bend index of 60° the drag factor D_f is 0.96.
Therefore, for the partially turbulent flow zone, the transmission factor is

$$F = 4 \times 0.96 \log_{10}\left(\frac{10{,}685{,}214}{1.4125 \times 22.13}\right) = 21.25$$

From the above two values of F, using the smaller number, we get the AGA transmission factor as $F = 20.01$.

Therefore, the corresponding friction factor f is found from Eqn (2.42) as

$$\frac{2}{\sqrt{f}} = 20.01 \quad \text{or} \quad f = 0.0100$$

Example 5.21
Using the AGA method, calculate the transmission factor and friction factor for gas flow in a DN 500 pipeline with 12-mm wall thickness. The flow rate is 6 Mm³/day and gas gravity = 0.6 and viscosity = 0.00012 poise. The pipe absolute pipe roughness is 0.02 mm. Assume a bend index of 60°, base pressure = 101 kPa, and base temperature = 15 °C. For a 60-km pipe length, calculate the upstream pressure needed to hold a downstream pressure of 5 MPa (absolute). Assume flow temperature = 20 °C and compressibility factor $Z = 0.85$. Neglect elevation effects.

Solution
Inside diameter of pipe = $500 - 2 \times 12 = 476$ mm.
The base temperature = $15 + 273 = 288$ K.
Gas flowing temperature = $20 + 273 = 293$ K.
We first calculate the Reynolds number.

$$R = 0.5134 \left(\frac{101}{288}\right)\left(\frac{0.6 \times 6 \times 10^6}{0.00012 \times 476}\right) = 11{,}347{,}470$$

Next calculate the two transmission factors as follows:
The fully turbulent transmission factor is

$$F = 4 \log_{10}\left(\frac{3.7 \times 476}{0.02}\right) = 19.78$$

For the smooth pipe zone, the Von Karman transmission factor is

$$F_t = 4 \log_{10}\left(\frac{11{,}347{,}470}{F_t}\right) - 0.6$$

Solving by successive iteration we get $F_t = 22.23$.
From Table 5.3, for a bend index of $60°$, the drag factor is 0.96.
Therefore, for the partially turbulent flow zone, the transmission factor is

$$F = 4 \times 0.96 \log_{10}\left(\frac{11{,}347{,}470}{1.4125 \times 22.23}\right) = 21.34$$

Using the smaller of the two values of F, the AGA transmission factor is

$$F = 19.78$$

Therefore, the corresponding friction factor is

$$\frac{2}{\sqrt{f}} = 19.78 \quad \text{or} \quad f = 0.0102$$

Using the general flow equation, we calculate the upstream pressure P_1 as follows:

$$6 \times 10^6 = 5.747 \times 10^{-4} \times 19.78 \times \left(\frac{288}{101}\right)\left[\frac{P_1^2 - 5000^2}{0.6 \times 293 \times 60 \times 0.85}\right]^{0.5} \times 476^{2.5}$$

Solving for P_1 we get

$$P_1 = 6130 \text{ kPa} = 6.13 \text{ MPa}$$

27. WEYMOUTH EQUATION

The Weymouth equation is used for high-pressure, high flow rate, and large-diameter gas-gathering systems. This formula directly calculates the flow rate through a pipeline for given values of gas gravity, compressibility, inlet and outlet pressures, pipe diameter, and length. In USCS units, the Weymouth equation is stated as follows:

$$Q = 433.5E \left(\frac{T_b}{P_b}\right) \left(\frac{P_1^2 - e^s P_2^2}{GT_f L_e Z}\right)^{0.5} D^{2.667} \quad (5.115)$$

where
Q – volume flow rate, standard ft^3/day (SCFD)
E – pipeline efficiency, a decimal value less than or equal to 1.0
P_b – base pressure, psia
T_b – base temperature, °R (460 + °F)
P_1 – upstream pressure, psia
P_2 – downstream pressure, psia
G – gas gravity (air = 1.00)
T_f – average gas flow temperature, °R (460 + °F)
L_e – equivalent length of pipe segment, miles
Z – gas compressibility factor, dimensionless
D – pipe inside diameter, in

Where the equivalent length L_e and s were defined earlier. By comparing the Weymouth equation with the general flow equation, we can isolate an equivalent transmission factor as follows.

The Weymouth transmission factor in USCS units is

$$F = 11.18(D)^{1/6} \quad (5.116)$$

In SI units, the Weymouth equation is as follows:

$$Q = 3.7435 \times 10^{-3} E \left(\frac{T_b}{P_b}\right) \left(\frac{P_1^2 - e^s P_2^2}{GT_f L_e Z}\right)^{0.5} D^{2.667} \quad (5.117)$$

where
Q – gas flow rate, standard condition m^3/day
T_b – base temperature, K (273 + °C)
P_b – base pressure, kPa
T_f – average gas flow temperature, K (273 + °C)
P_1 – upstream pressure, kPa

P_2 – downstream pressure, kPa
L_e – equivalent length of pipe segment, km
Other symbols are as defined previously.
The Weymouth transmission factor in SI units is

$$F = 6.521(D)^{1/6} \qquad (5.118)$$

You will notice that a pipeline efficiency factor E is used in the Weymouth equation so that we can compare the throughput performance of a pipeline using the general flow equation that does not include an efficiency factor.

Example 5.22
Calculate the flow rate using Weymouth equation in a gas pipeline system, 15 miles long, NPS 12 pipe with 0.250-in wall thickness at an efficiency of 0.95. The upstream pressure is 1200 psia and the delivery pressure required at the end of the pipe segment is 750 psia. Use gas gravity = 0.59 and viscosity = 0.000008 lb/ft-sec. Flowing temperature of gas = 75 °F. Base pressure = 14.7 psia and base temperature = 60 °F. Assume compressibility factor to be 0.94.

Neglect elevation difference along pipe. How does this compare with the flow rate calculated using the general flow equation with the Colebrook friction factor? Assume a pipe roughness of 700 micro in.

Solution
Using Eqn (5.115), we get the flow rate for Weymouth equation as follows:

$$Q = 433.5 \times 0.95 \left(\frac{60+460}{14.7}\right) \left(\frac{1200^2 - 750^2}{0.59 \times (75+460) \times 15 \times 0.94}\right)^{0.5} 12.25^{2.667}$$

$$Q = 163{,}255{,}858 \text{ SCFD}$$

Or

$$Q = 163.26 \text{ MMSCFD}$$

Next we will calculate the Reynolds number.

$$R = \frac{0.0004778 \times Q \times 0.59 \times 14.7}{12.25 \times 0.000008 \times 520}$$

where Q is the flow rate in SCFD.

Simplifying, we get R = 0.0813 Q.

Because Q is unknown, we will first assume a transmission factor F = −20 and calculate the flow rate from the general flow equation.

$$Q = 38.77 \times 20 \left(\frac{520}{14.7}\right) \left[\frac{1200^2 - 750^2}{0.59 \times 535 \times 15 \times 0.94}\right]^{0.5} \times 12.25^{2.5}$$

$$= 202{,}284{,}747 \text{ SCFD} \quad \text{Or} \quad Q = 202.28 \text{ MMSCFD}$$

Next, we will calculate the Reynolds number and the transmission factor based on this flow rate as

$$R = 0.0813 \times 202{,}284{,}747 = 16.45 \text{ million}$$

And using transmission factor equation.

$$F = -4 \log_{10} \left(\frac{700 \times 10^{-6}}{3.7 \times 12.25} + \frac{1.255F}{16.45 \times 10^6}\right)$$

Solving for F, we get:

$$F = 19.09$$

Using this value the revised flow rate is found by proportion as

$$Q = 202.28 \times \frac{19.09}{20} = 193.08 \text{ MMSCFD}$$

Repeating calculation of R and F, we get:

$$R = 16.45 \times \frac{193.08}{202.28} = 15.7 \text{ million}$$

and

$$F = -4 \log_{10} \left(\frac{700 \times 10^{-6}}{3.7 \times 12.25} + \frac{1.255F}{15.7 \times 10^6}\right)$$

Therefore, F = 19.08.

This is fairly close to the previous value of F = 19.09; therefore, we will use this value and calculate the flow rate as

$$Q = 202.28 \times \frac{19.08}{20} = 192.98 \text{ MMSCFD}$$

Comparing this result using the general flow equation with that calculated using the Weymouth equation, we see that the latter equation is quite conservative.

28. PANHANDLE A EQUATION

The Panhandle A equation was developed for use in natural gas pipeline incorporating an efficiency factor, for Reynolds numbers in the range of 5 million to 11 million. In this equation, the pipe roughness is not used. Sometimes the compressibility factor is also not used. The general form of the Panhandle A equation is expressed in USCS units as follows:

$$Q = 435.87 E \left(\frac{T_b}{P_b}\right)^{1.0788} \left(\frac{P_1^2 - e^s P_2^2}{G^{0.8539} T_f L_e Z}\right)^{0.5394} D^{2.6182} \quad (5.119)$$

where
  Q – volume flow rate, standard ft^3/day (SCFD)
  E – pipeline efficiency, a decimal value less than 1.0
  P_b – base pressure, psia.
  T_b – base temperature, °R (460 + °F)
  P_1 – upstream pressure, psia
  P_2 – downstream pressure, psia
  G – gas gravity (air = 1.00)
  T_f – average gas flow temperature, °R (460 + °F)
  L_e – equivalent length of pipe segment, miles
  Z – gas compressibility factor, dimensionless
  D – pipe inside diameter, in
Other symbols are as defined previously.
In SI units, the Panhandle A equation is

$$Q = 4.5965 \times 10^{-3} E \left(\frac{T_b}{P_b}\right)^{1.0788} \left(\frac{P_1^2 - e^s P_2^2}{G^{0.8539} T_f L_e Z}\right)^{0.5394} D^{2.6182}$$

$$(5.120)$$

where
  Q – gas flow rate, standard condition m^3/day
  E – pipeline efficiency, a decimal value less than 1.0
  T_b – base temperature, K (273 + °C)
  P_b – base pressure, kPa
  T_f – average gas flow temperature, K (273 + °C)
  P_1 – upstream pressure, kPa (absolute)
  P_2 – downstream pressure, kPa (absolute)
  L_e – equivalent length of pipe segment, km
Other symbols are as defined previously.

Because of the exponents involved in this equation, all pressures must be in kPa.

By comparing the Panhandle A equation with the general flow equation, we can calculate an equivalent transmission factor in USCS units as follows:

$$F = 7.2111 E \left(\frac{QG}{D}\right)^{0.07305} \quad \text{(USCS)} \quad (5.121)$$

And in SI Units, it is

$$F = 11.85 E \left(\frac{QG}{D}\right)^{0.07305} \quad \text{(SI)} \quad (5.122)$$

Sometimes the transmission factor is used to compare the results of calculations using the general flow equation and the Panhandle A equation.

Example 5.23

Using the Panhandle A Equation, calculate the outlet pressure in a natural gas pipeline, NPS 16 with a 0.250-in wall thickness, 15 miles long. The gas flow rate is 100 MMSCFD at an inlet pressure of 1000 psia. The gas gravity = 0.6 and viscosity = 0.000008 lb/ft-sec. The average gas temperature is 80 °F. Assume base pressure = 14.73 psia and base temperature = 60 °F. For compressibility factor Z, use the CNGA method. Assume pipeline efficiency of 0.92.

Solution

The average pressure P_{avg} needs to be calculated before the compressibility factor Z can be determined. Because the inlet pressure $P_1 = 1000$ psia and the outlet pressure P_2 is unknown, we will have to assume a value of P_2 (such as 800 psia) and calculate P_{avg} and then calculate the value of Z. Once Z is known, from the Panhandle A equation we can calculate the outlet pressure P_2. Using this value of P_2, a better approximation for Z is calculated from a new P_{avg}. This process is repeated until successive values of P_2 are within allowable limits, such as 0.1 psia.

Assume $P_2 = 800$ psia. The average pressure is

$$P_{avg} = \frac{2}{3}\left(1000 + 800 - \frac{1000 \times 800}{1000 + 800}\right) = 903.7 \text{ psia}$$

Next we calculate the compressibility factor Z using CNGA method.

$$Z = \frac{1}{1 + \dfrac{(903.7 - 14.73) \times 3.444 \times 10^5 \times (10)^{1.785 \times 0.6}}{(80 + 460)^{3.825}}}$$

Or

$$Z = 0.8869$$

From the Panhandle A Eqn (5.119), substituting given values, neglecting elevations, we get

$$100 \times 10^6 = 435.87 \times 0.92 \left(\frac{60+460}{14.73}\right)^{1.0788}$$
$$\times \left(\frac{1000^2 - P_2^2}{(0.6)^{0.8539}(540 \times 15 \times 0.8869)}\right)^{0.5394} (15.5)^{2.6182}$$

Solving for P_2, we get

$$P_2 = 968.02 \text{ psia}$$

Because this is different from the assumed value of $P_2 = 800$, we recalculate the average pressure.and Z using $P_2 = 968.02$ psia

Revised average pressure is

$$P_{avg} = \frac{2}{3}\left(1000 + 968.02 - \frac{1000 \times 968.02}{1000+968.02}\right) = 984.10 \text{ psia}$$

Using this value of P_{avg}, we recalculate Z as

$$Z = \frac{1}{1 + \frac{(984.10-14.73)\times 3.444 \times 10^5 \times (10)^{1.785 \times 0.6}}{(80+460)^{3.825}}}$$

Or

$$Z = 0.8780$$

Recalculating P_2, we get

$$100 \times 10^6 = 435.87 \times 0.92 \left(\frac{60+460}{14.73}\right)^{1.0788}$$
$$\times \left(\frac{1000^2 - P_2^2}{(0.6)^{0.8539}(540 \times 15 \times 0.8780)}\right)^{0.5394} (15.5)^{2.6182}$$

Solving for P_2, we get

$$P_2 = 968.35 \text{ psia}$$

This is within 0.5 psi of the previously calculated value. Hence, we will not continue the iteration any further. Therefore, the outlet pressure is 968.35 psia.

Example 5.24

Using the Panhandle A equation, calculate the inlet pressure required in a natural gas pipeline, DN 300 with a 6-mm wall thickness, 24 km long, for a gas flow rate of 3.5 Mm^3/day.

The gas gravity $= 0.6$ and viscosity $= 0.000119$ poise. The average gas temperature is 20 °C. The delivery pressure is 6000 kPa (absolute). Assume base pressure $= 101$ kPa, base temperature $= 15$ °C, and compressibility factor $Z = 0.90$ with a pipeline efficiency of 0.92.

Solution

Pipe inside diameter $D = 300 - 2 \times 6 = 288$ mm.

Gas flow temperature $= 20 + 273 = 293$ K

Using Panhandle A Eqn (5.120) and neglecting elevation effect, we substitute:

$$3.5 \times 10^6 = 4.5965 \times 10^{-3} \times 0.92 \left(\frac{15 + 273}{101}\right)^{1.0788}$$

$$\times \left(\frac{P_1^2 - 6000^2}{(0.6)^{0.8539}(293 \times 24 \times 0.9)}\right)^{0.5394} (288)^{2.6182}$$

Solving for inlet pressure, we get
$P_1^2 - (6000)^2 = 19{,}812{,}783$
Or
$P_1 = 7471$ kPa (absolute)

29. PANHANDLE B EQUATION

The Panhandle B equation also known as the revised Panhandle equation is used in large-diameter, high-pressure transmission lines. In fully turbulent flow, it is found to be accurate for values of Reynolds number in the range of 4 million to 40 million. This equation in USCS units is as follows:

$$Q = 737 E \left(\frac{T_b}{P_b}\right)^{1.02} \left(\frac{P_1^2 - e^s P_2^2}{G^{0.961} T_f L_e Z}\right)^{0.51} D^{2.53} \qquad (5.123)$$

where
 Q – volume flow rate, standard ft^3/day (SCFD)
 E – pipeline efficiency, a decimal value less than 1.0
 P_b – base pressure, psia
 T_b – base temperature, °R (460 + °F)
 P_1 – upstream pressure, psia

P_2 – downstream pressure, psia
G – gas gravity (air = 1.00)
T_f – average gas flow temperature, °R (460 + °F)
L_e – equivalent length of pipe segment, miles
Z – gas compressibility factor, dimensionless
D – pipe inside diameter, in
Other symbols are as defined previously.
In SI units, the Panhandle B equation is

$$Q = 1.002 \times 10^{-2} E \left(\frac{T_b}{P_b}\right)^{1.02} \left(\frac{P_1^2 - e^s P_2^2}{G^{0.961} T_f L_e Z}\right)^{0.51} D^{2.53} \qquad (5.124)$$

where
Q – gas flow rate, standard condition m³/day
E – pipeline efficiency, a decimal value less than 1.0
T_b – base temperature, K (273 + °C)
P_b – base pressure, kPa
T_f – average gas flow temperature, K (273 + °C)
P_1 – upstream pressure, kPa (absolute)
P_2 – downstream pressure, kPa (absolute)
L_e – equivalent length of pipe segment, km
Z – gas compressibility factor at the flowing temperature, dimensionless
Other symbols are as defined previously.
The equivalent transmission factor for Panhandle B equation in USCS is given by.

$$F = 16.7 E \left(\frac{QG}{D}\right)^{0.01961} \qquad (5.125)$$

In SI units, it is

$$F = 19.08 E \left(\frac{QG}{D}\right)^{0.01961} \qquad (5.126)$$

Example 5.25
Using the Panhandle B equation, calculate the outlet pressure in a natural gas pipeline, NPS 16 with a 0.250-in wall thickness, 15 miles long. The gas flow rate is 100 MMSCFD at 1000 psia inlet pressure. The gas gravity = 0.6 and viscosity = 0.000,008 lb/ft-sec. The average gas temperature is 80 °F. Assume base pressure = 14.73 psia and base temperature = 60 °F. Compressibility factor Z = 0.90 and pipeline efficiency is 0.92.

Solution
Inside diameter of pipe $= 16 - 2 \times 0.25 = 15.5$ in
Gas flow temperature $= 80 + 460 = 540\,°R$
Using the Panhandle B equation, substituting given value, we get

$$100 \times 10^6 = 737 \times 0.92 \left(\frac{60+460}{14.73}\right)^{1.02}$$

$$\times \left(\frac{1000^2 - P_2^2}{(0.6)^{0.961}(540 \times 15 \times 0.90)}\right)^{0.51} (15.5)^{2.53}$$

Solving for P_2, we get

$$1000^2 - P_2^2 = 60{,}778$$

$$P_2 = 969.13 \text{ psia}$$

Compare this with the results of Panhandle A equation in example (5.23), where the outlet pressure $P_2 = 968.35$ psia. Therefore, the Panhandle B equation gives a slightly lower pressure drop compared with that from Panhandle A equation. In other words, Panhandle A is more conservative and will give a lower flow rate for the same pressures compared with Panhandle B. In this example, we use the constant value of $Z = 0.9$, whereas in example 5.23, Z was calculated using the CNGA equation as $Z = 0.8780$. If we factor this in, the result for the outlet pressure in this example will be 969.9 psia, which is not too different from the calculated value of 969.13 psia.

Example 5.26
Using the Panhandle B equation, calculate the inlet pressure in a natural gas pipeline, DN 300 with 6-mm wall thickness, 24 km long. The gas flow rate is 3.5 Mm³/day and the gas gravity $= 0.6$ and viscosity $= 0.000119$ poise. The average gas temperature is 20 °C and the delivery pressure is 6000 kPa (absolute). Assume base pressure $= 101$ kPa, base temperature $= 15\,°C$, and compressibility factor $Z = 0.90$. The pipeline efficiency is 0.92.

Solution
Inside diameter of pipe $= 300 - 2 \times 6 = 288$ mm.
Gas flow temperature $= 20 + 273 = 293$ K.
Neglecting elevations, using Panhandle B equation, we get

$$3.5 \times 10^6 = 1.002 \times 10^{-2} \times 0.92 \left(\frac{15+273}{101}\right)^{1.02}$$

$$\times \left(\frac{P_1^2 - 6000^2}{(0.6)^{0.961}(293 \times 24 \times 0.9)}\right)^{0.51} 288^{2.53}$$

Solving for the inlet pressure P_1, we get
$P_1^2 - (6000)^2 = 19{,}945{,}469$
$P_1 = 7480$ kPa (absolute)

Compare this with the results of Panhandle A equation in example (5.24), where the inlet pressure $P_1 = 7471$ kPa (absolute). Again, we see that Panhandle B equation gives a slightly lower pressure drop compared with that obtained from Panhandle A equation.

30. INSTITUTE OF GAS TECHNOLOGY EQUATION

The Institute of Gas Technology (IGT) equation proposed by the IGT is also known as the IGT distribution equation and is stated as follows for USCS units.

$$Q = 136.9E \left(\frac{T_b}{P_b}\right) \left(\frac{P_1^2 - e^s P_2^2}{G^{0.8} T_f L_e \mu^{0.2}}\right)^{0.555} D^{2.667} \quad (5.127)$$

where
  Q – volume flow rate, standard ft^3/day (SCFD)
  E – pipeline efficiency, a decimal value less than 1.0
  P_b – base pressure, psia
  T_b – base temperature, °R (460 + °F)
  P_1 – upstream Pressure, psia
  P_2 – downstream Pressure, psia
  G – gas gravity (air = 1.00)
  T_f – average gas flow temperature, °R (460 + °F)
  L_e – equivalent length of pipe segment, miles
  Z – gas compressibility factor, dimensionless
  D – pipe inside diameter, in
  μ – gas viscosity, lb/ft-s
  Other symbols are as defined previously.

In SI units, the IGT equation is expressed as follows:

$$Q = 1.2822 \times 10^{-3} E \left(\frac{T_b}{P_b}\right) \left(\frac{P_1^2 - e^s P_2^2}{G^{0.8} T_f L_e \mu^{0.2}}\right)^{0.555} D^{2.667} \quad (5.128)$$

where
  Q – gas flow rate, standard condition m^3/day
  E – pipeline efficiency, a decimal value less than 1.0
  T_b – base temperature, K (273 + °C)

Fluid Flow in Pipes

P_b – base pressure, kPa
T_f – average gas flow temperature, K (273 + °C)
P_1 – upstream pressure, kPa (absolute)
P_2 – downstream pressure, kPa (absolute)
L_e – equivalent length of pipe segment, km
μ – gas viscosity, poise
Other symbols are as defined previously.

Example 5.27
Using the IGT equation, calculate the flow rate in a natural gas pipeline, NPS 16 with a 0.250-in wall thickness, 15 miles long. The inlet and outlet pressure are 1000 psig and 800 psig, respectively. The gas gravity = 0.6 and viscosity = 0.000,008 lb/ft-s. The average gas temperature is 80 °F, the base pressure = 14.7 psia, and base temperature = 60 °F. Compressibility factor Z = 0.90 and the pipeline efficiency is 0.95.

Solution
Inside diameter of pipe = 16 − 2 × 0.25 = 15.5 in
 The pressures given are in psig and they must be converted to absolute pressures.
 Therefore
$P_1 = 1000 + 14.7 = 1014.7$ psia
$P_2 = 800 + 14.7 = 814.7$ psia
$T_b = 60 + 460 = 520\ °R$
$T_f = 80 + 460 = 540\ °R$
 Substituting in IGT equation, we get

$Q = 136.9$

$$\times 0.95 \left(\frac{520}{14.7}\right) \left(\frac{1014.7^2 - 814.7^2}{(0.6)^{0.8} \times 540 \times 15 \times (8 \times 10^{-6})^{0.2}}\right)^{0.555} 15.5^{2.667}$$

$Q = 263.1 \times 10^6\ \text{ft}^3/\text{day} = 263.1$ MMSCFD

Therefore the flow rate is 263.1 MMSCFD.

Example 5.28
A natural gas pipeline, DN 400 with 6-mm wall thickness, 24 km long is used to transports gas at an inlet pressures of 7000 kPa (gauge) and an outlet pressure of 5500 kPa (gauge). The gas gravity = 0.6 and viscosity = 0.000,119 poise. The average gas temperature is 20 °C. Assume

base pressure = 101 kPa and base temperature = 15 °C. Compressibility factor Z = 0.90 and pipeline efficiency is 0.95.
1. Calculate the flow rate using the IGT equation
2. What are the gas velocities at inlet and outlet?
3. If the velocity must be limited to 10 m/s, what should be the minimum pipe size be, assuming the flow rate and inlet pressure remain constant?

Solution
Inside diameter of pipe D = 400 − 2 × 6 = 388 mm.
All pressures are given in gauge values and must be converted to absolute values.
Inlet pressure P_1 = 7000 + 101 = 7101 kPa (absolute)
Outlet pressure P_2 = 5500 + 101 = 5601 kPa (absolute)
Base temperature T_b = 15 + 273 = 288 K
Flowing temperature T_f = 20 + 273 = 293 K
From the IGT equation, we get the flow rate in m^3/day as

$$Q = 1.2822 \times 10^{-3} \times 0.95 \left(\frac{288}{101}\right) \left(\frac{7101^2 - 5601^2}{(0.6)^{0.8} \times 293 \times 24 \times (1.19 \times 10^{-4})^{0.2}}\right)^{0.555} (388)^{2.667}$$

Or

$$Q = 7{,}665{,}328 \ m^3/day = 7.67 \ Mm^3/day$$

1. Therefore, the flow rate is 7.67 Mm^3/day
2. Next, we calculate the average velocity of the gas at the inlet and outlet pressures as

$$\text{Inlet velocity } u_1 = 14.7349 \left(\frac{7.67 \times 10^6}{388^2}\right) \left(\frac{101}{288}\right) \left(\frac{0.9 \times 293}{7101}\right)$$

$$= 9.78 \ m/s$$

In the preceding we assumed a constant compressibility factor, Z = 0.9. Similarly, at the outlet pressure, the average gas velocity is

$$\text{Outlet velocity } u_2 = 14.7349 \left(\frac{7.67 \times 10^6}{388^2}\right) \left(\frac{101}{288}\right) \left(\frac{0.9 \times 293}{5601}\right)$$

$$= 12.4 \ m/s$$

3. Because the velocity must be limited to 10 m/s, the pipe diameter must be increased. Increasing the pipe diameter will also increase the outlet pressure, if we keep both the flow rate and inlet pressure the same as

before. The increased outlet pressure will also reduce the gas velocity. We will try DN 450 pipe with 10-mm wall thickness.

Inside diameter of pipe $D = 450 - 2 \times 10 = 430$ mm.

Assuming P_1 and Q are the same as before, we calculate the new outlet pressure P_2 as

$$7.67 \times 10^6 = 1.2822 \times 10^{-3} \times 0.95 \left(\frac{288}{101}\right)$$
$$\times \left(\frac{7101^2 - P_2^2}{(0.6)^{0.8} \times 293 \times 24 \times (1.19 \times 10^{-4})^{0.2}}\right)^{0.555} (430)^{2.667}$$

Solving for P_2, we get

$$P_2 = 6228 \text{ kPa}$$

The new velocity at the outlet will be

$$u_2 = 14.7349 \left(\frac{7.67 \times 10^6}{430^2}\right)\left(\frac{101}{288}\right)\left(\frac{0.9 \times 293}{6228}\right) = 9.08 \text{ m/s}$$

Because this is less than 10 m/s specified, the DN 450 pipe is satisfactory.

In the preceding calculations, we assumed the same compressibility factor for both inlet and outlet pressures. Actually, a more correct solution would be to calculate Z using the CNGA equation at both inlet and outlet conditions and using these values in the calculation of gas velocities. This is left as an exercise for the reader.

31. SPITZGLASS EQUATION

The Spitzglass equation has been around for many years and originally was used in fuel gas piping calculations. There are two versions of the Spitzglass equation. One equation is for low pressure (≤ 1 psig) and another is for high pressure (>1 psig). These equations have been modified to include a pipeline efficiency and compressibility factor.

The low-pressure (less than or equal to one psig) version of Spitzglass equation is

$$Q = 3.839 \times 10^3 E \left(\frac{T_b}{P_b}\right) \left(\frac{P_1 - P_2}{GT_f L_e Z \left(1 + \frac{3.6}{D} + 0.03D\right)}\right)^{0.5} D^{2.5}$$

(5.129)

where

Q – volume flow rate, standard ft^3/day (SCFD)
E – pipeline efficiency, a decimal value less than 1.0
P_b – base pressure, psia
T_b – base temperature, °R (460 + °F)
P_1 – upstream pressure, psia
P_2 – downstream pressure, psia
G – gas gravity (air = 1.00)
T_f – average gas flow temperature, °R (460 + °F)
L_e – equivalent length of pipe segment, miles
D – pipe inside diameter, in
Z – gas compressibility factor, dimensionless
Other symbols are as defined previously.

The low-pressure (less than 6.9 kPa) version of Spitzglass equation in SI units is

$$Q = 5.69 \times 10^{-2} E \left(\frac{T_b}{P_b}\right) \left(\frac{P_1 - P_2}{GT_f L_e Z \left(1 + \frac{91.44}{D} + 0.0012D\right)}\right)^{0.5} D^{2.5}$$

(5.130)

where

Q – gas flow rate, standard condition m^3/day
E – pipeline efficiency, a decimal value less than 1.0
T_b – base temperature, K (273 + °C)
P_b – base pressure, kPa
P_1 – upstream pressure, kPa (absolute)
P_2 – downstream pressure, kPa (absolute)
G – gas gravity (air = 1.00)
T_f – average gas flow temperature, K (273 + °C)
L_e – equivalent length of pipe segment, km
Z – gas compressibility factor, dimensionless
Other symbols are as defined previously.

The high-pressure (>1 psig) version in USCS units is as follows:

$$Q = 729.6087 E \left(\frac{T_b}{P_b}\right) \left(\frac{P_1^2 - e^s P_2^2}{GT_f L_e Z \left(1 + \frac{3.6}{D} + 0.03D\right)}\right)^{0.5} D^{2.5} \quad (5.131)$$

where
- Q – volume flow rate, standard ft^3/day (SCFD)
- E – pipeline efficiency, a decimal value less than 1.0
- P_b – base pressure, psia
- T_b – base temperature, °R (460 + °F)
- P_1 – upstream pressure, psia
- P_2 – downstream pressure, psia.
- G – gas gravity (air = 1.00)
- T_f – average gas flow temperature, °R (460 + °F)
- L_e – equivalent length of pipe segment, miles
- D – pipe inside diameter, in
- Z – gas compressibility factor, dimensionless

Other symbols are as defined previously.
In SI units, the high-pressure (>6.9 kPa) version of Spitzglass equation is

$$Q = 1.0815 \times 10^{-2} E \left(\frac{T_b}{P_b}\right) \left(\frac{P_1^2 - e^s P_2^2}{GT_f L_e Z \left(1 + \frac{91.44}{D} + 0.0012 D\right)}\right)^{0.5} D^{2.5} \tag{5.132}$$

where
- Q – gas flow rate, standard condition m^3/day
- E – pipeline efficiency, a decimal value less than 1.0
- T_b – base temperature, K (273 + °C)
- P_b – base pressure, kPa
- P_1 – upstream pressure, kPa (absolute)
- P_2 – downstream pressure, kPa (absolute)
- G – gas gravity (air = 1.00)
- T_f – average gas flow temperature, K (273 + °C)
- L_e – equivalent length of pipe segment, km
- Z – gas compressibility factor, dimensionless

Other symbols are as defined previously.

32. MUELLER EQUATION

The Mueller equation is another form of flow rate versus pressure relationship in gas pipelines. In USCS units, it is expressed as follows:

$$Q = 85.7368 E \left(\frac{T_b}{P_b}\right) \left(\frac{P_1^2 - e^s P_2^2}{G^{0.7391} T_f L_e \mu^{0.2609}}\right)^{0.575} D^{2.725} \tag{5.133}$$

where

Q – volume flow rate, standard ft³/day (SCFD)
E – pipeline efficiency, a decimal value less than 1.0
P_b – base pressure, psia
T_b – base temperature, °R (460 + °F)
P_1 – upstream pressure, psia
P_2 – downstream pressure, psia
G – gas gravity (air = 1.00)
T_f – average gas flow temperature, °R (460 + °F)
L_e – equivalent length of pipe segment, miles
D – pipe inside diameter, in
μ – gas viscosity, lb/ft-s
Other symbols are as defined previously.
In SI units, the Mueller equation is as follows:

$$Q = 3.0398 \times 10^{-2} E \left(\frac{T_b}{P_b}\right) \left(\frac{P_1^2 - e^s P_2^2}{G^{0.7391} T_f L_e \mu^{0.2609}}\right)^{0.575} D^{2.725} \quad (5.134)$$

where

Q – gas flow rate, standard condition m³/day
E – pipeline efficiency, a decimal value less than 1.0
T_b – base temperature, K (273 + °C)
P_b – base pressure, kPa
P_1 – upstream pressure, kPa (absolute)
P_2 – downstream pressure, kPa (absolute)
G – gas gravity (air = 1.00)
T_f – average gas flow temperature, K (273 + °C)
L_e – equivalent length of pipe segment, km
μ – gas viscosity, cP
Other symbols are as defined previously.

33. FRITZSCHE EQUATION

Fritzsche formula developed in Germany in 1908 has found extensive use in compressed air and gas piping. In USCS units, it is expressed as follows:

$$Q = 410.1688 E \left(\frac{T_b}{P_b}\right) \left(\frac{P_1^2 - P_2^2}{G^{0.8587} T_f L_e}\right)^{0.538} D^{2.69} \quad (5.135)$$

All symbols are as defined before.
In SI units:

$$Q = 2.827E \left(\frac{T_b}{P_b}\right) \left(\frac{P_1^2 - e^s P_2^2}{G^{0.8587} T_f L_e}\right)^{0.538} D^{2.69} \quad (5.136)$$

All symbols are as defined before.

34. EFFECT OF PIPE ROUGHNESS

In the preceding sections, we used the pipe roughness as a parameter in the friction factor and transmission factor calculations. Both the AGA and the Colebrook–White equations use the pipe roughness, whereas the Panhandle and Weymouth equations do not use the pipe roughness directly in the calculations. Instead, these equations use a pipeline efficiency to compensate for the internal conditions and age of the pipe. Therefore, when comparing the predicted flow rates or pressures using the AGA or Colebrook–White equations with the Panhandle or Weymouth equations, we can adjust the pipeline efficiency to correlate with the pipe roughness used in the former equations.

Because most gas pipelines operate in the turbulent zone, the laminar flow friction factor, which is independent of pipe roughness, is of little interest to us. Concentrating therefore on turbulent flow, we see that the Colebrook–White equation is affected by variation in pipe internal roughness. For example, suppose we want to compare an internally coated pipeline with an uncoated pipeline. The internal roughness of the coated pipe may be in the range of 100–200 micro in, whereas the uncoated pipe may have a roughness of 600–800 micro in or more. If the pipe is NPS 20 with a 0.500-in wall thickness, the relative roughness using the lower roughness value are as follows:

$$\text{For coated pipe} \frac{e}{D} = \frac{100 \times 10^{-6}}{19} = 5.263 \times 10^{-6}$$

and

$$\text{For uncoated pipe} \frac{e}{D} = \frac{600 \times 10^{-6}}{19} = 1.579 \times 10^{-5}$$

Substituting these values of relative roughness in Colebrook–White equation and using a Reynolds number of 10 million, we calculate the following transmission factors.

$$F = 21.54 \quad \text{for coated pipe}$$

And

$$F = 20.65 \quad \text{for uncoated pipe}$$

Because the flow rate is directly proportional to the transmission factor F, from the general flow equation, we see that the coated pipe will be able to transport.

$\frac{21.54-20.65}{20.65} = 0.043 = 4.3\%$, more flow rate than the uncoated pipe, if all other parameters remain the same. This is true in the fully turbulent zone where Reynolds number has little effect on friction factor f and the transmission factor F. However, in the smooth pipe zone, pipe roughness has less effect on the friction factor and the transmission factor. This is evident from the Moody diagram Figure 5.9.

Using a Reynolds number of 10^6, we find from Moody diagram Figure 5.9, for coated pipe, that

$$f = 0.0118 \quad \text{and} \quad F = 18.41$$

And for the uncoated pipe

$$f = 0.0122 \quad \text{and} \quad F = 18.10$$

Therefore, the increase in flow rate in this case will be

$$\frac{18.41 - 18.10}{18.10} = 0.017 = 1.7 \text{ percent}$$

Thus the impact of pipe roughness is less in the smooth pipe zone or for lower Reynolds number. A similar comparison can be made using the AGA equation.

Figure 5.10 shows the effect of pipe roughness on the pipeline flow rate considering the AGA and Colebrook–White equations. The graph is based on NPS 20 pipe, 0.500-in wall thickness, 120-miles long with 1200 psig upstream pressure and 800 psig downstream pressure. The flowing temperature of gas is 70 °F.

It can be seen that as the pipe roughness is increased from 200 micro in to 800 micro in, the flow rate decreases as follows.

224 MMSCFD to 206 MMSCFD for the Colebrook–White equation.

Fluid Flow in Pipes

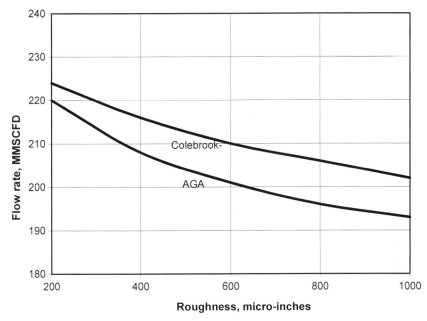

Figure 5.10 Effect of pipe roughness.

and

220 MMSCFD to 196 MMSCFD for the AGA equation.

We can therefore conclude that decreasing the pipe roughness directly results in throughput increase in a pipeline. However, the cost of internally coating a pipe to reduce the pipe roughness must be weighed against the revenue increase because of enhanced flow rate.

35. COMPARISON OF FLOW EQUATIONS

In the preceding sections, we calculated the flow rates and pressures in gas pipelines using the various flow equations. Each equation is slightly different from the other and some equations consider the pipeline efficiency while others use an internal pipe roughness value. How do these equations compare when predicting flow rates through a given pipe size when the upstream or downstream pressure is held constant? Obviously, some equations will predict higher flow rates for the same pressures than others. Similarly, if we start with a fixed upstream pressure in a pipe segment at a given flow rate, these equations will predict different downstream pressures. This indicates that some equations calculate higher pressure drops for the same flow rate

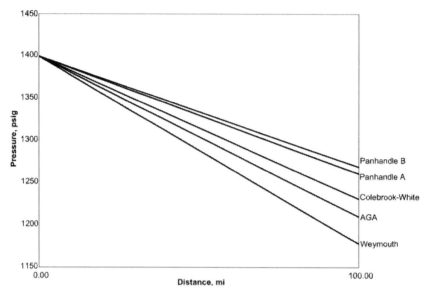

Figure 5.11 Comparison of flow equations.

than others. Figure 5.11 and Figure 5.12 show some of these comparisons when using the AGA, Colebrook–White, Panhandle, and Weymouth equations.

In Figure 5.11, we consider a pipeline 100 miles long, NPS 16 with 0.250-in wall thickness operating at a flow rate of 100 MMSCFD.

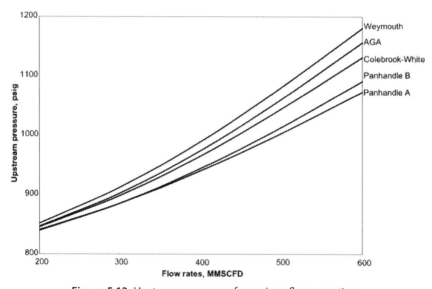

Figure 5.12 Upstream pressures for various flow equations.

The gas flowing temperature is 80 °F. With the upstream pressure fixed at 1400 psig, the downstream pressure was calculated using the different flow equations. By examining Figure 5.11, it is clear that the highest pressure drop is predicted by the Weymouth equation and the lowest pressure drop is predicted by the Panhandle B equation. We used a pipe roughness of 700 micro in for both the AGA and Colebrook equations, whereas a pipeline efficiency of 0.95 was used in the Panhandle and Weymouth equations.

Figure 5.12 shows a comparison of the flow equations from a different perspective. In this case, we calculated the upstream pressure required for NPS 30 pipeline, 100 miles long holding the delivery pressure constant at 800 psig. The upstream pressure required for various flow rates ranging from 200 MMSCFD to 600 MMSCFD were calculated using the five flow equations. Again, it can be seen that Weymouth equation predicts the highest upstream pressure at any flow rate, whereas the Panhandle A equation calculates the least pressure. We therefore conclude that the most conservative flow equation that predicts the highest pressure drop is the Weymouth equation and the least conservative flow equation is the Panhandle A.

36. SUMMARY

In this chapter, we defined pressure and how it is measured in both static and dynamic context. The velocity and Reynolds number calculations for pipe flow were introduced and the use of Reynolds number in classifying liquid flow as laminar, critical, and turbulent were explained. Existing methods of calculating the pressure drop because of friction in a pipeline using the Darcy–Weisbach equation were discussed and illustrated using examples. The importance of the Moody diagram was explained. Also, the trial-and-error solutions of friction factor from the Colebrook–White equation were covered. The use of Hazen–Williams and MIT pressure drop equations were discussed. Minor losses in pipelines from valves, fittings, pipe enlargements, and pipe contractions were analyzed. The concept of drag reduction as a means of reducing frictional head loss was also introduced.

Also, we introduced the various methods of calculating the pressure drop in a pipeline transporting gas and gas mixtures. The more commonly used equations for pressure drop in gas pipelines versus flow rate and pipe size were discussed and illustrated using examples. The effect of elevation changes was explained and the concept of Reynolds number, friction factor, and transmission factor were introduced. We compared the more commonly used pressure drop equations such as AGA, Colebrook–White, Weymouth, and Panhandle equations. The use of pipeline efficiency in comparing various equations was illustrated using an example. The average velocity of gas flow was introduced and the limiting value of erosional velocity was discussed.

CHAPTER SIX

Pressure Required to Transport

In this chapter, we will use the concepts of pressure drop calculations developed in an earlier chapter to determine the total pressure required for transporting a fluid (liquid or gas) in a pipeline that is constructed and operated in various configurations, such as series and parallel pipelines. We will discuss the various components that make up this total pressure required and how the pressure is influenced by the flow rate, fluid properties, and the pipeline elevation profile. Transmission pipelines may have a fixed flow rate throughout the pipeline, or an initial flow rate at the origin, followed by several additional injections or deliveries along the length of the line. The intermediate injections may be fluids of different compositions (specific gravity and viscosity) and therefore it will be necessary to calculate the combined properties of the fluid mixture for pressure drop calculations. The effect of intermediate delivery volumes and injection rates along a gas pipeline, the impact of contract delivery pressures, and the necessity of regulating pressures using a control valve or pressure regulators will also be analyzed.

In liquid pipelines, we will introduce the concept of system head curves and how they are developed. In a subsequent chapter, the interaction between the system head curve and the pump head curve will be examined. Thermal effects resulting from heat transfer between the fluid in the pipe and the surrounding soil in a buried pipeline, soil temperatures, soil thermal conductivities, and the Joule–Thompson effect (in gases) will be explained. In series and parallel piping systems, we will discuss the concepts of equivalent length and equivalent diameters that will help simplify calculations. To increase pipeline throughput, pipe loops are sometimes installed in liquid and gas pipelines. We will compare different pipe looping scenarios to accomplish this task and review the concept of the hydraulic pressure gradient. In gas pipeline systems, the important concept of line pack and how it is calculated will be explained.

1. TOTAL PRESSURE DROP REQUIRED TO PUMP A GIVEN VOLUME OF FLUID THROUGH A PIPELINE

In general, the pressure required at the origin of the pipeline to transport a specified volume of fluid from point A to point B will consist of the following components.

1. Pressure required to compensate for pipe friction.
2. Pressure required to account for the elevation difference between the ends of the pipeline.
3. Certain minimum pressure required to prevent vaporization of the product in a liquid pipeline.
4. Pressure required to ensure some minimum contract delivery pressure at the end of the pipeline.

In some cases where the pipeline traverses mountainous regions with considerable elevation differences, we must also take into account the minimum pressure in a pipeline so that vaporization of liquid does not occur. For example, in a liquefied petroleum gas (LPG) pipeline, we may be required to maintain a minimum pipeline pressure of 250–300 psig depending upon the flowing temperature. If a high vapor pressure liquid vaporizes in a pipeline, this may result in two-phase flow, which causes higher pressure drop and therefore more pumping power requirement, in addition to possible damage to pumping equipment. Thus, single-phase incompressible fluids must be pumped so that the pressure at any point in the pipeline does not drop below the vapor pressure of the liquid. Contractual requirements in a gas pipeline may also dictate that the gas pressure not drop below a certain minimum value.

Suppose the total frictional pressure drop in a liquid pipeline segment is 500 psig and the elevation difference contributes an additional 300 psig. If the delivery pressure at the end of the pipeline is set at 100 psig, the total pressure required at the origin of the pipeline is 500 + 300 + 100 = 900 psig.

When pumping gases, the pressure component from elevation may be a small quantity because the gas gravity is very small compared with that of a liquid. As explained in Chapter 5, the pipe segment elevation is taken into account in gas pipelines slightly differently compared with liquids. We will discuss this aspect in more detail in this chapter.

We will now discuss each of these components that make up the total pressure required by comparing the situation between a liquid pipeline and a gas pipeline.

2. FRICTIONAL COMPONENT

The frictional pressure drop results from the flow rate, fluid viscosity, and pipe roughness. It is similar in liquid and gas flow. The effect of friction was discussed in Chapter 5, where we introduced the internal roughness of pipe and how the friction factor was calculated using the Moody diagram, Colebrook–White and other equations. In gas flow, we discussed how the Weymouth and Panhandle equations took into account the internal conditions and age of the pipe by the pipeline efficiency factor rather than a friction factor. As we saw in Chapter 5, the pressure drop in a liquid pipeline increases with flow rate. The relationship is not linear. In fact, it can be seen from the Darcy equation and the other more practical equations, the pressure drop from friction varies as the square of the flow rate. Thus doubling the flow rate will result in four times the pressures drop from friction. Similarly in gas pipelines, the pressure drop varies very nearly as the square of the gas flow rate. The magnitude of the pressure drop from friction in a gas pipeline is generally smaller compared with that in a liquid pipeline. This is because efficient gas pipeline transportation requires keeping the average gas pressure in a pipe segment as high as possible. As pressure drops from expansion of gas, there is loss in efficiency. In a pipeline with multiple compressor stations, the lower the pressure at the downstream end of the first pipe segment, the higher will be the compression ratio at the downstream station (hence higher the horsepower (HP)) to boost the pressure for shipment downstream to the next compressor station in a long-distance gas pipeline. In this chapter, we will illustrate how to calculate the pressure drop resulting from friction in various pipe configurations that include flow injection, deliveries, and series and parallel piping.

3. EFFECT OF PIPELINE ELEVATION

The elevation component referred to earlier results from the difference in elevation along the pipeline that necessitates additional pressure for raising the fluid in the pipeline from one point to another. Of course, a drop in elevation will have the opposite effect of a rise in elevation. Thus, if the pipeline elevation increases from the origin to the delivery terminus, the effect of pipeline elevation and the frictional component are the same. In other words, they are additive. However, a pipeline that has elevation decrease from the beginning of the pipeline to the delivery terminus will provide pressure from gravity to assist the flow, compared with the frictional component that tries to impede the flow. Thus the friction component and the elevation component are of opposite sign.

In Section 1, the elevation component of 300 psig discussed depends on the static elevation difference between the beginning of the pipeline A and the delivery point B and the liquid specific gravity. In the case of a gas pipeline, the elevation component will depend on the static elevation differences between A and B as well as the gas gravity. However, the relationship between these parameters is more complex in a gas pipeline compared with a liquid pipeline. If the pipeline profile is a rolling terrain, the rise and fall in elevations between the origin A and the terminus B must be accounted for separately and summed up. Also, compared with a liquid, the density of a gas is several orders of magnitude lower and hence the influence of elevation is smaller in a gas pipeline.

Generally, if we were to breakdown the total pressure required in a gas pipeline into the components discussed earlier, we will find that the elevation component is very small. This is illustrated in the next example.

Problem 6.1

A gas pipeline Nominal Pipe Size (NPS) 36, with 0.500-in wall thickness, 100 miles long transports natural gas (specific gravity = 0.6 and viscosity = 0.000008 lb/ft-s) at a flow rate of 250 million standard cubic feet per day (MMSCFD) at an inlet temperature of 60 °F. Assuming isothermal flow, calculate the inlet pressure required, if the required delivery pressure at the pipeline terminus is 870 psig. The base pressure and base temperature are 14.7 psig and 60 °F, respectively. Use the Colebrook equation with pipe roughness of 0.0007 in.

Case A—Consider no elevation changes along the pipeline length.

Case B—Consider elevation changes as follows. Inlet elevation of 100 ft and elevation at delivery point 450 ft with elevation at the midpoint of 250 ft.

Solution

Inside diameter of pipe $D = 16 - 2 \times 0.250 = 15.5$ in.

First, we calculate the Reynolds number from Eqn (5.89).

$$R = 0.0004778 \left(\frac{14.7}{60 + 460}\right)\left(\frac{0.6 \times 100 \times 10^6}{0.000008 \times 15.5}\right) = 6,535,664$$

Next, using the Colebrook Eqn (2.39), we calculate the friction factor as

$$\frac{1}{\sqrt{f}} = -2Log_{10}\left(\frac{0.0007}{3.7 \times 15.5} + \frac{2.51}{6535664\sqrt{f}}\right)$$

Solving by trial and error we get
$f = 0.0109$

Therefore, the transmission factor is, using Eqn (5.97),

$$F = \frac{2}{\sqrt{0.0109}} = 19.1954$$

To calculate the compressibility factor, Z, the average pressure is required. Because the inlet pressure is unknown, we will calculate an approximate value of Z using a value of 110% of the delivery pressure for the average pressure.

The average pressure is

$$P_{avg} = 1.1 \times (870 + 14.7) = 973.17 \text{ psia}$$

Using the California Natural Gas Association (CNGA) Eqn (1.34), we calculate the value of compressibility factor as

$$Z = \frac{1}{\left[1 + \left(\frac{(973.17 - 14.7) \times 344400(10)^{1.785 \times 0.6}}{520^{3.825}}\right)\right]} = 0.8629$$

Case A

Because there is no elevation difference between the beginning and the end of the pipeline, the elevation component in Eqn (2.7) can be neglected and $e^s = 1$.

Outlet pressure is

$$P_2 = 870 + 14.7 = 884.7 \text{ psia.}$$

From the general flow Eqn (5.57), substituting given values we get

$$100 \times 10^6 = 38.77$$

$$\times 19.1954 \left(\frac{520}{14.7}\right) \left(\frac{P_1^2 - 884.7^2}{0.6 \times 520 \times 50 \times 0.8629}\right)^{0.5} (15.5)^{2.5}$$

Therefore, the upstream pressure is

$$P_1 = 999.90 \text{ psia} = 985.20 \text{ psig}$$

Using this value of P_1, we calculate the new average pressure using Eqn (5.69)

$$P_{avg} = \frac{2}{3}\left(999.9 + 884.7 - \frac{999.9 \times 884.7}{999.9 + 884.7}\right) = 943.47 \text{ psia}$$

compared with 973.17 we used for calculating Z. Recalculating Z using the new value of P_{avg}, we get

$$Z = \frac{1}{\left[1 + \left(\frac{(943.47 - 14.7) \times 344400(10)^{1.785 \times 0.6}}{520^{3.825}}\right)\right]} = 0.8666$$

This compares with 0.8629 we calculated earlier for Z. We will now recalculate the inlet pressure using this value of Z. From the general flow Eqn (5.57), we get

$$100 \times 10^6 = 38.77$$

$$\times\ 19.1954 \left(\frac{520}{14.7}\right) \left(\frac{P_1^2 - 884.7^2}{0.6 \times 520 \times 50 \times 0.8666}\right)^{0.5} (15.5)^{2.5}$$

Solving for the upstream pressure, we get

$$P_1 = 1000.36 \text{ psia} = 985.66 \text{ psig}$$

This is close enough to the previously calculated value 985.20 psig that no further iteration is needed. Therefore, the pressure required at the beginning of the pipeline in Case A is 985.66 psig when elevation difference is zero.

We will now calculate the pressure required taking into account the given elevations at the beginning, midpoint, and end of the pipeline.

Case B

We will use $Z = 0.8666$ throughout as in Case A.

Using Eqn (2.10), the elevation adjustment factor is first calculated for each of the two segments.

For the first segment, from milepost 0.0 to milepost 25.0, we get

$$s_1 = 0.0375 \times 0.6 \left(\frac{250 - 100}{520 \times 0.8666}\right) = 0.0075$$

Similarly, for the second segment, from milepost 25.0 to milepost 50.0 we get

$$s_2 = 0.0375 \times 0.6 \left(\frac{450 - 100}{520 \times 0.8666}\right) = 0.0175$$

Therefore, the adjustment for elevation is, using Eqn (2.12),

$$j = \frac{e^{0.0075} - 1}{0.0075} = 1.0038 \text{ for first segment}$$

and

$$j = \frac{e^{0.0175} - 1}{0.0175} = 1.0088 \text{ for second segment}$$

For the entire length

$$s_2 = 0.0375 \times 0.6 \left(\frac{450 - 100}{520 \times 0.8666}\right) = 0.0175$$

The equivalent length from Eqn (2.13) is then

$$L_e = 1.0038 \times 25 + 1.0088 \times 25 \times e^{0.0075} = 50.5049 \text{ mi}.$$

Therefore, we see that the effect of the elevation is taken into account partly by increasing the pipe length from approximately 50—50.50 miles.

Substituting in Eqn (5.7), we get

$$100 \times 10^6 = 38.77$$

$$\times 19.1954 \left(\frac{520}{14.7}\right) \left(\frac{P_1^2 - e^{0.0175} 884.7^2}{0.6 \times 520 \times 50.50 \times 0.8666}\right)^{0.5} 15.5^{2.5}$$

Solving for inlet pressure P_1:

$$P_1 = 1008.34 \text{ psia} = 993.64 \text{ psig}$$

Thus, the pressure required at the beginning of the pipeline in Case B is 993.64 psig, taking into account elevation difference along the pipeline. Compare this with the 985.66 calculated, ignoring the elevation differences.

For simplicity, we assume the same value of Z in the preceding calculations as in the previous case. To be correct, we should recalculate Z based on the average pressure and repeat calculations until the results are within 0.1 psi. This is left as an exercise for the reader.

It can be seen from the preceding calculations that because of an elevation difference of (450—100) or 350 ft between the delivery point and the beginning of the pipeline, the required pressure is (993.64—985.66 psig) or approximately 8 psig more. In a liquid line, the effect of elevation would have been more. The elevation difference of 350 ft in a water line would result in an increased pressure of $350 \times 0.433 = 152$ psi approximately at the upstream end.

4. EFFECT OF CHANGING PIPE DELIVERY PRESSURE

The delivery pressure component discussed in Section 1 is also similar between liquid and gas pipelines. The higher the pressure desired at the delivery end or terminus of the pipeline, the higher the total pressure required will be at the upstream end of the pipeline. Suppose a 50-mile liquid pipeline requires an origin pressure of 1120 psig based on a terminus delivery pressure of 50 psig. If the delivery pressure is increased to 100 psig, the origin pressure will increase exactly to 1170 psig. In a gas pipeline, it is not so simple. The impact of changing the delivery pressure is not linear in the case of a compressible fluid such as natural gas. We will now explore the effect of changing the contract delivery pressure at the end of a gas pipeline. In a gas pipeline, the increase or decrease in the upstream pressure will not be proportionate from the nonlinear nature of the gas pressure drop.

Consider the previous problem. All parameters in Case A are the same except for the delivery pressure, which is increased from 870 to 950 psig. The increased delivery pressure will cause the compressibility factor to change slightly because of the change in average pressure. However, for simplicity we will assume $Z = 0.8666$ as calculated before.

New delivery pressure is

$$P_2 = 950 + 14.7 = 964.7 \text{ psia}$$

Substituting in general flow Eqn (5.7) we get

$$100 \times 10^6 = 38.77$$

$$\times 19.1954 \left(\frac{520}{14.7}\right) \left(\frac{P_1^2 - 964.7^2}{0.6 \times 520 \times 50 \times 0.8666}\right)^{0.5} 15.5^{2.5}$$

Therefore:

$$P_1 = 1071.77 \text{ psia} = 1057.07 \text{ psig}$$

Thus, the pressure required at the beginning of the pipeline is approximately 1057 psig. This compares with a value of 1066 psig we calculated if the pressure variation were linear. In general, for a gas pipeline, if the delivery pressure is increased by ΔP, the inlet pressure will increase by less than ΔP. Similarly, if the delivery pressure is decreased by ΔP, the inlet pressure will decrease by less than ΔP. We will illustrate this using the preceding example.

5. PIPELINE WITH INTERMEDIATE INJECTIONS AND DELIVERIES

A pipeline in which the fluid enters at the beginning of the pipeline and the same volume exits at the end of the pipeline is a pipeline with no intermediate injection or deliveries. When portions of the inlet volume are delivered at various points along the pipeline and the remaining volume delivered at the end of the pipeline, we call this system a pipeline with intermediate delivery points. A more complex case with fluid flow into the pipeline (injection) at various points along its length combined with deliveries at other points is shown in Figure 6.1. In such a pipeline system the pressure required at the beginning, point A will be calculated by considering the pipeline divided into segments AB, BC, and so on.

Figure 6.1 Pipeline with injection and deliveries.

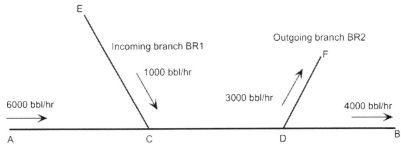

Figure 6.2 Pipeline with branches.

Another piping system may consist of fluid flow at the inlet of the pipeline along with multiple pipe branches (incoming and outgoing) making deliveries of the fluid, as shown in Figure 6.2.

In the pipeline system depicted in Figure 6.2, the pipe segment AB has a certain volume Q_1 flowing through it. At point B, another pipeline CB (incoming branch) brings in additional volume Q_2, resulting in a total volume of $(Q_1 + Q_2)$ flowing through pipe segment BD. At D, an outgoing branch pipe DE delivers a volume of Q_3 to a customer location E. The remaining volume $(Q_1 + Q_2 - Q_3)$ then flows from D to F through the pipe segment DF to the final delivery terminus at F.

In the subsequent sections, we will analyze pipelines with intermediate flow deliveries, injections, and branch pipes as shown in Figures 6.1 and 6.2. The objectives in all cases will be to calculate the pressures and flow rates through the various pipe sections and to determine pipe sizes required to limit pressure drop in certain pipe segments.

Problem 6.2: Gas Pipeline

A 150-mile-long natural gas pipeline consists of several injections and deliveries as shown in Figure 6.3. The pipeline is NPS 20, 0.500-in wall thickness and has an inlet volume of 250 MMSCFD. At points B (milepost 20) and C (milepost 80), 50 MMSCFD and 70 MMSCFD, respectively, are delivered.

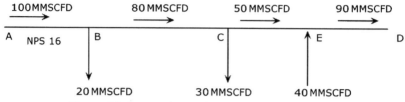

Figure 6.3 Gas pipeline with injections and deliveries.

At D (milepost 100), gas enters the pipeline at 60 MMSCFD. All streams of gas may be assumed to have a specific gravity of 0.65 and a viscosity of 8.0×10^{-6} lb/ft-s. The pipe is internally coated (to reduce friction), resulting in an absolute roughness of 150 μ in. Assume a constant gas flow temperature of 60 °F and base pressure and base temperature of 14.7 psia and 60 °F, respectively. Use a constant compressibility factor of 0.85 throughout. Neglect elevation differences along the pipeline.

1. Using the American Gas Association (AGA) equation, calculate the pressures along the pipeline at points A, B, C, and D for a minimum delivery pressure of 300 psig at the terminus E. Assume drag factor = 0.96.
2. What diameter pipe will be required for section DE if the required delivery pressure at E is increased to 500 psig? The inlet pressure at A remains the same as calculated above.

Solution

We will start calculations beginning with the last segment DE.
Pipe inside diameter $D = 20 - 2 \times 0.500 = 19.00$ in
The flow rate in pipe DE is 190 MMSCFD.
Using Eqn (5.89), the Reynolds number is

$$R = 0.0004778 \left(\frac{14.7}{520} \right) \left(\frac{0.65 \times 190 \times 10^6}{8 \times 10^{-6} \times 19} \right) = 10,974,469$$

Next calculate the two transmission factors required per AGA.
1. The fully turbulent transmission factor, using Eqn (5.97) is

$$F = 4 Log_{10} \left(\frac{3.7 \times 19}{150 \times 10^{-6}} \right) = 22.68$$

2. The smooth pipe zone Von Karman transmission factor, using Eqn (5.x) is

$$F_t = 4 Log_{10} \left(\frac{10,974,469}{F_t} \right) - 0.6$$

Solving for F_t by iteration, we get

$$F_t = 22.18$$

Therefore, for partly turbulent flow zone the transmission factor, using Eqn (2.49) is

$$F = 4 \times 0.96 Log_{10}\left(\frac{10,974,469}{1.4125 \times 22.18}\right) = 21.29$$

Using the smaller of the two values, the AGA transmission factor is

$$F = 21.29$$

Next we use the general flow Eqn (5.57) to calculate the upstream pressure P_1 at D based on given downstream pressure of 300 psig at E.

$$190 \times 10^6 = 38.77 \times 21.29 \left(\frac{520}{14.7}\right)\left(\frac{P_1^2 - 314.7^2}{0.65 \times 520 \times 50 \times 0.85}\right)^{0.5} 19^{2.5}$$

Solving for P_1, we get the pressure at D as

$$P_1 = 587.11 \text{ psia} = 572.41 \text{ psig}$$

Next we consider the pipe segment CD, which has a flow rate of 130 MMSCFD. We calculate the pressure at C using the downstream pressure at D calculated previously.

To simplify calculation, we will use the same AGA transmission factor we calculated for segment DE. A more correct solution will be to calculate the Reynolds number and the two transmission factors as we did for the segment DE. However, for simplicity, we will use $F = 21.29$ for all pipe segments.

Applying the general flow Eqn (2.4), we calculate the pressure P_1 at C as follows:

$$130 \times 10^6 = 38.77$$
$$\times 21.29 \left(\frac{520}{14.7}\right)\left(\frac{P_1^2 - 587.11^2}{0.65 \times 520 \times 20 \times 0.85}\right)^{0.5} (19.0)^{2.5}$$

Solving for P_1 we get the pressure at C as

$$P_1 = 625.06 \text{ psia} = 610.36 \text{ psig}$$

Similarly, we calculate the pressure at B considering the pipe segment BC that flows 200 MMSCFD.

$$200 \times 10^6 = 38.77$$
$$\times 21.29 \left(\frac{520}{14.7}\right)\left(\frac{P_1^2 - 625.06^2}{0.65 \times 520 \times 60 \times 0.85}\right)^{0.5} (19.0)^{2.5}$$

Solving for P_1, we get the pressure at B as

$$P_1 = 846.95 \text{ psia} = 832.25 \text{ psig}$$

Finally, for pipe segment AB, which flows 250 MMSCFD, we calculate the pressure P_1 at A as follows:

$$250 \times 10^6 = 38.77$$

$$\times 21.29 \left(\frac{520}{14.7}\right) \left(\frac{P_1^2 - 846.95^2}{0.65 \times 520 \times 20 \times 0.85}\right)^{0.5} (19.0)^{2.5}$$

Solving for P_1 we get the pressure at A as

$$P_1 = 942.04 \text{ psia} = 927.34 \text{ psig}$$

If we maintain the same inlet pressure 927.34 psig at A and increase the delivery pressure at E to 500 psig, we can determine the pipe diameter required for section DE by considering the same upstream pressure of 572.41 psig at D as we calculated before.

Therefore for segment DE:
Upstream pressure $P_1 = 572.41 + 14.7 = 587.11$ psia
Downstream pressure $P_2 = 500 + 14.7 = 514.7$ psia
Using the general flow Eqn (2.4) with the same AGA transmission factor as before, we get

$$190 \times 10^6 = 38.77 \times 21.29 \left(\frac{520}{14.7}\right) \left(\frac{587.11^2 - 514.7^2}{0.65 \times 520 \times 50 \times 0.85}\right)^{0.5} (D)^{2.5}$$

Solving for the inside diameter D of pipe DE, we get

$$D = 23.79 \text{ in.}$$

The nearest standard pipe size is NPS 26 with 0.500-in wall thickness. This will give an inside diameter of 25 in, which is slightly more than the required minimum of 23.79 in calculated previously.

The wall thickness required for this pipe diameter and pressure will be dictated by the pipe material and is the subject of Chapter 6.

Problem 6.3: Liquid Pipeline

A crude oil pipeline shown in Figure 6.2 consists of injections and deliveries. The pipeline segment from point A to point B is 48 miles long and is NPS 18, 0.281-in wall thickness. It is constructed of 5LX-65 grade steel. At A, crude oil of specific gravity 0.85 and 10 cSt viscosity enters the pipeline at a flow rate of 6000 barrels (bbl)/h. At C (milepost 22), a new stream of crude oil with a specific gravity of 0.82 and 3.5 cSt viscosity enters the pipeline at a flow rate of 1000 bbl/h. The mixed stream then continues to point D (milepost 32), where 3000 bbl/h is stripped off the pipeline. The remaining volume continues to the end of the pipeline at point B. The flowing temperature may be assumed to be constant at 60 °F.

1. Calculate the pressure required at A and the composition of the crude oil arriving at terminus B at a minimum delivery pressure of 50 psi. Assume elevations at A, C, D, and B to be 100, 150, 250, and 300 ft, respectively. Use Colebrook−White equations for pressure drop calculations and assume a pipe roughness of 0.002 in.
2. How much pump power (HP) will be required to maintain this flow rate at the origin point A, assuming a 50-psi pump suction pressure at A and 80% pump efficiency?
3. If a positive displacement (PD) pump is used to inject the stream at C, what pressure and HP are required at C?

Solution

The pressure drop from friction for segment AC is calculated using Eqn (3.27) as follows:

Reynolds number $= 92.24 \times 6000 \times 24/(17.438 \times 10) = 76{,}170$
Friction factor $= 0.02$
Pressure drop $= 13.25$ psi/mi
Frictional pressure drop between A and C $= 13.25 \times 22 = 291.5$ psi

Next, we calculate the blended properties of the liquid stream after mixing two streams at point C, by blending 6000 bbl/h of crude A (specific gravity of 0.85 and viscosity of 10 cSt) with 1000 bbl/h of crude B (specific gravity of 0.82 and viscosity of 3.5 cSt) using Eqns (2.4) and (2.21) as follows:

Blended specific gravity at C $= 0.8457$
Blended viscosity at C $= 8.366$ cSt

For pipe segment CD, we calculate the pressure drop by using the previous properties at a flow rate of 7000 bbl/h.

Reynolds number $= 92.24 \times 7000 \times 24/(17.438 \times 8.366) = 106{,}222$
Friction Factor $= 0.0188$
Pressure drop $= 16.83$ psi/mi
Frictional pressure drop between C and D $= 16.83 \times 10 = 168.3$ psi

Finally, we calculate for pipe segment DB the pressure drop by using above liquid properties at a flow rate of 4000 bbl/h.

Reynolds number $= 92.24 \times 4000 \times 24/(17.438 \times 8.366) = 60{,}698$
Friction factor $= 0.021$
Pressure drop $= 6.13$ psi/mi
Frictional pressure drop between D and B $= 6.13 \times 16 = 98.08$ psi

Therefore, the total frictional pressure drop between point A and point B is

$$291.5 + 168.3 + 98.08 = 557.9 \text{ psi}$$

The elevation head between A and B consists of (150−100) ft between A and C and (300−150) between C and B. We need to separate the total elevation head in this fashion because of differences in liquid properties in pipe segments AC and CB. Therefore, the total elevation head is

$$[(150-100) \times 0.85/2.31] + [(300-150) \times 0.8457/2.31] = 73.32 \text{ psi}$$

Adding the delivery pressure of 50 psi, the total pressure required at A is therefore

$$557.9 + 73.32 + 50 = 681.22 \text{ psi}$$

Therefore, the pressure required at A is 681.22 psi and the composition of the crude oil arriving at terminus B is: specific gravity of 0.8457 and viscosity of 8.366 cSt.

The power calculations will be introduced in Chapter 8. However, here we will simply calculate the power required using Eqn (8.x)

The HP required at A is calculated using Eqn (5.16) as follows:

$$\text{HP} = 6000 \times (681.22 - 50)/(0.8 \times 2449) = 1,933 \text{ HP}$$

To calculate the injection pump requirement at point C, we must first calculate the pressure in the pipeline at point C that the PD pump has to overcome.

The pressure at C = pressure at A − pressure drop from A to C − elevation head A to C.

$$P_C = 681.22 - 291.5 - (150 - 100) \times 0.85/2.31 = 371.3 \text{ psi}$$

The HP required for the PD pump at C is calculated using Eqn (5.16) as follows:

$$\text{PD Pump HP required} = (371.3 - 50) \times 1000/(0.8 \times 2449) = 164$$

Assuming 50 psi suction pressure and 80% pump efficiency.

Problem 6.4: Gas Pipeline

A pipeline 150 miles long transports natural gas from Corona to Beaumont. The gas has a specific gravity of 0.60 and a viscosity of 8×10^{-6} lb/ft-s. What is the minimum pipe diameter required to flow 150 MMSCFD from Corona to Beaumont for a delivery pressure of 800 psig at Beaumont and inlet pressure of 1400 psig at Corona? The gas may be assumed constant at 60 °F and the base pressure and base temperature are 14.7 psia and 60 °F, respectively. Use constant value of 0.90 for compressibility factor and a pipe roughness of 700 μ in. Compare results using the AGA, Colebrook–White, Panhandle B, and Weymouth equations. Use 95% pipeline efficiency. Neglect elevation differences along the pipeline. How will the result change if the elevation at Corona is 100 ft and is 500 ft at Beaumont?

Solution

We will first use AGA equation to determine the pipe diameter. Because the transmission factor F depends on Reynolds number, which depends on the unknown pipe diameter, we will first assume a value of $F = 20$.

From the general flow Eqn (2.4), we get

$$100 \times 10^6 = 38.77 \times 20.0 \left(\frac{520}{14.7}\right) \left[\frac{1414.7^2 - 814.7^2}{0.6 \times 520 \times 100 \times 0.9}\right]^{0.5} \times (D)^{2.5}$$

Solving for diameter D

$D = 12.28$ in or NPS 12 with a 0.250 in wall thickness approximately.

Next we will recalculate the transmission factor using this pipe size.
Using NPS 12 with a 0.250-in wall thickness.
The inside pipe diameter $D = 12.75 - 2 \times 0.250 = 12.25$ in
Calculating the Reynolds number from Eqn (2.34), we get

$$R = 0.0004778 \left(\frac{14.7}{520}\right) \left(\frac{0.6 \times 100 \times 10^6}{8 \times 10^{-6} \times 12.25}\right) = 8{,}269{,}615$$

The fully turbulent transmission factor, using Eqn (2.48) is

$$F = 4Log_{10} \left(\frac{3.7 \times 12.25}{0.0007}\right) = 19.25$$

For the smooth pipe zone, using Eqn (2.50), the Von Karman transmission factor is

$$F_t = 4Log_{10} \left(\frac{8{,}269{,}615}{F_t}\right) - 0.6$$

Solving for F_t by iteration, we get

$$F_t = 21.72$$

Using a drag factor of 0.96, for partly turbulent flow, the transmission factor is from Eqn (2.49)

$$F = 4 \times 0.96 \times Log_{10} \left(\frac{8{,}269{,}615}{1.4125 \times 21.72}\right) = 20.85$$

Using the lower of the two values, the AGA transmission factor is

$$F = 19.25$$

Using this value of F, we recalculate the minimum pipe diameter from the general flow Eqn (2.4) as follows:

$$100 \times 10^6 = 38.77 \times 19.25 \left(\frac{520}{14.7}\right) \left[\frac{1414.7^2 - 814.7^2}{0.6 \times 520 \times 100 \times 0.9}\right]^{0.5} \times (D)^{2.5}$$

Solving for diameter D

$$D = 12.47 \text{ in}$$

We will not continue iteration any further, because the new diameter will not change the value of F appreciably.

Therefore, based on the AGA equation, the pipe inside diameter required is 12.47 in.

Next we calculate the transmission factor based on Colebrook–White equation assuming an inside diameter of 12.25 in and a Reynolds number = 8,269,615 calculated earlier.

Using Colebrook–White Eqn (2.45), we get

$$F = -4 Log_{10} \left[\frac{0.0007}{3.7 \times 12.25} + \frac{1.255 F}{8,269,615} \right]$$

Solving for F by successive iteration, we get the Colebrook–White transmission factor as

$$F = 18.95$$

Using the general flow equation with this Colebrook–White transmission factor, we calculate the diameter as follows:

$$100 \times 10^6 = 38.77 \times 18.95 \left(\frac{520}{14.7} \right) \left[\frac{1414.7^2 - 814.7^2}{0.6 \times 520 \times 100 \times 0.9} \right]^{0.5} \times (D)^{2.5}$$

Solving for diameter D

$$D = 12.55 \text{ in}$$

Recalculating the Reynolds number and transmission factor using the pipe inside diameter of 12.55 in we get

$$R = 8,071,935 \text{ and } F = 18.94$$

Therefore, the new diameter required is by proportions, using the general flow equation

$$\left(\frac{D}{12.55} \right)^2 = \left(\frac{18.95}{18.94} \right)$$

Or $D = 12.55$ approximately. There is no appreciable change in the diameter required.

Therefore, based on Colebrook–White equation, the pipe inside diameter required is $D = 12.55$ in.

Next we determine the diameter required using the Panhandle B Eqn (2.59) and a pipeline efficiency of 0.95

$$100 \times 10^6 = 737$$

$$\times 0.95 \left(\frac{520}{14.7}\right)^{1.02} \left(\frac{1414.7^2 - 814.7^2}{0.6^{0.961} \times 520 \times 100 \times 0.9}\right)^{0.51} D^{2.53}$$

Solving for diameter D, we get

$$D = 11.93 \text{ in.}$$

Therefore, based on the Panhandle B equation, the pipe inside diameter required is 11.93 in.

Next we calculate the diameter required, using the Weymouth Eqn (2.52) and a pipeline efficiency of 0.95

$$100 \times 10^6 = 433.5 \times 0.95 \left(\frac{520}{14.7}\right) \left(\frac{1414.7^2 - 814.7^2}{0.6 \times 520 \times 100 \times 0.9}\right)^{0.5} D^{2.667}$$

Solving for diameter D, we get

$$D = 13.30 \text{ in.}$$

Therefore, based on the Weymouth equation, the pipe inside diameter required is 13.30 in.

In summary, the minimum pipe inside diameter required based on the various flow equations is as follows

AGA	$D = 12.47$ in
Colebrook–White	$D = 12.55$ in
Panhandle B	$D = 11.93$ in
Weymouth equation	$D = 13.30$ in

It can be seen that the Weymouth equation is the most conservative equation. The AGA and Colebrook–White equations predict almost the same pipe size, whereas Panhandle B predicts the smallest pipe size. To further illustrate the comparison of various pressure drop equations, refer to discussion in Chapter 2 and Figure 2.5. This figure shows how the delivery pressure varies for a fixed flow rate and inlet pressure. Table 3.1 also summarizes the various pressure drop equations used in the gas pipeline industry.

Considering elevation effects, with a single slope from Corona (100 ft) to Beaumont (500 ft), the elevation adjustment parameter is from Eqn (2.10):

$$s = 0.0375 \times 0.6 \left(\frac{500 - 100}{520 \times 0.9}\right) = 0.0192$$

Therefore, the equivalent length from Eqn (2.12) is

$$L_e = 100 \times \frac{e^{0.0192} - 1}{0.0192} = 100.97 \text{ mi}$$

We will apply the elevation correction factor for the extreme cases (Weymouth and Panhandle B equations) that produce the largest and the smallest diameters, respectively.

From the Weymouth Eqn (2.52), we see that keeping all other items same, the diameter and pipe length are related by the following equation

$$\frac{D^{2.667}}{\sqrt{L}} = \text{Constant}$$

$$\left(\frac{D}{13.3}\right)^{2.667} = \left(\frac{100.97}{100}\right)^{0.5}$$

Solving for pipe inside diameter D, we get

$$D = 13.32 \text{ in.}$$

Not an appreciable change from the previous value of 13.30 in.

Similarly, from the Panhandle B Eqn (2.59), we see that the pipe diameter and length are related by

$$\frac{D^{2.53}}{L^{0.51}} = \text{Constant}$$

$$\left(\frac{D}{11.93}\right)^{2.53} = \left(\frac{100.97}{100}\right)^{0.51}$$

Solving for pipe inside diameter D, we get

$$D = 11.95$$

Not an appreciable change from the previous value of 11.93 in.

Therefore, considering elevation difference between Corona and Beaumont, the minimum pipe size required are as follows

Panhandle B	$D = 11.95$ in
Weymouth equation	$D = 13.32$ in

We thus see that even with 400-ft elevation difference, the pipe diameter does not change appreciably.

Problem 6.5: Gas Pipeline

A natural gas distribution piping system consists of NPS 12 with 0.250-in wall thickness, 24 miles long as shown in Figure 6.4.

At Yale, an inlet flow rate of 65 MMSCFD of natural gas enters the pipeline at 60 °F. At the Compton terminus, gas must be supplied at a flow rate of

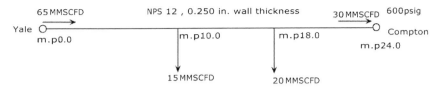

Figure 6.4 Yale to Compton gas distribution pipeline.

30 MMSCFD at a minimum pressure of 600 psig. There are intermediate deliveries of 15 MMSCFD at milepost 10 and 20 MMSCFD at milepost 18. What is the required inlet pressure at Yale? Use a constant friction factor of 0.01 throughout. The compressibility factor may be assumed to be 0.94. The gas gravity and viscosity are 0.6 and 7×10^{-6} lb/ft-s, respectively. Assume isothermal flow at 60 °F. The base temperature and base pressure are 60 °F and 14.7 psia, respectively. If the delivery volume at B is increased to 30 MMSCFD and other deliveries remain the same, what increased pressure is required at Yale to maintain the same flow rate and delivery pressure at Compton? Neglect elevation differences along the pipeline.

Solution

For each section of piping such as AB, we must calculate the pressure drop from friction at the appropriate flow rate and then determine the total pressure drop for the entire pipeline.

Inside diameter of pipe = $12.75 - 2 \times 0.250 = 12.25$ in
Friction factor $f = 0.01$
Therefore, transmission factor using Eqn (2.42) is

$$F = \frac{2}{\sqrt{0.01}} = 20.00$$

Using the general flow Eqn (2.7) for the last pipe segment from milepost 18 to milepost 24, we get

$$30 \times 10^6 = 38.77 \times 20.0 \left(\frac{520}{14.7}\right) \left[\frac{P_C^2 - 614.7^2}{0.6 \times 520 \times 6 \times 0.94}\right]^{0.5} \times (12.25)^{2.5}$$

Solving for the pressure at C

$$P_C = 620.88 \text{ psia}$$

Next we will use this pressure P_C to calculate the pressure P_B for the 8-mile section of pipe segment BC flowing 50 MMSCFD.

Using the general flow Eqn (2.7)

$$50 \times 10^6 = 38.77 \times 20 \left(\frac{520}{14.7}\right) \left[\frac{P_B^2 - 620.88^2}{0.6 \times 520 \times 8 \times 0.94}\right]^{0.5} \times (12.25)^{2.5}$$

Solving for P_B, we get

$$P_B = 643.24 \text{ psia}$$

Finally, we calculate the pressure P_1 at Yale by considering the 10-mile pipe segment from Yale to point B that flows 65 MMSCFD.

$$65 \times 10^6 = 38.77 \times 20 \left(\frac{520}{14.7}\right) \left[\frac{P_1^2 - 643.24^2}{0.6 \times 520 \times 10 \times 0.94}\right]^{0.5} \times (12.25)^{2.5}$$

Solving for the pressure at Yale, we get

$$P_1 = 688.09 \text{ psia} = 673.39 \text{ psig}$$

Therefore, the required inlet pressure at Yale is 673.39 psig.

When the delivery volume at B is increased from 15 MMSCFD to 30 MMSCFD and all other delivery volumes remain the same, the inlet flow rate at Yale will increase to $65 + 15 = 80$ MMSCFD. If the delivery pressure at Compton is to remain the same as before, the pressures at B and C will also be the same as calculated before because the flow rates in BC and CD are the same as before. Therefore, we can recalculate the inlet pressure for the pipe section from Yale to point B considering a flow rate of 80 MMSCFD that causes a pressure of 643.24 psia at B.

Using the general flow Eqn (2.7), the pressure P_1 at Yale is

$$80 \times 10^6 = 38.77 \times 20.0 \left(\frac{520}{14.7}\right) \left[\frac{P_1^2 - 643.24^2}{0.6 \times 520 \times 10 \times 0.94}\right]^{0.5}$$
$$\times (12.25)^{2.5}$$

Solving for pressure at Yale

$$P_1 = 710.07 \text{ psia} = 695.37 \text{ psig}$$

Therefore, increasing the delivery volume at B by 15 MMSCFD causes the pressure at Yale to increase by approximately 22 psig.

Next we use this pressure as the inlet pressure for the last pipe segment EF and calculate the outlet pressure at F using general flow equation as follows:

$$100 \times 10^6 = 77.54 \left(\frac{1}{\sqrt{0.015}}\right) \left(\frac{520}{14.73}\right) \left[\frac{(1145.63^2 - P_2^2)}{0.6 \times 540 \times 20 \times 0.92}\right]^{0.5} 15.5^{2.5}$$

Solving for the outlet pressure at F, we get

$$P_2 = 1085.85 \text{ psia} = 1071.12 \text{ psig}$$

In summary, the calculated results are as follows:
Pressure at beginning of pipe loops = 1166.6 psig
Pressure at the end of pipe loops = 1130.9 psig
Outlet pressure at the end of pipeline = 1071.12 psig
Flow rate in NPS 14 loop = 51 MMSCFD
Flow rate in NPS 12 loop = 49 MMSCFD

6. SYSTEM HEAD CURVES: LIQUID PIPELINES

A system head curve, also abbreviated to system curve, for a pipeline is a graphic representation of how the pressure to pump a liquid in a pipeline varies with the flow rate. Figure 6.5 shows a typical system head curve. As the flow rate increases, the head (pressure) required increases. From Chapter 5, Eqn (5.x), we know that the pressure drop from friction varies as the square of the flow rate. If the head loss at 1000 gal/min is 10 psi/mi, at 2000 gal/min the head loss is approximately $(2000/1000)^2 \times 10 = 40$ psi/mi. Therefore, doubling the flow rate causes the head loss to be four times as before.

Therefore, if we neglect the elevation profile, and consider 50 psi terminus delivery pressure, the total pressure P_T required at the origin of a 30-mile pipeline is 350 psi $(30 \times 10 + 50)$ at 1000 gal/min and 1250 psi at 2000 gal/min. Thus, we can calculate the head loss and hence the total

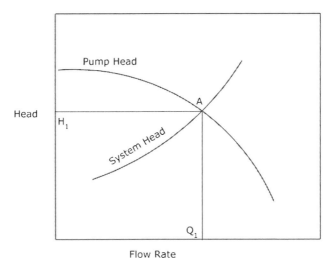

Figure 6.5 System head curve.

pressure P_T required for various flow rates and plot a curve representing the system curve. We can also take into account the elevations along the pipeline and using the method described in Section 3, obtain the values of P_T and plot the system head curve. Usually, when plotting system curves, the pressures are converted to feet of head as indicated in Figure 6.5. However, sometimes the system head curve may be plotted in psi as well. In US customary system units, the horizontal scale is the flow rate in gal/min or bbl/h. In SI units, the head is plotted in meters of liquid and the flow rates in m^3/h or L/min.

To recap, consider a pipeline of inside diameter D and length L that transports a liquid of specific gravity Sg and viscosity v from a pump station at A to delivery terminus located at B. We calculate the pressure required at A to transport the liquid at a particular flow rate Q. By varying flow rate Q, we can determine the pressure required at A for each flow rate so that a given delivery pressure at B is maintained. For each flow rate Q, we can calculate the pressure drop from friction for the length L of the pipeline, add the head required to account for elevation difference between A and B, and, finally, add the delivery pressure required at B as follows, using Eqn (6.1).

Pressure at A = Friction pressure drop + Elevation head + Delivery pressure

Once the pressure at A is calculated for each flow rate, we can plot a system head curve as shown in Figure 6.5.

In Chapter 11, system head curves along with pump head curves will be reviewed in detail. We will see how the system head curve in conjunction with the pump head curve will determine the operating point for a particular pump—pipeline configuration. Because a system head curve represents the pressure required to pump various flow rates through a given pipeline, we can plot a family of such curves for different liquids as shown in Figure 6.6. The higher specific gravity and viscosity of diesel fuel requires greater pressures (psi or kPa) compared with gasoline. Hence the diesel system head curve is located above that of gasoline as shown in Figure 6.6.

The shape of the system curve varies depending on the amount of friction head component compared to the elevation head. Figure 6.6 show two system curves that illustrate this. In Figure 6.7, the friction component is higher than the elevation component. Most of the system head required is from the friction in the pipe.

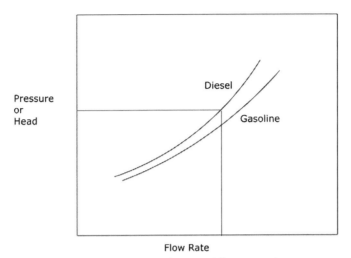

Figure 6.6 System head curve: different products.

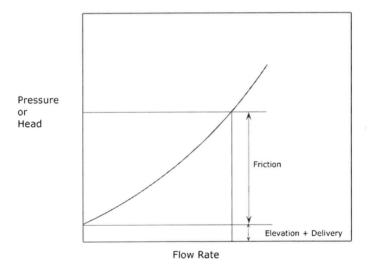

Figure 6.7 System head curve: high friction.

Alternatively, if the elevation differences are much higher than the frictional head loss, then system curve is less sensitive to flow rate changes.

Figure 6.8 shows a system head curve that consists mostly of the static head resulting from the pipe elevations.

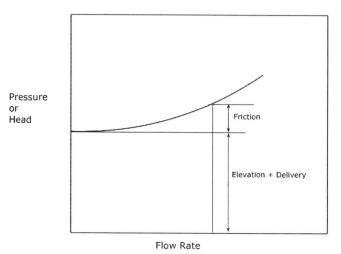

Figure 6.8 System head curve: high elevation.

7. HYDRAULIC PRESSURE GRADIENT: LIQUID PIPELINE

From frictional losses, the pressure in a liquid pipeline decreases continuously from the pipe inlet to the pipe delivery terminus. If there is no elevation difference between the two ends of the pipeline and the pipe elevation profile is essentially flat, the inlet pressure at the beginning of the pipeline will decrease continuously by the friction loss at a particular flow rate. When there are elevation differences along the pipeline, the decrease in pipeline pressure along the pipeline will be from the combined effect of pressure drop from friction and the algebraic sum of pipeline elevations. Thus starting at 1000 psi pressure at the beginning of the pipeline, assuming 15 psi/mi pressure drop from friction in a practically flat pipeline (no elevation difference) with constant diameter throughout, at distance of 20 miles from the beginning of the pipeline the pressure would decrease to

$$1000 - 15 \times 20 = 700 \text{ psi}$$

If the pipeline is 60 miles long, the pressure drop from friction in the entire line will be

$$15 \times 60 = 900 \text{ psi}$$

Therefore, the pressure at the end of the pipeline will be $1000 - 900 = 100$ psi.

Thus the liquid pressure in the pipeline has uniformly dropped from 1000 psi in the beginning to 100 psi at the end of the 60-mile length.

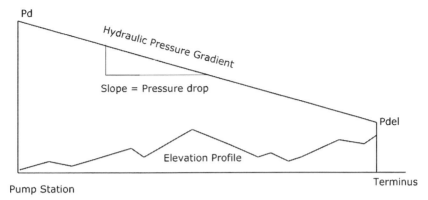

Figure 6.9 Hydraulic pressure gradient: liquid pipeline.

This pressure profile is referred to as the hydraulic pressure gradient in the pipeline (Figure 6.9).

The hydraulic pressure gradient is a graphical representation of the variation of pressures along the pipeline. It is shown along with the pipeline elevation profile. Because elevation is plotted in feet, it is convenient to represent the pipeline pressures also in feet of liquid head. This is shown in Figures 6.6–6.8.

In the problem discussed in Section 6, we calculated the pressure required at the beginning of the pipeline to be 1034 psi for pumping crude oil at a flow rate of 4000 bbl/h. This pressure requires one pump station at the origin of the pipeline at point A.

Suppose the pipe length is 100 miles and the maximum allowable operating pressure (MAOP) of the pipeline is limited to 1200 psi. Assume the total pressure required at A is calculated to be 1600 psi at a flow rate of 4000 bbl/h. Because this is higher than the MAOP, we would require an additional intermediate pump station between A and B to limit the maximum pressure to 1200 psi. Because of the MAOP limit, the total pressure required at A will be provided in steps. The first pump station at A will provide approximately half the pressure followed by the second pump station located at some intermediate point providing the other half. This results in a sawtooth-like hydraulic gradient as shown in Figure 6.10.

The actual discharge pressure at each pump station will be calculated considering pipeline elevations between A and B and the required minimum suction pressures at the second pump station. An approximate calculation is described next, referring to Figure 6.10.

Let P_s and P_d represent the common suction and discharge pressure, respectively, for each pump station and P_{del} is the required delivery pressure

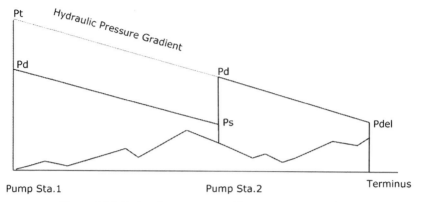

Figure 6.10 Hydraulic pressure gradient: two pump stations.

at the pipe terminus B. The total pressure P_t required at A can be stated as follows:

$$P_t = P_{friction} + P_{elevation} + P_{del} \quad (6.1)$$

where

P_t—total pressure required at A
$P_{friction}$—total frictional pressure drop between A and B
$P_{elevation}$—elevation head between A and B
P_{del}—required delivery pressure at B

Also, from Figure 6.10, based on geometry, we can state that:

$$P_t = P_d + P_d - P_s \quad (6.2)$$

Solving for P_d we get

$$P_d = (P_t + P_s)/2 \quad (6.3)$$

where

P_d—pump station discharge pressure
P_s—pump station suction pressure

For example, if the total pressure calculated is 1600 psi and the maximum pipeline pressure allowed is 1200 psi, we would require two pump stations. Considering minimum suction pressure of 50 psi, each pump station would have a discharge pressure of

$$P_d = (1600 + 50)/2 = 825 \text{ psi using Eqn (6.3)}$$

Each pump station operates at 825 psi discharge pressure and the pipeline MAOP is 1200 psi. It is clear that based on pipeline pressures, we have the capability of increasing pipeline throughput further to fully use the 1200 psi MAOP at each pump station. Of course, this would correspondingly require enhancing the pumping equipment at each pump station, since more HP will be required at the higher flow rate. We can estimate the increased throughput possible if we were to operate the pipeline at 1200 psi MAOP level at each pump station as discussed next.

Assume that the pipeline elevation difference between A and B contributes 300 psi to the total pressure required. This simply represents the station elevation head between A and B converted to psi. This component of the total pressure required ($P_t = 1600$ psi) depends only on the pipeline elevation and liquid specific gravity and therefore does not vary with flow rate. Similarly, the delivery pressure of 50 psi at B is also independent of flow rate. We can then calculate the frictional component (which depends on flow rate) of the total pressure P_t using Eqn (6.1) as follows:

Frictional pressure drop $= 1600 - 300 - 50 = 1,250$ psi.

Assuming a 100-mile length, the frictional pressure drop per mile of pipe is

$$P_m = 1250/100 = 12.5 \text{ psi/mi}$$

This pressure drop occurs at 4000 bbl/h flow rate. From previous chapters, we know that the pressure drop per mile, P_m, varies as the square of the flow rate as long as the liquid properties and pipe size do not change. Therefore, we can write

$$P_m = K(Q)^2 \qquad (6.4)$$

where

P_m—frictional pressure drop per mile of pipe
K—a constant for this pipeline that depends on liquid properties and pipe diameter
Q—pipeline flow rate

Note that K is not the same as the head loss coefficient discussed in Chapter 5. Strictly speaking, K also includes a friction factor f, which varies with flow rate. However, for simplicity, we will assume that K is the constant that encompasses liquid specific gravity and pipe diameter. A more rigorous approach will require an additional parameter in Eqn (6.4) that would include the friction factor f, which in turn depends on Reynolds number, pipe roughness, and so on.

Therefore, using Eqn (6.4), we can state that at the initial flow rate of 4000 bbl/h.

$$12.5 = K(4000)^2 \qquad (6.5)$$

In a similar manner, we can estimate the frictional pressure drop per mile when flow rate is increased to some value Q to fully use the 1200 psi MAOP of the pipeline.

Using Eqn (6.3), if we allow each pump station to operate at 1200 psi discharge pressure, we can write

$$1200 = (P_t + 50)/2$$

or

$$P_t = 2400 - 50 = 2350 \text{ psi}$$

This total pressure will now consist of friction, elevation, and delivery pressure components at the higher flow rate Q.

From Eqn (5.1) we can write

$$2350 = P_{friction} + P_{elevation} + P_{del} \text{ at the higher flow rate Q}$$

or

$$2350 = P_{friction} + 300 + 50$$

Therefore:

$$P_{friction} = 2350 - 300 - 50 = 2000 \text{ psi at the higher flow rate Q}$$

Thus the pressure drop per mile at the higher flow rate Q is

$$P_m = 2000/100 = 20 \text{ psi/mi}$$

From Eqn (6.4), we can write

$$20 = K(Q)^2 \tag{6.6}$$

where Q is the unknown higher flow rate in bbl/h.

By dividing Eqn (6.3) by Eqn (6.4), we get the following:

$$20/12.5 = (Q/4000)^2$$

Solving for Q we get

$$Q = 4000(20/12.5)^{1/2} = 5059.64 \text{ bbl/h}$$

Therefore, by fully using the 1200 psi MAOP of the pipeline with the two pump stations, we are able to increase the flow rate to approximately 5060 bbl/h. As previously mentioned, this will definitely require additional pumps at both pump stations to provide the higher discharge pressure. Pumps will be discussed in a later chapter.

In the preceding sections, we have considered a pipeline to be of uniform diameter and wall thickness for its entire length. In reality, pipe diameter and wall thickness change, depending on the service requirement, design code, and the local regulatory requirements. Pipe wall thickness may have to be increased from different specified minimum yield strength of pipe because a higher or lower grade of pipe was used at some locations. As mentioned previously, some cities or counties through which the pipeline traverses may require different design factors (0.66 instead of 0.72) to be used, thus necessitating a different wall thickness. If there are significant elevation changes along the pipeline, the low elevation points may require higher wall thickness to withstand the higher pipe operating pressures. If the pipeline has intermediate flow delivery or injections, the pipe diameter may be reduced or increased for certain portions to optimize pipe use. In all these cases, we can conclude that the pressure drop from friction will not be the same uniform value throughout the entire pipeline length. Injections and delivery along the pipeline and their impact on pressure required is discussed later in this chapter.

When pipe diameter and wall thickness change along a pipeline, the slope of the hydraulic gradient, as shown in Figure 6.10 will no longer be uniform. Because of varying frictional pressure drop (because of pipe diameter and wall thickness change), the slope of the hydraulic gradient will vary along the pipe length.

8. TRANSPORTING HIGH VAPOR PRESSURE LIQUIDS

As mentioned previously, transportation of high vapor pressure liquids such as LPG requires that a certain minimum pressure be maintained throughout the pipeline. This minimum pressure must be greater than the liquid vapor pressure at the flowing temperature. Otherwise, liquid may vaporize causing two phase flow in the pipeline, which the pumps cannot handle. If the vapor pressure of LPG at the flowing temperature is 250 psi, the minimum pressure anywhere in the pipeline must be greater than 250 psi. To be safe, at high elevation points or peaks along the pipeline profile, we must ensure that more than the minimum pressure is maintained. This is illustrated in Figure 6.11. Additionally, the delivery pressure at the end of the pipeline must also satisfy the minimum pressure requirements.

Thus, the delivery pressure at the pipeline terminus for LPG may be 300 psi or higher to account for any meter station and manifold piping losses at the delivery point. Also, sometimes, with high vapor pressure liquids, the delivery point may be a pressure vessel or a pressurized sphere maintained at 500 to 600 psi and therefore may require even higher minimum pressures

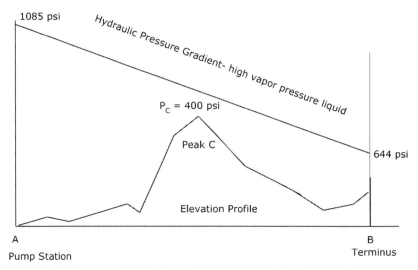

Figure 6.11 Transporting high vapor pressure liquid in a pipeline.

compared to the vapor pressure of the liquid. Therefore, both the delivery pressure and the minimum pressure must be considered when analyzing pipelines transporting high vapor pressure liquids.

9. HYDRAULIC PRESSURE GRADIENT: GAS PIPELINE

The hydraulic pressure gradient is a graphical representation of the gas pressures along the pipeline as shown in Figure 6.12. The horizontal axis shows the distance along the pipeline starting at the upstream end. The vertical axis represents the pipeline pressures.

Because pressure in a gas pipeline is nonlinear compared with liquid pipelines, the hydraulic gradient for a gas pipeline will appears to be a slightly curved line instead of a straight line. The slope of the hydraulic gradient at any point represents the pressure loss from friction per unit length of pipe. As discussed earlier, even with constant flow rate and uniform pipe diameter, for gas pipelines, this slope is more pronounced as we move toward the downstream end of the pipeline because the pressure drop is larger toward the end of the pipeline. In a liquid pipeline, under the same conditions, the slope of the hydraulic pressure gradient will be the same throughout. If the gas flow rate through the pipeline is a constant value (no intermediate injections or deliveries) and pipe size is uniform throughout, the hydraulic gradient appears to be a slightly curved line as shown in Figure 6.12 with

Pressure Required to Transport

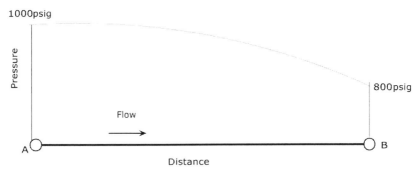

Figure 6.12 Hydraulic pressure gradient for uniform flow.

no appreciable breaks. If there are intermediate deliveries or injections along the pipeline, the hydraulic gradient will be a series of broken lines as indicated in Figure 6.13.

A similar broken hydraulic gradient may also be seen in the case of a pipeline with variable pipe diameters and wall thicknesses, even if the flow rate is constant. Unlike liquid pipelines, the breaks in hydraulic pressure gradient are not as conspicuous in gas pipelines.

In a long-distance gas pipeline, because of limitations of pipe pressure (MAOP), intermediate compressor stations will be installed to boost the gas pressure to the required value so that the gas can be delivered at the contract delivery pressure at the end of the pipeline. For example, consider a 200-mile-long, NPS 16 pipeline that transports 150 MMSCFD of gas from Compton to a delivery point at Beaumont as shown in Figure 6.14.

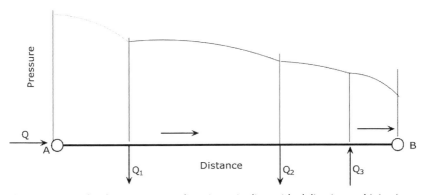

Figure 6.13 Hydraulic pressure gradient in a pipeline with deliveries and injections.

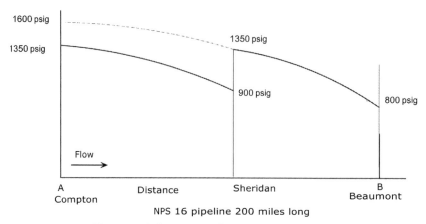

Figure 6.14 Compton to Beaumont gas pipeline.

Suppose calculations show that 1600 psig pressure is required at Compton to deliver gas to Beaumont at 800 psig. If the MAOP of this pipeline is limited to 1350 psig, obviously we will need more than one compressor station. The first compressor station will be located at Compton and will provide a pressure of 1350 psig. As gas flows from Compton toward some intermediate location, such as Sheridan, the gas pressure would have dropped to some value such as 900 psig. At Sheridan, a second compressor station would boost the gas pressure to 1350 psig on its way to the terminus at Beaumont. By installing the second compressor station at Sheridan, pipeline pressures are maintained within maximum operating pressure limits. The actual location of the intermediate compressor station at Sheridan will depend on many factors including pipeline elevation profile, the gas pressure at Sheridan, and the delivery pressure required at Beaumont. The hydraulic pressure gradient in this case is as shown in Figure 6.14. In the preceding discussion, we picked an arbitrary pressure of 900 psig at Sheridan. This gives us an approximate compression ratio of

$$\frac{1350 + 14.7}{914.7} = 1.492$$

which is a reasonable number for centrifugal compressors used in gas pipelines. In reference to Figure 6.14, we will now outline the method of locating the intermediate compressor station at Sheridan.

Starting at Compton with an inlet pressure $P_1 = 1350$ psig, we calculate the pipe length L that will cause the pressure to drop to 900 psig, using the general flow equation. Assuming a flow rate of 150 MMSCFD and a friction factor $f = 0.01$, we get

$$150 \times 10^6 = 77.54 \left(\frac{1}{\sqrt{0.01}}\right) \left(\frac{520}{14.7}\right) \left[\frac{1364.7^2 - 914.7^2}{0.6 \times 520 \times L \times 0.9}\right]^{0.5} \times (15.5)^{2.5}$$

Solving for L, we get

$$L = 109.28 \text{ mi}$$

Thus the approximate location of the second compressor station Sheridan is 109.28 miles from Compton. If we allow the compressor at Sheridan to boost the gas pressure to 1350 psig, the compressor ratio is

$$r = \frac{1350 + 14.7}{914.7} = 1.492$$

which is a reasonable ratio for a centrifugal compressor. Therefore, starting at 1350 psig at Sheridan, we calculate the delivery pressure at Beaumont using the general flow equation for $(200 - 109.28) = 90.7$ miles of pipe as follows

$$150 \times 10^6 = 77.54 \left(\frac{1}{\sqrt{0.01}}\right) \left(\frac{520}{14.7}\right) \left[\frac{1364.7^2 - P_2^2}{0.6 \times 520 \times 90.7 \times 0.9}\right]^{0.5} \times (15.5)^{2.5}$$

Solving for P_2, we get

$$P_2 = 1005.5 \text{ psia} = 990.82 \text{ psig}$$

This is more than the required delivery pressure at Beaumont of 800 psig. We could go back and repeat these calculations considering slightly lower pressure at Compton, say 1300 psig, to get the correct delivery pressure of 800 psig at Beaumont. This is left as an exercise for the reader.

Alternatively, we could start with the required delivery pressure of 800 psig at Beaumont and work backwards to determine the distance at which the upstream pressure reaches 1350 psig. That will be the location for the Sheridan compressor station. Next, we will determine the pressure at Sheridan beginning with the 1350 psig upstream pressure at Compton. This will establish the suction pressure of the Sheridan compressor station. Knowing the suction pressure and the discharge pressure at Sheridan we can calculate the compression ratio required. In Chapter 9, we will discuss multiple compressor station in more detail.

10. PRESSURE REGULATORS AND RELIEF VALVES

In a long-distance gas pipeline with intermediate delivery points, there may be a need to regulate the fluid pressure at certain delivery points in order to satisfy the customer requirements. Suppose the pressure at the delivery point B in Figure 6.14 is 800 psig, whereas the customer requirement is only 500 psig. Obviously, some means of reducing the pressure must be provided so that the customer may use the fluid for his or her requirements at the correct pressure. This is achieved by means of a pressure regulator that will ensure a constant pressure downstream of the delivery point, regardless of the pressure on the upstream side of the pressure regulator. This concept is further illustrated using the example pipeline shown in Figure 6.14.

The main pipeline from A to C is shown along with a branch pipe BE. The flow rate from A to B is 100 MMSCFD with an inlet pressure of 1200 psig at A. At B, gas is delivered into a branch line BE at the rate of 30 MMSCFD. The remaining volume of 70 MMSCFD is delivered to the pipeline terminus at C at a delivery pressure of 600 psig. Based on the delivery pressure requirement of 600 psig at C and a takeoff of 30 MMSCFD at point B, the calculated pressure at B is 900 psig. Starting with 900 psig on the branch line at B at 30 MMSCFD, gas is delivered to point E at 600 psig. If the actual requirement at E is only 400 psig, a pressure regulator will be installed at E to reduce the delivery pressure by 200 psig.

It can be seen from Figure 6.15 that at point D, immediately upstream of the pressure regulator, the gas pressure is approximately 600 psig and is

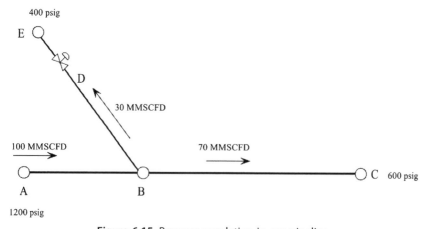

Figure 6.15 Pressure regulation in gas pipeline.

regulated to 400 psig downstream at E. If the mainline flow rate changes from 100 to 90 MMSCFD and the delivery at B is maintained the same at 30 MMSCFD, the gas pressure at B will reduce to a value below 900 psig. Accordingly, the pressure at point D in the branch pipe BE will also reduce to some value below 600 psig. Regardless, because of the pressure regulator, the pressure at E will be maintained at the required 400 psig. However, if for some reason the pressure upstream of the regulator at D falls below 400 psig, the downstream pressure at E cannot be maintained at the original value of 400 psig. The pressure regulator can only reduce the pressure downstream to the required value. It cannot increase the pressure beyond the pressure on the upstream side. If the pressure at D drops to 300 psig, the pressure regulator is ineffective and will remain fully open and the delivery pressure at E will be 300 psig as well.

Problem 6.6

A natural gas pipeline (NPS 16, 0.250-in wall thickness, 50 miles long) with a branch pipe (NPS 8, 0.250-in wall thickness, 15 miles long) as shown in Figure 6.16 is used to transport 100 MMSCFD gas (gravity = 0.6 and viscosity = 0.000,008 lb/ft-s) from A to B. At B (milepost 20), a delivery of 30 MMSCFD occurs into the branch pipe BE. The delivery pressure at E must be maintained at 300 psig. The remaining volume of 70 MMSCFD is shipped to the terminus C at a delivery pressure of 600 psig. Assume constant gas temperature of 60 °F and a pipeline efficiency of 0.95. The base temperature and base pressure are 60 °F and 14.7 psia, respectively.
Compressibility factor $Z = 0.88$.
1. Using Panhandle A equation, calculate the inlet pressure required at A.
2. Is a pressure regulator required at E?
3. If the inlet flow at A drops to 60 MMSCFD, what is the impact in the branch pipeline BE, if the flow rate of 30 MMSCFD is maintained?

Solution

Pipe inside diameter for pipe segment AB and BC = $16 - 2 \times 0.25 = 15.5$ in

First we will consider the pipe segment BC and calculate the pressure P_1 at B for 70 MMSCFD flow rate to deliver gas at 600 psig at C. Using the Panhandle A Eqn (2.55), neglecting elevation effects

$$70 \times 10^6 = 435.87 \times 0.95 \left(\frac{60 + 460}{14.7}\right)^{1.0788}$$
$$\times \left(\frac{P_1^2 - 614.7^2}{0.6^{0.8539} \times 520 \times 30 \times 0.88}\right)^{0.5394} (15.5)^{2.6182}$$

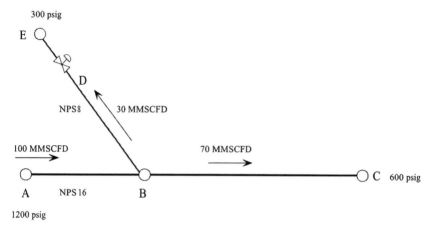

Figure 6.16 Problem: pressure regulation.

Solving for pressure at B, we get

$$P_1 = 660.39 \text{ psia} = 645.69 \text{ psig}$$

Next considering pipe segment AB, flowing 100 MMSCFD, we calculate the inlet pressure P_1 at A using the outlet pressure 660.39 psia we calculated at B.

From the Panhandle A Eqn (2.55)

$$100 \times 10^6 = 435.87 \times 0.95 \left(\frac{520}{14.7}\right)^{1.0788} \left(\frac{P_1^2 - 660.39^2}{0.6^{0.8539} \times 520 \times 20 \times 0.88}\right)^{0.5394}$$
$$\times (15.5)^{2.6182}$$

Solving for the pressure at A, we get

$$P_1 = 715.08 \text{ psia} = 700.38 \text{ psig}$$

Next using the pressure 660.39 psia at B, we calculate the outlet pressure of branch BE that flows 30 MMSCFD through the 15-mile NPS 8 pipe.

Using the Panhandle A Eqn (2.55),

$$30 \times 10^6 = 435.87 \times 0.95 \left(\frac{520}{14.7}\right)^{1.0788} \left(\frac{660.39^2 - P_2^2}{0.6^{0.8539} \times 520 \times 15 \times 0.88}\right)^{0.5394}$$
$$\times (8.125)^{2.6182}$$

Solving for the pressure at E

$$P_2 = 544.90 \text{ psia} = 530.2 \text{ psig}$$

Because the required delivery pressure at E is 300 psig, a pressure regulator will be required at E.

If the flow rate at A drops to 60 MMSCFD and the branch BE flow rate is maintained constant at 30 MMSCFD, we will calculate the junction pressure at B by using the Panhandle A Eqn (2.55) for the pipe segment BC considering a flow rate of 30 MMSCFD and a delivery pressure of 600 psig at C.

$$30 \times 10^6 = 435.87 \times 0.95 \left(\frac{520}{14.7}\right)^{1.0788}$$
$$\times \left(\frac{P_1^2 - 614.7^2}{0.6^{0.8539} \times 520 \times 30 \times 0.88}\right)^{0.5394} (15.5)^{2.6182}$$

Solving for the pressure at B, we get

$$P_1 = 624.47 \text{ psia} = 609.77 \text{ psig}$$

Using the pressure at B, we calculate the outlet pressure at E on branch BE for 30 MMSCFD flow rate using the Panhandle A Eqn (2.55):

$$30 \times 10^6 = 435.87 \times 0.95 \left(\frac{520}{14.7}\right)^{1.0788}$$
$$\times \left(\frac{624.47^2 - P_2^2}{0.6^{0.8539} \times 520 \times 15 \times 0.88}\right)^{0.5394} (8.125)^{2.6182}$$

Solving for the pressure P_2 we get

$$P_2 = 500.76 \text{ psia} = 486.06 \text{ psig}$$

This is the new pressure at E. Comparing this pressure with the previously calculated pressure of 530.2 psig, we see that the delivery pressure at E has dropped by approximately 44 psig. To maintain the delivery pressure of 300 psig at E, a pressure regulator is still required.

Therefore, the answers are
1. Inlet pressure at A = 700.38 psig.
2. A pressure regulator is required at E to reduce the pressure from 530.2 to 300 psig.
3. Finally, a pressure regulator is required at E, to reduce the pressure from 486.1 to 300 psig.

11. SUMMARY

In this chapter, we continued to look at the application of the pressure drop equations introduced in Chapter 5. Several piping configurations such as pipes in series, pipes in parallel, and gas pipelines with injections and deliveries were analyzed to determine pressures required and pipe size needed to satisfy certain requirements. The concepts of equivalent length in series piping and equivalent diameter in pipe loops were explained and illustrated using problems. The hydraulic pressure gradient and the need for intermediate pump/compressor stations to transport given volumes of liquid/gas without exceeding allowable pipeline MAOP were also covered. In liquid pipelines, the concept of system head curves was introduced and illustrated using examples. The method of calculating the line pack volume in a gas pipeline was also explained.

CHAPTER SEVEN

Thermal Hydraulics

In this chapter, we will discuss thermal hydraulics in which the temperature variation in a pipeline and its effect on the pressure drop from friction are taken into account. In many instances, it is important to take into account the variation of the liquid or gas temperature along a transmission pipeline resulting from its impact on pressure drop and flow rate. Isothermal hydraulics, which formed the majority of calculations in previous chapters, will be compared with thermal hydraulics. Because manual calculation of liquid or gas pipeline hydraulics, considering thermal effects is quite laborious and time-consuming, we will use examples of pipeline simulation cases using two popular commercial pipeline hydraulics software programs.

In isothermal flow, there is no significant temperature variation in the fluid being pumped. Examples of isothermal flow in liquid pipelines include water pipelines, refined petroleum products (gasoline, diesel, etc.), and other light crude oil pipelines in which the liquid temperature is close to ambient temperatures. In many cases where heavy crude oil and other liquids of high viscosity have to be pumped, the liquid is heated to some temperature (such as 150°F–180°F) before being pumped through the pipeline. The heated liquid loses heat to the surrounding soil in a buried pipeline as it flows from the origin to the delivery terminus. This results in temperature drop in the liquid which in turn affects the specific gravity and viscosity and hence the friction loss. The overall effect is change in total pressure required to transport the liquid compared with isothermal flow. We will explore how thermal hydraulics is performed in both liquid and gas pipelines by taking into consideration the ambient soil properties in a buried pipeline.

1. TEMPERATURE-DEPENDENT FLOW

In the preceding chapters, we concentrated on steady-state liquid and gas flow in pipelines without paying much attention to temperature variations along the pipeline, resulting from heat transfer between the fluid in the pipeline and the medium surrounding the pipe. We assumed in a liquid pipeline that the liquid entered the pipeline inlet at some temperature such as 70 °F. The liquid properties such as specific gravity and viscosity at the inlet temperature were used to calculate the Reynolds number and friction factor and finally the pressure drop from friction. Similarly, we also used the specific gravity at inlet temperature to calculate the elevation head based on the pipeline topography. In all cases, the liquid properties were considered at some constant flowing temperature. These calculations are therefore based on isothermal (constant temperature) flow.

All of the above may be valid in most cases in which the liquid transported such as water, gasoline, diesel, or light crude oil is at ambient temperatures. As the liquid flows through the pipeline, heat may be transferred to or from the liquid to the surrounding soil (buried pipeline) or the ambient air (above-ground pipeline). Significant changes in liquid temperatures from heat transfer with surroundings will affect liquid properties such as specific gravity and viscosity. This in turn will affect pressure drop calculations. So far we have ignored this heat transfer effect, assuming minimal temperature variations along the pipeline. However, there are instances when the liquid has to be heated to a much higher temperature than ambient conditions to reduce the viscosity and make it flow easily. Pumping a higher viscosity liquid that is heated will also require less pump horsepower.

For example, a high-viscosity crude oil (200 cSt to 800 cSt or more at 60 °F) may be heated to 160 °F before it is pumped into the pipeline. This high-temperature liquid loses heat to the surrounding soil as it flows through the pipeline by conduction of heat from the interior of the pipe, to the soil through the pipe wall. The ambient soil temperature may be 40 °F–50 °F during the winter and 60 °F–80 °F during the summer. Therefore, considerable temperature difference exists between the hot liquid in the pipe and the surrounding soil.

The temperature difference of about 120 °F in winter and 100 °F during summer will cause significant heat transfer between the crude oil and surrounding soil. This will result in temperature drop of the liquid and variation in liquid specific gravity and viscosity as it flows through the pipeline. Therefore, in such instances, we will be wrong in assuming a constant flowing temperature to calculate pressure drop as we do in isothermal flow. Such heated liquid pipeline may be bare or insulated.

To illustrate further, consider a 20-inch buried pipeline transporting 8000 barrels (bbl)/h of a heavy crude oil that enters the pipeline at an inlet temperature of 160 °F. Assume that the liquid temperature has dropped to 124 °F at a location 50 miles from the pipeline inlet. Suppose the crude oil properties at 160 °F inlet conditions and at 124 °F at milepost 50 are as follows:

Temperature (°F)	Specific gravity	Viscosity (cSt)
160	0.9179	40.55
124	0.9306	103.69

Using the 160 °F inlet temperature, we calculate the frictional pressure drop at inlet conditions to be 7.6 psi/mi. At milepost 50, at 124 °F, using the given liquid properties, we find that the frictional pressure drop has increased to 23.97 psi/mi. Thus, in a 50-mile section of pipe the pressure drop from friction varies from 7.6 psi/mi to 23.97 psi/mi as shown in Figure 7.1.

We could use an average value of the pressure drop per mile to calculate the total frictional pressure drop in the first 50 miles of the pipe. However, this will be a very rough estimate.

A better approach would be to subdivide the 50-mile section of the pipeline into smaller segments of 5 or 10 miles each and calculate the pressure drop per mile for each segment. We will then add the individual pressure drops for each 5- or 10-mile segment to get the total friction drop in the 50-mile length of pipeline. Of course, this assumes that we know the temperature of the liquid at 5- or 10-mile increments up to milepost 50. The liquid properties and pressure drop from friction can then be calculated at the boundaries of each 5- or 10-mile segment as illustrated in Figure 7.2.

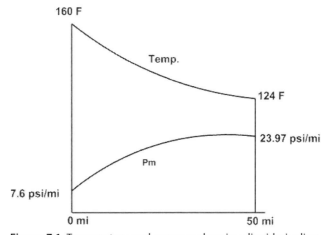

Figure 7.1 Temperature and pressure drop in a liquid pipeline.

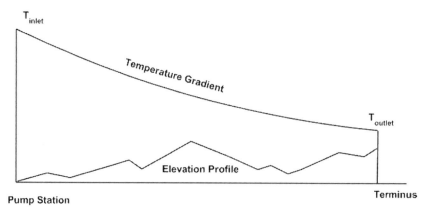

Figure 7.2 Thermal temperature gradient: liquid pipeline.

To calculate the temperature variation along the pipeline, we must resort to a more complex series of calculation, taking into account the soil temperatures along the pipeline, thermal conductivity of pipe material, pipe insulation if any, thermal conductivity of soil, and pipe burial depth. We will present a simplified approach to calculating the temperature profile in a buried pipeline. The method and formulas used were developed originally for the Trans Alaska Pipeline system. These have been found to be quite accurate over the range of temperatures and pressures encountered in heated liquid pipelines today.

The hydraulic gradient showing the pressure profile in a heated buried liquid pipeline as shown in Figure 7.3. Note the curved shape of the gradient compared with the straight line gradient in isothermal flow.

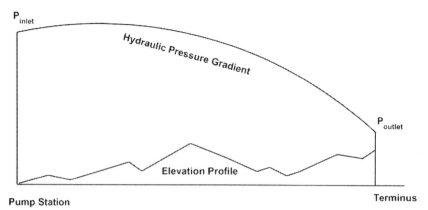

Figure 7.3 Thermal hydraulic pressure gradient: liquid pipeline.

An example will illustrate the use of these formulas. More accurate methods include using a computer software program that will subdivide the pipeline into small segments and compute the heat balance and pressure drop calculations to develop the pressure and temperature profile for the entire pipeline. One such commercially available program is LIQTHERM developed by SYSTEK (www.systek.us).

2. FORMULAS FOR THERMAL HYDRAULICS: LIQUID PIPELINES

2.1 Thermal Conductivity

Thermal conductivity is the property used in heat conduction through a solid. In English units, it is measured in Btu/h/ft/°F. In SI units, thermal conductivity is expressed in W/m/°C.

For heat transfer through a solid of area A and thickness dx, with a temperature difference dT, the formula in English units is:

$$H = K(A)(dT/dx) \qquad (7.1)$$

where

H = heat flux perpendicular to the surface area, Btu/h
K = thermal conductivity of solid, Btu/h/ft/°F
A = area of heat flux, ft^2
Dx = thickness of solid, ft
dT = temperature difference across the solid, °F.

The term dT/dx represents the temperature gradient in °F/ft. Equation (7.1) is also known as the Fourier heat conduction formula.

It can be seen from Eqn (7.1) that the thermal conductivity of a material is numerically equal to the amount of heat transferred across a unit area of the solid material with unit thickness, when the temperature difference between the two faces of the solid is maintained at 1°.

The thermal conductivity for steel pipe and soil are as follows:
K for steel pipe = 29 Btu/h/ft/°F
K for soil = 0.2 to 0.8 Btu/h/ft/°F.

Sometimes, heated liquid pipelines are insulated on the outside with an insulating material. The K value for insulation may range from 0.01 to 0.05 Btu/h/ft/°F.

In SI units, Eqn (7.1) becomes

$$H = K(A)(dT/dx) \qquad (7.2)$$

where
 H = heat flux, W
 K = thermal conductivity of solid, W/m/°C.
 A = area of heat flux, m^2
 dx = thickness of solid, m
 dT = temperature difference across the solid, °C

In SI units, the thermal conductivity for steel pipe, soil, and insulation are as follows:
 K for steel pipe = 50.19 W/m/°C
 K for soil = 0.35 W/m/°C to 1.4 W/m/°C
 K for insulation = 0.02–0.09 W/m/°C.

As an example, heat transfer occurs across a flat steel plate of 8-in thickness and 100-ft^2 area, where the temperature difference across the plate thickness is 20 °F. From Eqn (7.1):
 Heat transfer = 29 × (100) × 20 × 12/8 = 87,000 Btu/h.

2.2 Overall Heat Transfer Coefficient

The overall heat transfer coefficient is also used in heat flux calculations. Eqn (7.1) for heat flux can be written in terms of overall heat transfer coefficient as follows:

$$H = U(A)(dT) \tag{7.3}$$

where
 U = overall heat transfer coefficient, Btu/h/ft^2/°F
 Other symbols in Eqn (7.3) are the same as in Eqn (7.1).
 In SI units, Eqn (7.3) becomes

$$H = U(A)(dT) \tag{7.4}$$

where
 U = overall heat transfer coefficient, W/m^2/°C
 Other symbols in Eqn (7.4) are the same as in Eqn (7.2).

The value of U may range from 0.3 to 0.6 Btu/h/ft^2/°F in English units and 1.7–3.4 W/m^2/°C in SI units.

When analyzing heat transfer between the liquid in a buried pipeline and the outside soil, we consider flow of heat through the pipe wall and pipe insulation, (if any) to the soil. If U represents the overall heat transfer coefficient, we can write from Eqn (7.3):

$$H = U(A)(T_L - T_S) \tag{7.5}$$

where

A = area of pipe under consideration, ft²
T_L = liquid temperature, °F
T_S = soil temperature, °F
U = overall heat transfer coefficient, Btu/h/ft²/°F

Because we are dealing with temperature variation along the pipeline length, we must consider a small section of pipeline at a time when applying Eqn (7.5) for heat transfer.

For example, consider a 100-ft length of 16-in pipe carrying a heated liquid at 150 °F. If the outside soil temperature is 70 °F and the overall heat transfer coefficient:

$$U = 0.5 \text{ Btu/hr/ft}^2/°F$$

We can calculate the heat transfer using Eqn (7.5) as follows:

$$H = 0.5A(150 - 70)$$

where A is the area through which heat flux occurs.

$$A = \pi \times (16/12) \times 100 = 419 \text{ ft}^2$$

Therefore:

$$H = 0.5 \times 419 \times 80 = 16760 \text{ Btu/h}$$

2.3 Heat Balance

The pipeline is subdivided and, for each segment, the heat content balance is computed as follows:

$$H_{in} - \text{DeltaH} + H_w = H_{out} \qquad (7.6)$$

where

H_{in} = heat content entering line segment, Btu/h
DeltaH = heat transferred from line segment to surrounding medium (soil or air), Btu/h
H_w = heat content from frictional work, Btu/h
H_{out} = heat content leaving line segment, Btu/h

Here we have included the effect of frictional heating in the term H_w. With viscous liquids, the effect of friction is to create additional heat, which would raise the liquid temperature. Therefore in thermal hydraulic analysis frictional heating is included to improve the calculation accuracy.

In SI units, Eqn (7.6) will be the same, with each term expressed in watts instead of Btu/h.

The heat balance in Eqn (7.6) forms the basis for computing the outlet temperature of the liquid in a segment starting with its inlet temperature and taking into account the heat loss (or gain) with the surroundings and accounting for frictional heating. In the following sections, we will formulate the method of calculating each term in Eqn (7.6).

2.4 Logarithmic Mean Temperature Difference

In heat transfer calculations, from varying temperatures it is customary to use a slightly different concept of temperature difference called logarithmic mean temperature difference (LMTD). The LMTD between the liquid in the pipeline and the surrounding medium is calculated as follows.

Consider a pipeline segment of length Δx with liquid temperatures T_1 at the upstream end and T_2 at the downstream end of the segment. If T_s represents the average soil temperature (or ambient air temperature, if above ground pipeline) surrounding this pipe segment, the logarithmic mean temperature of the pipe (T_m) segment is calculated as follows:

$$T_m - T_S = \frac{(T_1 - T_S) - (T_2 - T_S)}{\text{Log}_e[(T_1 - T_S)/(T_2 - T_S)]} \tag{7.7}$$

where

T_m = logarithmic mean temperature of pipe segment, °F
T_1 = temperature of liquid entering pipe segment, °F
T_2 = temperature of liquid leaving pipe segment, °F
T_S = sink temperature (soil or surrounding medium), °F

In SI units, Eqn (7.7) will be the same, with all temperatures expressed in °C instead of °F.

For example, if the average soil temperature is 60 °F and the temperature of a pipe segment upstream and downstream are 160 °F and 150 °F, respectively, the logarithmic mean temperature of the pipe segment is

$$T_m = 60 + \frac{(160 - 60) - (150 - 60)}{\text{Log}_e[(160 - 60)/(150 - 60)]} = 60 + 94.88 = 154.88\ °F$$

Thus we calculated the logarithmic mean temperature of the pipe segment to be 154.88 °F. If we had used a simple arithmetic average, we would get the following for mean temperature of pipe segment.

Arithmetic Mean temperature = $(160 + 150)/2 = 155\ °F$

This is not too far off from the logarithmic mean temperature T_m calculated previously. It can be seen that the logarithmic mean temperature approach gives a slightly more accurate representation of the average liquid temperature in the pipe segment. Note that the use of natural logarithm in Eqn (7.7) signifies an exponential decay of the liquid temperature in the pipeline segment. In this example, the LMTD for the pipe segment is

$$\text{LMTD} = 154.88 - 60 = 94.88\ °F$$

If we assume an overall heat transfer coefficient $U = 0.5\ \text{Btu/h/ft}^2/°F$, we can estimate the heat flux from this pipe segment to the surrounding soil using Eqn (7.4) as follows:

Heat flux $= 0.5 \times 1 \times 94.88 = 47.44\ \text{Btu/h}$ per ft^2 of pipe area.

2.5 Heat Entering and Leaving Pipe Segment

The heat content of the liquid entering and leaving a pipe segment is calculated using the mass flow rate of the liquid, its specific heat, and the temperatures at the inlet and outlet of the segment. The heat content of the liquid entering the pipe segment is calculated from

$$H_{in} = w(C_{pi})(T_1) \qquad (7.8)$$

The heat content of the liquid leaving the pipe segment is calculated from

$$H_{out} = w(C_{po})(T_2) \qquad (7.9)$$

where

H_{in} = heat content of liquid entering pipe segment, Btu/h
H_{out} = heat content of liquid leaving pipe segment, Btu/h
C_{pi} = specific heat of liquid at inlet, Btu/lb/°F
C_{po} = specific heat of liquid at outlet, Btu/lb/°F
w = liquid flow rate, lb/h
T_1 = temperature of liquid entering pipe segment, °F
T_2 = temperature of liquid leaving pipe segment, °F.

The specific heat C_p of most liquid range between 0.4 and 0.5 Btu/lb/°F (0.84–2.09 kJ/kg/°C) and increases with liquid temperature. For petroleum fluids, C_p can be calculated if the specific gravity or API gravity and temperatures are known.

In SI units, Eqns (7.8) and (7.9) become

$$H_{in} = w(C_{pi})(T_1) \qquad (7.10)$$

$$H_{out} = w(C_{po})(T_2) \qquad (7.11)$$

where
H_{in} = heat content of liquid entering pipe segment, J/s (W)
H_{out} = heat content of liquid leaving pipe segment, J/s (W)
C_{pi} = specific heat of liquid at inlet, kJ/kg/°C
C_{po} = specific heat of liquid at outlet, kJ/kg/°C
w = liquid flow rate, kg/s
T_1 = temperature of liquid entering pipe segment, °C
T_2 = temperature of liquid leaving pipe segment, °C.

2.6 Heat Transfer: Buried Pipeline

Consider a buried pipeline, with insulation, that transports a heated liquid. If the pipeline is divided into segments of length L, we can calculate the heat transfer between the liquid and the surrounding medium using the following equations.
In English units:

$$H_b = 6.28(L)(T_m - T_S)/(Parm1 + Parm2) \qquad (7.12)$$

$$Parm1 = (1/K_{ins})\text{Log}_e(R_i/R_p) \qquad (7.13)$$

$$Parm2 = (1/K_s)\text{Log}_e\left[2S/D + ((2S/D)^2 - 1)^{1/2}\right] \qquad (7.14)$$

where
H_b = heat transfer, Btu/h
T_m = log mean temperature of pipe segment, °F
T_S = ambient soil temperature, °F
L = pipe segment length, ft
R_i = pipe insulation outer radius, ft
R_p = pipe wall outer radius, ft
K_{ins} = thermal conductivity of insulation, Btu/h/ft/°F
K_s = thermal conductivity of soil, Btu/h/ft/°F
S = depth of cover (pipe burial depth) to pipe centerline, ft
D = pipe outside diameter, ft

Parm1 and Parm2 are intermediate values that depend on parameters indicated.

In SI units, Eqns (7.12), (7.13) and (7.14) become

$$H_b = 6.28(L)(T_m - T_S)/(Parm1 + Parm2) \qquad (7.15)$$

$$Parm1 = (1/K_{ins})\text{Log}_e(R_i/R_p) \qquad (7.16)$$

$$Parm2 = (1/K_s)\text{Log}_e\left[2S/D + ((2S/D)^2 - 1)^{1/2}\right] \qquad (7.17)$$

where

H_b = heat transfer, W
T_m = log mean temperature of pipe segment, °C
T_S = ambient soil temperature, °C
L = pipe segment length, m
R_i = pipe insulation outer radius, mm
R_p = pipe wall outer radius, mm
K_{ins} = thermal conductivity of insulation, W/m/°C
K_s = thermal conductivity of soil, W/m/°C
S = depth of cover (pipe burial depth) to pipe centerline, mm
D = pipe outside diameter, mm

2.7 Heat Transfer: Above-Ground Pipeline

Similar to the buried pipeline discussed in the earlier section, an above-ground insulated pipeline is used to transports a heated liquid. If the pipeline is divided into segments of length L, we can calculate the heat transfer between the liquid and the ambient air using the following equations.

In English units:

$$H_a = 6.28(L)(T_m - T_S)/(Parm1 + Parm3) \qquad (7.18)$$

$$Parm3 = 1.25/[R_i(4.8 + 0.008(T_m - T_S))] \qquad (7.19)$$

$$Parm1 = (1/K_{ins})\text{Log}_e(R_i/R_p) \qquad (7.20)$$

where

H_a = heat transfer, Btu/h
T_m = log mean temperature of pipe segment, °F
T_S = ambient soil temperature, °F
L = pipe segment length, ft

R_i = pipe insulation outer radius, ft
R_p = pipe wall outer radius, ft
K_{ins} = thermal conductivity of insulation, Btu/h/ft/°F
K_s = thermal conductivity of soil, Btu/h/ft/°F
S = depth of cover (pipe burial depth) to pipe centerline, ft
D = pipe outside diameter, ft

In SI units, Eqns (7.18), (7.19) and (7.20) become

$$H_a = 6.28(L)(T_m - T_S)/(Parm1 + Parm3) \qquad (7.21)$$

$$Parm3 = 1.25/[R_i(4.8 + 0.008(T_m - T_S))] \qquad (7.22)$$

$$Parm1 = (1/K_{ins})\text{Log}_e(R_i/R_p) \qquad (7.23)$$

where
H_a = heat transfer, W
T_m = log mean temperature of pipe segment, °C
T_S = ambient soil temperature, °C
L = pipe segment length, m
R_i = pipe insulation outer radius, mm
R_p = pipe wall outer radius, mm
K_{ins} = thermal conductivity of insulation, W/m/°C
K_s = thermal conductivity of soil, W/m/°C
S = depth of cover (pipe burial depth) to pipe centerline, mm
D = pipe outside diameter, mm

2.8 Frictional Heating

The frictional pressure drop causes heating of the liquid. The heat gained by the liquid from friction is calculated using the following equations.

$$H_w = 2545(HHP) \qquad (7.24)$$

$$HHP = (1.7664 \times 10^{-4})(Q)(Sg)(h_f)(L_m) \qquad (7.25)$$

where
H_w = frictional heat gained, Btu/h
HHP = hydraulic horsepower required for pipe friction
Q = liquid flow rate, bbl/h
Sg = liquid specific gravity
h_f = frictional head loss, ft/mi
L_m = pipe segment length, miles

In SI units, Eqns (7.24) and (7.25) become

$$H_w = 1000(\text{Power}) \tag{7.26}$$

$$\text{Power} = (0.00272)(Q)(Sg)(h_f)(L_m) \tag{7.27}$$

where
H_w = frictional heat gained, W
Power = power required for pipe friction, kW
Q = liquid flow rate, m³/h
Sg = liquid specific gravity
h_f = friction loss, m/km
L_m = pipe segment length, km

2.9 Pipe Segment Outlet Temperature

Using the formulas developed in the preceding sections and referring to the heat balance Eqn (7.6), we can now calculate the temperature of the liquid at the outlet of the pipe segment as follows.

For buried pipe:

$$T_2 = (1/wC_p)\left[2545(\text{HHP}) - H_b + (wC_p)T_1\right] \tag{7.28}$$

For above-ground pipe:

$$T_2 = (1/wC_p)\left[2545(\text{HHP}) - H_a + (wC_p)T_1\right] \tag{7.29}$$

where
H_b = heat transfer for buried pipe, Btu/h from Eqn (7.12)
H_a = heat transfer for above ground pipe, Btu/h from Eqn (7.18)
C_p = average specific heat of liquid in pipe segment

For simplicity, we have used the average specific heat as previously for the pipe segment based on C_{pi} and C_{po} discussed earlier in Eqns (7.8) and (7.9).

In SI units, Eqns (7.28) and (7.29) can be expressed as follows.
For buried pipe:

$$T_2 = (1/wC_p)\left[1000(\text{Power}) - H_b + (wC_p)T_1\right] \tag{7.30}$$

For above-ground pipe:

$$T_2 = (1/wC_p)\left[1000(\text{Power}) - H_a + (wC_p)T_1\right] \tag{7.31}$$

where
- H_b = heat transfer for buried pipe, W
- H_a = heat transfer for above ground pipe, W
- Power = frictional power defined in Eqn (7.27), kW

2.10 Liquid Heating from Pump Inefficiency

Because a centrifugal pump is not 100% efficient, the difference between the hydraulic horsepower and the brake horsepower (BHP) represents power lost. Most of this power lost is converted to heating the liquid being pumped. The temperature rise of the liquid from pump inefficiency may be calculated from the following equation.

$$\Delta T = (H/778 C_P)(1/E - 1) \quad (7.32)$$

where
- ΔT = temperature rise, °F
- H = pump head, ft
- C_P = specific heat of liquid, Btu/lb/°F
- E = pump efficiency as a decimal value, less than 1.0

When considering thermal hydraulics, these temperature rises as the liquid moves through a pump station should be included in the temperature profile calculation. For example, if the liquid temperature has dropped to 120 °F at the suction side of a pump station and the temperature rise from pump inefficiency causes a 3 °F rise, the liquid temperature at the pump discharge will be 123 °F.

Example Problem 7.1
Calculate the temperature rise of a liquid (specific heat = 0.45 Btu/lb/°F) as it flows through a pump from pump inefficiency.
Pump head = 2450 ft and pump efficiency = 75%

Solution
From Eqn (7.32), the temperature rise is

$$\Delta T = (2450/(778 \times 0.45))(1/0.75 - 1) = 2.33 \,°F$$

We will now use the equations discussed in this chapter to calculate the thermal hydraulic temperature profile of a crude oil pipeline.

Example Problem 7.2

A 16-in, 0.250-in wall thickness, 50-mile long buried pipeline transports 4000 bbl/h of heavy crude oil that enters the pipeline at 160 °F. The crude oil has a specific gravity and viscosity as follows:

Temperature (°F)	100	140
Specific gravity	0.967	0.953
Viscosity (cSt)	2277	348

Assume a pipe burial depth of 36 in to the top of pipe and 1.5-in insulation thickness with a thermal conductivity (K value) of 0.02 Btu/h/ft/°F. Also assume a uniform soil temperature of 60 °F with a K value of 0.5 Btu/h/ft/°F. Using the heat balance equation, calculate the outlet temperature of the crude oil at the end of the first mile segment. Assume average specific heat of 0.45 for the crude oil.

Solution

First calculate the heat transfer for buried pipe using Eqns (7.12) through (7.14).

$$\text{Parm1} = (1/0.02)\text{Log}_e(9.5/8) = 8.5925$$

$$\text{Parm2} = (1/0.5)\text{Log}_e\left[2 \times 44/16 + \left((2 \times 44/16)^2 - 1\right)^{1/2}\right] = 4.7791$$

$$H_b = 6.28(5280)(T_m - 60)/(8.5925 + 4.7791)$$

or

$$H_b = 2479.76(T_m - 60) \text{Btu/hr} \qquad (7.33)$$

The log mean temperature T_m of this 1-mile pipe segment has to be approximated first because it depends on the inlet temperature, soil temperature, and the unknown liquid temperature at the outlet of the 1-mile segment.

As a first approximation, assume the outlet temperature at the end the first 1-mile segment to be $T_2 = 150$. Calculate T_m using Eqn (7.7).

$$T_m = 60 + \frac{(160 - 60) - (150 - 60)}{\text{Log}_e\left[(160 - 60)/(150 - 60)\right]} \qquad (7.34)$$

or

$$T_m = 154.91 \text{ °F}$$

Therefore, Hb from Eqn (7.32) becomes

$$H_b = 2479.76(154.91 - 60) = 235,354 \text{ Btu/hr}$$

The frictional heating component H_w will be calculated using Eqns (7.24) and (7.25). The friction drop h_f depends on specific gravity and viscosity at the calculated mean temperature T_m. Using the viscosity temperature relationship from Chapter 2, we calculate the specific gravity and viscosity at 154.91 °F to be 0.9478 and 200.22 cSt. respectively.

Reynolds number:

$$R = \frac{92.24 \times (4000 \times 24)}{15.5 \times 200.22} = 2853$$

Using the Colebrook—White equation, the friction factor is

$$f = 0.034$$

Friction drop h_f will be calculated from Darcy—Weisbach Eqn (3.26) as follows:

$$h_f = 0.034 (5280 \times 12/15.5) (V^2/64.4)$$

The velocity V is calculated using Eqn (3.12) as follows:

$$V = 0.2859 (4000)/(15.5)^2 = 4.76 \text{ ft/s}$$

Therefore, friction pressure drop is

$$h_f = 0.034 (5280 \times 12/15.5)(4.76 \times 4.76/64.4) = 48.89 \text{ ft}$$

From Eqn (7.25), the frictional horsepower (HP) is

$$HHP = (1.7664 \times 10^{-4}) \times 4000 \times 0.9478 \times 48.89 \times 1.0 = 32.74$$

Therefore, frictional heating from Eqn (7.24) is

$$H_w = 2545 \times 32.74 = 83323 \text{ Btu/h}$$

The mass flow rate is

$$w = 4000 \times 5.6146 \times 0.9478 \times 62.4 = 1.328 \times 10^6 \text{ lb/h}$$

From Eqn (7.30), the liquid temperature at the outlet of the 1-mile segment is

$$T_2 = (1/(1.328 \times 10^6 \times 0.45))$$
$$\times [83,323 - 235,354 + 1.328 \times 10^6 \times 0.45 \times 160]$$

$$T_2 = [-0.255 + 160] = 159.75 \text{ °F}$$

This value of T_2 is used as a second approximation in Eqn (7.34) to calculate a new value of T_m and subsequently the next approximation for T_2. Calculations are repeated until successive values of T_2 are within close agreement. This is left as an exercise for the reader.

It can be seen from the foregoing that manual calculation of temperatures and pressures along a heated oil pipeline is definitely a laborious process that can be eased using programmable calculators and personal computers.

Thermal hydraulics is very complex and calculations require use of some type of a computer program to generate quick results. Such a program can subdivide the pipeline into short segments and calculate the temperatures, liquid properties, and pressure drops as we have seen in the examples in this chapter. Several commercial software packages are available to perform thermal hydraulics. One such software is LIQTHERM developed by SYSTEK Technologies, Inc. (www.systek.us). For a sample output report from a liquid pipeline thermal hydraulics analysis using LIQTHERM software, refer to Appendix A.15.

3. ISOTHERMAL VERSUS THERMAL HYDRAULICS: GAS PIPELINES

In the previous chapters, the hydraulic analysis of gas flow through pipelines was mainly done based upon isothermal or constant temperature flow. This assumption is fairly good in long-distance pipelines where the gas temperature reaches a constant value equal to or close to the surrounding soil (or ambient) temperature at large distances from the compressor stations. However, upon compressing the gas, depending upon the compression ratio, the outlet temperature of the gas from the compressor station may be considerably higher than that of the ambient air or surrounding soil.

Recalling from a previous chapter, we found that when gas is compressed adiabatically from 60 °F suction temperature and a compression ratio of 2.0, the discharge temperature was 278.3 °F. Because pipe coating limitations restrict temperatures to about 140–150 °F, cooling of the compressed gas is necessary at the downstream side of the compressor station. In this example, assuming gas cooling results in a discharge temperature of 140 °F as gas enters the pipeline, we find that the temperature difference between the gas at 140 °F and the surrounding soil at 70 °F will cause heat transfer to take place between the pipeline gas and the surrounding soil. The gas temperature drops off rapidly for the first few miles and eventually reaches a temperature close to the soil temperature. Additionally, in a long transmission pipeline, the soil temperature may vary along the pipeline as well, causing different heat transfer rates at locations along the pipeline. This is illustrated in Figure 7.4.

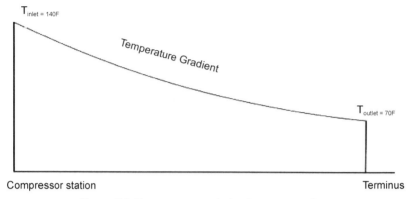

Figure 7.4 Temperature variation in a gas pipeline.

In some instances, the expansion of gas as it flows along a pipeline may result in gas temperature reaching slightly lower temperatures than the surrounding soil. This is called the Joule–Thompson cooling effect. Thus if the soil temperature is fairly constant at 70 °F, from the Joule–Thompson effect, the final temperature of the gas at the terminus of the pipeline may drop to 60 °F or 65 °F. This is illustrated in Figure 7.5.

This cooling of gas below the surrounding soil temperature depends upon the pressure differential and the Joule–Thompson coefficient. Ignoring this cooling effect will result in a more conservative (lower flow rate for a given pressure drop) flow rate calculation, because cooler temperature means less pressure drop in a gas pipeline and hence higher flow rate.

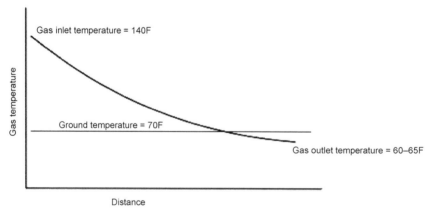

Figure 7.5 Joule Thompson cooling effect in a gas pipeline.

Thermal hydraulics is the study of gas pipeline pressures, temperatures, and flow rates taking into account the thermal properties of the soil, pipe, and pipe insulation, if any. From such variation in gas temperature, calculation of pressure drop must be made by considering short lengths of pipe that make up the total pipeline. For example, if the pipeline is 50 miles long we would subdivide it into short segments of 1- or 2-mile lengths and apply the general flow equation for each pipe segment, considering an average gas temperature and an average ambient soil temperature. Starting with the upstream pressure of segment 1, the downstream pressure will be calculated assuming an average temperature for segment 1. Next, using the calculated downstream pressure as the upstream pressure for segment 2, we calculate the downstream pressure for segment 2. The process is continued until all segments of the pipeline are covered. The variation of temperature from segment to segment must be taken into account to calculate the compressibility factor used in the general flow equation. The following equation is the general flow equation that we used extensively in previous chapters.

$$Q = 77.54 \left(\frac{T_b}{P_b}\right) \left(\frac{P_1^2 - P_2^2}{GT_f LZf}\right)^{0.5} D^{2.5} \qquad (7.34a)$$

From the general flow equation, we see that if the gas flowing temperature T_f is constant (isothermal flow) throughout the length of the pipeline L_e, the compressibility factor Z of the gas will depend on the average pressure of the pipe segment. If the gas temperature also varies along the pipeline, based on the preceding discussions, the compressibility factor Z will change in a different manner because it is a function of both the gas temperature and gas pressure. Therefore, it is seen that calculation of P_1 or P_2 for a given flow rate Q from Eqn (7.34a) will yield different results if isothermal conditions did not exist.

The calculation of gas temperature at any point along the pipeline by taking into account the heat transfer between the gas and surrounding soil is quite complicated. It does not lend itself easily to manual calculations. We will discuss the method of calculation for thermal hydraulics in this section. To accurately take into account the temperature variations, a suitable gas pipeline hydraulics simulation program must be used, because, as indicated previously, manual calculation is quite laborious and time-consuming. Several commercial simulation programs are available to model steady state gas pipeline hydraulics. These programs calculate the gas temperature and pressures by taking into consideration variations of soil temperature, pipe burial depth, and thermal conductivities of pipe, insulation, and soil. One such software program is GASMOD, marketed by SYSTEK Technologies, Inc. (http://www.systek.us). The Appendix includes an example simulation of a gas pipeline using the GASMOD software. In this chapter, we will use GASMOD to illustrate thermal hydraulics analysis.

4. TEMPERATURE VARIATION AND GAS PIPELINE MODELING

Consider a buried pipeline transporting gas from point A to point B. We will analyze a short segment of length ΔL of this pipe as shown in Figure 7.6 and apply the principles of heat transfer to determine how the gas temperature varies along the pipeline.

The upstream end of the pipe segment of length ΔL is at a temperature T_1 and the downstream end at temperature T_2. The average gas temperature in this segment is represented by T. The outside soil temperature at this location is T_s. Assume steady-state conditions and the mass flow rate of gas to be m. The gas flow from the upstream end to the downstream end of the segment causes a temperature drop of ΔT. The heat loss from the gas may be represented by

$$\Delta H = -mCp\Delta T \qquad (7.35)$$

where

ΔH = heat transfer rate, Btu/h
m = mass flow rate of gas, lb/h
Cp = average specific heat of gas, Btu/h/lb/F
ΔT = temperature difference = $T_1 - T_2$, °F

The negative sign in Eqn (7.35) indicates loss of heat from upstream temperature T_1 to downstream temperature T_2

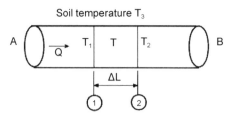

Figure 7.6 Analysis of temperature variation.

Next we consider the heat transfer from the gas to the surrounding soil in terms of the overall heat transfer coefficient U and the difference in temperature between the gas and surrounding soil represented by $(T - T_s)$. Therefore, we can write the following equation for heat transfer:

$$\Delta H = U \Delta A (T - T_s) \quad (7.36)$$

where
- U = overall heat transfer coefficient, Btu/h/ft^2/°F
- ΔA = surface area of pipe for heat transfer = $\pi D \Delta L$
- T = average gas temperature in pipe segment, °F
- T_s = average soil temperature surrounding pipe segment, °F
- D = pipe inside diameter, ft

Equating the two values of heat transfer rate ΔH from Eqns (7.35) and (7.36), we get

$$-mCp\Delta T = U\Delta A(T - T_s)$$

Simplifying, we get

$$\frac{\Delta T}{T - T_s} = -\left(\frac{\pi U D}{mCp}\right)\Delta L \quad (7.37)$$

Rewriting Eqn (7.37) in differential form and integrating, we get

$$\int_1^2 \frac{dT}{T - T_s} = \int_2^1 -\left(\frac{\pi U D}{mCp}\right) dL \quad (7.38)$$

Integrating and simplifying, we get

$$\frac{T_2 - T_s}{T_1 - T_s} = e^{-\theta} \quad (7.39)$$

where e = base of natural logarithms (e = 2.718…) and

$$\theta = \frac{\pi U D \Delta L}{mCp} \quad (7.40)$$

Simplifying Eqn (7.40) further, we get the downstream temperature of the pipe segment of length ΔL as

$$T_2 = T_s + (T_1 - T_s)e^{-\theta} \quad (7.41)$$

It can be seen from Eqn (7.41) that as the pipe length increases, the term approaches zero and the temperature T_2 becomes equal to soil temperature T_s. Therefore, in a long gas pipeline, the gas temperature ultimately equals the surrounding soil temperature. This is illustrated in Figure 7.4 (temperature variation in a gas pipeline).

In the preceding analysis, we made several simplifying assumptions. We assumed that the soil temperature and the overall heat transfer coefficient remained constant and ignored the Joule–Thompson effect as gas expands through a pipeline. In a long pipeline, the soil temperature may actually vary along the pipeline and therefore must be taken into account in these calculations. One approach would be to subdivide the pipeline into segments that have constant soil temperatures and perform calculations for each segment separately. The Joule–Thompson effect causes the gas to cool slightly from expansion. Therefore, in a long pipeline, the gas temperature at the delivery point may fall below that of the ground or soil temperature as indicated in Figure 7.5 (Joule–Thompson cooling effect).

5. REVIEW OF SIMULATION MODEL REPORTS

To illustrate the effect of thermal effects in a gas pipeline, we will analyze a gas transmission pipeline first using the method outlined in Chapter 5. Next we will analyze the same pipeline taking into account the thermal conductivity and soil temperatures. The latter method requires the use of a computer simulation model. To do this, we have chosen the commercially available software known as GASMOD. We will compare the results of the isothermal hydraulics of chapter 5 with the thermal hydraulics using GASMOD. Examples will be used to illustrate the comparison.

Example Problem 7.3
A natural gas pipeline system is being built from Rockford to Concord, a distance of 240 miles. The pipeline is constructed of NPS 30 with 0.500-in wall thickness, API 5L-X60 pipe. The maximum operating pressure (MOP) is 1400 psig. Gas enters the Rockford compressor station at 70 °F and 800 psig pressure. The soil temperature may be assumed to be 60 °F throughout. The gas flow rate is 420 million standard cubic feet per day (MMSCFD) and the gas specific gravity and viscosity are 0.6 and 0.000008 lb/ft.s. The contract delivery pressure required at Concord is 500 psig. Assume isothermal flow at 70 °F and gas-specific heat ratio is 1.29. Use compressor adiabatic efficiency of 80% and mechanical efficiency of 98%. Use the general flow equation with the Colebrook friction factor,

assuming a pipe internal roughness of 700 micro in. Calculate the pressure profile and the compressor horsepower required at Rockford. Compare these results with thermal hydraulic analysis using GASMOD. Assume a base pressure of 14.7 psia and base temperature of 60 °F. The pipeline elevation profile is essentially flat.

Solution
Inside diameter of pipe $D = 30 - 2 \times 0.500 = 29$ in
First we calculate the Reynolds number:

$$R = 0.0004778 \left(\frac{14.7}{520}\right) \left(\frac{0.6 \times 420 \times 10^6}{0.000008 \times 29}\right) = 14.67 \times 10^6$$

Next, using the Colebrook equation we calculate the friction factor as

$$\frac{1}{\sqrt{f}} = -2 \text{Log}_{10} \left(\frac{0.0007}{3.7 \times 29} + \frac{2.51}{14.67 \times 10^6 \sqrt{f}}\right)$$

Solving by trial and error, we get

$$f = 0.0097$$

Therefore, the transmission factor is

$$F = \frac{2}{\sqrt{0.0097}} = 20.33$$

To calculate the compressibility factor Z, the average pressure is required. Because the inlet pressure is unknown, we will calculate an approximate value of Z using a value of 110% of the delivery pressure for the average pressure.
The average pressure is

$$P_{avg} = 1.1 \times (500 + 14.7) = 566.17 \text{ psia} = 551.47 \text{ psig}$$

Using the California Natural Gas Association (CNGA) equation we calculate the value of compressibility factor as

$$Z = \frac{1}{\left[1 + \left(\frac{551.47 \times 344400 (10)^{1.785 \times 0.6}}{520^{3.825}}\right)\right]} = 0.9217$$

Because there is no elevation difference between the beginning of the pipeline and the end of the pipeline, the elevation component can be neglected and $e^s = 1$.
Outlet pressure is

$$P_2 = 500 + 14.7 = 514.7 \text{ psia}.$$

From the general flow equation, we get

$$420 \times 10^6 = 38.77 \times 20.33 \left(\frac{520}{14.7}\right) \left(\frac{P_1^2 - 514.7^2}{0.6 \times 530 \times 240 \times 0.9217}\right)^{0.5} (29.0)^{2.5}$$

Solving for the upstream pressure, we get

$$P_1 = 1021.34 \text{ psia} = 1006.64 \text{ psig}$$

Using this value of P_1, we calculate the new average pressure.

$$P_{avg} = \frac{2}{3}\left(1021.34 + 514.7 - \frac{1021.34 \times 514.7}{1021.34 + 514.7}\right) = 795.87 \text{ psia}$$

Compare this with the value of 566.17 psia we assumed initially for calculating Z. Obviously, we were way off. Recalculating Z using the recently calculated value of P_{avg}, we get

$$Z = \frac{1}{\left[1 + \left(\frac{(795.87 - 14.7) \times 344400(10)^{1.785 \times 0.6}}{530^{3.825}}\right)\right]} = 0.8926$$

This compares with 0.9217 we calculated earlier for Z. We will now recalculate the inlet pressure using this value of Z.

$$420 \times 10^6 = 38.77 \times 20.33 \left(\frac{520}{14.7}\right) \left(\frac{P_1^2 - 514.7^2}{0.6 \times 530 \times 240 \times 0.8926}\right)^{0.5} (29.0)^{2.5}$$

Solving for the upstream pressure, we get

$$P_1 = 1009.24 \text{ psia} = 994.54 \text{ psig}$$

This compares with 1021.34 psia calculated earlier. This is almost 1% different. We could repeat the process and get a better approximation.

Using this recently calculated value of P_1, we calculate the new average pressure as

$$P_{avg} = = 788.72 \text{ psia}$$

this compares with the previous approximation of 795.87 psia. The error is less than a percent.

Recalculating Z using this value of P_{avg}, we get $= 0.8935$.

We will now recalculate the inlet pressure using this value of Z, in which we get

$$420 \times 10^6 = 38.77$$

$$\times 20.33 \left(\frac{520}{14.7}\right) \left(\frac{P_1^2 - 514.7^2}{0.6 \times 530 \times 240 \times 0.8935}\right)^{0.5} (29.0)^{2.5}$$

Solving for the upstream pressure, we get

$$P_1 = 1009.62 \text{ psia} = 994.92 \text{ psig}$$

This compares with 1009.24 psia calculated earlier. The difference is less than 0.04% and therefore we can stop iterating any further.

The HP required at Rockford compressor station will be calculated as follows:

$$\text{HP} = 0.0857 \times 420 \left(\frac{1.29}{0.29}\right)(70 + 460)$$

$$\times \left(\frac{1 + 0.8935}{2}\right) \left(\frac{1}{0.8}\right) \left[\left(\frac{1009.62}{814.7}\right)^{\frac{0.29}{1.29}} - 1\right] = 4962$$

Next, we calculate the driver horsepower required based on a mechanical efficiency of 0.98.

The BHP required = 4962/0.98 = 5063.

The final results are:

Inlet pressure at Rockford = 994.92 psig

Delivery pressure at Concord = 500 psig at a flow rate of 420 MMSCFD

BHP required at Rockford compressor station = 5063 HP.

It must be noted that the preceding calculations ignored any elevation changes along the pipeline. If we had considered the pipe elevations at Rockford and Concord, the result would have been different.

This pressure of 994.92 psig required at the Rockford compressor station was calculated assuming a constant gas flowing temperature of 70 °F and considering the pipeline as a single segment 240 miles long. As explained in earlier chapters, the calculation accuracy is improved if we subdivide the pipeline into short segments. By doing so, we calculate the upstream pressure of each segment starting with the last segment near Concord. If the pipeline is divided into 100 equal pipe segments of 2.4 miles each, the pressure P_{100} at the upstream end of the last segment is calculated using the general flow equation, considering 500 psig downstream pressure. Next,

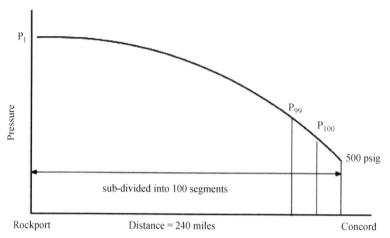

Figure 7.7 Subdividing pipe into segments.

using this calculated pressure P_{100}, we calculate the upstream pressure P_{99} of the 99th segment. The process is repeated until all segments are covered, and the value of the pressure P_1 at Rockford is calculated. This is illustrated in Figure 7.7.

By subdividing the pipeline in this fashion, we are improving the accuracy of calculations. Of course, manual calculation in this manner is going to be quite laborious and time-consuming and we should use some form of a computer program to perform this task.

Next we will compare the isothermal calculation results with thermal hydraulics using the GASMOD program. Following is the output report from the GASMOD program.

```
*********** GASMOD - GAS PIPELINE HYDRAULIC SIMULATION ***********

*********** GASMOD - GAS PIPELINE HYDRAULIC SIMULATION ***********
*********** Version 6.00.780 ************
DATE:                                    3-December-2013    TIME:  13:24:58
PROJECT DESCRIPTION:
Pipeline from Rockford to Concord
Case Number:                             1001
Pipeline data file:                      C:\Users\My Documents\Gasmod\Problem73.TOT

Pressure drop formula:                   Colebrook-White
Pipeline efficiency:                     1.00
Compressibility Factor Method:           CNGA

Inlet Gas Gravity(Air=1.0):              0.60000
Inlet Gas Viscosity:                     0.0000080(lb/ft-sec)
Gas specific heat ratio:                 1.29
Polytropic compression index:            1.30
```

Thermal Hydraulics

**** Calculations Based on Specified Thermal Conductivities of Pipe, Soil and Insulation ****

```
Base temperature:                         60.00(degF)
Base pressure:                            14.70(psia)

Origin suction temperature:               70.00(degF)
Origin suction pressure:                  800.00(psig)
Pipeline Terminus Delivery  pressure:     500.35(psig)
Minimum pressure:                         100.0(psig)
Maximum gas velocity:                     50.00(ft/sec)

Inlet  Flow rate:                         420.00(MMSCFD)
Outlet Flow rate:                         420.00(MMSCFD)

CALCULATION OPTIONS:
Polytropic compression considered:        YES
Branch pipe calculations:                 NO
Loop pipe calculations:                   NO
Compressor Fuel Calculated:               NO
Joule Thompson effect included :          NO
Customized Output:                        NO
Holding Delivery Pressure at terminus
```

ALL PRESSURES ARE GAUGE PRESSURES, UNLESS OTHERWISE SPECIFED AS ABSOLUTE PRESSURES

**************** PIPELINE PROFILE DATA ************

Distance (mi)	Elevation (ft)	Diameter (in)	Thickness (in)	Roughness (in)
0.00	250.00	30.000	0.500	0.000700
10.00	250.00	30.000	0.500	0.000700
20.00	250.00	30.000	0.500	0.000700
30.00	250.00	30.000	0.500	0.000700
40.00	250.00	30.000	0.500	0.000700
50.00	250.00	30.000	0.500	0.000700
60.00	250.00	30.000	0.500	0.000700
70.00	250.00	30.000	0.500	0.000700
100.00	250.00	30.000	0.500	0.000700
120.00	250.00	30.000	0.500	0.000700
140.00	250.00	30.000	0.500	0.000700
150.00	250.00	30.000	0.500	0.000700
170.00	250.00	30.000	0.500	0.000700
190.00	250.00	30.000	0.500	0.000700
200.00	250.00	30.000	0.500	0.000700
220.00	250.00	30.000	0.500	0.000700
240.00	250.00	30.000	0.500	0.000700

*************** THERMAL CONDUCTIVITY AND INSULATION DATA ****************

Distance (mi)	Cover (in)	Thermal Conductivity (Btu/hr/ft/degF)			Insul.Thk (in)	Soil Temp (degF)
		Pipe	Soil	Insulation		
0.000	36.000	29.000	0.800	0.020	0.000	60.00
10.000	36.000	29.000	0.800	0.020	0.000	60.00
20.000	36.000	29.000	0.800	0.020	0.000	60.00
30.000	36.000	29.000	0.800	0.020	0.000	60.00
40.000	36.000	29.000	0.800	0.020	0.000	60.00
50.000	36.000	29.000	0.800	0.020	0.000	60.00
60.000	36.000	29.000	0.800	0.020	0.000	60.00
70.000	36.000	29.000	0.800	0.020	0.000	60.00
100.000	36.000	29.000	0.800	0.020	0.000	60.00
120.000	36.000	29.000	0.800	0.020	0.000	60.00
140.000	36.000	29.000	0.800	0.020	0.000	60.00
150.000	36.000	29.000	0.800	0.020	0.000	60.00
170.000	36.000	29.000	0.800	0.020	0.000	60.00
190.000	36.000	29.000	0.800	0.020	0.000	60.00
200.000	36.000	29.000	0.800	0.020	0.000	60.00
220.000	36.000	29.000	0.800	0.020	0.000	60.00
240.000	36.000	29.000	0.800	0.020	0.000	60.00

**************** LOCATIONS AND FLOW RATES ****************

Location	Distance (mi)	Flow in/out (MMSCFD)	Gravity	Viscosity (lb/ft-sec)	Pressure (psig)	GasTemp. (degF)
Rockford	0.00	420.0000	0.6000	0.00000800	800.00	70.00
Concord	240.00	-420.0000	0.6000	0.00000800	500.35	60.00

**************** COMPRESSOR STATION DATA ***************

FLOW RATES, PRESSURES AND TEMPERATURES:

Name MaxPipe Temp (degF)	Flow Rate (MMSCFD)	Suct. Press. (psig)	Disch. Press. (psig)	Compr. Ratio	Suct. Loss. (psia)	Disch. Loss. (psia)	Suct. Temp. (degF)	Disch. Temp (degF)
Rockford 140.00	420.00	795.00	1006.42	1.2611	5.00	10.00	70.00	105.47

************* COMPRESSOR EFFICIENCY, HP AND FUEL USED ****************

Name Installed (HP)	Distance (mi)	Compr Effy. (%)	Mech. Effy. (%)	Overall Effy. (%)	Horse Power	Fuel Factor (MCF/day/HP)	Fuel Used (MMSCFD)
Rockford 5000	0.00	80.00	98.00	78.40	5,163.89	0.2000	------

Total Compressor Station Horsepower: 5,163.89
5,000.

WARNING!
Required HP exceeds the installed HP at compressor station: Rockford

************** REYNOLD'S NUMBER AND HEAT TRANSFER COEFFICIENT **************

Distance (mi)	Reynold'sNum.	FrictFactor (Darcy)	Transmission Factor	HeatTransCoeff (Btu/hr/ft2/degF)	CompressibilityFactor (CNGA)
0.000	14,671,438.	0.0099	20.13	0.3361	0.8879
10.000	14,671,438.	0.0099	20.13	0.3361	0.8801
20.000	14,671,438.	0.0099	20.13	0.3361	0.8754
30.000	14,671,438.	0.0099	20.13	0.3361	0.8729
40.000	14,671,438.	0.0099	20.13	0.3361	0.8721
50.000	14,671,438.	0.0099	20.13	0.3361	0.8724
60.000	14,671,438.	0.0099	20.13	0.3361	0.8734
70.000	14,671,438.	0.0099	20.13	0.3361	0.8767
100.000	14,671,438.	0.0099	20.13	0.3361	0.8821
120.000	14,671,438.	0.0099	20.13	0.3361	0.8870
140.000	14,671,438.	0.0099	20.13	0.3361	0.8909
150.000	14,671,438.	0.0099	20.13	0.3361	0.8952
170.000	14,671,438.	0.0099	20.13	0.3361	0.9012
190.000	14,671,438.	0.0099	20.13	0.3361	0.9061
200.000	14,671,438.	0.0099	20.13	0.3361	0.9114
220.000	14,671,438.	0.0099	20.13	0.3361	0.9192
240.000	14,671,438.	0.0099	20.13	0.3361	0.9192

******************* PIPELINE TEMPERATURE AND PRESSURE PROFILE *******************

Distance Location (mi)	Diameter (in)	Flow (MMSCFD)	Velocity (ft/sec)	Press. (psig)	GasTemp. (degF)	SoilTemp. (degF)	MAOP (psig)
0.00 Rockford	30.000	420.0000	15.43	996.42	105.47	60.00	1400.00
10.00	30.000	420.0000	15.69	979.75	89.05	60.00	1400.00
20.00	30.000	420.0000	15.96	963.35	78.31	60.00	1400.00
30.00	30.000	420.0000	16.23	947.03	71.43	60.00	1400.00
40.00	30.000	420.0000	16.51	930.64	67.09	60.00	1400.00
50.00	30.000	420.0000	16.80	914.09	64.38	60.00	1400.00
60.00	30.000	420.0000	17.11	897.31	62.70	60.00	1400.00
70.00	30.000	420.0000	17.44	880.24	61.66	60.00	1400.00
100.00	30.000	420.0000	18.54	826.86	60.39	60.00	1400.00
120.00	30.000	420.0000	19.42	789.12	60.15	60.00	1400.00
140.00	30.000	420.0000	20.43	749.30	60.05	60.00	1400.00
150.00	30.000	420.0000	21.00	728.50	60.03	60.00	1400.00
170.00	30.000	420.0000	22.31	684.85	60.01	60.00	1400.00
190.00	30.000	420.0000	23.91	637.95	60.00	60.00	1400.00
200.00	30.000	420.0000	24.86	613.06	60.00	60.00	1400.00
220.00	30.000	420.0000	27.17	559.71	60.00	60.00	1400.00
240.00 Concord	30.000	420.0000	30.30	500.35	60.00	60.00	1400.00

```
****************** LINE PACK VOLUMES AND PRESSURES ********************

Distance    Pressure    Line Pack
(mi)        (psig)      (million std.cu.ft)

0.00        996.42      0.0000
10.00       979.75      17.3641
20.00       963.35      17.6640
30.00       947.03      17.7542
40.00       930.64      17.6905
50.00       914.09      17.5188
60.00       897.31      17.2744
70.00       880.24      16.9816
100.00      826.86      48.8875
120.00      789.12      30.7385
140.00      749.30      29.1384
150.00      728.50      13.9433
170.00      684.85      26.5757
190.00      637.95      24.7439
200.00      613.06      11.6489
220.00      559.71      21.7583
240.00      500.35      19.5575

Total line pack in main pipeline =   349.2396(million std.cubic ft)

Started simulation at:  13:23:10
Finished simulation at: 13:24:59
Time elapsed            :  109 seconds
Time elapsed            :  0 hours  1 minutes  49 seconds
DATE:   3-December-2013
```

It can be seen from the GASMOD thermal hydraulic analysis report, the inlet pressure at Rockford is 996.42 psig, whereas the manual calculation considering isothermal flow yielded an inlet pressure of 994.92 psig at the Rockford compressor station. Thus taking into account the temperature variation of the gas along the pipeline, the pressure required at Rockford is approximately 4 psig higher. This does not seem to be very significant. However, in many cases, the temperature variation altrun -1ong the pipeline will cause pressures calculated to be significantly different. To recap, the manual calculations were based on isothermal gas flow temperature of 70 °F, whereas the thermal hydraulics shows variation of the gas temperature ranging from 102.92 °F at the Rockford compressor discharge to 60 °F at Concord. The gas temperature reaches the soil temperature of 60 °F at approximately milepost 190, after which it remains constant at 60 °F. Figure 7.8 shows the temperature (Figure 7.9) variation in this case.

Next we will illustrate calculation of the pressure and temperature profile considering pipeline elevation difference between Rockford and Concord

Thermal Hydraulics

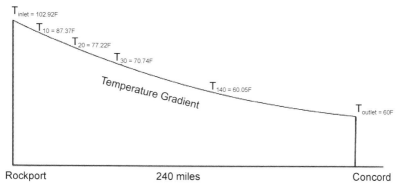

Figure 7.8 Gas temperature variation: Rockford to Concord pipeline.

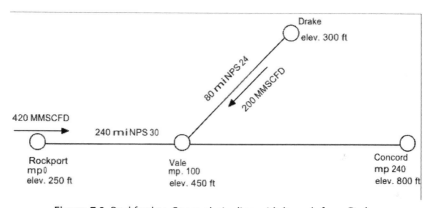

Figure 7.9 Rockford to Concord pipeline with branch from Drake.

and considering a branch pipeline bringing in an additional 200 MMSCFD as illustrated in the next example.

Example Problem 7.4
A natural gas pipeline system from Rockford (elevation 250 ft) to Concord (elevation 800 ft), a distance of 240 miles. The pipeline is constructed of NPS 30 with 0.500-in wall thickness, API 5L-X60 pipe. The MOP is 1400 psig.

The pipeline elevation profile is listed below.

Milepost	Elevation	Location
0.00	250.00	Rockford
10.00	300.00	
20.00	200.00	
30.00	320.00	
40.00	400.00	
50.00	375.00	
60.00	410.00	
70.00	430.00	
100.00	450.00	Vale
120.00	500.00	
140.00	400.00	
150.00	600.00	
170.00	700.00	
190.00	710.00	
200.00	720.00	
220.00	750.00	
240.00	800.00	Concord

Gas enters the Rockford compressor station at 70 °F and 800 psig pressure. The soil temperature may be assumed to be 60 °F throughout. The gas flow rate is 420 MMSCFD and the gas-specific gravity and viscosity are 0.6 and 0.000,008 lb/ft.s. At Vale (milepost 100, elevation 450 ft), a branch pipeline 80 miles long, NPS 24, 0.375-in wall thickness brings in an additional 200 MMSCFD gas from a gathering facility at Drake. The elevation at Drake is 300 ft and that at Vale is 450 ft. The MAOP is 1000 psig.

The pipeline elevation profile for the branch pipe from Drake to Vale is as follows:

Milepost	Elevation	Location
0.00	300.00	Drake
10.00	100.00	
20.00	125.00	
40.00	200.00	
50.00	250.00	
70.00	300.00	
80.00	450.00	Vale

The inlet temperature at the beginning of the branch is 70 °F.

Thermal Hydraulics

The contract delivery pressure required at Concord is 500 psig. Assume isothermal flow at 70 °F and gas-specific heat ratio is 1.29. Use compressor adiabatic efficiency of 80% and mechanical efficiency is 98%. Use the general flow equation with the Colebrook friction factor, assuming a pipe internal roughness of 700 micro in. Calculate the pressure profile and the compressor horsepower required at Rockford. Compare these results with thermal hydraulic analysis using GASMOD. Assume base pressure of 14.7 psia and base temperature of 60 °F.

Solution

Inside diameter of pipe $D = 30 - 2 \times 0.500 = 29$ in.

The calculation of the pressure at milepost 100 will be done first. This is because we know the delivery pressure at Concord and the 140-mile segment from Vale, at milepost 100 to the pipeline terminus at Concord flows 620 MMSCFD. In comparison, the first 100 miles from Rockford to Vale carries only 420 MMSCFD and both upstream and downstream pressures are unknown. After finding the pressure at Vale, we can calculate the upstream pressure at Rockford considering the 100-mile segment at the lower flow rate.

First we calculate the Reynolds number for 620 MMSCFD flow rate.

$$R = 0.0004778 \left(\frac{14.7}{60 + 460}\right) \left(\frac{0.6 \times 620 \times 10^6}{0.000008 \times 29}\right) = 21{,}657{,}837$$

Next, using the Colebrook equation, we calculate the friction factor as:

$$\frac{1}{\sqrt{f}} = -2\text{Log}_{10}\left(\frac{0.0007}{3.7 \times 29} + \frac{2.51}{21657837\sqrt{f}}\right)$$

Solving by trial and error, we get

$$f = 0.0096$$

Therefore, the transmission factor is, using

$$F = \frac{2}{\sqrt{0.0096}} = 20.41$$

To calculate the compressibility factor Z, the average pressure is required. Because the inlet pressure at Vale is unknown, we will calculate an approximate value of Z using a value of 110% of the delivery pressure for the average pressure.

The average pressure is

$$P_{\text{avg}} = 1.1 \times (500 + 14.7) = 566.17 \text{ psia} = 551.47 \text{ psig}$$

Using the CNGA equation, we calculate the value of compressibility factor as

$$Z = \frac{1}{\left[1 + \left(\frac{(566.17-14.7) \times 344400(10)^{1.785 \times 0.6}}{530^{3.825}}\right)\right]} = 0.9217$$

Because there is an elevation difference of (800 − 450) or 350 ft between Vale and Concord, we must apply the elevation correction.

Therefore, the elevation adjustment parameter is

$$s = 0.0375 \times 0.6 \left(\frac{350}{520 \times 0.9217}\right) = 0.0161$$

The equivalent length L_e is

$$L_e = \frac{140(1.0163 - 1)}{0.0161} = 141.74 \text{ mi}$$

Next we calculate the pressure at Vale, milepost 100, as follows

$$620 \times 10^6 = 38.77$$

$$\times 20.41 \left(\frac{520}{14.7}\right) \left(\frac{P_1^2 - 1.0163 \times 514.7^2}{0.6 \times 530 \times 141.32 \times 0.9217}\right)^{0.5} 29^{2.5}$$

Solving for the upstream pressure at Vale, we get

$$P_1 = 1122.49 \text{ psia} = 1107.79 \text{ psig}$$

Using this calculated value of P_1, we calculate the new average pressure as

$$P_{avg} = \frac{2}{3}\left(1122.49 + 514.7 - \frac{1122.49 \times 514.7}{1122.49 + 514.7}\right) = 856.2 \text{ psia}$$

$$= 841.5 \text{ psig}$$

Compare this with the previous approximation of 551.47 psig. Obviously, the assumed value was way off. Recalculating the compressibility factor using the recently calculated average pressure:

$$Z = \frac{1}{\left[1 + \left(\frac{551.47 \times 344400(10)^{1.785 \times 0.6}}{520^{3.825}}\right)\right]} = 0.8852$$

Recalculating the pressure at Vale, we get

$$620 \times 10^6 = 38.77$$

$$\times 20.41 \left(\frac{520}{14.7}\right) \left(\frac{P_1^2 - 1.0163 \times 514.7^2}{0.6 \times 530 \times 141.32 \times 0.8852}\right)^{0.5} 29^{2.5}$$

Solving for the upstream pressure at Vale, we get

$$P_1 = 1104.88 \text{ psia} = 1090.18 \text{ psig}$$

Compared with the last calculated value the difference is: $1090.18 - 1107.79 = -17.61$ psig or 1.6%. One more iteration would get us closer to the correct value. Recalculate the new average pressure and we get

$$P_{avg} = 845.63 \text{ psia} = 830.93 \text{ psig}$$

Recalculating the compressibility factor using recently calculated average pressure:

$$Z = \frac{1}{\left[1 + \left(\frac{551.47 \times 344400(10)^{1.785 \times 0.6}}{520^{3.825}}\right)\right]} = 0.8865$$

Recalculating the pressure at Vale, we get

$$620 \times 10^6 = 38.77 \times 20.41 \left(\frac{520}{14.7}\right) \left(\frac{P_1^2 - 1.0163 \times 514.7^2}{0.6 \times 530 \times 141.32 \times 0.8865}\right)^{0.5} 29^{2.5}$$

Solving for the upstream pressure at Vale, we get

$$P_1 = 1106.15 \text{ psia} = 1091.45 \text{ psig}$$

Compared with the last calculated value, the difference is: $1090.18 - 1091.45 = -1.27$ psig or 0.12%. This is considered close enough.

Therefore the pressure at Vale = 1106.15 psia = 1091.45 psig

Next, using this as the downstream pressure for the pipe segment from Rockford to Vale, we calculate the upstream pressure at Rockford as follows for a flow rate of 420 MMSCFD.

From previous calculations at 420 MMSCFD,

The Reynolds number is 14,671,438

The friction factor was calculated as $f = 0.0097$ and the transmission factor was $F = 20.33$

To calculate the compressibility factor, Z the average pressure is required. Because the inlet pressure at Rockford is unknown, we will calculate an approximate value of Z using a value of 110% of the downstream pressure at Vale for the average pressure.

The average pressure is

$$P_{avg} = 1.1 \times 1106.15 = 1216.77 \text{ psia} = 1202.07 \text{ psig}$$

Using this, we calculate the compressibility factor as

$$Z = \frac{1}{\left[1 + \left(\frac{(1216.77 - 14.7) \times 344400(10)^{1.785 \times 0.6}}{530^{3.825}}\right)\right]} = 0.8437$$

Because there is an elevation difference of $(450 - 250)$ or 200 ft between Rockford and Vale, applying the elevation correction as before, we get the elevation adjustment parameter as

$$s = 0.0375G\left(\frac{H_2 - H_1}{T_f Z}\right) = 0.0101$$

The equivalent length L_e is

$$L_e = \frac{L(e^s - 1)}{s} = 100 \text{ mi}$$

Calculating the pressure at Rockford, we get

$$420 \times 10^6 = 38.77 \times 20.33 \left(\frac{520}{14.7}\right) \left(\frac{P_1^2 - 1.0101 \times 1106.15^2}{0.6 \times 530 \times 100 \times 0.8437}\right)^{0.5} 29^{2.5}$$

Solving for the upstream pressure at Rockford, we get

$$P_1 = 1238.04 \text{ psia} = 1223.34 \text{ psig}$$

Next we will compare the isothermal calculation results with thermal hydraulics using the GASMOD program. Given below is the output report from GASMOD program.

```
*********** GASMOD - GAS PIPELINE HYDRAULIC SIMULATION ***********
*********** Version 6.00.780 ************
DATE:                                   3-December-2013    TIME:  14:56:03
PROJECT DESCRIPTION:
Pipeline from Rockford to Concord with injection at Vale
Case Number:                            1001
Pipeline data file:                     C:\Users\My Documents\Gasmod\Problem74.TOT

Pressure drop formula:                  Colebrook-White
Pipeline efficiency:                    1.00
Compressibility Factor Method:          CNGA

Inlet Gas Gravity(Air=1.0):             0.60000
Inlet Gas Viscosity:                    0.0000080(lb/ft-sec)
Gas specific heat ratio:                1.29
```

**** Calculations Based on Specified Thermal Conductivities of Pipe, Soil and Insulation ****

```
Base temperature:                          60.00 (degF)
Base pressure:                             14.70 (psia)

Origin suction temperature:                70.00 (degF)
Origin suction pressure:                  800.00 (psig)
Pipeline Terminus Delivery  pressure:     499.66 (psig)
Minimum pressure:                         100.0  (psig)
Maximum gas velocity:                      50.00 (ft/sec)

Inlet  Flow rate:                         420.00 (MMSCFD)
Outlet Flow rate:                         620.00 (MMSCFD)

CALCULATION OPTIONS:
Polytropic compression considered:        NO
Branch pipe calculations:                 YES
Loop pipe calculations:                   NO
Compressor Fuel Calculated:               NO
Joule Thompson effect included :          NO
Customized Output:                        NO
Holding Delivery Pressure at terminus
```

ALL PRESSURES ARE GAUGE PRESSURES, UNLESS OTHERWISE SPECIFED AS ABSOLUTE PRESSURES

*************** PIPELINE PROFILE DATA ***********

Distance (mi)	Elevation (ft)	Diameter (in)	Thickness (in)	Roughness (in)
0.00	250.00	30.000	0.500	0.000700
10.00	300.00	30.000	0.500	0.000700
20.00	200.00	30.000	0.500	0.000700
30.00	320.00	30.000	0.500	0.000700
40.00	400.00	30.000	0.500	0.000700
50.00	375.00	30.000	0.500	0.000700
60.00	410.00	30.000	0.500	0.000700
70.00	430.00	30.000	0.500	0.000700
100.00	450.00	30.000	0.500	0.000700
120.00	500.00	30.000	0.500	0.000700
140.00	400.00	30.000	0.500	0.000700
150.00	600.00	30.000	0.500	0.000700
170.00	700.00	30.000	0.500	0.000700
190.00	710.00	30.000	0.500	0.000700
200.00	720.00	30.000	0.500	0.000700
220.00	750.00	30.000	0.500	0.000700
240.00	800.00	30.000	0.500	0.000700

************** THERMAL CONDUCTIVITY AND INSULATION DATA ****************

Distance (mi)	Cover (in)	Thermal Conductivity (Btu/hr/ft/degF)			Insul.Thk (in)	Soil Temp (degF)
		Pipe	Soil	Insulation		
0.000	36.000	29.000	0.800	0.020	0.000	60.00
10.000	36.000	29.000	0.800	0.020	0.000	60.00
20.000	36.000	29.000	0.800	0.020	0.000	60.00
30.000	36.000	29.000	0.800	0.020	0.000	60.00
40.000	36.000	29.000	0.800	0.020	0.000	60.00
50.000	36.000	29.000	0.800	0.020	0.000	60.00
60.000	36.000	29.000	0.800	0.020	0.000	60.00
70.000	36.000	29.000	0.800	0.020	0.000	60.00
100.000	36.000	29.000	0.800	0.020	0.000	60.00
120.000	36.000	29.000	0.800	0.020	0.000	60.00
140.000	36.000	29.000	0.800	0.020	0.000	60.00
150.000	36.000	29.000	0.800	0.020	0.000	60.00
170.000	36.000	29.000	0.800	0.020	0.000	60.00
190.000	36.000	29.000	0.800	0.020	0.000	60.00
200.000	36.000	29.000	0.800	0.020	0.000	60.00
220.000	36.000	29.000	0.800	0.020	0.000	60.00
240.000	36.000	29.000	0.800	0.020	0.000	60.00

***************** LOCATIONS AND FLOW RATES ****************

Location	Distance (mi)	Flow in/out (MMSCFD)	Gravity	Viscosity (lb/ft-sec)	Pressure (psig)	GasTemp. (degF)
Rockford	0.00	420.0000	0.6000	0.00000800	800.00	70.00
Vale	100.00	200.0000	0.6000	0.00000800	1085.66	60.46
Concord	240.00	-620.0000	0.6000	0.00000800	499.66	60.00

**************** COMPRESSOR STATION DATA **************

FLOW RATES, PRESSURES AND TEMPERATURES:

Name MaxPipe Temp (degF)	Flow Rate (MMSCFD)	Suct. Press. (psig)	Disch. Press. (psig)	Compr. Ratio	Suct. Loss. (psia)	Disch. Loss. (psia)	Suct. Temp. (degF)	Disch. Temp (degF)
Rockford 140.00	420.00	800.00	1223.93	1.5204	0.00	0.00	70.00	132.24

************* COMPRESSOR EFFICIENCY, HP AND FUEL USED ****************

Name Installed (HP)	Distance (mi)	Compr Effy. (%)	Mech. Effy. (%)	Overall Effy. (%)	Horse Power	Fuel Factor (MCF/day/HP)	Fuel Used (MMSCFD)
Rockford 10000	0.00	80.00	98.00	78.40	9,512.55	0.2000	------

Total Compressor Station Horsepower: 9,512.55   10,000.

************** REYNOLD'S NUMBER AND HEAT TRANSFER COEFFICIENT **************

Distance (mi)	Reynold'sNum.	FrictFactor (Darcy)	Transmission Factor	HeatTransCoeff (Btu/hr/ft2/degF)	CompressibilityFactor (CNGA)
0.000	14,671,438.	0.0099	20.13	0.3361	0.8820
10.000	14,671,438.	0.0099	20.13	0.3361	0.8678
20.000	14,671,438.	0.0099	20.13	0.3361	0.8578
30.000	14,671,438.	0.0099	20.13	0.3361	0.8516
40.000	14,671,438.	0.0099	20.13	0.3361	0.8479
50.000	14,671,438.	0.0099	20.13	0.3361	0.8460
60.000	14,671,438.	0.0099	20.13	0.3361	0.8454
70.000	14,671,438.	0.0099	20.13	0.3361	0.8465
100.000	21,657,838.	0.0098	20.25	0.3365	0.8515
120.000	21,657,838.	0.0098	20.25	0.3365	0.8588
140.000	21,657,838.	0.0098	20.25	0.3365	0.8650
150.000	21,657,838.	0.0098	20.25	0.3365	0.8721
170.000	21,657,838.	0.0098	20.25	0.3365	0.8821
190.000	21,657,838.	0.0098	20.25	0.3365	0.8903
200.000	21,657,838.	0.0098	20.25	0.3365	0.8998
220.000	21,657,838.	0.0098	20.25	0.3365	0.9148
240.000	21,657,838.	0.0098	20.25	0.3365	0.9148

Thermal Hydraulics

```
***************** PIPELINE TEMPERATURE AND PRESSURE PROFILE ********************
```

Distance Location (mi)	Diameter (in)	Flow (MMSCFD)	Velocity (ft/sec)	Press. (psig)	GasTemp. (degF)	SoilTemp. (degF)	MAOP (psig)
0.00 Rockford	30.000	420.0000	12.60	1223.93	132.24	60.00	1400.00
10.00	30.000	420.0000	12.76	1208.55	107.14	60.00	1400.00
20.00	30.000	420.0000	12.87	1197.93	90.16	60.00	1400.00
30.00	30.000	420.0000	13.05	1181.33	79.02	60.00	1400.00
40.00	30.000	420.0000	13.22	1166.01	71.88	60.00	1400.00
50.00	30.000	420.0000	13.36	1153.82	67.38	60.00	1400.00
60.00	30.000	420.0000	13.52	1139.84	64.56	60.00	1400.00
70.00	30.000	420.0000	13.68	1126.20	62.81	60.00	1400.00
100.00 Vale	30.000	620.0000	20.94	1085.66	60.46	60.00	1400.00
120.00	30.000	620.0000	22.18	1023.93	60.24	60.00	1400.00
140.00	30.000	620.0000	23.59	961.87	60.12	60.00	1400.00
150.00	30.000	620.0000	24.57	922.83	60.09	60.00	1400.00
170.00	30.000	620.0000	26.73	847.14	60.05	60.00	1400.00
190.00	30.000	620.0000	29.53	765.53	60.02	60.00	1400.00
200.00	30.000	620.0000	31.32	720.84	60.02	60.00	1400.00
220.00	30.000	620.0000	36.25	620.90	60.01	60.00	1400.00
240.00 Concord	30.000	620.0000	44.79	499.66	60.00	60.00	1400.00

```
******************* LINE PACK VOLUMES AND PRESSURES ********************
```

Distance (mi)	Pressure (psig)	Line Pack (million std.cu.ft)
0.00	1223.93	0.0000
10.00	1208.55	20.6236
20.00	1197.93	21.5215
30.00	1181.33	22.0920
40.00	1166.01	22.3347
50.00	1153.82	22.4188
60.00	1139.84	22.3750
70.00	1126.20	22.2208
100.00	1085.66	65.2595
120.00	1023.93	41.3944
140.00	961.87	38.6798
150.00	922.83	18.2378
170.00	847.14	34.0286
190.00	765.53	30.7115
200.00	720.84	14.0359
220.00	620.90	25.1636
240.00	499.66	20.7997

Total line pack in main pipeline = 441.8972 (million std.cubic ft)

```
************ PIPE BRANCH CALCULATION SUMMARY ***********

Number of Pipe Branches = 1

BRANCH TEMPERATURE AND PRESSURE PROFILE:

Incoming Branch File: C:\Users\Shashi\My Documents\Gasmod\VALEBRANCH.TOT

Branch Location: Vale at 100 (mi)

Distance   Elevation  Diameter  Flow      Velocity  Press.    Gas Temp.  Amb Temp.
Location
(mi)       (ft)       (in)      (MMSCFD)  (ft/sec)  (psig)    (degF)     (degF)

0.00       150.00     24.000    200.000   9.76      1170.51   130.57     60.00
Drake
10.00      100.00     24.000    200.000   9.76      1161.56   90.53      60.00
20.00      125.00     24.000    200.000   9.92      1151.22   72.50      60.00
40.00      200.00     24.000    200.000   10.10     1130.29   61.97      60.00
50.00      200.00     24.000    200.000   10.18     1120.90   60.77      60.00
70.00      200.00     24.000    200.000   10.35     1101.91   60.12      60.00
80.00      450.00     24.000    200.000   10.51     1085.16   60.05      60.00
Vale

Total line pack in branch pipeline C:\Users\Shashi\My Documents \Gasmod\VALEBRANCH.TOT
= 111.6601(million std.cubic ft)

Compressor Power reqd. at the beginning of branch: 4,199.20 HP
Compression ratio: 1.48
Suction temperature: 70.00 (degF)
Suction pressure: 800.00 (psig)
Suction piping loss: 5.00 (psig)
Discharge piping loss: 10.00 (psig)
Started simulation at:  14:54:03
Finished simulation at: 14:56:04
Time elapsed            : 121 seconds
Time elapsed            : 0 hours  2 minutes  1 seconds
DATE:   3-December-2013
```

It can be seen from the GASMOD thermal hydraulic analysis report that the inlet pressure at Rockford is approximately 1224 psig, whereas the manual calculation considering isothermal flow yielded an inlet pressure of approximately 1223 psig at the Rockford compressor station. This difference is not very significant. However, in many cases the temperature variation along the pipeline will cause pressures calculated to be significantly different, especially in short pipelines. As an example, if we had a pipeline 100 miles long, similar to the pipe section between Rockford and Vale, the thermal hydraulics will show a drastic temperature variation, from 132.24 to 60.00 °F. Therefore an isothermal analysis at 70 °F for the entire 100-mile length will show considerable discrepancy in pressures. This is left as an exercise for the reader.

To recap, the manual calculations were based on isothermal gas flow temperature of 70 °F, whereas the thermal hydraulics shows variation of the gas temperature ranging from 132.24 °F at the Rockford compressor discharge to 60.01 °F at Concord, which is very close to the surrounding soil temperature of 60 °F.

The compression ratio is 1.52 at the Rockford compressor station, where the 800 psig inlet pressure of the gas is increased to the discharge pressure of 1224 psig. This, in accordance with the analysis in Chapter 10 under compressors, causes the discharge temperature of the gas to increase to 132.24 °F. If the compression ratio were higher, the discharge temperature of the gas from compression would have been still higher. The pipeline coating temperature limitation is 140 °F and would then require gas cooling to avoid damage to the pipe coating. It can be seen from the GASMOD report that the gas flow temperature starts off at 132.24 °F at milepost 0 (Rockford) and quickly drops to 67.38 °F at milepost 50. This is the exponential temperature decay we discussed in an earlier section. Beginning at milepost 50, the gas temperature starts dropping off more gradually until it almost attains the soil temperature at milepost 240 (Concord). Also the section of pipe between milepost 170 and Concord is at a fairly constant temperature close to the soil temperature of 60 °F. Therefore, the 70-mile pipeline section between milepost 170 and Concord may be considered to be in isothermal flow for all practical purposes. A manual calculation of this last 70-mile pipe segment flowing at 620 MMSCFD will yield a pressure profile very close to the pressures shown in the GASMOD report. This will be demonstrated next.

First, we calculate the Reynolds number, $R = 21{,}657{,}837$

Next, using the Colebrook equation, we calculate the friction factor as

$$\frac{1}{\sqrt{f}} = -2\text{Log}_{10}\left(\frac{0.0007}{3.7 \times 29} + \frac{2.51}{21657837\sqrt{f}}\right)$$

Solving by trial and error we get $f = 0.0096$.

Therefore, the transmission factor is

$$F = \frac{2}{\sqrt{0.0096}} = 20.41$$

To calculate the compressibility factor Z, the average pressure is required. Because the pressure at milepost 170 is unknown, we will calculate an approximate value of Z using a value of 850 psig for the average pressure.

Using the CNGA equation, we calculate the value of compressibility factor as

$$Z = \cfrac{1}{\left[1 + \left(\cfrac{850 \times 344400 (10)^{1.785 \times 0.6}}{520^{3.825}}\right)\right]} = 0.8765$$

Because there is an elevation difference of (800 − 700) or 100 ft between milepost 170 and Concord, we must apply the elevation correction and the elevation adjustment parameter is

$$s = 0.0375 \times 0.6 \left(\frac{800 - 700}{520 \times 0.8765}\right) = 0.0049$$

The equivalent length L_e is 70.17 miles.

Next, we calculate the pressure at milepost 70 as follows:

$$620 \times 10^6 = 38.77$$

$$\times\, 20.41 \left(\frac{520}{14.7}\right) \left(\frac{P_1^2 - 1.0049 \times 514.7^2}{0.6 \times 520 \times 70.17 \times 0.8765}\right)^{0.5} 29^{2.5}$$

$$P_1 = 851.59 \text{ psia} = 836.89 \text{ psig}$$

This compares with 850 psig we assumed earlier. Recalculating the average pressure based on $P_1 = 851.59$ psia and $P_2 = 514.7$ psia, we get

$$P_{avg} = \frac{2}{3}\left(851.59 + 514.7 - \frac{851.59 \times 514.7}{851.59 + 514.7}\right) = 696.99 \text{ psia}$$

Next, using the CNGA equation, we recalculate the compressibility factor as

$$Z = \cfrac{1}{\left[1 + \left(\cfrac{682.29 \times 344400 (10)^{1.785 \times 0.6}}{520^{3.825}}\right)\right]} = 0.8984$$

Next, we calculate the pressure at milepost 70 as follows:

$$620 \times 10^6 = 38.77$$

$$\times\, 20.41 \left(\frac{520}{14.7}\right) \left(\frac{P_1^2 - 1.0049 \times 514.7^2}{0.6 \times 520 \times 70.17 \times 0.8984}\right)^{0.5} 29^{2.5}$$

$$P_1 = 858.30 \text{ psia} = 843.6 \text{ psig}$$

This value is less than 1% different from the previously calculated value of 836.39 psig. Therefore, we do not have to iterate any further. Comparing the value of 843.6 psig with the pressure of approximately 847 psig from the GASMOD report, we see that we are less than 0.5% apart. Thus the assumption of isothermal flow in the last 70-mile section of the pipeline is a valid one. We would have been closer still if the 70-mile section had been subdivided into two or more segments and the upstream pressures calculated as discussed in an earlier section.

In conclusion, we can state that calculating the pressures and HP in a gas pipeline based on the assumption of constant temperature throughout the pipeline will yield satisfactory answers if the pipeline is long. For shorter pipelines, calculations must be performed by subdividing the pipeline into short segments and taking into account heat transfer between pipeline gas and the surrounding soil.

6. SUMMARY

In this chapter, we explored thermal effects of pipeline hydraulics compared with isothermal flow. To transport viscous liquids, they have to be heated to a temperature, sometimes much higher than the ambient conditions. This temperature differential between the pumped liquid and the surrounding soil (buried pipeline) or ambient air (above-ground pipeline) causes heat transfer to occur, resulting in temperature variation of the liquid along the pipeline. Unlike isothermal flow, where the liquid temperature is uniform throughout the pipeline, heated pipeline hydraulics requires subdividing the pipeline into short segments and calculating pressure drops based on liquid properties at the average temperature of each segment. We illustrated this using an example that showed how the temperature varies along the pipeline. Also, a method to compute the heat transfer between the liquid and the surrounding medium was shown, taking into account thermal conductivities of pipe, soil, and insulation, if present. The heating of liquid from friction was also quantified, as was the liquid heating associated with pump inefficiency. A popular liquid pipeline hydraulic simulation software was used to illustrate the calculation methodology.

In gas pipelines, we reviewed the thermal effects of pressure drop and HP required.

We pointed out the differences between the results obtained from an isothermal versus thermal hydraulic analysis. This was illustrated with example problems using an isothermal analysis compared with a more rigorous approach considering heat transfer between the pipeline gas and the surrounding soil. A popular gas pipeline hydraulic simulation software was used to illustrate the calculation methodology.

7. PRACTICE PROBLEMS

1. Apply the technique discussed in the temperature variation calculation section to calculate the temperature profile of a gas pipeline 4 miles long, NPS 20, with 0.375-in wall thickness at a flow rate of 200 MMSCFD.
2. A 200-mile, NPS 24, 0.500-in wall thickness, pipeline from Mobile to Savannah, is used for transporting 300 MMSCFD of natural gas (gravity = 0.65 and viscosity = 0.000008 lb/ft.s). The MOP is 1400 psig. Gas inlet temperature and pressure at Mobile are 80 °F and 1200 psig, respectively. The soil temperature may be assumed to be 60 °F throughout. The delivery pressure required at Savannah is 900 psig. Assume isothermal flow at 70 °F. Using the Panhandle B equation with an efficiency of 0.95, calculate the free flow volume with no compressor stations. Compare these results with thermal hydraulic analysis using subdivided pipe segments and heat transfer calculations. The base pressure is 14.7 psia and base temperature is 60 °F.

CHAPTER EIGHT

Power Required to Transport

This chapter will discuss the horsepower or power required to transport liquid or gas through transmission pipelines.

1. HORSEPOWER REQUIRED

So far we have examined the pressure required to transport a given amount of liquid through a pipeline system. Depending on the flow rate and maximum allowable operating pressure of the pipeline, we may need one or more pump stations to safely transport the specified throughput. The pressure required at each pump station will generally be provided by centrifugal or positive displacement pumps. Pump operation and performance will be discussed in a later chapter In this section, we will calculate the horsepower (HP) required to pump a given volume of liquid through the pipeline regardless of the type of pumping equipment used.

1.1 Hydraulic Horsepower

Power required is defined as energy or work performed per unit time. In English units, energy is measured in ft-lb and power is expressed in HP. One HP is defined as 33,000 ft lb/min or 550 ft lb/s.

In SI units, energy is measured in Joules and power is measured in Joules/second (watts). The larger unit kilowatts (kW) is more commonly used. One HP is equal to 0.746 kW.

To illustrate the concept of work, energy, and power required, imagine a situation that requires 150,000 gal of water to be raised 500 ft to supply the needs of a small community. If this requirement is on a 24-h basis, we can state that the work done in lifting 150,000 gal of water by 500 ft is

$$(150,000/7.48) \times 62.34 \times 500 = 625,066,845 \text{ ft lb}$$

where the specific weight of water is assumed to be 62.34 lb/ft^3 and 1 ft^3 = 7.48 gal.

Thus we need to expend 6.25×10^8 ft lb of energy over a 24-h period to accomplish this task. Because 1 HP equals 33,000 ft lb/min, the power required in this case is

$$HP = \frac{6.25 \times 10^8}{24 \times 60 \times 33,000} = 13.2$$

This is also known as the hydraulic HP (HHP) because we have not considered pumping efficiency.

As a liquid flows through a pipeline, pressure loss occurs because of friction. The pressure needed at the beginning of the pipeline to account for friction and any elevation changes is then used to calculate the amount of energy required to transport the liquid. Factoring in the time element, we get the power required to transport the liquid.

Problem 8.1
Consider 4000 barrels (bbl)/h of liquid being transported through a pipeline with one pump station operating at 1000 psi discharge pressure. If the pump station suction pressure is 50 psi, the pump has to produce (1000−50) or 950 psi differential pressure to pump 4000 bbl/h of the liquid. If the liquid specific gravity is 0.85 at flowing temperature, calculate the HP required at this flow rate.

Solution
Flow rate of liquid in lb/min is calculated as follows:

$$M = 4000 \text{ bbl/hr} \left(5.6146 \text{ ft}^3/\text{bbl}\right) \left(1 \text{ hr}/60 \text{ min}\right) (0.85) \left(62.34 \text{ lb/ft}^3\right)$$

where 62.34 is the specific weight of water in lb/ft^3.

or

$$M = 19,834.14 \text{ lb/min}$$

Therefore the HP required is

$$HP = \frac{(\text{lb/min})(\text{ft.head})}{33,000}$$

or

$$HP = ((19,834.14)(950)(2.31)/0.85)/33,000 = 1,552$$

In the above calculation, no efficiency value has been considered. In other words, we have assumed 100% pumping efficiency. Therefore, the previous HP calculated is referred to as hydraulic HP, based on 100% efficiency.

$$HHP = 1552$$

1.2 Brake Horsepower

The Brake HP (BHP) takes into account the pump efficiency. If a pump efficiency of 75% is used, we can calculate the BHP in the previous example as follows:

Brake Horsepower = Hydraulic Horsepower/Pump Efficiency

$$BHP = HHP/0.75 = 1552/0.75 = 2070$$

If an electric motor is used to drive the above pump, the actual motor HP required would be calculated as

Motor HP = BHP/Motor Efficiency

Generally, induction motors used for driving pumps have fairly high efficiencies, ranging from 95–98%. Using 98% for motor efficiency, we can calculate the motor HP required as follows:

$$\text{Motor HP} = 2070/0.98 = 2112$$

Because the closest standard size electric motor is 2500 HP, this application will require a pump that can provide a differential pressure of 950 psi at a flow rate of 4000 bbl/h and will be driven by a 2500 HP electric motor.

Pump companies measure pump flow rates in gal/min and pump pressures are expressed in terms of feet of liquid head. We can therefore convert the flow rate from bbl/h to gal/min and the pump differential pressure of 950 psi can be converted to liquid head in feet.

$$4000 \text{ bbl/hr} = \frac{4000 \times 42}{60} = 2800 \text{ gal/min}$$

$$950 \text{ psi} = \frac{950 \times 2.31}{0.85} = 2582 \text{ ft}$$

This statement for the pump requirement can then be reworded as follows:

This application will use a pump that can provide a differential pressure of 2582 ft of head at 2800 gal/min and will be driven by a 2500 HP electric motor. We will discuss pump performance in more detail in a later chapter.

The formula for BHP required in terms of customary pipeline units is as follows:

$$BHP = QP/(2449E) \qquad (8.1)$$

where
Q = flow rate, bbl/h
P = differential pressure, psi
E = efficiency, expressed as a decimal value less than 1.0

Two additional formulas for BHP are expressed in terms of flow rate in gal/min and pressure in psi or feet of liquid are as follows:

$$BHP = (GPM)(H)(Spgr)/(3960E) \qquad (8.2)$$

and

$$BHP = (GPM)P/(1714E) \qquad (8.3)$$

where
GPM = flow rate, gal/min
H = differential head, ft
P = differential pressure, psi
E = efficiency, expressed as a decimal value less than 1.0
Spgr = liquid specific gravity, dimensionless

In SI units, power in kW can be calculated as follows:

$$\text{Power (kW)} = \frac{QH\,Spgr}{367.46\,(E)} \qquad (8.4)$$

where
Q = flow rate, m^3/h
H = differential head, m
Spgr = liquid specific gravity
E = efficiency, expressed as a decimal value less than 1.0

and

$$\text{Power (kW)} = \frac{QP}{3600\,(E)} \qquad (8.5)$$

where
P = pressure, kPa
Q = flow rate, m^3/h
E = efficiency, expressed as a decimal value less than 1.0

Problem 8.2

A water distribution system requires a pump that can produce 2500 ft head pressure to transport a flow rate of 5000 gal/min. Assuming a centrifugal pump driven by an electric motor, calculate the hydraulic HP, pump BHP, and the motor HP required. Pump efficiency = 82%. Motor efficiency = 96%.

Solution

$$\text{Hydraulic HP (HHP)} = \frac{(5000)}{7.48} \times 62.34 \times \frac{2500}{33000}$$

$$\text{HHP} = 3157$$

Pump BHP required = $3157/0.82 = 3850$ HP

Motor HP required = $3850/0.96 = 4010$ HP

Problem 8.3

A water pipeline is used to move 320 L/s and requires a pump pressure of 750 m. Calculate the power required at 80% pump efficiency and 98% motor efficiency.

Solution
Using Eqn (8.4)

$$\text{Pump power required} = \frac{(320 \times 60 \times 60)}{1000} \times \frac{750 \times 1.0}{367.46 \times 0.80}$$

$$= 2939 \text{ KW}$$

Motor power required = $2939/0.98 = 3000$ KW

2. EFFECT OF GRAVITY AND VISCOSITY

It can be seen from the previous discussions that the pump BHP is directly proportional to specific gravity of liquid. Therefore, if the HP for pumping water is 1000, the HP required when pumping a crude oil of specific gravity 0.85 is

Crude Oil HP = 0.85 (Water HP) = $0.85 \times 1000 = 850$ HP

Similarly, when pumping a liquid of specific gravity greater than 1.0, the HP required will be higher. This can be seen from examining Eqn (8.4).

We can therefore conclude that, for same pressure and flow rate, HP required increases with specific gravity of liquid pumped. The HP required is also affected by viscosity of the liquid pumped. Consider water with a viscosity of 1.0 cSt. If a particular pump generates a head of 2500 ft at a flow rate of 3000 gal/min and has an efficiency of 85%, we can calculate the water HP using Eqn (8.2) as follows:

$$\text{BHP} = \frac{(3000)(2500)(1.0)}{(0.85)(3960)} = 2228.16$$

If liquid with a viscosity of 1000 SSU is used with this pump we must correct the pump head, flow rate and efficiency values using the Hydraulic Institute viscosity correction charts for centrifugal pumps. These will be discussed in more detail in the pumps and compressor chapter. The net result is that the BHP required when pumping the high-viscosity liquid will be higher than the previously calculated value. It has been found that, with high-viscosity liquids the pump efficiency degrades a lot faster than flow rate or head. Also considering pipeline hydraulics, we can say that high viscosities increase pressure required to transport a liquid and therefore increase HP required.

3. GAS: HORSEPOWER

3.1 Horsepower Required

The amount of energy input to the gas by the compressors is dependent on the pressure of the gas and flow rate. The HP that represents the energy per unit time also depends on the gas pressure and the flow rate. As the flow rate increases, the pressure also increases and hence the HP needed will also increase. Because energy is defined as work done by a force, we can state the power required in terms of the gas flow rate and the discharge pressure of the compressor station.

Suppose the gas flow rate is Q measured in standard ft^3 per day (SCFD) and the suction and discharge pressures of the compressor station are P_s and P_d, respectively. The compressor station adds the differential pressure of $(P_d - P_s)$ psia to the gas flowing at Q SCFD. Therefore, the rate at which energy is supplied to the gas is

$$(P_d - P_s) \times Q \times \text{Const1}$$

Where Const1 is a constant depending on the units used. This is a simplistic approach, because the gas properties vary with temperature and pressure. Also the compressibility factor and the type of gas compression (adiabatic or polytropic) must be taken into account. Therefore, the calculation for HP will be approached from another angle in what follows.

The head developed by the compressor is defined as the amount of energy supplied to the gas per unit mass of gas. Therefore, by multiplying the mass flow rate of gas by the compressor head we can calculate the total energy supplied to the gas. Dividing this by compressor efficiency, we will get the HP required to compress the gas. The equation for HP can be expressed as follows.

$$\text{HP} = \frac{M \times \Delta H}{\eta} \tag{8.6}$$

where

HP = compressor horsepower
M = mass flow rate of gas, lb/min
ΔH = compressor head, ft-lb/lb
η = compressor efficiency, %

Another more commonly used formula for compressor HP that takes into account the compressibility of gas is as follows.

$$\text{HP} = 0.0857 \left(\frac{\gamma}{\gamma - 1}\right) QT_1 \left(\frac{Z_1 + Z_2}{2}\right) \left(\frac{1}{\eta_a}\right) \left[\left(\frac{P_2}{P_1}\right)^{\frac{\gamma-1}{\gamma}} - 1\right] \tag{8.7}$$

where

HP = compression horsepower
γ = ratio of specific heats of gas, dimensionless
Q = gas flow rate, million SCFD
T_1 = suction temperature of gas, °R
P_1 = suction pressure of gas, psia
P_2 = discharge pressure of gas, psia
Z_1 = compressibility of gas at suction conditions, dimensionless
Z_2 = compressibility of gas at discharge conditions, dimensionless
η_a = compressor adiabatic (isentropic) efficiency, decimal value

In SI units, the power equation is as follows.

$$\text{Power} = 4.0639 \left(\frac{\gamma}{\gamma-1}\right) Q T_1 \left(\frac{Z_1+Z_2}{2}\right)\left(\frac{1}{\eta_a}\right)\left[\left(\frac{P_2}{P_1}\right)^{\frac{\gamma-1}{\gamma}} - 1\right] \quad (8.8)$$

where
  Power = compression Power, kW
  γ = ratio of specific heats of gas, dimensionless
  Q = gas flow rate, Mm^3/day
  T_1 = suction temperature of gas, K
  P_1 = suction pressure of gas, kPa
  P_2 = discharge pressure of gas, kPa
  Z_1 = compressibility of gas at suction conditions, dimensionless
  Z_2 = compressibility of gas at discharge conditions, dimensionless
  η_a = compressor adiabatic (isentropic) efficiency, decimal value

The adiabatic efficiency η_a generally ranges from 0.75 to 0.85. By considering a mechanical efficiency η_m of the compressor driver, we can calculate the BHP required to run the compressor as follows.

$$\text{BHP} = \frac{\text{HP}}{\eta_m} \quad (8.9)$$

Where HP is the horsepower calculated from the preceding equations, taking into account the adiabatic efficiency η_a of the compressor. The mechanical efficiency η_m of the driver may range from 0.95–0.98. The overall efficiency η_o is defined as the product of the adiabatic efficiency η_a and the mechanical efficiency η_m.

$$\eta_o = \eta_a \times \eta_m \quad (8.10)$$

From the adiabatic compression equation, eliminating the volume V, the discharge temperature of the gas is related to the suction temperature and the compression ratio by means of the following equation.

$$\left(\frac{T_2}{T_1}\right) = \left(\frac{P_2}{P_1}\right)^{\frac{\gamma-1}{\gamma}} \quad (8.11)$$

The adiabatic efficiency η_a may also be defined as the ratio of the adiabatic temperature rise to the actual temperature rise. Thus if the gas temperature, because of compression, increases from T_1 to T_2 the actual temperature rise is (T_2-T_1).

The theoretical adiabatic temperature rise is obtained from the adiabatic pressure, temperature relationship as follows, considering the gas compressibility factors similar to Eqn (8.11)

$$\left(\frac{T_2}{T_1}\right) = \left(\frac{Z_1}{Z_2}\right)\left(\frac{P_2}{P_1}\right)^{\frac{\gamma-1}{\gamma}} \tag{8.12}$$

or

$$T_2 = T_1 \left(\frac{Z_1}{Z_2}\right)\left(\frac{P_2}{P_1}\right)^{\frac{\gamma-1}{\gamma}} \tag{8.13}$$

Therefore, the theoretical adiabatic temperature rise is

$$T_1 \left(\frac{Z_1}{Z_2}\right)\left(\frac{P_2}{P_1}\right)^{\frac{\gamma-1}{\gamma}} - T_1$$

Therefore, the adiabatic efficiency is

$$\eta_a = \frac{T_1 \left(\frac{Z_1}{Z_2}\right)\left(\frac{P_2}{P_1}\right)^{\frac{\gamma-1}{\gamma}} - T_1}{T_2 - T_1} \tag{8.14}$$

Where T_2 is the actual discharge temperature of the gas. Simplifying, we get:

$$\eta_a = \left(\frac{T_1}{T_2 - T_1}\right)\left[\left(\frac{Z_1}{Z_2}\right)\left(\frac{P_2}{P_1}\right)^{\frac{\gamma-1}{\gamma}} - 1\right] \tag{8.15}$$

For example, if the inlet gas temperature is 80 °F and the suction and discharge pressures are 800 psia and 1400 psia, respectively. We can calculate the adiabatic efficiency if the outlet temperature is given as 200 °F. Using $\gamma = 1.4$ and from Eqn (8.15), the adiabatic efficiency is, assuming compressibility factors to be equal to 1.0.

$$\eta_a = \left(\frac{80 + 460}{200 - 80}\right)\left[\left(\frac{1400}{800}\right)^{\frac{1.4-1}{1.4}} - 1\right] = 0.7802 \tag{8.16}$$

Thus, the adiabatic compression efficiency is 0.7802.

Problem 8.4
Calculate the compressor HP required for an adiabatic compression of 106 million SCFD gas with inlet temperature of 68 °F and 725 psia pressure. The discharge pressure is 1305 psia. Assume compressibility factor at suction and discharge condition to be $Z_1 = 1.0$ and $Z_2 = 0.85$ and adiabatic exponent $\gamma = 1.4$ with the adiabatic efficiency $\eta_a = 0.8$. If the mechanical efficiency of the compressor driver is 0.95, what BHP is required? Calculate the outlet temperature of the gas.

Solutions
From Eqn (8.7), the HP required is

$$HP = 0.0857 \times 106 \left(\frac{1.40}{0.40}\right)(68 + 460)\left(\frac{1 + 0.85}{2}\right)\left(\frac{1}{0.8}\right)\left[\left(\frac{1305}{725}\right)^{\frac{0.40}{1.40}} - 1\right]$$

$$= 3550$$

Using Eqn (8.9), we calculate the driver HP required based on a mechanical efficiency of 0.95.

$$\text{The BHP required} = \frac{3550}{0.95} = 3737$$

The outlet temperature of the gas is found from Eqn (8.13) after transposing as follows.

$$T_2 = (68 + 460) \times \left[\frac{\left(\frac{1}{0.85}\right)\left(\frac{1305}{725}\right)^{\frac{0.4}{1.4}} - 1}{0.8}\right] + (68 + 460)$$

$$= 786.46\text{R} = 326.46 \text{ F}$$

The discharge temperature of the gas is 326.46 °F.

Problem 8.5
Natural gas at 3 Mm^3/day and 20 °C is compressed isentropically ($\gamma = 1.4$) from a suction pressure of 5 MPa absolute to a discharge pressure of 9 MPa absolute in a centrifugal compressor with an isentropic efficiency of 0.80. Calculate the compressor power required assuming the compressibility factors at suction and discharge condition to be $Z_1 = 0.95$ and $Z_2 = 0.85$, respectively. If the mechanical efficiency of the compressor driver is 0.95, what is the driver power required? Calculate the outlet temperature of the gas.

Solutions

From Eqn (8.8), the power required is

$$\text{Power} = 4.0639 \times 3\left(\frac{1.40}{0.40}\right)(20+273)\left(\frac{0.95+0.85}{2}\right)\left(\frac{1}{0.8}\right)\left[\left(\frac{9}{5}\right)^{\frac{0.40}{1.40}} - 1\right]$$

$$= 2572 \text{ kW}$$

$$\text{Power} = 2572 \text{ kW}$$

Using Eqn (8.9), we calculate the driver power required as follows.

$$\text{The driver power required} = \frac{2572}{0.95} = 2708 \text{ kW}$$

The outlet temperature of the gas is found from Eqn (8.13) as follows.

$$T_2 = \frac{20+273}{0.8} \times \left[\left(\frac{0.95}{0.85}\right)\left(\frac{9}{5}\right)^{\frac{0.4}{1.4}} - 1\right] + (20+273)$$

$$= 410.94 \text{ K} = 137.94 \text{ C}$$

4. SUMMARY

In this chapter, we discussed the hydraulic HP, BHP, and motor HP calculations for a given liquid pipeline system using examples. A more comprehensive analysis of pumps and HP required will be covered in a later chapter. We also covered HP required for a pipeline system transporting gas.

CHAPTER NINE

Pump Stations

1. INTRODUCTION

In this chapter, we will discuss the pump stations and pumping configurations used in liquid pipelines. The optimum locations of pump stations for hydraulic balance will be reviewed. Centrifugal pumps and positive displacement pumps will be discussed along with their performance characteristics. We will introduce affinity laws for centrifugal pumps, the importance of net positive suction head (NPSH), and how to calculate power requirements when pumping different liquids. Also, viscosity corrected pump performance using the Hydraulic Institute chart will be explained. The performance of two or more pumps in series or parallel configuration will be examined and how to estimate the operating point for a pump in conjunction with the pipeline system head curve. Next, we will look at some of the major components in a pump station, such as pumps and drivers and control valves. The use of variable speed pumps to save pumping power under different operating conditions, such as batching will be reviewed.

2. LIQUID-PUMP STATIONS
2.1 Multipump Station Pipelines

In Chapter 6, we explained how the total pressure required to pump a liquid through a pipeline from point A to point B at a specified flow rate was determined. We found that depending upon the flow rate and maximum allowable operating pressure (MAOP), one or more pump stations located along the pipeline may be required to handle the throughput. Suppose calculations show that the total pressure required at the pipeline origin, taking into account the frictional pressure loss, elevation profile of the pipeline, and the terminus delivery pressure is 1950 psig. If the MAOP of the pipeline is 1200 psig, we conclude that we need two pump stations to handle the throughput without exceeding the MAOP. The first pump station located at the beginning of the pipeline will discharge at the maximum pressure of 1200 psig and, as the liquid flows through the pipeline, the pressure will reduce because of friction as well as decrease (or increase) resulting from the elevation profile and finally reach some point at the minimum pressure, such as 50 psig. The second pump station at this location will boost the liquid pressure to some value (below MAOP), which will then be sufficient to take the product all the way to the terminus.

The question is then: Where do we locate this intermediate booster pump station? If the pipeline is of uniform diameter and wall thickness, the pipeline elevation profile is fairly flat, and there are no intermediate injections or deliveries, the second pump station will be located at the midpoint along the pipeline. If the pipeline is 100 miles long, the first pump station will be at the origin (milepost 0.0) and the second booster pump station will be at milepost 50.0.

2.2 Hydraulic Balance and Pump Stations Required

Suppose calculations indicate that at a flow rate of 5000 gal/min, a 100-mile pipeline requires a pressure of 1950 psig at the beginning of the pipeline to deliver the liquid to the pipeline terminus at some minimum pressure, 50 psig. This 1950 psig pressure may be provided in two steps of 975 psig each or three steps of approximately 650 psig each. In fact, because of the MAOP limit of the pipe, we may not be able to operate with just one pump station at the beginning of the pipeline, discharging at 1950 psig. Therefore, in long pipelines, the total pressure required to pump the liquid is provided in two or more stages by installing intermediate booster pumps along the pipeline.

In the example case with 1950 psig requirement and an MAOP of 1200 psig, we would provide this pressure as follows. The pump station at the beginning of the pipeline will provide a discharge pressure of 975 psig, which would be consumed by friction loss in the pipeline and, at some point (roughly halfway) along the pipeline, the pressure will drop to zero. At this location, we boost the liquid pressure to 975 psig using an intermediate booster pump station. We have, of course, neglected the elevation profile of the pipeline and assumed an essentially flat profile.

This pressure of 975 psig will be sufficient to take care of the friction loss in the second half of the pipeline length. The liquid pressure will reduce to zero at the end of the pipeline. However, the liquid pressure at any point along the pipeline must be higher than the vapor pressure of the liquid at the flowing temperature. In addition, the intermediate pump station requires certain minimum suction pressure. Therefore we cannot allow the pressure at any point to drop to zero. Accordingly, we will locate the second pump station at a point where the pressure has dropped to a suitable minimum suction pressure, such as 50 psig. The minimum suction pressure required is also dictated by the particular pump and may have to be higher than 50 psi, to account for any restrictions and suction piping losses at the pump station. For the present, we will assume 50 psig suction pressure is adequate for each pump station. Hence, starting with a discharge pressure of 1025 psig (975 + 50), we will locate the second pump station along the pipeline where the pressure has dropped to 50 psig. This pump station will then boost the liquid pressure back up to 1025 psig and will deliver the liquid to the pipeline terminus at 50 psig. Thus each pump station provides 975 psig differential pressure (discharge pressure less suction pressure) to the liquid, together matching the total pressure requirement of 1950 psig at 5000 gal/min flow rate.

Note that in this analysis, if we considered the pipeline elevations, the location of the intermediate booster pump will be different from that of a pipeline along a flat terrain.

When each pump station supplies the same amount of energy to the liquid, we say that the pump stations are in hydraulic balance. This will result in the same power added to the liquid at each pump station. For a single flow rate at the inlet of the pipeline (no intermediate injections or deliveries), the hydraulic balance will also result in identical discharge pressures at each pump station. Because of the topography of the pipeline route, it may not be possible to locate the intermediate pump station at the theoretical locations for hydraulic balance. It is possible that the hydraulically balanced pump station location of milepost 50 may actually be in the middle of a swamp or a river. Hence we will have to relocate this pump station to a more suitable location after field investigation. If the revised location of the second pump station were at milepost 52, the hydraulic balance would no longer be valid. Recalculating the hydraulics with the revised pump station location will show that the pump stations are not in hydraulic balance and will not be operating at the same discharge pressure or providing the same amount of power. Therefore, although it is desirable to have all pump stations balanced, it may not be practical. Hydraulically balanced pump station locations afford the advantage of using identical pumps and motors and the convenience of maintaining a common set of spare parts (pump rotating elements, mechanical seal) at a central operating district location.

In Chapter 7, we introduced the hydraulic pressure gradient and how to locate an intermediate pump station. We presented a method to calculate the discharge pressure for a two pump station pipeline system,

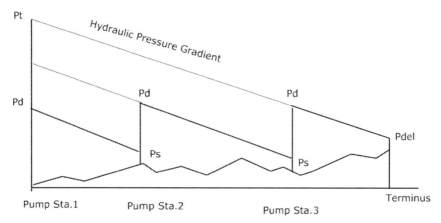

Figure 9.1 Hydraulic gradient: multiple pump stations.

knowing the total pressure required for a particular flow rate. We will now discuss a method to calculate the pump station pressures for hydraulic balance.

Figure 9.1 shows a pipeline with varying elevation profile, but no significant controlling peaks along the pipeline. First, the total pressure, P_T, is calculated for a given flow rate and liquid properties. If we used only one pump station at the beginning, the hydraulic gradient with the pump station discharging at pressure P_T is as shown, delivering the liquid at the terminus delivery pressure, P_{del}. Because P_T may be higher than the MAOP of the pipeline, let's assume that three pump stations are required to provide the pressures needed without exceeding the MAOP. Each pump station will discharge at some common pressure, P_D, that is just below the MAOP. If P_S represents the common pump station suction pressure, from the geometry of the hydraulic gradient, we see that:

$$P_D + (P_D - P_S) + P_D - P_S = P_T \tag{9.1}$$

Because this is based on one origin pump station and two intermediate pump stations, we can extend the above equation for N pump stations as follows:

$$P_D + (N-1)(P_D - P_S) = P_T \tag{9.2}$$

Solving for the number of pump stations, N, we get

$$N = (P_T - P_S)/(P_D - P_S) \tag{9.3}$$

Equation (9.3) can be used to estimate the number of pump stations required for hydraulic balance, knowing the discharge pressure limit P_D at each pump station and the calculated total pressure P_T.

Solving Eqn (9.3) for the pump station discharge pressure, we get

$$P_D = (P_R - P_S)/N + P_S \qquad (9.4)$$

For example, suppose we calculated the total pressure required as 2950 psig and the pump station suction pressure is 50 psig, the number of pump station required with 1200 psig discharge pressure is, using Eqn (9.3)

$$N = \frac{(2950 - 50)}{(1200 - 50)} = 2.52$$

Rounding up to the nearest whole number, $N = 3$.

Therefore, three pump stations are required, to limit the maximum discharge pressure to 1200 psig. From Eqn (9.4), with $N = 3$, each pump station will operate at a discharge pressure of

$$P_D = (2950 - 50)/3 + 50 = 1017 \text{ psig}$$

After calculating the discharge pressure required for hydraulic balance, as explained previously, the pump stations can be located along the pipeline profile, graphically. Let's assume we need three pump stations, as shown in Figure 9.1. First the pipeline profile (milepost versus elevation) is plotted and the hydraulic gradient superimposed upon it by drawing the sloped line starting at P_T at A and ending at P_{del} at D as shown in Figure 9.1. Because the pipeline elevations are in feet, the pressures must also be plotted in feet of liquid head. Thus P_T and P_{del} must be converted to feet of head first. Next, starting at the first pump station A at discharge pressure P_D, a line is drawn parallel to the hydraulic gradient. The location B of the second pump station will be established at a point where the hydraulic gradient between A and B meet the vertical line at the required suction pressure P_S. The process is continued to determine the location C of the third pump station.

In the preceding analysis, we made some simplifying assumptions. We assumed that the pressure drop per mile (slope of the hydraulic gradient) was constant throughout the pipeline, based on constant pipe inside diameter and flow rate throughout the entire pipeline. With variable pipe diameter or wall thickness, the slope of the hydraulic gradient may not be constant, as will be explained shortly.

2.3 Telescoping Pipe Wall Thickness

Reviewing a typical hydraulic gradient as shown in Figure 9.2, we note that because of friction, the liquid pressure decreases from the pump station to the terminus in the direction of flow. The pipeline segment immediately downstream of a pump station will be at a higher pressure such as 1200 psig or more, depending on MAOP, whereas the tail end of that segment before the next pump station (or the terminus) will be subject to lower pressures in the range of 50–100 psig. If we use the same wall thickness throughout the pipeline, we will be underusing the downstream portion of the piping that is subject to a lower pressure. Therefore, a more efficient approach would be to reduce the pipe wall thickness as we move away from a pump station toward the suction side of the next pump station or the delivery terminus. For example, the pipe wall thickness immediately downstream of the pump station may be 0.375 in for some length and then the wall thickness gradually reduces to 0.250 in closer the suction of the next pump station.

From the discussions on pipe strength, the higher pipe wall thickness immediately adjacent to the pump station will be able to withstand the higher discharge pressure. As the pressure reduces down the length of the pipeline, the wall thickness would be reduced to some value just enough to withstand the lower pressures as we approach the next pump station or delivery terminus. This process of varying the wall thickness to compensate for reduced pipeline pressures is referred to as telescoping pipe wall thickness.

However, telescoping pipe wall thickness must be done cautiously. Suppose a pipeline has two pump stations and the second pump station is

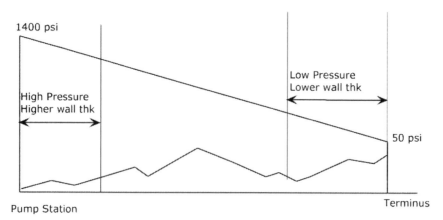

Figure 9.2 Telescoping pipe wall thickness.

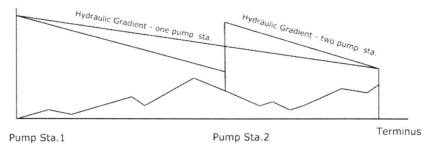

Figure 9.3 Hydraulic gradient: pump station shutdown.

shut down for some reason, flow rate will reduce and the hydraulic gradient will be flatter as shown in Figure 9.3.

It can be seen that portions of the pipeline on the upstream side of the second pump station will be subject to higher pressure than when the second pump station was online. Therefore, wall thickness reductions (telescoping) implemented upstream of a pump station must be able to handle the higher pressures because of the shutdown of an intermediate pump station.

2.4 Change of Pipe Grade: Grade Tapering

Similar to reducing the pipe wall thickness to compensate for lower pressures as we approach the next pump station or delivery terminus, the pipe material grade may also be varied. Thus the high pressure sections immediately downstream of a pump station may be constructed of API 5L X-52 grade steel, whereas the lower pressure section may be constructed of API 5L X-42 grade pipe, thereby reducing the total cost. This process of varying the pipe grade is referred to as grade tapering. Sometimes a combination of telescoping and grade tapering is used to reduce pipe material cost. Note that such wall thickness variation and pipe grade reduction to match the requirements of steady state pressures may not always be feasible. Consideration must be given to increased pipeline pressures under intermediate pump station shut down and upset conditions such as pump start up, valve closure, and so on. These transient conditions cause surge pressures in a pipeline and therefore must be taken into account when selecting optimum wall thickness and pipe grade.

2.5 Slack Line and Open Channel Flow

Most pipelines typically flow full, with no vapor space or a free liquid surface. However, under certain topographic conditions with drastic elevation changes, we may encounter pipeline sections that are partially full, resulting in open channel flow or slack line flow. Slack line operation may be unavoidable in some liquid pipelines. However, when pumping high vapor pressure liquids and in batched pipelines, slack line flow cannot be allowed. When pumping high vapor pressure products, the liquid must not vaporize so as to prevent pump damage. In batched pipelines, slack line flow will cause intermingling of batches, which is unacceptable for product quality reasons.

2.6 Batching Different Liquids

Batching is the process of transporting multiple liquids simultaneously through a pipeline with minimal mixing. Some comingling of the batches is unavoidable at the boundary or interface between contiguous batches. For example, gasoline, diesel, and kerosene may be shipped through a pipeline in a batched mode from a refinery to a storage terminal. Batched pipelines must operate in fully turbulent mode or the velocities should be sufficiently high to ensure Reynolds number higher than 4000. If the flow were laminar ($R < 2000$), the product batches would intermingle, thereby contaminating or degrading the products. Also, batched pipelines must be run in packed conditions (with no slack line or open channel flow) to avoid contamination or intermingling of batches, in pipelines that have significant elevation changes.

In a batched pipeline, the total frictional pressure drop for a given flow rate is calculated by adding the individual pressure drops for each product, based on its specific gravity, viscosity, and the batch length.

In batched pipelines, we must first calculate the line fill of the pipeline as follows, in US customary system (USCS) units:

$$\text{Line fill volume, bbl} = 5.129 L(D)^2 \quad (\text{USCS}) \qquad (9.5a)$$

Where D is pipe inside diameter in inches and L is pipe length in miles. In SI units, line fill is calculated from.

$$\text{Line fill volume, m}^3 = 7.855 \times 10^4 L(D)^2 \quad (\text{SI}) \qquad (9.5b)$$

Where D is pipe inside diameter in mm and L is pipe length in km.

The Appendix shows the line fill volumes for various pipe diameters and wall thickness in USCS and SI units.

Problem 9.1

A refined products pipeline, NPS 12 in, 0.250-in wall thickness, 120 miles long from Douglas Refinery to Hampton Terminal is used to ship three products at a flow rate of 1500 gal/min as shown in Figure 9.4. The instantaneous condition shows the three batches within the pipeline.

Figure 9.4 Batched pipeline.

Assuming the following physical properties and batch sizes for the three products, calculate the pressure drop for each liquid batch at the given flow rate. Use Hazen–Williams head loss with C factors shown. The batch size for diesel and kerosene are 50,000 barrels (bbl) and 30,000 bbl, respectively. Gasoline fills the remainder of the pipeline.

Product	Sp.Grav	C factor
Diesel	0.85	125
Kerosene	0.82	135
Gasoline	0.74	140

Solution

First, calculate the total liquid volume in the pipeline, also known as the line fill volume, using Eqn (9.5a).

$$\text{Line fill} = 5.129(120)(12.25)^2 = 92,361 \text{ bbl}$$

The gasoline batch size = 92,361 bbl − 80,000 bbl = 12,361 bbl.

Using Eqn (9.5), the batch lengths are calculated as 64.96 miles for diesel, 38.98 miles for kerosene, and 16.06 miles for gasoline.

The frictional head loss in the different batch segments are calculated from Table 8.7(a), Chapter 8, adjusting for the C values, and flow rates.

For example, for water with C = 120, at 3000 gal/min in NPS 12 pipe, the Table 8.7(a) shows a head loss of 20.39 ft per 1000 ft of pipe.

For diesel multiplying by $\left(\frac{120}{C} \times \frac{Q}{3000}\right)^{1.852}$ results in the head loss as:

$$\text{Diesel head loss} = 20.39 \times \left(\frac{120}{125} \times \frac{1500}{3000}\right)^{1.852}$$

$$= 5.24 \text{ ft per 1000 ft of pipe}$$

Converting to psi/mi:
diesel pressure loss = (5.24 × 0.85/2.31) × 5.28 psi/mi = 10.18 psi/mi.

Similarly, the pressure loss in kerosene and gasoline are calculated as:

Diesel:	10.18 psi/mi	Batch length: 64.96 miles
Kerosene:	9.82 psi/mi	Batch length: 38.98 miles
Gasoline:	8.86 psi/mi	Batch length: 16.06 miles

In the preceding, for each liquid, the pipeline length that represents the batch volume is shown. Thus the diesel batch will start at milepost 0.0 and end at milepost 64.96. Similarly the kerosene batch will start at milepost 64.96 and end at milepost (64.96 + 38.98) = 103.94. Finally, the gasoline batch will start at milepost 103.94 and end at milepost (103.94 + 16.06) = 120.0 for the snapshot configuration shown in Figure 9.4.

The batch lengths calculated previously are based on 92,361/120 = 769.68 bbl per mile of 12-in pipe calculated using Eqn (9.5). The total frictional pressure drop for the entire 100-mile pipeline is obtained by adding up the individual frictional pressure drops for each product as follows:

Total pressure drop = 10.18 (64.96) + 9.82 (38.98) + 8.86 (16.06) = 1186.4 psig

In addition, the elevation head and the delivery pressure at Hampton are added to the frictional pressure loss of 1186.4 psig to determine the total pressure required at Douglas Refinery.

When batching different products such as gasoline and diesel, flow rates vary as the batches move through the pipeline because of the changing composition of liquid in the pipeline. To economically operate the batched pipeline, by minimizing pumping cost, there is an optimum batch size for the various products in a pipeline system. An analysis needs to be made over a finite period, such as a week or a month, to determine the flow rates and pumping costs considering various batch sizes. The combination of batch sizes that result in the least total pumping cost, consistent with shipper and market demands, will then be the optimum batch sizes for the particular pipeline system.

2.7 Centrifugal Pumps versus Reciprocating Pumps

Centrifugal and reciprocating pumps are used to pump liquids through a pipeline from an originating point to the delivery terminus at the required flow rate and pressure. To increase the flow rate, more pump pressure will be required. The majority of liquid pipelines use centrifugal pumps because of their flexibility and less operating cost compared to reciprocating pumps.

Reciprocating pumps belong to the category of positive displacement (PD) pumps and are typically used for liquid injection lines in oil pipeline gathering systems.

A centrifugal pump increases the kinetic energy of a liquid because of centrifugal velocity from the rotation of the pump impeller. This kinetic energy is converted to pressure energy in the pump volute. The higher the impeller speed, the higher the pressure developed. Larger impeller diameter increases the velocity and hence the pressure generated by the pump. Compared with PD pumps, centrifugal pumps have a lower efficiency. However, centrifugal pumps can operate at higher speeds to generate higher flow rates and pressures. Centrifugal pumps also have lower maintenance requirements than PD pumps.

PD pumps, such as reciprocating pumps, operate by forcing a fixed volume of liquid from the inlet to outlet of the pump. These pumps operate at lower speeds than centrifugal pumps. Reciprocating pumps cause intermittent flow. Rotary screw pumps and gear pumps are also PD pumps, but operate continuously compared to reciprocating pumps.

Modern liquid pipelines are mostly designed with centrifugal pumps because of their flexibility in volumes and pressures. In petroleum pipeline installations where liquid from a field gathering system is injected into a main pipeline, PD pumps may be used. Figures 9.5 and 9.6 shows typical centrifugal pumps and reciprocating pumps used in the pipeline industry.

Figure 9.5 Typical centrifugal pump.

Figure 9.6 Reciprocating pump.

Centrifugal pumps are generally classified as radial flow, axial flow, and mixed flow pumps. Radial flow pumps develop head by centrifugal force. Axial flow pumps, on the other hand, develop the head from the propelling or lifting action of the impeller vanes on the liquid. Radial flow pumps are used when high heads are required, whereas the axial flow and mixed flow pumps are mainly used for low head–high capacity applications.

The performance of a centrifugal pump is represented by a series of curves, collectively called the pump characteristic curves. These shows how the pump head, efficiency, and pump power varies with flow rate (aka capacity), as shown in Figure 9.7.

The head curve shows the pump head on the left vertical axis, whereas the flow rate is shown on the horizontal axis. This curve may be referred to as the H-Q curve or the head-capacity curve. The term capacity is used interchangeably with flow rate when dealing with pumps. The efficiency curve is called the E-Q curve and shows how the pump efficiency varies with capacity. The power curve, such as brake horsepower (BHP) versus capacity curve, indicates the power required to run the pump at various flow rates. Another important pump characteristic is the NPSH versus flow rate curve. NPSH, or net positive suction head, is important when pumping high vapor pressure liquids and will be discussed later in this chapter.

The performance curves for a particular model pump are typically plotted for a particular pump impeller size and speed (example: 10 in.

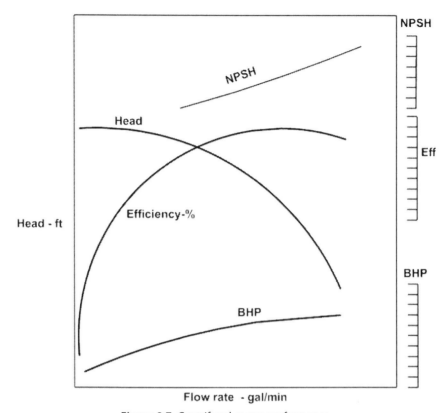

Figure 9.7 Centrifugal pump performance.

impeller, 3560 rpm). The manufacturer's pump curves are always based on water as the liquid pumped. When pumping liquids other than water, these curves may require adjustments for specific gravity and viscosity of the liquid. In USCS units, the pressure generated by the pump is measured in feet of water and flow rate is shown in gal/min. In SI units, the head is stated in meters of water and the flow rate may be in m^3/h or L/s. In USCS units, pump power is always stated as BHP, whereas kW is used in SI units.

In addition to the four characteristic curves, pump vendors provide pump head curves drawn for different impeller diameters and iso-efficiency curves.

An example of this is seen in Figure 9.8.

Another set of curves provided by centrifugal pump vendors is called a composite rating chart, and is shown in Figure 9.9.

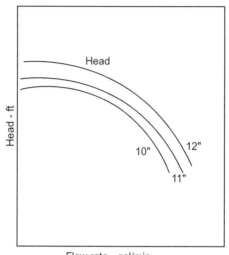

Figure 9.8 Centrifugal pump performance for different impeller sizes.

A PD pump continuously pumps a fixed volume at various pressures. It is able to provide any pressure required at a fixed flow rate, within the limits of structural design. This fixed flow rate depends on the geometry of the pump such as bore, stroke, etc. A typical PD pump pressure volume curve is shown in Figure 9.10.

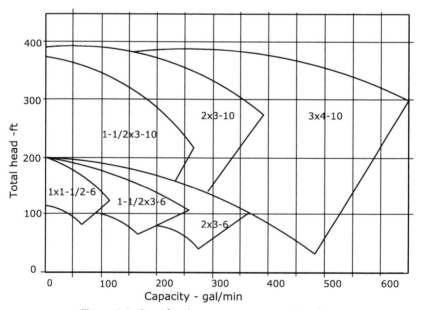

Figure 9.9 Centrifugal pump composite rating chart.

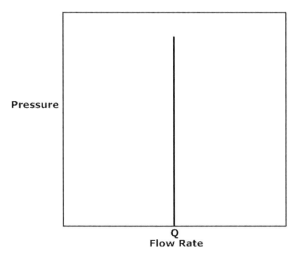

Figure 9.10 Positive displacement pump performance.

2.8 Centrifugal Pump Head and Efficiency versus Flow Rate

A typical pump manufacturer's performance curve is shown in Figure 9.7.

It shows the head versus capacity and the efficiency versus capacity as well as the BHP versus capacity curves for a specific model pump at a pump speed of 3560 rpm and an impeller diameter of 16-15/16 inches.

The solid curves are for water and the dashed curves represent the performance when pumping a heavy liquid, with a viscosity of 1075 SSU. It can be seen that the H-Q curve is a gradually drooping curve that starts of at the highest value (shutoff head) of head at zero flow rate. The head decreases as the flow rate through the pump increases. The trailing point of the curve represents the maximum flow and the corresponding head the pump can generate. The shutoff head for this pump is approximately 1320 ft and the maximum capacity is 7000 gal/min.

Pump head is always plotted in feet of head of water. Therefore the pump is said to develop the same head in feet of liquid, regardless of the liquid. Because this particular pump develops 1280 ft head at 4000 gal/min flow rate, the corresponding pump pressure in psig depends on the specific gravity of the liquid. Using Eqn (9.10):

When pumping water, pump pressure = $1280 \times 1/2.31 = 554$ psi.

When pumping gasoline, pump pressure = $1280 \times 0.74/2.31 = 410$ psi.

The head curve will be the same for water or a petroleum liquid, if its viscosity is less than 10 cSt. At higher viscosities, the pump does not produce the same head as that produced when pumping water and the H-Q curve will have to be corrected downward, as explained later. In fact, Figure 9.7 shows that when pumping a crude oil with a viscosity of 1075 SSU (236 cSt), the head generated at 4000 gal/min flow rate drops to 1260 ft. The effect of the viscosity is more pronounced on the pump efficiency. With water, the pump efficiency is approximately 82.5% at 4000 gal/min. However, this drops to 70% with the viscous liquid. The maximum efficiency of this pump is 86% when pumping water and occurs at a capacity of 5500 gal/min. The flow rate and head corresponding to the maximum pump efficiency is called the best efficiency point (BEP). In this case at BEP, Q = 5500 gal/min, H = 1180 ft and E = 86%.

The head versus flow rate curve is shown for this particular pump at an impeller diameter of 16-15/16 in operating at 3560 rpm. Within the same pump body, a certain range of impeller sizes can be accommodated. The maximum impeller size for this pump is 18-1/2 in. Similarly, the minimum impeller size will be specified by the vendor. The pump head versus capacity curve in Figure 9.11 shows a series of head curves at different impeller sizes for a fixed pump speed. It can be seen that the 9-in curve and 12-in curve are parallel to the 10-in curve. The variation of the pump H-Q curves with impeller diameter follow the affinity laws, discussed later in this chapter.

Similar to H-Q variation with pump impeller diameter, we can generate curves for different impeller speeds. If the impeller diameter is kept constant at 10 in and the initial H-Q curve was based on a pump speed of 3560 rpm, by varying the pump speed, we can generate a family of parallel curve as shown in Figure 9.12.

A pump may develop the head in stages. A single-stage pump may generate 200 ft head at 2500 gal/min. A three-stage pump of this design will generate 600 ft head at the same flow rate and pump speed. An application that requires 2400 ft head at 3500 gal/min will be handled by a six-stage pump with each stage providing 400 ft of head. Destaging is the process of reducing the active number of stages in a pump to reduce the total head developed. The six-stage pump discussed previously may be destaged to four stages if we need only 1600 ft head at 3500 gal/min.

When choosing a centrifugal pump for a particular application, we try to get the operating point as close as possible to the BEP. To allow for future increase in flow rates, the initial operating point is chosen slightly to the left of the BEP. This will ensure that, with increase in the pipeline throughput, the operating point on the pump curve will move to the right, which would result in a slightly better efficiency.

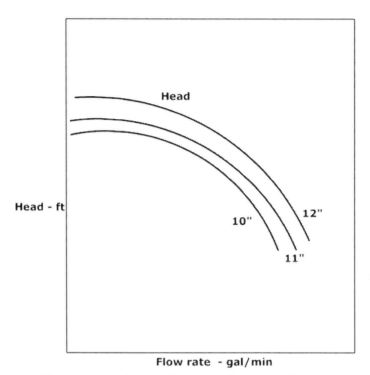

Figure 9.11 Head versus flow rate: different impeller sizes.

2.9 BHP versus Flow Rate

From the H-Q curve and efficiency versus flow rate curves, we can calculate the BHP required at every flow rate as follows:

$$\text{Pump BHP} = \frac{Q \times H \times Sg}{3960 \times E} \quad \text{(USCS)} \tag{9.6}$$

Where Q is pump flow rate, gal/min; H is pump head, ft; E is pump efficiency as a decimal value, less than 1.0, and Sg is liquid specific gravity (for water Sg = 1.0)

In SI units, power in kW can be calculated as follows:

$$\text{Power kW} = \frac{Q \times H \times Sg}{367.46 \times E} \quad \text{(SI)} \tag{9.7}$$

Where Q is pump flow rate, m^3/h; H is pump head, m; E is pump efficiency as a decimal value, less than 1.0, and Sg is liquid specific gravity (for water Sg = 1.0).

For example, from Figure 9.7, at the BEP the flow rate, head, and efficiency for the water curve are: Q = 5500 gal/min, H = 1180 ft and E = 86%.

The BHP at this flow rate for water is calculated from Eqn (9.6) as:

$$\text{Pump BHP} = \frac{5500 \times 1180 \times 1.0}{3960 \times 0.86} = 1906$$

Similarly, BHP can be calculated at various flow rates from zero to 7000 gal/min by reading the corresponding head and efficiency values from the H-Q and E-Q curves and using Eqn (9.6) as previously. The BHP versus flow rate curve can be seen below the H-Q curve in Figure 9.7.

Note that there is a dashed curve above the solid BHP curve. The dashed curve is based on the crude oil with Sg = 0.943 and viscosity of 1075 SSU. Because of the higher viscosity, the BHP required for crude oil is higher than that for water.

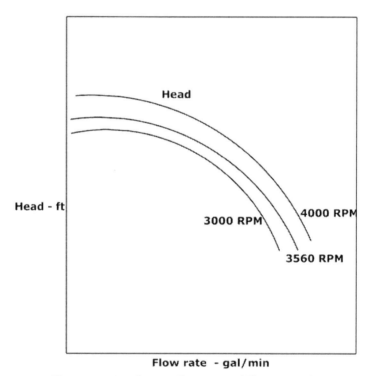

Figure 9.12 Head versus flow rate at different speeds.

The BHP calculated previously is the BHP demanded by the pump. An electric motor driving the pump will have an efficiency that ranges from 95% to 98%. Therefore the motor HP required is equal to the pump BHP divided by the motor efficiency as follows:

Motor HP = Pump BHP/Motor Efficiency = 1906/0.95

= 2006 HP, based on 95% motor efficiency.

2.10 NPSH versus Flow Rate

In addition to the H-Q, E-Q, and BHP versus capacity curves discussed in the preceding sections, the pump performance data will include a fourth curve for NPSH versus capacity. This curve is generally located above the head, efficiency, and BHP curves as shown in Figure 9.13.

The NPSH curve shows the variation of the minimum net positive suction head at the impeller suction versus the flow rate. The NPSH increases as the flow rate increases. NPSH represents the resultant positive pressure at pump suction after accounting for frictional loss and liquid vapor pressure. NPSH is discussed in detail later in this chapter.

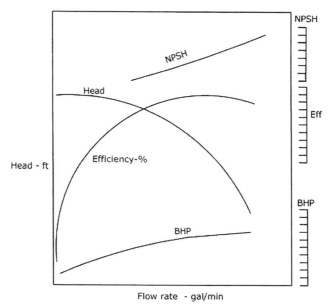

Figure 9.13 NPSH versus flow rate.

2.11 Specific Speed

The specific speed of a centrifugal pump is used for comparing geometrically similar pumps and for classifying the different types of centrifugal pumps.

Specific speed may be defined as the speed at which a geometrically similar pump must be run such that it produces a head of 1 ft at a flow rate of 1 gal/min. It is calculated as follows:

$$N_S = NQ^{1/2}/H^{3/4} \qquad (9.8)$$

Where N_S is pump specific speed; N is pump impeller speed, rpm; Q is flow rate or capacity, gal/min; and H is head, ft.

Both Q and H are measured at the BEP for the maximum impeller diameter. The head H is measured per stage for a multistage pump.

Another related term, called the suction specific speed is defined as follows:

$$N_{SS} = NQ^{1/2}/(NPSH_R)^{3/4} \qquad (9.9)$$

Where N_{SS} is suction specific speed; N is pump impeller speed, rpm; Q is flow rate or capacity, gal/min; and $NPSH_R$ is NPSH required at BEP.

When applying these equations to calculate the pump-specific speed and suction-specific speed, use the full Q value for single or double suction pumps for N_S calculation. For N_{SS} calculation, use one-half of the Q value for double suction pumps.

Table 9.1 lists the specific speed range for centrifugal pumps.

Table 9.1 Specific speeds of centrifugal pumps

Description	Application	Specific Speed, Ns
Radial vane	Low capacity/high head	500–1000
Francis screw type	Medium capacity/medium head	1000–4000
Mixed flow type	Medium to high capacity, low to medium head	4000–7000
Axial flow type	High capacity/low head	7000–20,000

Problem 9.2

Calculate the specific speed of a five-stage double suction centrifugal pump, 12-in diameter impeller that when operated at 3560 rpm generates a head of 2200 ft at a capacity of 3000 gal/min at the BEP on the head capacity curve. If the NPSH required is 25 ft, calculate the suction specific speed.

Solution

$$N_S = NQ^{1/2}H^{3/4} = 3560(3000)^{1/2}/(2200/5)^{3/4} = 2030$$

The suction specific speed is

$$N_{SS} = NQ^{1/2}/NPSH_R^{3/4} = 3560(3000/2)^{1/2} = 12,332$$

2.12 Affinity Laws for Centrifugal Pumps

The affinity laws for centrifugal pumps are used to predict pump performance for changes in impeller diameter and impeller speed.

The affinity laws are represented as follows:

For impeller diameter change:

$$Q_2/Q_1 = D_2/D_1 \qquad (9.10)$$

$$H_2/H_1 = (D_2/D_1)^2 \qquad (9.11)$$

Where Q_1, Q_2 are the initial and final flow rates; H_1, H_2 are the initial and final heads; and D_1, D_2 are the initial and final impeller diameters.

Similarly, affinity laws state that for the same impeller diameter, if pump speed is changed, flow rate is directly proportional to the speed, whereas the head is directly proportional to the square of the speed. As with diameter change, the BHP is proportional to the cube of the impeller speed. This is represented mathematically as follows.

For impeller speed change:

$$Q_2/Q_1 = N_2/N_1 \qquad (9.12)$$

$$H_2/H_1 = (N_2/N_1)^2 \qquad (9.13)$$

Where Q_1, Q_2 is are the initial and final flow rates; H_1, H_2 are the initial and final heads; and N_1, N_2 are the initial and final impeller speeds.

Note that the affinity laws for speed change are exact. However, the affinity laws for impeller diameter change are only approximate and valid for small changes in impeller sizes. The pump vendor must be consulted to verify that the predicted values using affinity laws for impeller size changes are accurate or if any correction factors are needed. With speed and impeller size changes, the efficiency versus flow rate can be assumed to be the same.

Problem 9.3

The head and efficiency versus capacity data for a centrifugal pump with a 10-in impeller is as shown below.

Q, gal/min	0	800	1600	2400	3000
H, ft	3185	3100	2900	2350	1800
E, %	0	55.7	78	79.3	72

The pump is driven by a constant speed electric motor at a speed of 3560 rpm.
1. Determine the performance of this pump with an 11-in impeller using affinity laws.
2. If the pump drive were changed to a variable frequency drive motor with a speed range of 3000 to 4000 rpm, calculate the new H-Q curve for the maximum speed of 4000 rpm with the original 10-in impeller.

Solution
1. Using affinity laws for impeller diameter changes, the multiplying factor for flow rate is:
Factor = 11/10 = 1.1 and the multiplier for head is $(1.1)^2 = 1.21$.
Therefore, we will generate a new set of Q and H values for the 11-in impeller by multiplying the given Q values by the factor 1.1 and the H values by the factor 1.21 as follows:

Q, gal/min	0	880	1760	2640	3300
H, ft	3854	3751	3509	2844	2178

These flow rates and head values represent the predicted performance of the 11-in impeller. The efficiency versus flow rate curve for the 11-in impeller will approximately be the same as that of the 10-in impeller.
2. Using affinity laws for speeds, the multiplying factor for the flow rate is:
Factor = 4000/3560 = 1.1236 and
the multiplier for head is $(1.1236)^2 = 1.2625$.

Therefore we will generate a new set of Q and H values for the pump at 4000 rpm by multiplying the given Q values by the factor 1.1236 and the H values by the factor 1.2625 as follows:

Q, gal/min	0	899	1798	2697	3371
H, ft	4021	3914	3661	2967	2273

These flow rates and head values represent the predicted performance of the 10-in impeller at 4000 rpm. The new efficiency versus flow rate curve will approximately be the same as the given curve for 3560 rpm.

2.13 Effect of Specific Gravity and Viscosity on Pump Performance

As mentioned previously, the pump vendor's performance curves are always based on water as the pumped liquid. When a pump is used to pump a viscous liquid, with a viscosity greater than 10 cSt, the H-Q and E-Q curves must be corrected for viscosity, using the Hydraulic Institute standards. Because BHP is a function of the specific gravity, the BHP calculated for water must also be adjusted for specific gravity of the viscous liquid. Therefore for a viscous liquid, capacity, head, and efficiency must be corrected as well as the specific gravity used to adjust the water BHP. The Hydraulic Institute has published viscosity correction charts that can be applied to correct the water performance curves to produce viscosity corrected curves. Figure 9.14 shows a typical chart from the Hydraulic Institute Engineering Data book. For any application involving high-viscosity liquids, the pump vendor should be given the liquid properties. The viscosity corrected performance curves will be supplied by the vendor as part of the pump proposal. For preliminary analysis, you may also use these charts to generate the viscosity-corrected pump curves.

The Hydraulic Institute method of viscosity correction requires determining the BEP values for Q, H, and E from the water performance curve. This is called the 100% BEP point. Three additional sets of Q, H, and E values are obtained at 60%, 80%, and 120% of the BEP flow rate from the water performance curve. From these four sets of data, the Hydraulic Institute chart can be used to obtain the correction factors C_q, C_h, and C_e for flow, head, and efficiency for each set of data. These factors are used to multiply the Q, H, and E values from the water curve, thus generating corrected values of Q, H, and E for 60%, 80%, 100%, and 120% BEP values. Problem 9.4 illustrates the Hydraulic Institute method of viscosity correction. Note that for multistage pumps, the values of H must be per stage.

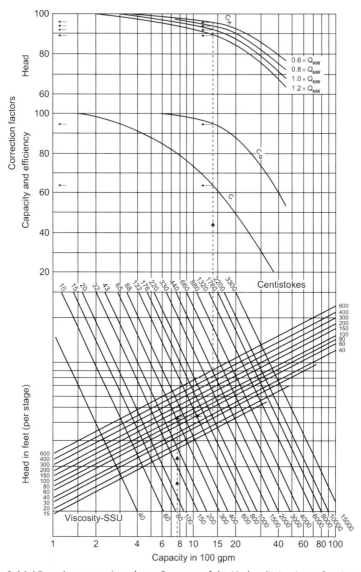

Figure 9.14 Viscosity correction chart. *Courtesy of the Hydraulic Institute, Parsippany, NJ, www.Pumps.org.*

Problem 9.4

The water performance of a single stage centrifugal pump for 60%, 80%, 100%, and 120% of the BEP is as shown below:

Q, gal/min	450	600	750	900
H, ft	114	108	100	86
E, %	72.5	80.0	82.0	79.5

1. Calculate the viscosity corrected pump performance when pumping oil with a specific gravity of 0.90 and a viscosity of 1000 SSU at pumping temperature.

Solution

1. By inspection, the BEP for this pump curve is

$$Q = 750 \quad H = 100 \quad \text{and} \quad E = 82$$

We first establish the four sets of capacities to correspond to 60%, 80%, 100%, and 120%. These have already been given as 450, 600, 750, and 900 gal/min. Because the head values are per stage, we can directly use the BEP value of head along with the corresponding capacity to enter the Hydraulic Institute Viscosity Correction chart at 750 gal/min on the lower horizontal scale. Go vertically from 750 gal/min to the intersection point on the line representing the 100-ft head curve and then horizontally to intersect the 1000 SSU viscosity line and finally vertically up to intersect the three correction factor curves C_e, C_q, and C_h.

From the Hydraulic Institute chart, Figure 9.14, we obtain the values of C_q, C_h, and C_e for flow rate, head, and efficiency as follows:

C_q	0.95	0.95	0.95	0.95
C_h	0.96	0.94	0.92	0.89
C_e	0.635	0.635	0.635	0.635

Corresponding to Q values of 450 (60% of Q_{NW}), 600 (80% of Q_{NW}), 750 (100% of Q_{NW}), and 900 (120% of Q_{NW}). The term Q_{NW} is the BEP flow rate from the water performance curve.

Using these correction factors, we generate the Q, H, and E values for the viscosity-corrected curves by multiplying the water performance value of Q by C_q, H by C_h, and E by C_e and obtain the following results.

Q_v	427	570	712	855
H_v	109.5	101.5	92.0	76.5
E_v	46.0	50.8	52.1	50.5
BHP_v	23.1	25.9	28.6	29.4

The last row of values for viscous BHP was calculated using Eqn (9.17) in Chapter 9.

The Hydraulic Institute chart, for obtaining the correction factors, consists of two separate charts. One chart applies to small pumps up to 100 gal/min capacity and head per stage of 6–400 ft. The other chart applies to larger pumps with capacity between 100 gal/min and 10,000 gal/min and head range of 15–600 ft per stage. Note that when data are taken from a water performance curve, the head has to be corrected per stage because the Hydraulic Institute charts are based on head in feet per stage rather than the total pump head. Therefore, if a six-stage pump has a BEP at 2500 gal/min and 3000 ft of head with an efficiency of 85%, the head per stage to be used with the chart will be 3000/6 = 500 ft. The total head (not per stage) from the water curve can then be multiplied by the correction factors from the Hydraulic Institute charts to obtain the viscosity corrected head for the six-stage pump.

2.14 Pump Configuration: Series and Parallel

In the preceding section, we discussed the performance of a single pump. To transport liquid through a pipeline, we may need to use more than one pump at a pump station to provide the necessary flow rate or head requirement. These pumps may be operated in series or parallel configurations. Series pumps are generally used for higher heads and parallel pumps for increased flow rates. When pumps are operated in series, the same flow rate goes through each pump and the resultant head is the sum of the heads generated by each pump. In parallel operation, the flow rate is split between the pumps, whereas each pump produces the same head. Series and parallel pump configuration are illustrated in Figures 9.15 and 9.16(a) and (b).

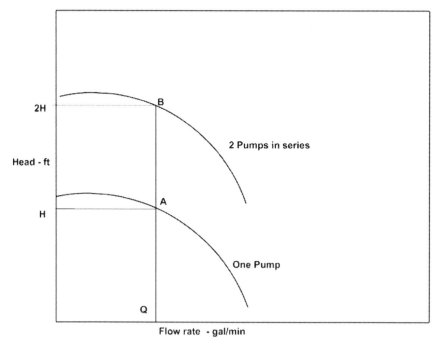

Figure 9.15 Pumps in series.

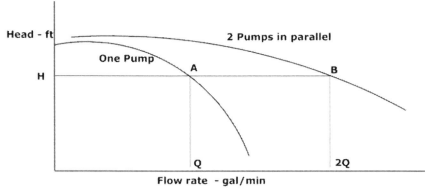

Figure 9.16 Pumps in parallel.

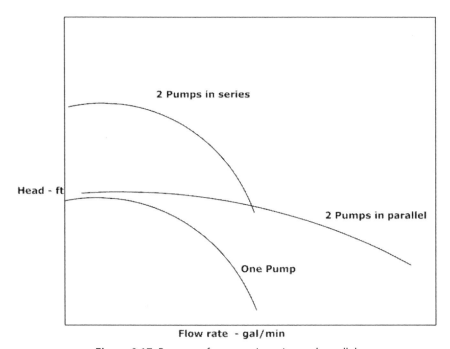

Figure 9.17 Pump performance in series and parallel.

The choice of series or parallel pumps for a particular application depends on many factors, including pipeline elevation profile as well as operational flexibility. Figure 9.17 shows the combined performance of two identical pumps in series and parallel configuration. It can be seen that parallel pumps are used when we need larger flows. Series pumps are used when we need higher heads than each individual pump.

If the pipeline elevation profile is essentially flat, the pump pressure required is mainly to overcome the pipeline friction. On the other hand, if the pipeline has drastic elevation changes, the pump head generated is mainly for the static lift and to a lesser extent for pipe friction. In the latter case, if two pumps are used in series and one shuts down, the remaining pump alone will only be able to provide half the head and therefore will not be able to provide the necessary head for the static lift at any flow rate. If the pumps were configured in parallel, then shutting down one pump will still allow the other pump to provide the necessary head for the static lift at half the previous flow rate. Thus parallel pumps are generally used when elevation differences are considerable. Series pumps are used where pipeline elevations are not significantly high.

Problem 9.5
One large pump and one small pump are operated in series. The H-Q characteristic of the pumps is defined as follows:

Q, gal/min	0	800	1600	2400	3000
Pump 1					
H, ft	2389	2325	2175	1763	1350
Pump 2					
Q, gal/min	0	800	1600	2400	3000
H, ft	796	775	725	588	450

1. Calculate the combined performance of pump 1 and pump 2 in series configuration.
2. Can these pumps be configured to operate in parallel?

Solution
1. Pumps in series have the same flow through each pump and the heads are additive. We can therefore generate the total head produced in series configuration by adding the head of each pump for each flow rate given as follows:
Combined performance of pump 1 and pump 2 in series:

Q, gal/min	0	800	1600	2400	3000
H, ft	3185	3100	2900	2351	1800

2. To operate satisfactorily in a parallel configuration, the two pumps must have a common range of heads so that at each common head, the corresponding flow rates can be added to determine the combined performance. Pump 1 and pump 2 are mismatched for parallel operation. Therefore, they cannot be operated in parallel.

2.15 Pump Head Curve versus System Head Curve

In Chapter 9, we discussed development of the pipeline system head curves. In this chapter, we will see how the system curve and the pump head curve together determine the operating point (Q, H) on the pump curve.

The system head curve for the pipeline represents the pressure required to pump a liquid through the pipeline at various flow rates (increasing pressure with increasing flow rate) and the pump H-Q curve shows the pump head available at various flow rates. When the head requirements of the pipeline match the available pump head, we have a point of intersection of the system head curve with the pump head curve as shown in Figure 9.18. This is the operating point for this pipeline and pump combination.

This figure shows system head curves for diesel and gasoline. The point of intersection of the pump head curve and the system head curve for diesel (point A) indicates the operating point for this pump when transporting diesel. Similarly, when pumping gasoline, the corresponding operating point (point B) is as shown in Figure 9.18. Therefore, with 100% diesel in the pipeline, the flow rate would be Q_A and the corresponding pump head would be H_A as shown. Similarly, with 100% gasoline in the pipeline, the flow rate would be Q_B and the corresponding pump head would be H_B as shown in Figure 9.18.

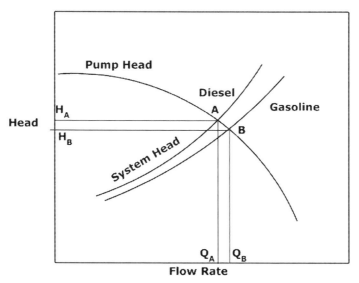

Figure 9.18 Pump curve–system head curve.

2.16 Multiple Pumps versus System Head Curve

Figure 9.19 shows the pipeline system head curve superimposed on the pump head curves to show the operating point with one pump, two pumps in series, and the same two pumps in parallel configurations. The operating points are shown as A, C, and B with flow rates of Q_A, Q_C, and Q_B, respectively.

In certain pipeline systems, depending upon the flow requirements, we may be able to obtain higher throughput by switching from a series pump configuration to a parallel pump configuration. From Figure 9.19, it can be seen that a steep system curve would be better with series pumps. If the system curve is relatively flat, parallel pumps operation will result in increased flow.

2.17 NPSH Required versus NPSH Available

The NPSH of a centrifugal pump is defined as the NPSH required at the pump impeller suction to prevent pump cavitation at any flow rate. Cavitation occurs when the suction pressure at the impeller falls below the liquid vapor pressure. This will damage the pump impeller and render it useless.

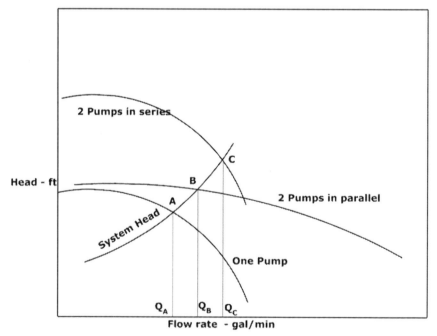

Figure 9.19 Multiple pumps and system head curve.

NPSH required for a pump at any flow rate is given by the pump vendor's NPSH curve, as in Figure 9.14. For that pump curve, it can be seen that the NPSH required ranges from 38 ft at 1200 gal/min to 64 ft at 2600 gpm. To obtain satisfactory performance with this pump, the actual *available* NPSH ($NPSH_A$) must be more than the *required* NPSH ($NPSH_R$). The $NPSH_A$ is calculated for a piping system by taking into account the positive tank head, including atmospheric pressure and subtracting the pressure drop resulting from friction in the suction piping and the liquid vapor pressure at the pumping temperature. The resulting value of NPSH for this piping configuration will represent the net pressure of the liquid at pump suction, above its vapor pressure. Shortly, we will review this calculation for a typical pump and piping system.

Consider a centrifugal pump with the suction and delivery tanks and interconnecting piping as shown in Figure 9.20.

The vertical distance from the liquid level on the suction side of the pump center line is defined as the static suction head. More correctly it is the static suction lift (H_S) when the center line of the pump is higher than that of the liquid supply level as in Figure 9.20. If the liquid supply level were higher than the pump center line, it will be called the static suction head on the pump. Similarly, the vertical distance from the pump center line to the liquid level on the delivery side is the static discharge head (H_d) as shown in Figure 9.20. The total static head on a pump is defined as the sum of the static suction head and the static discharge head. It represents the vertical distance between the liquid supply level and the liquid discharge level. The static suction head, static discharge head, and the total static head on a pump are all measured in feet of liquid in USCS or meters of liquid in the SI system.

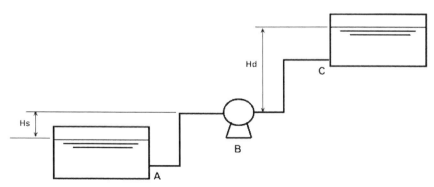

Figure 9.20 Suction and discharge heads for a centrifugal pump.

The friction head, measured in feet of liquid, is the head loss resulting from friction in both suction and discharge piping. It represents the pressure required to overcome the frictional resistance of all piping, fittings, and valves on the suction and discharge side of the pump as shown in Figure 9.20.

On the suction side of the pump, the available suction head H_S will be reduced by the friction loss in the suction piping. This net suction head on the pump will be the available suction head at the pump center line.

$$\text{Suction Head} = H_S - H_{fs} \qquad (9.14)$$

where H_{fs} is the friction loss in suction piping.

The NPSH available is calculated by adding the suction head to the atmospheric pressure on the liquid surface in the suction tank and subtracting the vapor pressure of the liquid at the flowing temperature, as follows:

$$\text{NPSHA} = (P_a - P_v)(2.31/S_g) + H + E1 - E2 - h_f \qquad (9.15)$$

where:
P_a is atmospheric pressure, psi;
P_v is liquid vapor pressure at flowing temperature, psi
S_g = liquid specific gravity at flowing temperature
H = tank head, ft
$E1$ = elevation of tank bottom, ft
$E2$ = elevation of pump suction, ft
h_f = friction loss in suction piping, ft

Problem 9.6

A centrifugal pump is used to pump a liquid from a storage tank through 500 ft of NPS 16 suction piping as shown in Figure 9.21. The head loss in the suction piping is estimated to be 12.5 ft.
1. Calculate the NPSH available at a flow rate of 3500 gal/min.
2. The pump data indicate $\text{NPSH}_R = 24$ ft at 3500 gal/min and 52 ft at 4500 gal/min. Can this piping system handle the higher flow rate without the pump cavitating?
3. If cavitation is a problem in (b), what changes must be made to the piping system to prevent pump cavitation at 4500 gal/min?

Solution
1. NPSH available in ft of liquid head is from Eqn (9.15):

$$\text{NPSH}_A = (P_a - P_v)(2.31/S_g) + H + E1 - E2 - h_f$$

Pump Stations

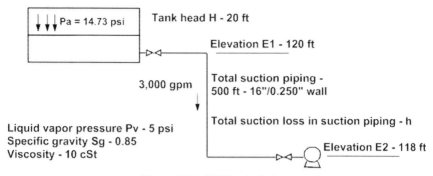

Figure 9.21 NPSH calculation.

Substituting given values we get

$$\text{NPSH}_A = (14.7 - 10) \times 2.31/0.85 + 20 + 125 - 115 - 12.5 = 30.27 \text{ ft}$$

2. Because $\text{NPSH}_R = 24$ ft at 3500 gal/min and $\text{NPSH}_A > \text{NPSH}_R$, the pump will not cavitate at this flow rate. Next, we will check for the higher flow rate.
   At 4500 gal/min flow rate, the head loss needs to be estimated.
   Using Chapter 8 concepts, at 4500 gal/min:

$$h_f = (4500/3500)^2 \times 12.5 = 20.7 \text{ ft}$$

   Recalculating NPSH_A at the higher flow rate we get:

$$\text{NPSH}_A = (14.7 - 10) \times 2.31/0.85 + 20 + 125 - 115 - 20.7 = 22.07 \text{ ft}$$

   Because the pump data indicate $\text{NPSH}_R = 52$ ft at 4500 gal/min, the pump does not have adequate NPSH ($\text{NPSH}_A < \text{NPSH}_R$) and therefore will cavitate at the higher flow rate.
3. The extra head required to prevent cavitation = $52 - 22.07 = 29.9$ ft
   One solution is to locate the pump suction at an additional 30 ft or more below the tank. This may not be practical. Another solution is to provide a small vertical can type pump that can serve as booster between the tank and the main pump. This booster pump will provide the additional head required to prevent cavitation.

2.18 Pump Station Configuration

A typical piping layout within a pump station in a simplified form is shown in Figure 9.22. In addition to the components shown, a pump station will contain other ancillary equipment such as strainers, flow meters, etc.

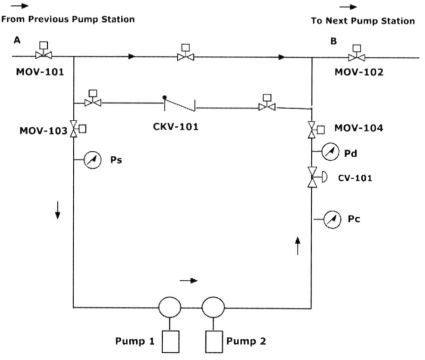

Figure 9.22 Typical pump station layout.

The pipeline enters the station boundary at point A, where the station block valve MOV-101 is located. The pipeline leaves the station boundary on the discharge side of the pump station at point B, where the station block valve MOV-102 is located.

Station bypass valves designated as MOV-103 and MOV-104 are used for bypassing the pump station in the event of pump station maintenance or other reasons where the pump station must be isolated from the pipeline. Along the main pipeline there is located a check valve, CKV-101, that prevents reverse flow through the pipeline. This typical station layout shows two pumps configured in series. Each pump pumps the same flow rate and the total pressure generated is the sum of the pressures developed by each pump. On the suction side of the pump station, the pressure is designated as P_s, whereas the discharge pressure on the pipeline side is designated as P_d. With constant-speed motor-driven pumps, there is always a control valve on the discharge side of the pump station shown as CV-101 in Figure 9.22. This control valve controls the pressure to the required value P_d by creating a pressure drop across it between the pump discharge pressure P_e and the station discharge pressure P_d. Because the pressure within the case of the second pump represents the sum of the suction pressure and the total pressure generated by both pumps, it is referred to as the case pressure P_c.

If the pump is driven by a variable speed drive (VSD) motor or an engine, the control valve is not needed because the pump may be slowed down or speeded up as required to generate the exact pressure P_d. In such a situation, the case pressure will equal the station discharge pressure P_d.

In addition to the valves shown, there will be additional valves on the suction and discharge of the pumps. Also, not shown in the figure, is a check valve located immediately after the pump discharge that prevents reverse flow through the pumps.

2.19 Control Pressure and Throttle Pressure

Mathematically, if ΔP_1 and ΔP_2 represent the differential head produced by pump 1 and pump 2 in series, we can write:

$$P_c = Ps + \Delta P_1 + \Delta P_2 \qquad (9.16)$$

Where P_c is the case pressure in pump 2 or upstream pressure at control valve.

The pressure throttled across the control valve is defined as

$$P_{thr} = P_c - P_d \qquad (9.17)$$

where

P_{thr} = control valve throttle pressure
P_d = pump station discharge pressure

The throttle pressure represents the mismatch that exists between the pump and the system pressure requirements at a particular flow rate. P_d is the pressure at the pump station discharge needed to transport liquid to the next pump station or delivery terminus. The case pressure P_c is the available pressure from the pumps. If the pumps were driven by variable speed motors, P_c would exactly match P_d and there would be no throttle pressure. In this case, no control valve is needed. The case pressure is also referred to as control pressure because it is the pressure upstream of the control valve. The throttle pressure represents unused pressure developed by the pump and hence results in wasted HP and dollars. The objective should be to reduce the amount of throttle pressure in any pumping situation. With VSD pumps there is no HP wasted since the pump case pressure exactly matches the station discharge pressure.

2.20 Variable Speed Pumps

If there are two or more pumps in series configuration, one of the pumps may be driven by a VSD motor or an engine. With parallel pumps all pumps will have to be VSD pumps, because parallel configurations require matching heads at the same flow rate. In the case of two pumps in series, we could convert one of the two pumps to be driven by a variable speed motor. This pump can then slow down to the required speed that would develop just the right amount of head, which when added to the head developed by the constant speed pump would provide exactly the total head required to match the pipeline system requirement.

2.21 VSD Pump versus Control Valve

In a single-pump station pipeline with one pump, a control valve is used to regulate the pressure for a given flow rate.

The pipeline from Essex pump station to the Kent delivery terminal is 120 miles long, NPS 16, 0.250-in wall thickness pipe with an MAOP of 1440 psi, as shown in Figure 9.23. The pipeline is designed to operate at 1400 psi discharge pressure, pumping 4000 bbl/h of liquid (specific gravity, 0.89; and viscosity, 30 cSt at 60 °F) on a continuous basis. The delivery pressure required at Kent is 50 psi. The pump suction pressure at Essex is 50 psi, and the pump differential pressure is $(1400-50) = 1350$ psi. Let's assume the single pump at Essex has the following design point:

$$Q = 2800 \text{ gal/min} \quad H = 3800 \text{ ft}$$

Converting the head of 3800 ft into psi, the pump pressure developed is

$$3800 \times 0.89/2.31 = 1464 \text{ psi}$$

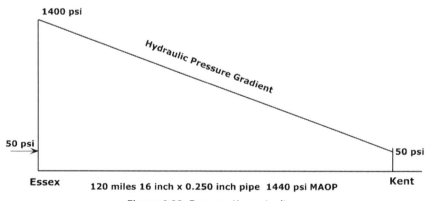

Figure 9.23 Essex to Kent pipeline.

We can see that this pressure combined with the 50-psi pump suction pressure would produce a pump discharge pressure (and hence a case pressure) of $1464 + 50 = 1514$ psi.

Because the MAOP is 1440 psi, this will overpressure the pipeline by 74 psi. The control valve located just downstream of the pump discharge will be used to reduce the discharge pressure to the required pressure of 1400 psi. This is shown by the modified system curve (2) in Figure 9.24.

The system head curve (1) in Figure 9.24 represents the pressure versus flow rate variation for our pipeline from Essex to Kent. At $Q = 2800$ gal/min (4000 bbl/h) flow rate, point C on the pipeline system head curve shows the operating point that requires a pipeline discharge pressure of 1400 psi at Essex. We have superimposed the pump H-Q curve and pressures are in feet of liquid head, so the pressure at C is

$$1400 \times 2.31/0.89 = 3634 \text{ ft}$$

Because the pump suction pressure is 50 psi, the vertical axis includes this suction head and the pump head curve in Figure 9.24 include this suction head as well. The pump H-Q curve shows that at a flow rate of

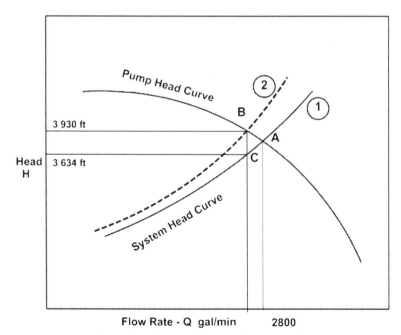

Figure 9.24 System curve and control valve.

2800 gal/min, the differential head generated is 3800 ft. Adding the 50 psi suction pressure point B on the H-Q curve corresponds to

$$3800 + 50 \times 2.31/0.89 = 3930 \text{ ft}$$

Therefore, the point of intersection of the pump H-Q curve and the pipeline system curve is the operating point A, which is at a higher flow rate than 2800 gal/min. Also, the pump discharge pressure at A will be higher than that at C. To limit this pressure within the MAOP, some form of pressure control is required. Thus using the control valve on the pump discharge will move the operating point from A to B on the pump curve corresponding to 2800 gal/min and a higher head of 3930 ft as calculated previously. The control valve pressure drop is the head difference BC as follows:

$$\text{Control valve pressure drop, BC} = 3930 - 3634 = 296 \text{ ft}$$

This pressure drop across the control valve, also called the pump throttle pressure is

$$\text{Throttle pressure} = 296 \times 0.89/2.31 = 114 \text{ psi}$$

Therefore, because of the slightly oversized pump and the pipe MAOP limit, we use a control valve on the pump discharge to limit the pipeline discharge pressure to 1400 psi at the required flow rate of 4000 bbl/h (2800 gal/min).

A dashed hypothetical system head curve designated as (2), passing through the point B on the pump head curve is the artificial system head curve because of the restriction imposed by the control valve.

The preceding analysis applies to a pump driven by a constant speed electric motor. If we had a VSD pump that can vary the pump speed from 60% to 100% rated speed and if the rated speed were 3560 rpm, the pump speed range will be $3560 \times 0.60 = 2136$ rpm to 3560 rpm.

Because the pump speed can be varied from 2136 rpm to 3560 rpm, the pump head curve will correspondingly vary according to the centrifugal pump affinity laws. Therefore, we can find some speed (less than 3560 rpm) at which the pump will generate the required head corresponding to point C in Figure 9.24.

From the preceding discussion, we can conclude that use of VSD pumps can provide the right amount of pressure required for a given flow rate, thus avoiding pump throttle pressures (and hence wasted HP), common with constant speed motor driven pumps with control valves. However, VSD pumps are expensive to install and operate compared with the use of a control valve. A typical control valve installation may cost $150,000, whereas a VSD may require $300,000–$500,000 incremental cost compared with the constant speed motor driven pump. We will have to factor in the increased operating cost of the VSD pump compared with the dollars lost in wasted HP from control valve throttling.

3. SUMMARY

In this section, we discussed pump stations and pumping configurations used in liquid pipelines. The optimum locations of pump stations for hydraulic balance were analyzed. Centrifugal pumps and positive displacement pumps were compared, along with their performance characteristics. We explained the affinity laws for centrifugal pumps, the importance of NPSH, and how to calculate power requirements when pumping different liquids. Viscosity-corrected pump performance using the Hydraulic Institute chart was explained. The performance of two or more pumps in series or parallel configuration was examined and the operating point for a pump in conjunction with the pipeline system head curve. We discussed some of the major components in a pump station, such as pumps and drivers and control valves. The use of variable speed pumps to save pumping power under different operating conditions was reviewed.

CHAPTER TEN

Compressor Stations

1. INTRODUCTION

In this chapter, we will discuss number and size of compressor stations required to transport gas in a transmission pipeline. The optimum locations and pressures at which compressor stations operate will be determined based upon the pipeline flow rate, allowable pipe operating pressures (MAOP) and pipeline topography. Centrifugal and positive displacement compressors used in gas transportation will be compared with reference to their performance characteristics and economics. Typical compressor station design and equipment used will be discussed. Isothermal, adiabatic, and polytropic compression processes and the compressor power required will be discussed. The discharge temperature of compressed gas, its impact on pipeline throughput, and gas cooling will also be explained.

2. COMPRESSOR STATION LOCATIONS

Compressor stations are installed on long-distance gas pipelines to provide the pressure needed to transport gas from one location to another. Because of the maximum allowable operating pressure (MAOP) limitations of pipeline, more than a single compressor station may be needed on a long pipeline. The locations and pressures at which these compressor stations operate are determined by the MAOP, compression ratio, power available, and environmental and geotechnical factors.

Consider a pipeline that is designed to transport 200 million standard ft^3/day (MMSCFD) of natural gas from Dover, UK, to an industrial plant at Grimsby, 70 miles away. First, we would calculate the pressure required at Dover to ensure delivery of the gas at a minimum contract pressure of 500 psig at Grimsby. Suppose the MAOP of the pipeline is 1200 psig and we calculated the required pressure at Dover to be 1130 psig. It is clear that there is no violation of MAOP and therefore a single compressor station at Dover would suffice to deliver gas to Grimsby at the required 500 psig delivery pressure. If the pipeline length had been 150 miles instead, calculations would show that, to deliver the same quantity of gas to Grimsby at the same 500 psig terminus pressure, the pressure required at Dover would have to be 1580 psig. Obviously, because this is greater than the MAOP, we would need more than the origin compressor station at Dover.

Having determined that we need two compressor stations, we need to locate the second compressor station between Dover and Grimsby. Where would this compressor station be installed? A logical site would be the midpoint between Dover and Grimsby, if the hydraulic characteristics (pipe size and ground profile) are uniform throughout the pipeline.

For simplicity, let us assume the pipeline elevation profile is fairly flat and therefore, elevation differences can be ignored. Based on this, we will locate the intermediate compressor station at the midpoint, Kent, as shown in Figure 10.1. Next we will determine the pressures at the two compressor stations.

Because the MAOP is limited to 1200 psig, assume that the compressor at Dover discharges at this pressure. Because of friction, the gas pressure drops from Dover to Kent, as indicated in Figure 10.1. Suppose the gas pressure reaches 900 psig at Kent and is then boosted back to 1200 psig by the compressor at Kent. Therefore, the compressor station at Kent has a suction pressure of 900 psig and a discharge pressure of 1200 psig. The gas continues to move from Kent to Grimsby starting at 1200 psig at Kent. As the gas reaches Grimsby, the pressure may or may not be equal to the desired terminus delivery pressure of 500 psig. Therefore, if the desired pressure at Grimsby is to be maintained, the discharge pressure of the Kent compressor stations may have to be adjusted. Alternatively, Kent could discharge at the same 1200 psig, but its location along the pipeline may have to be adjusted.

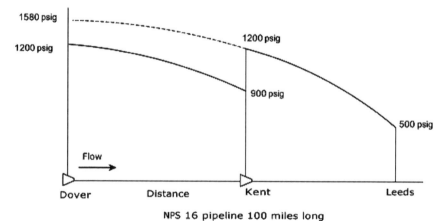

Figure 10.1 Gas pipeline with two compressor stations.

We assumed the 900 psig suction pressure at the Kent compressor station quite arbitrarily. It could well have been 700 psig or 1000 psig. The actual value depends on the so-called compression ratio desired. The compression ratio is simply the ratio of the compressor discharge pressure to its suction pressure, both pressures being expressed in absolute units.

$$\text{Compression ratio } r = \frac{P_d}{P_s} \tag{10.1}$$

where the compressor suction and discharge pressures P_s and P_d are in absolute units.

In the present case, using the assumed values, the compression ratio for Kent is

$$r = \frac{1200 + 14.7}{900 + 14.7} = 1.33$$

In this calculation, we assumed the base pressure to be 14.7 psia. If instead of the 900 psig, we had chosen a suction pressure of 700 psig, the compression ratio would be

$$r = \frac{1200 + 14.7}{700 + 14.7} = 1.7$$

A typical compression ratio for centrifugal compressors is around 1.5. A larger value compression ratio will mean more compressor power, whereas a smaller compression ratio means less power required. In gas pipelines, it is desirable to keep the average pipeline pressure as high as possible to reduce the total compression power. Therefore, if the suction pressure at Kent is allowed to drop to 700 psig or lower, the average pressure in the pipeline would be lower than if we had used 900 psig. Obviously, there is a tradeoff between the number of compressor stations, the suction pressure, and compression power required. We will discuss this in more detail later in this chapter.

Going back to the example problem previously, we concluded that we may have to adjust the location of the Kent compressor station or adjust its discharge pressure to ensure the 500 psig delivery pressure at Grimsby. Alternatively, we could leave the intermediate compressor station at the halfway point discharging at 1200 psig, eventually delivering gas to Grimsby at a pressure such as 600 psig. If calculations showed that by discharging out of Kent results in 600 psig at Grimsby, we have satisfied the minimum contract delivery pressure (500 psig) requirement at Grimsby. However, there is extra energy associated with the extra 100 psig delivery pressure. If the power plant can use this extra energy, then there is no waste. On the other hand, if the power plant requirement is 500 psig maximum, then a pressure regulator must be installed at the delivery point. The extra 100 psig would be reduced through the pressure regulator or pressure control valve at Grimsby and some energy will be wasted. Another option would be to keep the Kent compressor at the midpoint but reduce its discharge pressure so that it will result in the requisite 500 psig at Grimsby. Because pressure drop in gas pipelines is nonlinear, remembering our discussion in Chapter 8, the Kent discharge pressure may have to be reduced by less than 100 psig to provide the fixed 500 psig delivery pressure at Grimsby. This would mean that Dover will operate at 1200 psig discharge, whereas Kent would discharge at 1150 psig. This will, of course, not ensure maximum average pressure in the pipeline. However, this is still a solution and to choose the best option we must compare two or more alternative approaches, taking into account the total compressor power required as well as the cost involved. By moving the Kent compressor station slightly upstream or downstream, there will be changes in the suction and discharge pressures and therefore the compressor power required. The change in the capital cost may not be significant. However, the compressor power variation will result in change in energy cost and therefore in annual operating cost. We must therefore take into account the capital cost and annual operating cost to produce the optimum solution. An example will illustrate this method.

Problem 10.1

A natural gas pipeline 140 miles long from Danby to Leeds is constructed of NPS 16, 0.250-in wall thickness pipe with an MAOP of 1400 psig. The gas specific gravity and viscosity are 0.6 and 8×10^{-6} lb/ft-s, respectively. The pipe roughness may be assumed to be 700 μm in and the base pressure and base temperature are 14.7 psia and 60 °F, respectively. The gas flow rate is 175 MMSCFD at 80 °F and the delivery pressure required at Leeds is 800 psig. Determine the number and locations of compressor stations required, neglecting elevation difference along the pipeline. Assume $Z = 0.85$.

Solution

We will use the Colebrook–White equation to calculate the pressure drop.
    The Reynolds number is calculated from Eqn (8.47) as follows.

$$R = \frac{0.0004778 \times 175 \times 10^6 \times 0.6 \times 14.7}{15.5 \times 8 \times 10^{-6} \times 520} = 11{,}437{,}412$$

Compressor Stations

$$\text{Relative roughness} = e/D = \frac{700 \times 10^{-6}}{15.5} = 4.5161 \times 10^{-5}$$

Using the Colebrook–White Eqn (8.51), we get the friction factor.

$$\frac{1}{\sqrt{f}} = -2 Log_{10} \left[\frac{4.516 \times 10^{-5}}{3.7} + \frac{2.51}{11{,}437{,}412 \sqrt{f}} \right]$$

Solving for f by successive iteration, we get

$$f = 0.0107$$

Using the General flow Eqn (8.55), neglecting elevation effects, we calculate the pressure required at Danby as

$$175 \times 10^6 = 77.54 \left(\frac{1}{\sqrt{0.0107}} \right) \left(\frac{520}{14.7} \right) \left(\frac{P_1^2 - 814.7^2}{0.6 \times 540 \times 140 \times 0.85} \right)^{0.5}$$
$$\times (15.5)^{2.5}$$

Solving for the pressure at Danby, we get

$$P_1 = 1594 \text{ psia} = 1594 - 14.7 = 1579.3 \text{ psig}$$

It can be seen from Figure 10.2 that because MAOP is 1400 psig, we cannot discharge at 1579.3 psig at Danby.

We will need to reduce the discharge pressure at Danby to 1400 psig and install an intermediate compressor station between Danby and Leeds as shown in Figure 10.3.

Initially, let's assume that the intermediate compressor station will be located at Hampton, halfway between Danby and Leeds. For the pipe segment from Danby to Hampton, we calculate the suction pressure at the Hampton compressor station as follows.

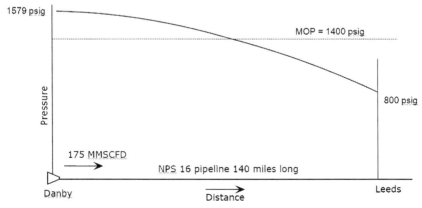

Figure 10.2 Pipeline with 1400 psig MAOP.

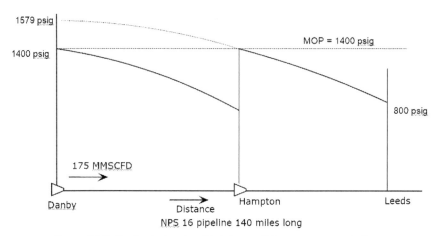

Figure 10.3 Danby to Leeds with intermediate compressor stations.

Using the general flow Eqn (8.55):

$$175 \times 10^6 = 77.54 \left(\frac{1}{\sqrt{0.0107}}\right)\left(\frac{520}{14.7}\right)\left(\frac{1414.7^2 - P_2^2}{0.6 \times 540 \times 70 \times 0.85}\right)^{0.5}$$
$$\times (15.5)^{2.5}$$

Solving for the pressure at Hampton (suction pressure):

$$P_2 = 1030.95 \text{ psia} = 1016.25 \text{ psig}$$

At Hampton, if we boost the gas pressure from 1016.25 to 1400 psig (MAOP) the compression ratio at Hampton is $\frac{1414.7}{1030.95} = 1.37$.

This is a reasonable compression ratio for a centrifugal compressor. Next, we will calculate the delivery pressure at Leeds, starting with the 1400 psig discharge pressure at Hampton.

For the 70-mile pipe segment from Hampton to Leeds, using the general flow equation, we get

$$175 \times 10^6 = 77.54 \left(\frac{1}{\sqrt{0.0107}}\right)\left(\frac{520}{14.7}\right)\left(\frac{1414.7^2 - P_2^2}{0.6 \times 540 \times 70 \times 0.85}\right)^{0.5}$$
$$\times (15.5)^{2.5}$$

Solving for P_2, the pressure at Leeds we get

$$P_2 = 1030.95 \text{ psia} = 1016.25 \text{ psig}$$

This is exactly the suction pressure at Hampton that we calculated earlier. This is because, hydraulically, Hampton is at the midpoint of the 140-mile pipeline discharging at the same pressure as Danby.

The calculated pressure at Leeds is higher than the 800 psig desired; therefore, we must move the location of Hampton compressor station slightly toward Danby so that the calculated delivery at Leeds will be

800 psig. We will calculate the distance L required between Hampton and Leeds; to achieve this, use the general flow equation.

$$175 \times 10^6 = 77.54 \left(\frac{1}{\sqrt{0.0107}}\right)\left(\frac{520}{14.7}\right)\left(\frac{1414.7^2 - 814.7^2}{0.6 \times 540 \times L \times 0.85}\right)^{0.5} \times (15.5)^{2.5}$$

Solving for length L, we get

$$L = 99.77 \text{ miles}$$

Therefore, the Hampton compressor station must be located at approximately 99.8 miles from Leeds, or 40.2 miles from Danby. Relocating Hampton from the midpoint (70 miles) results in a higher suction pressure at Hampton and therefor a different compression ratio. This can be calculated as follows.

Using the general flow Eqn (8.55) for the 40.2-mile pipe segment between Danby and Hampton, we get

$$175 \times 10^6 = 77.54 \left(\frac{1}{\sqrt{0.0107}}\right)\left(\frac{520}{14.7}\right)\left(\frac{1414.7^2 - P_2^2}{0.6 \times 540 \times 40.2 \times 0.85}\right)^{0.5} \times (15.5)^{2.5}$$

Solving for P_2 we get

$$P_2 = 1209.3 \text{ psia} = 1194.6 \text{ psig}$$

Therefore the suction pressure at Hampton = 1194.6 psig.
The compression ratio at Hampton = $\frac{1414.7}{1209.3} = 1.17$.

Figure 10.4 shows the revised location of the Hampton compressor station.

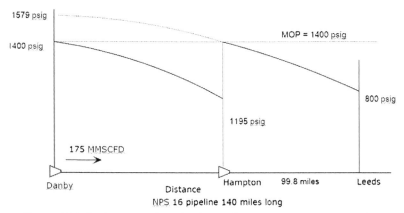

Figure 10.4 Danby to Leeds with relocated Hampton compressor station.

3. HYDRAULIC BALANCE

In the preceding discussions, we considered each compressor station operating at the same discharge pressure and the same compression ratio. However, to provide the desired delivery pressure at the terminus, we had to relocate the intermediate compressor station, thus changing its compression ratio. From the definition of compression ratio, Eqn (10.1), we can say that each compressor station operates at the same suction and discharge pressures, if doing so results in adequate delivery pressure at the terminus. If there are no intermediate gas flows in or out (injection or delivery) of the pipeline, other than at the beginning and end, as in Example 10.1, each compressor station is required to compress the same amount of gas. Therefore, with pressures and flow rates being the same, each compressor station will require the same amount of power. This is known as hydraulic balance. In a long pipeline with multiple compressor stations, if each compressor station adds the same amount of energy to the gas, this is a hydraulically balanced pipeline.

One of the advantages of a hydraulically balanced pipeline is that all compression equipment may be identical and result in minimum inventory of spare parts, thus reducing maintenance cost. Also, to pump the same gas volume through a pipeline, hydraulically balanced compressor stations will require less total power than if the stations were not in hydraulic balance.

Next we will review the different gas compression processes, such as isothermal, adiabatic, and polytropic compression. This will be followed by estimation of compressor power required.

4. ISOTHERMAL COMPRESSION

Isothermal compression occurs when the gas pressure and volume vary so that the temperature remains constant. Isothermal compression requires the least amount of work compared with other forms of compression. This process is of theoretical interest because, in practice, maintaining the temperature constant in a gas compressor is not practical.

Figure 10.5 shows the pressure versus volume variation for isothermal compression. Point 1 represents the inlet conditions of pressure P_1, volume V_1, and at temperature T_1. Point 2 represents the final compressed conditions of pressure P_2, volume V_2, and at constant temperature T_1.

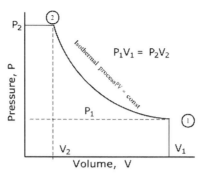
Figure 10.5 Isothermal compression.

The relationship between pressure P and volume V in an isothermal process is as follows:

$$PV = C \tag{10.2}$$

where C is a constant. Recall from Chapter 1 that this is basically Boyle's law. Applying subscripts 1 and 2, we can state that

$$P_1 V_1 = P_2 V_2 \tag{10.3}$$

The work done in compressingh 1 pound of natural gas isothermally is given by

$$Wi = \frac{53.28}{G} T_1 Log_e \left(\frac{P_2}{P_1}\right) \quad \text{(USCS units)} \tag{10.4}$$

where
  Wi: isothermal work done, ft-lb/lb of gas
  G: gas gravity, dimensionless
  T_1: suction temperature of gas, °R
  P_1: suction pressure of gas, psia
  P_2: discharge pressure of gas, psia
  Log_e: natural logarithm to base e ($e = 2.718$)
The ratio $\left(\frac{P_2}{P_1}\right)$ is also called the compression ratio
In SI units, the work done in isothermal compression of 1 kg of gas is

$$Wi = \frac{286.76}{G} T_1 Log_e \left(\frac{P_2}{P_1}\right) \quad \text{(SI units)} \tag{10.5}$$

where

Wi: isothermal work done, J/kg of gas
T_1: suction temperature of gas, K
P_1: suction pressure of gas, kPa absolute
P_2: discharge pressure of gas, kPa absolute
Other symbols are as defined earlier

Problem 10.2

Natural gas is compressed isothermally at 80 °F from an initial pressure of 800 psig to a pressure of 1000 psig. The gas gravity is 0.6. Calculate the work done in compressing 4 pounds of gas. Use 14.7 psia and 60 °F for the base pressure and temperature, respectively.

Solutions

Using Eqn (10.4), the work done per lb of gas is

$$Wi = \frac{53.28}{0.6}(80 + 460) Log_e \left(\frac{1000 + 14.7}{800 + 14.7}\right) = 10{,}527 \text{ ft.lb/lb}$$

Total work done in compressing 4 pounds of gas is

$$W_T = 10{,}527 \times 4 = 42{,}108 \text{ ft.lb}$$

5. ADIABATIC COMPRESSION

Adiabatic compression occurs when there is no heat transfer between the gas and the surroundings. The term adiabatic and isentropic are used synonymously, although isentropic really means constant entropy. Actually, an adiabatic process, without friction, is called an isentropic process. In adiabatic compression process, the gas pressure and volume are related as follows.

$$PV^\gamma = Const \qquad (10.6)$$

where

γ: ratio of specific heats of gas, $\frac{C_p}{C_v}$
C_p: specific heats of gas at constant pressure
C_v: specific heats of gas at constant volume
$Const$: a constant

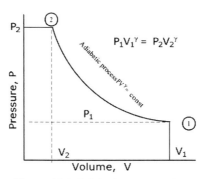

Figure 10.6 Adiabatic compression.

γ is sometimes called the adiabatic or isentropic exponent for the gas and has a value ranging between 1.2 and 1.4.

Therefore, considering subscript 1 and 2 as the beginning and ending conditions, respectively, of the adiabatic compression process, we can write:

$$P_1 V_1^\gamma = P_2 V_2^\gamma \tag{10.7}$$

Figure 10.6 shows adiabatic compression diagram showing the variation of pressure versus volume of the gas.

The work done in compressing 1 pound of natural gas adiabatically can be calculated as follows:

$$Wa = \frac{53.28}{G} T_1 \left(\frac{\gamma}{\gamma - 1}\right) \left[\left(\frac{P_2}{P_1}\right)^{\frac{\gamma-1}{\gamma}} - 1\right] \quad \text{(USCS units)} \tag{10.8}$$

where
  Wa: adiabatic work done, ft-lb/lb of gas
  G: gas gravity, dimensionless
  T_1: suction temperature of gas OR
  γ: ratio of specific heats of gas, dimensionless
  P_1: suction pressure of gas, psia
  P_2: discharge pressure of gas, psia

In SI units, the work done in adiabatic compression of 1 kg of gas is

$$Wa = \frac{286.76}{G} T_1 \left(\frac{\gamma}{\gamma - 1}\right) \left[\left(\frac{P_2}{P_1}\right)^{\frac{\gamma-1}{\gamma}} - 1\right] \quad \text{(SI units)} \tag{10.9}$$

where

   Wa: adiabatic work done, J/kg of gas
   T_1: suction temperature of gas, K
   P_1: suction pressure of gas, kPa absolute
   P_2: discharge pressure of gas, kPa absolute
   Other symbols are as defined earlier

Problem 10.3
Natural gas is compressed adiabatically from an initial temperature and pressure of 60 °F and 500 psig, respectively, to a final pressure of 1000 psig. The gas gravity is 0.6 and the ratio of specific heat is 1.3. Calculate the work done in compressing 5 pounds of gas. Use 14.7 psia and 60 °F for the base pressure and temperature, respectively.

Solutions
Using Eqn (10.8), the work done in adiabatic compression is

$$Wa = \frac{53.28}{0.6}(60+460)\left(\frac{1.3}{0.3}\right)\left[\left(\frac{1014.7}{514.7}\right)^{\frac{0.3}{1.3}} - 1\right] = 33{,}931 \text{ ft.lb/lb}$$

Therefore, total work done in compressing 5 pounds of gas is

$$W_T = 33{,}931 \times 5 = 169{,}655 \text{ ft.lb}$$

Problem 10.4
Calculate the work done in compressing 2 kg of gas (gravity = 0.65) adiabatically from an initial temperature of 20 °C and pressure of 700 kPa to a final pressure of 2000 kPa. The specific heat ratio of gas is 1.4 and the base pressure and base temperature are 101 kPa and 15 °C, respectively.

Solutions
Using Eqn (10.9), the work done in adiabatic compression of 1 kg of gas is

$$Wa = \frac{286.76}{0.65}(20+273)\left(\frac{1.4}{0.4}\right)\left[\left(\frac{2000+101}{700+101}\right)^{\frac{0.4}{1.4}} - 1\right] = 143{,}512 \text{ J/Kg}$$

Therefore, total work done in compressing 2 kg of gas is

$$W_T = 143{,}512 \times 2 = 287{,}024 \text{ J}$$

6. POLYTROPIC COMPRESSION

Polytropic compression is similar to adiabatic compression but there can be heat transfer, unlike in adiabatic compression. In a polytropic process, the gas pressure and volume are related as follows:

$$PV^n = Const \tag{10.10}$$

where

n: polytropic exponent
$Const$: a constant different than that in Eqn (10.6)
As before, from initial to final conditions, we can state that:

$$P_1 V_1^n = P_2 V_2^n \tag{10.11}$$

Because polytropic compression is similar to adiabatic compression, we can easily calculate the work done in polytropic compression by substituting n for γ in Eqns (10.10) and (10.11).

Problem 10.5
Natural gas is compressed polytropically from an initial temperature and pressure of 60 °F and 500 psig, respectively, to a final pressure of 1000 psig. The gas gravity is 0.6 and the base pressure and base temperature are 14.7 psia and 60 °F, respectively. Calculate the work done in compressing 5 pounds of gas using a polytropic exponent of 1.5.

Solutions
Polytropic compression is similar to adiabatic compression; therefore, the same equation can be used for work done, substituting the polytropic exponent n for the adiabatic exponent γ (the ratio of specific heat).

Using Eqn (10.8), the work done in polytropic compression of 1 pound of gas is

$$Wp = \frac{53.28}{0.6}(60 + 460)\left(\frac{1.5}{0.5}\right)\left[\left(\frac{1014.7}{514.7}\right)^{\frac{0.5}{1.5}} - 1\right] = 35{,}168 \text{ ft.lb/lb}$$

Therefore, total work done in compressing 5 pounds of gas is

$$W_T = 35{,}168 \times 5 = 175{,}840 \text{ ft.lb}$$

7. DISCHARGE TEMPERATURE OF COMPRESSED GAS

In adiabatic or polytropic compression of natural gas, the final temperature of the gas can be calculated, given the initial temperature and initial and final pressures, as follows. The initial conditions are called suction conditions and the final state is the discharge condition.

From Eqn (10.6) for adiabatic compression and the perfect gas law from Chapter 1, we eliminate the volume V, to get a relationship between pressure, temperature, and the compressibility factor:

$$\left(\frac{T_2}{T_1}\right) = \left(\frac{Z_1}{Z_2}\right)\left(\frac{P_2}{P_1}\right)^{\frac{\gamma-1}{\gamma}} \tag{10.12}$$

where
 T_1: suction temperature of gas, °R
 T_2: discharge temperature of gas, °R
 Z_1: gas compressibility factor at suction, dimensionless
 Z_2: gas compressibility factor at discharge, dimensionless
 Other symbols are as defined earlier

Replacing the adiabatic exponent γ in Equation (10.12) with the polytropic exponent n, we can similarly calculate the discharge temperature, for polytropic compression from the following equation.

$$\left(\frac{T_2}{T_1}\right) = \left(\frac{Z_1}{Z_2}\right)\left(\frac{P_2}{P_1}\right)^{\frac{n-1}{n}} \tag{10.13}$$

where all symbols are as defined earlier.

Problem 10.6
Gas is compressed adiabatically ($\gamma = 1.4$) from 60 °F suction temperature and a compression ratio of 2.0. Calculate the discharge temperature, assuming $Z_1 = 0.99$ and $Z_2 = 0.85$.

Solution
Using Eqn (10.12):

$$\left(\frac{T_2}{60+460}\right) = \left(\frac{0.99}{0.85}\right)(2.0)^{\frac{0.4}{1.4}} = 1.4198$$

$$T_2 = 1.4198 \times 520 = 738.3 \,°R = 278.3 \,°F$$

Therefore, the discharge temperature of the gas is 278.3 °F.

8. COMPRESSION POWER REQUIRED

We have seen from the preceding calculations that the amount of energy input to the gas by the compression process will depend on the gas pressure and the quantity of gas. The latter is proportional to the gas flow rate. The power (horsepower [HP] in US customary system units and kW in SI units) that represents the energy expended per unit time will also depend on the gas pressure and the flow rate. As the flow rate increases, the pressure also increases and therefore the power needed will also increase. For a gas compressor, the power required can be stated in terms of the gas flow rate and the discharge pressure of the compressor station as explained next.

If the gas flow rate is Q (standard ft^3/day) and the suction and discharge pressures of the compressor station are P_1 and P_2, respectively, the compressor station provides the differential pressure of $(P_2 - P_1)$ to the gas flowing at Q. This differential pressure is called the head. The discharge pressure P_d will depend on the type of compression (e.g., adiabatic, polytropic).

The head developed by the compressor is also defined as the amount of energy supplied to the gas per unit mass of gas (ft-lb/lb). Therefore, by multiplying the mass flow rate of gas by the compressor head, we can calculate the total energy supplied to the gas per unit time, or the power supplied to the gas. Dividing this power by the efficiency of the compressor, we get the compressor power input. The equation for the compressor power can be expressed as follows, in US customary system units.

$$\text{HP} = 0.0857 \left(\frac{\gamma}{\gamma - 1} \right) Q T_1 \left(\frac{Z_1 + Z_2}{2} \right) \left(\frac{1}{\eta_a} \right) \left[\left(\frac{P_2}{P_1} \right)^{\frac{\gamma-1}{\gamma}} - 1 \right] \quad (10.14)$$

where

HP: compression HP
γ: ratio of specific heats of gas, dimensionless
Q: gas flow rate, MMSCFD
T_1: suction temperature of gas OR
P_1: suction pressure of gas, psia
P_2: discharge pressure of gas, psia
Z_1: compressibility of gas at suction conditions, dimensionless
Z_2: compressibility of gas at discharge conditions, dimensionless
η_a: compressor adiabatic (isentropic) efficiency, less than 1.0

In SI units, the compressor power is as follows:

$$\text{Power} = 4.0639 \left(\frac{\gamma}{\gamma - 1}\right) Q T_1 \left(\frac{Z_1 + Z_2}{2}\right) \left(\frac{1}{\eta_a}\right) \left[\left(\frac{P_2}{P_1}\right)^{\frac{\gamma-1}{\gamma}} - 1\right] \quad (10.15)$$

where
Power: compression power, kW
γ: ratio of specific heats of gas, dimensionless
Q: gas flow rate, million m^3/day (Mm^3/day)
T_1: suction temperature of gas, K
P_1: suction pressure of gas, kPa
P_2: discharge pressure of gas, kPa
Z_1: compressibility of gas at suction conditions, dimensionless
Z_2: compressibility of gas at discharge conditions, dimensionless
η_a: compressor adiabatic (isentropic) efficiency, decimal value

The adiabatic efficiency η_a generally ranges from 0.75 to 0.85. By considering a mechanical efficiency η_m of the compressor driver we can calculate the brake HP (BHP) required to run the compressor as follows:

$$\text{BHP} = \frac{\text{HP}}{\eta_m} \quad (10.16)$$

where HP is the horsepower calculated from the preceding equations taking into account the adiabatic efficiency η_a of the compressor.

Similarly, in SI units, the brake power is calculated as follows.

$$\text{Brake Power} = \frac{\text{Power}}{\eta_m} \quad (10.17)$$

The mechanical efficiency η_m of the driver may range from 0.95 to 0.98. Therefore, the overall efficiency η_o is defined as the product of the adiabatic efficiency η_a and the mechanical efficiency η_m.

$$\eta_o = \eta_a \times \eta_m \quad (10.18)$$

From the adiabatic compression Eqn (10.6), using the perfect gas law, we eliminate the volume V, and the discharge temperature of the gas is then related to the suction temperature and the compression ratio as follows:

$$\left(\frac{T_2}{T_1}\right) = \left(\frac{P_2}{P_1}\right)^{\frac{\gamma-1}{\gamma}} \quad (10.19)$$

The adiabatic efficiency η_a may also be defined as the ratio of the adiabatic temperature rise to the actual temperature rise. Thus if the gas temperature, because of compression, increases from T_1 to T_2, the actual temperature rise is $(T_2 - T_1)$.

The theoretical adiabatic temperature rise is obtained from the adiabatic pressure and temperature relationship as follows, considering the gas compressibility factors similar to Eqn (10.12).

$$\left(\frac{T_2}{T_1}\right) = \left(\frac{Z_1}{Z_2}\right)\left(\frac{P_2}{P_1}\right)^{\frac{\gamma-1}{\gamma}} \qquad (10.20)$$

Simplifying and solving for T_2:

$$T_2 = T_1\left(\frac{Z_1}{Z_2}\right)\left(\frac{P_2}{P_1}\right)^{\frac{\gamma-1}{\gamma}} \qquad (10.21)$$

Therefore, the adiabatic efficiency is

$$\eta_a = \frac{T_1\left(\frac{Z_1}{Z_2}\right)\left(\frac{P_2}{P_1}\right)^{\frac{\gamma-1}{\gamma}} - T_1}{T_2 - T_1} \qquad (10.22)$$

Simplifying, we get

$$\eta_a = \left(\frac{T_1}{T_2 - T_1}\right)\left[\left(\frac{Z_1}{Z_2}\right)\left(\frac{P_2}{P_1}\right)^{\frac{\gamma-1}{\gamma}} - 1\right] \qquad (10.23)$$

where T_2 is the actual discharge temperature of the gas.

If the inlet gas temperature is 80 °F and the suction and discharge pressures are 800 psia and 1400 psia, respectively. In addition, if the discharge gas temperature is given as 200 °F, we can calculate the adiabatic efficiency from Eqn (10.23). Using $\gamma = 1.4$, the adiabatic efficiency is, assuming compressibility factors to be equal to 1.0 (approximately).

$$\eta_a = \left(\frac{80 + 460}{200 - 80}\right)\left[\left(\frac{1400}{800}\right)^{\frac{1.4-1}{1.4}} - 1\right] = 0.7802 \qquad (10.24)$$

Thus, the adiabatic compression efficiency is 0.7802.

Problem 10.7
Calculate the compressor power required for an adiabatic compression of 100 MMSCFD gas with inlet temperature of 70 °F and 725 psia pressure. The discharge pressure is 1305 psia. Assume compressibility factor at suction and discharge condition to be $Z_1 = 1.0$ and $Z_2 = 0.85$ and adiabatic exponent $\gamma = 1.4$ with the adiabatic efficiency $\eta_a = 0.8$. If the mechanical efficiency of the compressor driver is 0.95, what is the compressor BHP? Also estimate the discharge temperature of the gas.

Solutions
From Eqn (10.15), the HP required is

$$HP = 0.0857 \times 100 \left(\frac{1.40}{0.40}\right)(70+460)\left(\frac{1+0.85}{2}\right)\left(\frac{1}{0.8}\right)$$

$$\times \left[\left(\frac{1305}{725}\right)^{\frac{0.40}{1.40}} - 1\right] = 3362$$

Using Eqn (10.17), the compressor driver BHP is calculated based on a mechanical efficiency of 0.95.

$$\text{The BHP required} = \frac{3362}{0.95} = 3539$$

The outlet temperature of the gas is estimated using Eqn (10.23) and with some simplification as follows:

$$T_2 = (70+460) \times \left(\frac{\left(\frac{1}{0.85}\right)\left(\frac{1305}{725}\right)^{\frac{0.4}{1.4}} - 1}{0.8}\right) + (70+460)$$

$$= 789.44 \text{ R} = 329.44 \text{ °F}$$

Therefore, the discharge temperature of the gas is 329.44 °F.

Problem 10.8
Natural gas at 4 million m³/day (Mm³/day) and 20 °C is compressed isentropically ($\gamma = 1.4$) from a suction pressure of 5 MPa absolute to a discharge pressure of 9 MPa absolute in a centrifugal compressor with an isentropic efficiency of 0.82. Calculate the compressor power required assuming the compressibility factors at suction and discharge condition to be $Z_1 = 0.95$

and $Z_2 = 0.85$, respectively. If the mechanical efficiency of the compressor driver is 0.95, what is the driver power required? Calculate the discharge temperature of the gas.

Solutions
From Eqn (10.16), the power required is

$$\text{Power} = 4.0639 \times 4 \left(\frac{1.40}{0.40}\right)(20+273)\left(\frac{0.95+0.85}{2}\right)\left(\frac{1}{0.82}\right)$$

$$\times \left[\left(\frac{9}{5}\right)^{\frac{0.40}{1.40}} - 1\right] = 3346 \text{ kW}$$

Compressor power = 3346 kW.
Using Eqn (10.17), we calculate the driver power required as follows.
The driver power required $= \frac{3346}{0.95} = 3522$ kW.
The discharge temperature of the gas is estimated from Eqn (10.23) as

$$T_2 = \frac{20+273}{0.82} \times \left[\left(\frac{0.95}{0.85}\right)\left(\frac{9}{5}\right)^{\frac{0.4}{1.4}} - 1\right] + (20+273) = 408.07 \text{ K}$$

$$= 135.07 \, ^\circ\text{C}$$

9. OPTIMUM COMPRESSOR LOCATIONS

In the preceding sections, we discussed a pipeline with two compressor stations to deliver gas from Danby to a Leeds power plant. In this section, we will explore how to locate the intermediate compressor stations at the optimum location taking into account the overall HP required. In Section 3, we discussed hydraulic balance. In the next example, we will analyze optimum compressor locations by considering both hydraulically balanced and unbalanced compressor station locations.

Problem 10.9
A gas transmission pipeline is 240 miles long, NPS 30, 0.500-in wall thickness runs from Payson to Douglas. The origin compressor station is at Payson and two intermediate compressor stations tentatively located at Williams (milepost 80) and Snowflake (milepost 160) as shown in Figure 10.7. There are no intermediate flow deliveries or injections, and the inlet flow rate at Payson is 900 MMSCFD. The terminus delivery pressure required at

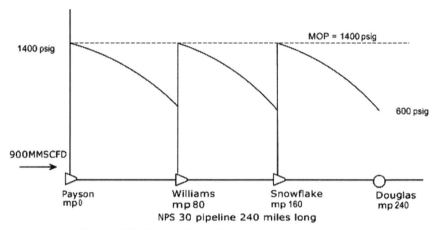

Figure 10.7 Gas pipeline with three compressor stations.

Douglas is 600 psig and the MAOP of the pipeline is 1400 psig throughout. Neglect the effects of elevation and assume a constant gas flow temperature of 80 °F and constant values of transmission factor $F = 20$ and compressibility factor $Z = 0.85$ throughout the pipeline. The gas gravity is 0.6, base pressure = 14.7 psia, and base temperature = 60 °F. Use polytropic compression coefficient of 1.38 and a compression efficiency of 0.9. The objective is to determine the locations of the intermediate compressor stations.

Solution

First, we calculate for each of the three segments, Payson to Williams, Williams to Snowflake, and Snowflake to Douglas, the downstream pressure starting with an upstream pressure of 1400 psig (MAOP). Accordingly, using the general flow equation for the Payson to Williams segment, we calculate the downstream pressure at Williams starting with a pressure of 1400 psig at Payson. This downstream pressure becomes the suction pressure at the Williams compressor station. Next, we repeat the calculations for the second segment from Williams to Snowflake to calculate the downstream pressure at Snowflake based on an upstream pressure of 1400 psig at Williams. This downstream pressure becomes the suction pressure at the Snowflake compressor station. Finally, we calculate the downstream pressure at Douglas, for the third and final pipe segment from Snowflake to Douglas, based on an upstream pressure of 1400 psig at Snowflake. This final pressure is the delivery pressure at the Douglas terminus. We have thus calculated the suction pressures at each of the two intermediate compressor stations at Williams and Snowflake and also calculated the final delivery pressure at Douglas. This pressure calculated at Douglas may or may not be equal to the desired delivery pressure of 600 psig because we performed a *forward calculation* going from Payson to Douglas. Therefore, because the delivery pressure is usually a desired or contracted value, we will have to adjust the location of the last compressor station at Snowflake to achieve the desired delivery pressure at Douglas, as we did before in the Danby to Leeds pipeline.

Another approach would be to perform a backward calculation starting at Douglas and proceeding toward Payson. In this case, we would start with segment three and calculate the location of the Snowflake compressor station that will result in an upstream pressure of 1400 psig at Snowflake. Thus we locate the Snowflake compressor station that will cause a discharge pressure of 1400 psig at Snowflake and a delivery pressure of 600 psig at Douglas. Having located the Snowflake compressor station, we can now recalculate the suction pressure at Snowflake by considering the pipe segment two and using an upstream pressure of 1400 psig at Williams. We will not have to repeat calculations for segment one because the location of Williams has not changed and therefore the suction pressure at Williams will remain the same as the previously calculated value. We have thus been able to determine the pressures along the pipeline with the given three compressor station configuration such that the desired delivery pressure at Douglas has been achieved and each compressor station discharges at the MAOP value of 1400 psig. But are these the optimum locations of the intermediate compressor stations Williams and Snowflake? In other words, are all compressor stations in hydraulic balance? We can say that these compressor stations are optimized and are in hydraulic balance only if each compressor station operates at the same compression ratio and therefore adds the same amount of HP to the gas at each compressor station. The locations of Williams and Snowflake may not result in the same suction pressures even though the discharge pressures are the same. Therefore, chances are that Williams may be operating at a lower compression ratio than Snowflake or Payson or vice versa, which will not result in hydraulic balance. However, if the compression ratios are close enough that the required compressor sizes are the same, we could still be in hydraulic balance and the station could be at optimum locations.

Next, let's perform the calculations and determine how much tweaking of the compressor station locations is required to optimize these stations. First we will perform the backward calculations for segment three, starting with a downstream pressure of 600 psig at Douglas and an upstream pressure of 1400 psig at Snowflake. With these constraints, we will calculate the pipe length L miles between Snowflake and Douglas.

Using the general flow Eqn (8.55) and neglecting elevations:

$$900 \times 10^6 = 38.77 \times 20.0 \left(\frac{520}{14.7}\right) \left(\frac{1414.7^2 - 614.7^2}{0.6 \times 540 \times L \times 0.85}\right)^{0.5} (29)^{2.5}$$

Solving for pipe length, we get $L = 112.31$ miles.

Therefore, to discharge at 1400 psig at Snowflake and deliver gas at 600 psig at Douglas, the Snowflake compressor station will be located at a distance of 112.31 miles upstream of Douglas or at milepost $(240 - 112.31) = 127.69$ measured from Payson.

Next, keeping the location of Williams compressor station at milepost 80, we calculate the downstream pressure at Snowflake for pipe segment two starting at 1400 psig at Williams. This calculated pressure will be the suction pressure of the Snowflake compressor station.

Using the general flow Eqn (8.55) and neglecting elevations:

$$900 \times 10^6 = 38.77 \times 20.0 \left(\frac{520}{14.7}\right) \left(\frac{1414.7^2 - P_2^2}{0.6 \times 540 \times 47.69 \times 0.85}\right)^{0.5} (29)^{2.5}$$

where the pipeline segment length between Williams and Snowflake was calculated as $127.69 - 80 = 47.69$ miles.

Solving for suction pressure at Snowflake, we get:

$$P_2 = 1145.42 \text{ psia} = 1130.72 \text{ psig}$$

Therefore, the compression ratio at Snowflake $= \frac{1414.7}{1145.42} = 1.24$

Next, for pipe segment one between Payson and Williams, we will calculate the downstream pressure at Williams starting at 1400 psig at Payson. This calculated pressure will be the suction pressure of the Williams compressor station.

Using the general flow Eqn (8.55) and neglecting elevations:

$$900 \times 10^6 = 38.77 \times 20.0 \left(\frac{520}{14.7}\right) \left(\frac{1414.7^2 - P_2^2}{0.6 \times 540 \times 80 \times 0.85}\right)^{0.5} (29)^{2.5}$$

Solving for suction pressure at Williams, we get:

$$P_2 = 919.20 \text{ psia} = 904.5 \text{ psig}$$

Therefore, the compression ratio at Williams $= \frac{1414.7}{919.2} = 1.54$

Therefore, from the preceding calculations, the compressor station at Williams requires a compression ratio $r = 1.54$, whereas the compressor station at Snowflake requires a compression ratio $r = 1.24$. Obviously, this is not a hydraulically balanced compressor station system. Further, we do not know what the suction pressure is at the Payson compressor station. If we assume that Payson receives gas at approximately the same suction pressure as Williams (905 psig), both Payson and Williams compressor stations will have the same compression ratio of 1.54. In this case, the Snowflake compressor will be the odd one, operating at the compression ratio of 1.24. How do we balance these compressor stations? One way would be to obtain the same compression ratios for all three compressor stations by simply relocating the Snowflake compressor station toward Douglas so that its suction pressure will drop from 1131 to 905 psig, while keeping the discharge at Snowflake at 1400 psig. This will then ensure that all three compressor stations will be operating at the following suction and discharge pressures and compression ratios.

Suction pressure, $P_s = 904.5$ psig.
Discharge pressure, $P_d = 1400$ psig.
Compression ratio, $r = \frac{1400+14.7}{904.5+14.7} = 1.54$.

However, because the Snowflake compressor is now located closer to Douglas than before (127.69), the discharge pressure of 1400 psig at Snowflake will result in a higher delivery pressure at Douglas than the required 600 psig as shown in Figure 10.8.

If the additional pressure at Douglas can be tolerated by the customer, then there will be no problem. But if the customer requires no more than 600 psig, we have to reduce the delivery pressure to 600 psig by installing a pressure regulator at Douglas, as shown in Figure 10.8. Therefore, by balancing the compressor station locations, we have also created a problem of getting rid of the extra pressure at the delivery point. Pressure regulation means wasted HP. The advantage of the balanced compressor stations versus the negative aspect of the pressure regulation must be factored in to the decision process.

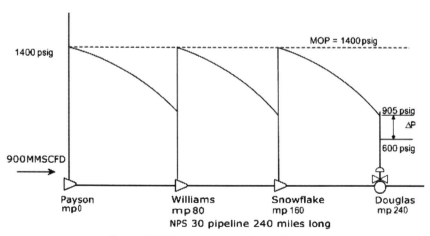

Figure 10.8 Pressure regulation at Douglas.

To illustrate this pressure regulation scenario, we will now determine the revised location of the Snowflake compressor station for hydraulic balance. We will calculate the length of pipe segment two by assuming 1400 psig discharge pressure at Williams and a suction pressure of 904.5 psig at Snowflake.

Using the general flow Eqn (8.55) and neglecting elevations:

$$900 \times 10^6 = 38.77 \times 20.0 \left(\frac{520}{14.7}\right) \left(\frac{1414.7^2 - 919.2^2}{0.6 \times 540 \times L \times 0.85}\right)^{0.5} (29)^{2.5}$$

Solving for pipe length for segment 2, we get:

$$L = 80 \text{ mi}$$

Therefore, the Snowflake compressor station should be located at a distance of 80 miles from Williams, or at milepost 160. We could have arrived at this without these calculations, because elevations are neglected and the Payson to Williams pressure profile will be the same as the pressure profile from Williams to Snowflake. With the Snowflake compressor station located at milepost 160 and discharging at 1400 psig, we conclude that the delivery pressure at Douglas will also be 904.5 psig, because all three pipe segments are hydraulically the same. We see that the delivery pressure at Douglas is approximately 305 psig more than the desired pressure. As indicated previously, a pressure regulator will be required at Douglas to reduce the delivery pressure to 600 psig. We can compare the hydraulically balanced scenario with the previously calculated case where Payson and Williams operate at a compression ratio of 1.54 and Snowflake operates at lower compression ratio of 1.24. By applying approximate cost per installed HP, we can compare these two cases. First using Eqn (10.14), calculate the HP required at each compressor station, assuming polytropic compression and compression ratio of 1.54 for balanced compressor station.

$$\text{HP} = 0.0857 \times 900 \times \left(\frac{1.38}{0.38}\right)(80 + 460)\left(\frac{1 + 0.85}{2}\right)\left(\frac{1}{0.9}\right)$$
$$\times \left[(1.54)^{\frac{0.38}{1.38}} - 1\right] = 19{,}627$$

Therefore, total HP required in the hydraulically balanced case is
Total HP = $3 \times 19{,}627 = 58{,}881$.
At a cost of $2000 per installed HP.
Total HP cost = $2000 \times 58{,}881 = \$117.76$ million.

In the hydraulically unbalanced case, the Payson and Williams compressor stations will operate at a compression ratio of 1.54 each, whereas the Snowflake compressor station will require a compression ratio of 1.24.

Using Eqn (10.14), the HP required at Snowflake compressor station is

$$HP = 0.0857 \times 900 \times \left(\frac{1.38}{0.38}\right)(80 + 460)\left(\frac{1 + 0.85}{2}\right)\left(\frac{1}{0.9}\right)$$
$$\times \left[(1.24)^{\frac{0.38}{1.38}} - 1\right] = 9487$$

Therefore, total HP required in the hydraulically unbalanced case is
Total HP = (2 × 19,627) + 9487 = 48,741.
At a cost of $2000 per installed HP.
Total HP cost = $2000 × 48,741 = $97.48 million.

The hydraulically balanced case requires (58,881 − 48,741) or 10,140 HP more and will cost approximately ($117.76 − $97.48) or $20.28 million more. In addition to the extra HP cost, the hydraulically balanced case will require a pressure regulator that will waste energy and result in extra equipment cost. Therefore, the advantages of using identical components, by reducing spare parts and inventory in the hydraulically balanced case, must be weighed against the additional cost. It may not be worth spending the extra $20 million to obtain this benefit. The preferred solution in this case is for the Payson and Williams compressor stations to be identical (compression ratio = 1.54) and the Snowflake compressor station to be a smaller one (compression ratio = 1.24) requiring the lesser compression ratio and HP, to provide the required 600 psig delivery pressure at Douglas.

10. COMPRESSORS IN SERIES AND PARALLEL

When compressors operate in series, each unit compresses the same amount of gas but at different compression ratios such that the overall pressure increase of the gas is achieved in stages as shown in Figure 10.9.

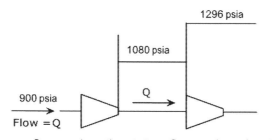

Figure 10.9 Compressors in series.

It can be seen from Figure 10.9 that the first compressor compresses gas from a suction pressure of 900 psia to 1080 psia at a compression ratio of 1.2. The second compressor takes the same volume and compresses it from 1080 psia to a discharge pressure of $1080 \times 1.2 = 1296$ psia. Thus, the overall compression ratio of the two identical compressors in series is $1296/900 = 1.44$. We have thus achieved the increase in pressure in two stages. At the end of each compression cycle, the gas temperature would rise to some value calculated in accordance with Eqn (10.23). Therefore, with multiple stages of compression, unless the gas is cooled between stages, the final gas temperature may be too high. High gas temperatures are not desirable because the throughput capability of a gas pipeline decreases with gas flow temperature. Therefore, with compressors in series, the gas is cooled to the original suction temperature between each stage of compression so that the final temperature at the end of all compressors in series is not exceedingly high. Suppose the calculated discharge temperature of a compressor is 232 °F, starting at 70 °F suction temperature and with a compression ratio of 1.4. If two of these compressors were in series and there was no cooling between compressions, the final gas temperature would reach approximately

$$\frac{(232 + 460)(232 + 460)}{70 + 460} = 903.5 \text{ R} = 443.5 \,°F$$

This is too high a temperature for pipeline transportation. On the other hand, if we cool the gas back to 70 °F before compressing it through the second compressor, the final temperature of the gas coming out of the second compressor will be approximately 232 °F.

Compressors are installed in parallel so that large volumes necessary can be provided by multiple compressors each producing the same compression ratio. Three identical compressors with a compression ratio of 1.4 may be used to provide 900 MMSCFD gas flow from a suction pressure of 900 psia. In this example, each compressor will compress 300 MMSCFD from 900 psia to a discharge pressure of

$$P_2 = 900 \times 1.4 = 1260 \text{ psia}.$$

This is illustrated schematically in Figure 10.10.

Unlike compressors in series, the discharge temperature of the gas coming out of the parallel bank of compressors will not be high because the gas does not undergo multiple compression ratios. The gas temperature on the

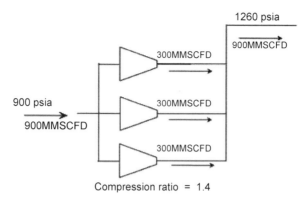

Figure 10.10 Compressors in parallel.

discharge side of each parallel compressor will be approximately the same as that of a single compressor with the same compression ratio. Therefore, three parallel compressors each compressing the same volume of gas at a compression ratio of 1.4 will have a final discharge temperature of 232 °F starting from a suction temperature of 70 °F. Gas cooling is required at these temperatures to achieve efficient gas transportation; also, operating at temperature not exceeding the limits of the pipe coating material. Generally, pipe coating require gas temperature not to exceed 140 °F–150 °F.

The compression ratio was defined earlier as the ratio of the discharge pressure to the suction pressure. The higher the compression ratio, the higher the gas discharge temperature will be, in accordance with Eqn (10.23).

Consider a suction temperature of 80 °F and the suction and discharge pressures of 900 psia and 1400 psia, respectively. The compression ratio is $1400/900 = 1.56$. Using Eqn (10.19), the approximate discharge temperature will be

$$\left(\frac{T_2}{80 + 460}\right) = \left(\frac{1400}{900}\right)^{\frac{1.3-1}{1.3}} = 598.36 \text{ R or } 138.36 \text{ °F}$$

If the compression ratio is increased from 1.56 to 2.0, the discharge temperature will become 173.67 °F. It can be seen that the discharge temperature of the gas increases considerably with the compression ratio. Because the throughput capacity of a gas pipeline decreases with increase in gas temperature, we must cool the discharge gas from a compressor to ensure maximum pipeline throughput.

Typically, centrifugal compressors are used in gas pipeline applications to have a compression ratio of 1.5–2.0, there may be instances where higher compression ratios are required due to lower gas receipt pressures and higher pipeline discharge pressures to enable a given volume of gas to be transported through pipeline. For this, reciprocating compressors are used. Manufacturers limit the maximum compression ratio of reciprocating compressors to a range of 5–6. This is due to high forces that are exerted on the compressor components that require expensive materials as well as complicated safety needs.

Suppose a compressor is required to provide gas at 1500 psia from a suction pressure 200 psia. This requires an overall compression ratio of 7.5. Because this is beyond the acceptable range of compression ratio, we will have to provide this compression in stages. If we provide the necessary pressure by using two compressors in series, each compressor will require to be at a compression ratio of $\sqrt{7.5}$ or approximately 2.74. The first compressor raises the pressure from 200 psia to $200 \times 2.74 = 548$ psia. The second compressor will then boost the gas pressure from 548 psia to approximately $548 \times 2.74 = 1500$ psia. In general, if n compressors are installed in series to achieve the required total compression ratio r_t, we can calculate the individual compression ratio of each compressor to be

$$r = (r_t)^{\frac{1}{n}} \tag{10.25}$$

where
   r: compression ratio of each compressor, dimensionless
   r_t: overall compression ratio, dimensionless
   n: number of compressors in series

By providing the overall compression ratio by means of identical compressors in series, power requirements will be minimized. Thus, in the preceding example, we assumed that two identical compressors in series each providing a compression ratio of 2.74 resulting in an overall compression ratio of 7.5 will be a better option than if we had a compressor with a compression ratio of 3.0 in series with another compressor with a compression ratio of 2.5. To illustrate this further, if an overall compression ratio of 20 is required and we were to use three compressors in series, the most economical option will be to use identical compressors each with a compression ratio of $(20)^{\frac{1}{3}} = 2.71$.

11. TYPES OF COMPRESSORS: CENTRIFUGAL AND POSITIVE DISPLACEMENT

Compressors used in natural gas transportation system are either positive displacement type or centrifugal type. Positive displacement compressors generate the pressure required by trapping a certain volume of gas within the compressor and increasing the pressure by reduction of volume. The high-pressure gas is then released through the discharge valve into the pipeline. Piston-operated reciprocating compressors fall within the category of positive displacement compressors. These compressors have a fixed volume and are able to produce high compression ratios. Centrifugal compressors, on the other hand, develop the pressure required by the centrifugal force resulting from rotation of the compressor wheel that translates the kinetic energy into pressure energy of the gas. Centrifugal compressors are more commonly used in gas transmission systems because of their flexibility. Centrifugal compressors have lower capital cost and lower maintenance expenses. They can handle larger volumes within a small area compared with positive displacement compressors. They also operate at high speeds and are of balanced construction. However, centrifugal compressors have less efficiency than positive displacement compressors.

Positive displacement compressors have flexibility in pressure range, have higher efficiency, and can deliver compressed gas at wide range of pressures. They are also not very sensitive to the composition of the gas. Positive displacement compressors have pressure ranges up to 30,000 psi and range from very low HP to more than 20,000 HP per unit. Positive displacement compressors may be single stage or multistage depending on the compression ratio required. The compression ratio per stage for positive displacement compressors is limited to 4.0 because a higher ratio causes a higher discharge pressure, which affects the valve life of positive displacement compressors. Heat exchangers are used between stages of compression so that the compressed heated gas is cooled to the original suction temperature before being compressed in the next stage. The HP required in a positive displacement compressor is usually estimated from charts provided by the compressor manufacturer. The following equation may be used for large, slow-speed compressors with compression ratios >2.5 and for gas specific gravity of 0.65.

$$BHP = 22rNQF \qquad (10.26)$$

where

BHP: brake horsepower
r: compression ratio per stage
N: number of stages
Q: gas flow rate, MMSCFD at suction temperature and 14.4 psia
F: factor that depends on the number of compression stages; 1.0 for single-stage compression, 1.08 for two stage compression, 1.10 for three stage compression

In Eqn (10.26) the constant 22 is changed to 20 when gas gravity is between 0.8 and 1.0. Also, for compression ratios between 1.5 and 2.0, the constant 22 is replaced with a number between 16 and 18.

Problem 10.10

Calculate the BHP required to compress a 5-MMSCFD gas at 14.4 psia and 70 °F, with an overall compression ratio of 7 considering two-stage compression.

Solution

Considering two identical stages, the compression ratio per stage = $\sqrt{7.0}$ = 2.65.

Using Eqn (10.26), we get

$$BHP = 22 \times 2.65 \times 2 \times 5 \times 1.08 = 629.64$$

Centrifugal compressors may be a single-wheel or single-stage compressor or multiwheel or multistage compressor. Single-stage centrifugal compressors have a volume range of 100–150,000 ft^3/min at actual conditions (actual cubic feet per minute [ACFM]). Multistage centrifugal compressors handle a volume range of 500–200,000 ACFM. The operational speeds of centrifugal compressors range from 3000 to 20,000 rpm. The upper limit of speed will be limited by the wheel-tip speed and stresses induced in the impeller. Advances in technology have produced compressor wheels operating at speeds in excess of 30,000 rpm. Centrifugal compressors are driven by electric motors, steam turbines, or gas turbines. Sometimes, speed increasers are used to increase the speeds necessary to generate the pressure.

12. COMPRESSOR PERFORMANCE CURVES

The performance curve of a centrifugal compressor that can be driven at varying speeds typically shows a graphic plot of the inlet flow rate in ACFM against the head or pressure generated at various percentages of

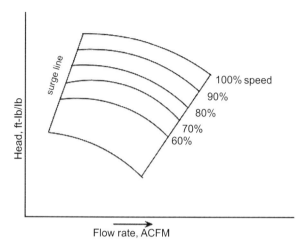

Figure 10.11 Typical centrifugal compressor performance curves.

the design speed. Figure 10.11 shows a typical centrifugal compressor performance curve or performance map.

The limiting curve on the left hand side is known as the surge line and the corresponding curve on the right side is known as the stonewall limit. The performance of a centrifugal compressor at different rotational speeds follows the so-called affinity laws. According to the affinity laws, as the rotational speed of centrifugal compressor is changed, the inlet flow and head vary as the speed and the square of the speed respectively, as indicated in the following equation.

For compressor speed change:

$$\frac{Q_2}{Q_1} = \frac{N_2}{N_1} \tag{10.27}$$

$$\frac{H_2}{H_1} = \left(\frac{N_2}{N_1}\right)^2 \tag{10.28}$$

where

Q_1, Q_2: initial and final flow rates
H_1, H_2: initial and final heads
N_1, N_2: initial and final compressor speeds

In addition, the HP for compression varies as the cube of the speed change as follows:

$$\frac{HP_2}{HP_1} = \left(\frac{N_2}{N_1}\right)^3 \tag{10.29}$$

Problem 10.11

The compressor head and volume flow rate for a centrifugal compressor at 18,000 rpm are as follows.

ACFM	Flow rate, Q Head, H
	ft-lb/lb
360	10,800
450	10,200
500	9700
600	8200
700	5700
730	4900

ACFM, Actual cubic feet per minute.

Using the affinity laws, determine the performance of this compressor at a speed of 20,000 rpm.

Solution

The ratio of speeds is

$$\frac{20000}{18000} = 1.11$$

The multiplier for the flow rate is 1.11 and the multiplier for the head is $(1.11)^2$ or 1.232.

Using the affinity laws, the performance of the centrifugal compressor at 20,000 rpm is as follows:

ACFM	Flow rate, Q Head, H
	ft-lb/lb
399.6	13,306
499.5	12,566
555.0	11,950
666.0	10,102
777.0	7022
810.0	6037

ACFM, Actual cubic feet per minute.

13. COMPRESSOR HEAD AND GAS FLOW RATE

The head developed by a centrifugal compressor is calculated from the suction and discharge pressures, the compressibility factor, and the polytropic or adiabatic exponent. We will explain how the ACFM is calculated from the standard gas flow rate. Knowing the maximum head that can be generated per stage, the number of stages needed can be calculated.

Consider a centrifugal compressor used to raise the gas pressure from 800 to 1440 psia starting at a suction temperature of 70 °F and gas flow rate of 80 MMSCFD. The average compressibility factor from the suction to the discharge side is 0.95. The compressibility factor at the inlet is assumed to be 1.0, the polytropic exponent is 1.3, and the gas gravity is 0.6. The head generated by the compressor is calculated from the previously introduced equation.

$$H = \frac{53.28}{0.6} \times 0.95 \times (70 + 460) \left(\frac{1.3}{0.3}\right) \left[\left(\frac{1440}{800}\right)^{\frac{0.3}{1.3}} - 1\right]$$

$$= 28{,}146 \text{ ft.lb/lb}$$

The actual flow rate at inlet conditions is calculated using the gas law as

$$Q_{act} = \frac{80 \times 14.7 \times 1.0}{800} \times \frac{70 + 460}{60 + 460} \times \frac{10^6}{24 \times 60}$$

$$= 1040.5 \text{ ft}^3/\text{min (ACFM)}$$

If this particular compressor, according to vendor data, can produce a maximum head per stage of 10,000 ft-lb/lb, the number of stages required to produce the required head is

$$n = \frac{28146}{10000} = 3$$

Rounding off to the nearest whole number. Next, suppose that this compressor has a maximum design speed of 16,000 rpm. The actual operating speed necessary for the three-stage compressor is, according to the affinity laws:

$$N_{act} = 16{,}000 \sqrt{\frac{28{,}146}{3 \times 10{,}000}} = 15{,}498$$

Therefore, to generate 28,146 ft-lb/lb of head at a gas flow rate of 1040.5 ACFM, this three-stage compressor must run at a speed of 15,498 rpm.

14. COMPRESSOR STATION PIPING LOSSES

As the gas enters the suction side of the compressor, it flows through a complex piping system within the compressor station. Similarly, the compressed gas leaving the compressor traverses the compressor station discharge

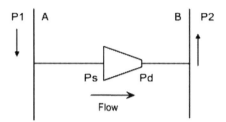

Figure 10.12 Compressor station suction and discharge piping.

piping system that consists of valves and fittings before entering the main pipeline on its way to the next compressor station or delivery terminus. This is illustrated in Figure 10.12.

It can be seen from Figure 10.12 that at the compressor station boundary A on the suction side, the gas pressure is P_1. This pressure drops to a value P_s at the compressor suction as the gas flows through the suction piping from A to B. This suction piping consisting of valves, fittings, filters, and meters causes a pressure drop of ΔP_s. Therefore, the actual suction pressure at the compressor is

$$P_s = P_1 - \Delta P_s \qquad (10.30)$$

where

P_s: compressor suction pressure, psia
P_1: compressor station suction pressure, psia
ΔP_s: pressure loss in compressor station suction piping, psi

At the compressor, the gas pressure is raised from P_s to P_d through a compression ratio r as follows:

$$r = \frac{P_d}{P_s} \qquad (10.31)$$

where

r: compression ratio, dimensionless
P_d: compressor discharge pressure, psia

The compressed gas then flows through the station discharge piping and loses pressure until it reaches the station discharge valve at the boundary D of the compressor station. If the station discharge pressure is P_2, we can write:

$$P_2 = P_d - \Delta P_d \qquad (10.32)$$

where

P_2: compressor station discharge pressure, psia
ΔP_d: pressure loss in compressor station discharge piping, psi

Typically, the values of ΔP_s and ΔP_d range from five to 15 psi.

Problem 10.12

A compressor station on a gas transmission pipeline has the following pressures at the station boundaries. Station suction pressure = 850 psig and station discharge pressure = 1430 psig. The pressure losses in the suction piping and discharge piping are 5 and 10 psi, respectively. Calculate the compression ratio of this compressor stations.

Solution

From Eqn (10.30), the compressor suction pressure is

$$P_s = 850 - 5 = 845 \text{ psia}$$

Similarly, the compressor discharge pressure is

$$P_d = 1430 + 10 = 1440 \text{ psia}$$

Therefore, compression ratio is

$$r = \frac{1440}{845} = 1.70$$

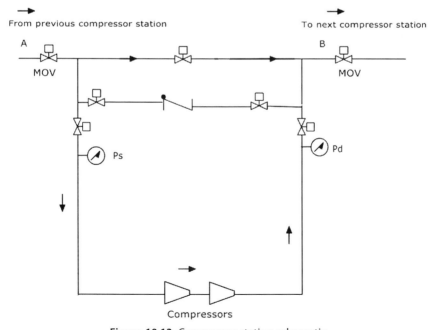

Figure 10.13 Compressor station schematic.

15. COMPRESSOR STATION SCHEMATIC

A typical compressor station schematic showing the arrangement of the valves, piping, and the compressor itself is shown in Figure 10.13.

16. SUMMARY

In this chapter, we discussed compressing a gas to generate the pressure needed to transport the gas from one point to another along a pipeline. The compression ratio and the power required to compress a certain volume of gas as well as the discharge temperature of the gas exiting the compressor were explained and illustrated using examples. In a long-distance gas transmission pipeline, locating intermediate compressor stations and minimizing energy lost were discussed. Hydraulically balanced and optimized compressor station locations were reviewed. Various compression processes, such as isothermal, adiabatic, and polytropic, were explained. The different types of compressors such as positive displacement and centrifugal were explained along with their advantages and disadvantages. The need for configuring compressors in series and parallel was explored. The centrifugal compressor performance curve was discussed and the effect of rotational speed on the flow rate and head using the affinity laws was illustrated with examples. Finally, the impact of the compressor station yard piping pressure drops and how they affect the compression ratio and HP were discussed.

CHAPTER ELEVEN

Series and Parallel Piping

1. SERIES PIPING

In the previous chapters, we assumed the pipeline to have the same diameter throughout its length. Pipes are said to be in series if different lengths of pipes of different diameters are joined end to end with the entire flow passing through all pipes.

Consider a pipeline consisting of two different lengths and pipe diameters joined together in series. A pipeline, 1000 ft long, 16-in diameter, connected in series with a pipeline, 500 ft long and 14-in diameter would be an example of a series pipeline. At the connection point, we will need to have a fitting known as a reducer, that will join the 16-in pipe with the smaller 14-in pipe. This fitting will be a 16-in × 14-in reducer. The reducer causes transition in the pipe diameter smoothly from 16 in to 14 in. We can calculate the total pressure drop through this 16-in/14-in pipeline system by adding the individual pressure drops in the 16-in and the 14-in pipe segment and accounting for the pressure loss in the 16-in × 14-in reducer.

There are situations where a pipeline may consist of different pipe diameter connected together in series to transport different volumes of fluid as shown in Figure 11.1.

In Figure 11.1, pipe section AB with a diameter of 16-in is used to transport natural gas volume of 100 million standard cubic feet per day (MMSCFD) and after making a delivery of 20 MMSCFD at B, the remainder of 80 MMSCFD flows through the 14-in diameter pipe BC.

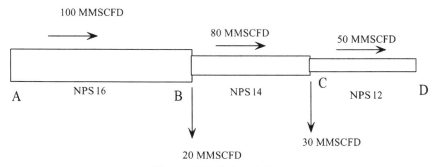

Figure 11.1 Series piping.

At C, a delivery of 30 MMSCFD is made and the balance volume of 50 MMSCFD is delivered to the terminus D through a 12-in pipeline CD.

It is clear that the pipe section AB flows the largest volume (100 MMSCFD), whereas the pipe segment CD transports the least volume (50 MMSCFD). Therefore, segments AB and CD for reasons of economy should be of different pipe diameters as indicated in the Figure 11.1. If we maintained the same pipe diameter of 16-in from A to D it would be a waste of pipe material and therefore cost. Constant diameter is used only when the same flow that enters the pipeline is also delivered at the end of the pipeline, with no intermediate injections or deliveries.

However, in reality there is no way of determining ahead what the future delivery volumes would be along the pipeline. Hence it is difficult to determine initially the different pipe sizes for each segment. Therefore, in many cases you will find that the same diameter pipe is used throughout the entire length of the pipeline even though there are intermediate deliveries. Even with the same nominal pipe diameter, different pipe sections may have different wall thicknesses; therefore, we have different pipe inside diameters for each pipe segment. Such wall thickness changes are made to compensate for varying pressures along the pipeline. The subject of pipe strength and its relation to pipe diameter and wall thickness were discussed in Chapter 5.

The pressure required to transport gas or liquid in a series pipeline from point A to point D in Figure11.1 is calculated by considering each pipe segment such as AB, BC, etc., and applying the appropriate pressure drop equation for each segment.

Another approach to calculating the pressures in series piping system is using the equivalent length concept. This method can be applied when the same uniform flow exists throughout the pipeline, with no intermediate deliveries or injections. We will explain this method of calculation for a series piping system with the same flow rate Q through all pipe segments. Suppose the first pipe segment has an inside diameter D_1 and length L_1 followed by the second segment of inside diameter D_2 and length L_2 and so on. We calculate the equivalent length of the second pipe segment based on the diameter D_1 such that the pressure drop in the equivalent length matches that in the original pipe segment of diameter D_2.

Pressure drop in diameter D_2 and Length L_2 =

Pressure drop in diameter D_1 and equivalent Length Le_2

Thus the second segment can be replaced with a piece of pipe of length Le$_2$ and diameter D$_1$ Similarly, the third pipe segment with diameter D$_3$ and length L$_3$ will be replaced with a piece of pipe of Le$_3$ and diameter D$_1$. Thus we have converted the three segments of pipe in terms of diameter D$_1$ as follows

Segment 1 – diameter D$_1$ and length L$_1$
Segment 2 – diameter D$_1$ and length Le$_2$
Segment 3 – diameter D$_1$ and length Le$_3$

For convenience, we picked the diameter D$_1$ of segment 1 as the base diameter to use, to convert from the other pipe sizes. We now have the series piping system reduced to one constant diameter (D$_1$) pipe of total equivalent length given by

$$Le = L_1 + Le_2 + Le_3 \qquad (11.1)$$

The pressure required at the inlet of this series piping system can then be calculated based on diameter D$_1$ and length Le. We will now explain how the equivalent length is calculated.

1.1 Equivalent Length of Pipes: Gas Pipelines

Upon examining the general flow equation, it can be seen that for the same flow rate and gas properties, neglecting elevation effects the pressure difference $(P_1^2 - P_2^2)$ is inversely proportional to the fifth power of the pipe diameter and directly proportional to the pipe length. Therefore, we can state that approximately

$$\Delta P_{sq} = \frac{CL}{D^5} \qquad (11.2)$$

where
ΔP_{sq} = difference in square of pressures $(P_1^2 - P_2^2)$ for pipe segment.
C = a constant
L = pipe length
D = pipe inside diameter

The value of C depends on the flow rate, gas properties, gas temperature, base pressure, and base temperature. Therefore, C will be the same for all pipe segments in a series pipeline with constant flow rate.

From Eqn (11.2), we conclude that the equivalent length for the same pressure drop is proportional to the fifth power of the diameter. Therefore, in the series piping example discussed earlier, the equivalent length of the second pipe segment of diameter D_2 and length L_2 is

$$\frac{CL_2}{D_2^5} = \frac{CLe_2}{D_1^5} \qquad (11.3)$$

Simplifying, we get

$$Le_2 = L_2 \left(\frac{D_1}{D_2}\right)^5 \qquad (11.4)$$

Similarly, for the third pipe segment of diameter D_3 and length L_3, the equivalent length is

$$Le_3 = L_3 \left(\frac{D_1}{D_3}\right)^5 \qquad (11.5)$$

Therefore, the total equivalent length Le for all three pipe segments in terms of diameter D_1 is

$$Le = L_1 + L_2 \left(\frac{D_1}{D_2}\right)^5 + L_3 \left(\frac{D_1}{D_3}\right)^5 \qquad (11.6)$$

It can be seen from Eqn (11.6) that if $D_1 = D_2 = D_3$, the total equivalent length becomes $(L_1 + L_2 + L_3)$ as expected.

We can now calculate the pressure drop for the series piping system considering a single pipe of length Le and uniform diameter D_1 flowing a constant volume Q. A problem will illustrate the equivalent length method.

Problem 11.1: Gas Pipeline

A series piping system, shown in Figure 11.2, consists of 12 miles of Nominal Pipe Size (NPS) 16, 0.375-in wall thickness connected to 24 miles of NPS 14, 0.250-in wall thickness and 8 miles of NPS 12, 0.250-in wall thickness pipes. Calculate the inlet pressure required at the origin A of this pipeline system for a gas flow rate of 100 MMSCFD. Gas is delivered to the terminus B at a de-livery pressure of 500 psig. The gas gravity and viscosity are 0.6 and 0.000008 lb/ft-s. The gas temperature is assumed constant at 60 °F. Use a compressibility factor of 0.90 and the general flow equation with Darcy

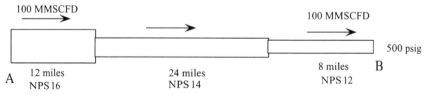

Figure 11.2 Series piping in a gas pipeline.

friction factor = 0.02. The base temperature and base pressure are 60 °F and 14.7 psia, respectively.

Compare results using the equivalent length method and with the more detailed method of calculating pressure for each pipe segment separately.

Solution

Inside diameter of first pipe segment = 16 − 2 × 0.375 = 15.25 in
Inside diameter of second pipe segment = 14 − 2 × 0.250 = 13.50 in
Inside diameter of third pipe segment = 12.75 − 21×10.250 = 12.25 in

Using Eqn (11.6), we calculate the equivalent length of the pipeline, considering NPS 16 as the base diameter.

$$Le = 12 + 24 \times \left(\frac{15.25}{13.5}\right)^5 + 8 \times \left(\frac{15.25}{12.25}\right)^5$$

or

$$Le = 12 + 44.15 + 23.92 = 80.07 \text{ mi}$$

Therefore, we will calculate the inlet pressure P_1 considering a single pipe from A to B having a length of 80.07 miles and inside diameter of 15.25 in.

Outlet pressure = 500 + 14.7 = 514.7 psia

Using the general flow, neglecting elevation effects and substituting given values, we get

$$100 \times 10^6 = 77.54 \left(\frac{1}{\sqrt{0.02}}\right)\left(\frac{520}{14.7}\right)\left[\frac{(P_1^2 - 514.7^2)}{0.6 \times 520 \times 80.07 \times 0.9}\right]^{0.5} 15.25^{2.5}$$

Transposing and simplifying, we get

$$P_1^2 - 514.7^2 = 724{,}642.99$$

Finally, solving for the inlet pressure P_1, we get

$$P_1 = 994.77 \text{ psia} = 980.07 \text{ psig}$$

Next we will compare the preceding result, using the equivalent length method, with the more detailed calculation of treating each pipe segment separately and adding the pressure drops.

Consider the 8-mile pipe segment 3 first because we know the outlet pressure at B is 500 psig. Therefore, we can calculate the pressure at the beginning of the segment 3 using the general flow equation as follows.

$$100 \times 10^6 = 77.54 \left(\frac{1}{\sqrt{0.02}}\right)\left(\frac{520}{14.7}\right)\left[\frac{(P_1^2 - 514.7^2)}{0.6 \times 520 \times 8 \times 0.9}\right]^{0.5} 12.25^{2.5}$$

Solving for the pressure P_1, we get

$$P_1 = 693.83 \text{ psia} = 679.13 \text{ psig}$$

This is the pressure at the beginning of the pipe segment 3, which is also the end of pipe segment 2.

Next consider pipe segment 2 (24 miles of NPS 14 pipe) and calculate the upstream pressure P_1 required for a downstream pressure of 679.13 psig calculated in the preceding section. Using the general flow equation for pipe segment 2, we get

$$100 \times 10^6 = 77.54 \left(\frac{1}{\sqrt{0.02}}\right)\left(\frac{520}{14.7}\right)\left[\frac{(P_1^2 - 693.83^2)}{0.6 \times 520 \times 24 \times 0.9}\right]^{0.5} 13.5^{2.5}$$

Solving for the pressure P_1, we get

$$P_1 = 938.58 \text{ psia} = 923.88 \text{ psig}$$

This is the pressure at the beginning of the pipe segment 2, which is also the end of pipe segment 1.

Next we calculate the inlet pressure P_1 of pipe segment 1 (12 miles of NPS 16 pipe) for an outlet pressure of 923.88 psig, we just calculated. Using the general flow equation for pipe segment 1, we get

$$100 \times 10^6 = 77.54 \left(\frac{1}{\sqrt{0.02}}\right)\left(\frac{520}{14.7}\right)\left[\frac{(P_1^2 - 938.58^2)}{0.6 \times 520 \times 12 \times 0.9}\right]^{0.5} 15.25^{2.5}$$

Solving for the pressure P_1, we get

$$P_1 = 994.75 \text{ psia} = 980.05 \text{ psig}$$

This compares well with the pressure of 980.07 psig we calculated earlier using the equivalent length method.

Problem 11.2

A natural gas pipeline consists of three different pipe segments connected in series, pumping the same uniform flow rate of 3.0 Mm^3/day at 20 °C. The first segment, DN 500 with a 12-mm wall thickness is 20-km long. The second segment is DN 400, with a 10-mm wall thickness and 25-km long. The last segment is DN 300, with a 6-mm wall thickness and 10 km long. The inlet pressure is 8500 kPa. Assuming flat terrain, calculate the delivery pressure, using the general flow equation and Colebrook friction factor of 0.02. Gas gravity = 0.65. Viscosity = 0.000119 poise. Compressibility factor Z = 0.9. Base temperature = 15 °C and base pressure = 101 kPa. Compare results using the equivalent length method as well as the method using individual pipe segment pressure drops.

Solution

Inside diameter of first pipe segment = 500 − 2 × 12 = 476 mm.
Inside diameter of second pipe segment = 400 − 2 × 10 = 380 mm.
Inside diameter of last pipe segment = 300 − 2 × 6 = 288 mm.
Equivalent length method:
Using Eqn (11.6), we calculate the total equivalent length of the pipeline system based on the first segment diameter DN 500 as follows.

$$Le = 20 + 25 \times \left(\frac{500 - 2 \times 12}{400 - 2 \times 10}\right)^5 + 10 \times \left(\frac{500 - 2 \times 12}{300 - 2 \times 6}\right)^5$$

or

$$Le = 20 + 77.10 + 123.33 = 220.43 \text{ km}$$

Thus the given pipeline system can be considered equivalent to a single pipe DN 500, 12-mm wall thickness, 220.43-km long.

The outlet pressure P_2 is calculated using the general flow equation as follows

$$3 \times 10^6 = 1.1494 \times 10^{-3} \left(\frac{15 + 273}{101}\right)$$
$$\times \left[\frac{(8500^2 - P_2^2)}{0.65 \times 293 \times 0.9 \times 0.02 \times 220.43}\right]^{0.5} (476)^{2.5}$$

Solving for P_2 we get

$$8500^2 - P_2^2 = 25,908,801$$

or

$$P_2 = 6807 \text{ kPa (absolute)}$$

We have assumed that given inlet pressure is in absolute value.
Therefore, the delivery pressure is 6807 kPa (absolute).
Next we calculate the delivery pressure considering the three pipe segments treated separately. For the first pipe segment, 20-km long, we calculate the outlet pressure P_2 at the end of the first segment as follows. Using the general flow equation, we get

$$3 \times 10^6 = 1.1494 \times 10^{-3} \left(\frac{15 + 273}{101}\right)$$
$$\times \left[\frac{(8500^2 - P_2^2)}{0.65 \times 293 \times 0.9 \times 0.02 \times 20}\right]^{0.5} (476)^{2.5}$$

Solving for P_2, we get

$$P_2 = 8361 \text{ kPa (absolute)}$$

Thus, the pressure at the end of the first pipe segment or the beginning of the second segment is 8361 kPa (absolute).

Next we repeat the calculation for the second pipe segment DN 400, 25-km long using

$P_1 = 8361$ kPa (absolute), to calculate P_2

$$3 \times 10^6 = 1.1494 \times 10^{-3} \left(\frac{15 + 273}{101}\right)$$

$$\times \left[\frac{(8361^2 - P_2^2)}{0.65 \times 293 \times 0.9 \times 0.02 \times 25}\right]^{0.5} (380)^{2.5}$$

Solving for P_2 we get

$$P_2 = 7800 \text{ kPa (absolute)}$$

This is the pressure at the end of the second pipe segment, which is also the inlet pressure for the third pipe segment.

Finally, we calculate the outlet pressure of the last pipe segment (DN 300, 10 km) using $P_1 = 7800$ kPa (absolute) as follows

$$3 \times 10^6 = 1.1494 \times 10^{-3} \left(\frac{15 + 273}{101}\right)$$

$$\times \left[\frac{(7800^2 - P_2^2)}{0.65 \times 293 \times 0.9 \times 0.02 \times 10}\right]^{0.5} (288)^{2.5}$$

Solving for P_2 we get.

$$P_2 = 6808 \text{ kPa (absolute)}$$

Therefore the delivery pressure is 6808 kPa (absolute).

This compares favorably with the values of 6807 kPa we calculated earlier using the equivalent length approach.

1.2 Equivalent Length of Pipes: Liquid Pipelines

A pipe is equivalent to another pipe or a pipeline system, when the same pressure loss from friction occurs in the equivalent pipe compared with that of the other pipe or pipeline system. Because the pressure drop can be caused by an infinite combination of pipe diameters and pipe lengths, we must specify a particular diameter to calculate the equivalent length.

Suppose a pipe A of length L_A and internal diameter D_A is connected in series with a pipe B of length L_B and internal diameter D_B. If we were to replace this two-pipe system with a single pipe of length L_E and diameter

D_E, we have what is known as the equivalent length of pipe. This equivalent length of pipe may be based on one of the two diameters (D_A or D_B) or a totally different diameter D_E.

The equivalent length L_E in terms of pipe diameter D_E can be written as

$$L_E/(D_E)^5 = L_A/(D_A)^5 + L_B/(D_B)^5 \qquad (11.7)$$

This formula for equivalent length is based on the premise that the total friction loss in the two-pipe system exactly equals that of the single equivalent pipe.

Because a pressure drop per unit length is inversely proportional to the fifth power of the diameter. If we refer to the diameter D_A as the basis, this equation becomes, after setting $D_E = D_A$.

$$L_E = L_A + L_B(D_A/D_B)^5 \qquad (11.8)$$

Thus, we have an equivalent length L_E that will be based on diameter D_A. This length L_E of pipe diameter D_A will produce the same amount of frictional pressure drop as the two lengths L_A and L_B in a series. We have thus simplified the problem by reducing it to one single pipe length of uniform diameter D_A.

The equivalent length method discussed previously is only approximate. Furthermore, if elevation changes are involved, it becomes more complicated, unless there are no controlling elevations along the pipeline system.

An example will illustrate this concept of equivalent pipe length.

Consider a pipeline 16-in × 0.281-in wall thickness pipeline, 20 miles long installed in series with a 14-in × 0.250-in wall thickness pipeline 10 miles long. The equivalent length of this pipeline is

$$20 + 10 \times (16 - 0.562)^5 / (14 - 0.50)^5 = 39.56 \text{ miles of 16 inch pipe.}$$

The actual physical length of 30 miles of 16-in and 14-in pipes is replaced with a single 16 in pipe 39.56 miles long, for pressure drop calculations. Note that we have left out the pipe fitting that would connect the 16-in pipe with the 14-in pipe. This would be a 16 × 14 reducer which would have its own equivalent length. To be precise, we should determine the equivalent length of the reducer from the table in Appendix A.10 and add it to the above length to obtain the total equivalent length, including the fitting.

After the equivalent length pipe is determined, we can calculate the pressure drop based on this pipe size.

Problem 11.3: Liquid Pipeline

A refined products pipeline consists of three pipe segments connected in series, pumping the same uniform flow rate of 60,000 barrels/day of diesel. The first segment, NPS 20, 0.500-in wall thickness is 20 miles long. The second segment is NPS 16, 0.250-in wall thickness and 15 miles long. The last segment is NPS 14, 0.250-in wall thickness and 10 miles long. The inlet pressure is 1400 psig. Assuming flat terrain, calculate the delivery pressure using the Colebrook–White equation for pressure drop and 0.002-in absolute pipe roughness throughout. Diesel gravity = 0.85. Viscosity = 5.0 cSt. Compare results using the equivalent length method as well as the method using individual pipe segment pressure drops.

Solution

Calculate the Reynolds number as follows:

$$R = \left(\frac{92.24 \times 60000}{5 \times (20 - 1.0)}\right) = 58,257 \quad \text{For a 20=in section}$$

and

$$R = \left(\frac{92.24 \times 60000}{5 \times 15.5}\right) = 71,412 \quad \text{For a 16=in section}$$

$$\text{Finally, } R = 81,991 \quad \text{for a 14=in section}$$

Next, calculate the friction factor f for each pipe size.

For a 20-in pipe section, using the Colebrook–White equation, we get f as follows:

$$\frac{1}{\sqrt{f}} = -2Log_{10}\left(\frac{0.002}{3.7 \times 19} + \frac{2.51 \times 1}{58257 f^{0.5}}\right)$$

Solving for f, we get $f = 0.02$ (corresponding $F = 14.14$)
Similarly, for a 16-in pipe section

$$\frac{1}{\sqrt{f}} = -2Log_{10}\left(\frac{0.002}{3.7 \times 15.5} + \frac{2.51 \times 1}{71412 f^{0.5}}\right)$$

and $f = 0.0198$ ($F = 14.21$)

Finally, for a 14-in pipe section, $f = 0.0194$ ($F = 14.36$)

Next we calculate the Pressure drop per mile Pm for 20 inch, 16 inch and 14 inch diameter pipe sizes as follows:

For 20-in pipe

$$Pm = 0.2421 \left(\frac{60000}{14.14}\right)^{2.0} \times \frac{0.85}{(19)^5} = 1.5 \text{ psi/mi}$$

For a 16-in pipe

$$Pm = 0.2421 \left(\frac{60000}{14.21}\right)^{2.0} \times \frac{0.85}{(15.5)^5} = 4.10 \text{ psi/mi}$$

For a 14-in pipe

$$Pm = 0.2421 \left(\frac{60000}{14.36}\right)^{2.0} \times \frac{0.85}{(13.5)^5} = 8.36 \text{ psi/mi}$$

The total pressure drop in the entire pipeline is

$$(1.5 \times 20) + (4.1 \times 15) + (8.36 \times 10) = 175.1 \text{ psi}$$

Therefore delivery pressure at the end of the pipe is $1400 - 175.1 = 1224.9$ psi.

Next we will calculate the results using the equivalent length method:

Converting each of the segment in terms of the base diameter of 20 in, we get the equivalent length of the middle segment (16 in) as $((20-1)/(15.5))^5 \times 15 = 41.5$ milesand similarly:

The equivalent length of the third segment (14 in) as $((20-1)/(13.5))^5 \times 10 = 55.22$ miles.

Therefore, the total equivalent length of the entire pipeline is

$$20.0 + 41.5 + 55.22 = 116.72 \text{ mi}$$

The total pressure drop is then $116.72 \times 1.5 = 175.08$ psi, which is the same as before.

2. PARALLEL PIPING

Sometimes two or more pipes are connected such that the fluid flow splits among the branch pipes and eventually combine downstream into a single pipe as illustrated in Figure 11.3. Such a piping system is referred to as parallel pipes. It is also called a looped piping system, where each parallel pipe is known as a loop. The reason for installing parallel pipes or loops is to reduce pressure drop in a certain section of the pipeline because of pipe pressure limitation or for increasing the flow rate in a bottleneck section. By

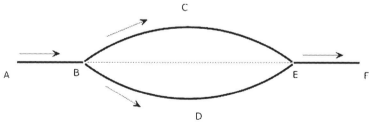

Figure 11.3 Parallel pipes.

installing a pipe loop from B to E in Figure 11.3, we are effectively reducing the overall pressure drop in the pipeline from A to F, because between B and E, the flow is split through two pipes.

In Figure 11.3, fluid flows through pipe AB and at point B, part of the flow branches off into pipe BCE, whereas the remainder flows through the pipe BDE. At point E, the flows recombine to the original value and the liquid flows through the pipe EF. We will assume that the entire pipeline system is in the horizontal plane with no changes in pipe elevations.

To solve for the pressures and flow rates in a parallel piping system, such as the one depicted in Figure 11.3, we use the following two principles of pipes in parallel.

1. Conservation of total flow.
2. Common pressure loss across each parallel pipe.

According to the principle (1), the total flow entering each junction of pipe must equal the total flow leaving the junction.

Therefore,

$$\text{Total Inflow} = \text{Total Outflow}$$

Thus, in Figure 11.3, all flows entering and leaving the junction B must satisfy the above principle. If the flow into the junction B is Q, the flow in branch BCE is Q_{BC} and flow in the branch BDE is Q_{BD}, we have from above conservation of total flow

$$Q = Q_{BC} + Q_{BD} \tag{11.9}$$

The second principle of parallel pipes, defined as (2), requires that the pressure drop across the branch BCE must equal the pressure drop across the branch BDE. This is simply because point B represents the common upstream pressure for each of these branches, whereas the pressure at point E represents the common downstream pressure. Referring to these pressures as P_B and P_E, we can state

$$\text{Pressure drop in branch BCE} = P_B - P_E \tag{11.10}$$

$$\text{Pressure drop in branch BDE} = P_B - P_E \tag{11.11}$$

Assuming that the flow Q_{BC} and Q_{BD} are in the direction of BCE and BDE, respectively. If we had a third pipe branch between B and E, such as that shown by the dashed line BE in Figure 11.3, we can state that the common pressure drop $P_B - P_E$ would be applicable to the third parallel pipe between B and E also.

We can rewrite Eqns (11.9) and (11.10) as follows for the three parallel pipe system.

$$Q = Q_{BC} + Q_{BD} + Q_{BE} \qquad (11.12)$$

and

$$\Delta P_{BCE} = \Delta P_{BDE} = \Delta P_{BE} \qquad (11.13)$$

where ΔP is the pressure drop in respective parallel pipes.

Similar to the equivalent length concept in series piping, we can calculate an equivalent pipe diameter for pipes connected in parallel.

2.1 Equivalent Diameter of Pipes: Liquid Pipelines

Because each of the parallel pipes in Figure 11.3 has a common pressure drop, we can replace all the parallel pipes between B and E with one single pipe of length L_E and diameter D_E so that the pressure drop through the single pipe at flow Q equals that of the individual pipes as follows:

Pressure drop in equivalent single pipe length L_E and diameter D_E at flow rate $Q = \Delta P_{BCE}$

Assuming now that we have only the two parallel pipes BCE and BDE in Figure 11.3, ignoring the dashed line BE, we can state that

$$Q = Q_{BC} + Q_{BD} \qquad (11.14)$$

and

$$\Delta P_{EQ} = \Delta P_{BCE} = \Delta P_{BDE} \qquad (11.15)$$

The pressure drop ΔP_{EQ} for the equivalent pipe can be written as follows, using Eqn (11.15)

$$\Delta P_{EQ} = K(L_E)(Q)^2/D_E^5 \qquad (11.16)$$

where K is a constant, that depends on the liquid properties.

Eqn (11.16) will then become

$$K\,L_E\,Q^2/D_E^5 = K\,L_{BC}\,Q_{BC}^2/D_{BC}^5 = K\,L_{BD}\,Q_{BD}^2/D_{BD}^5 \qquad (11.17)$$

Simplifying, we get

$$L_E\,Q^2/D_E^5 = L_{BC}\,Q_{BC}^2/D_{BC}^5 = L_{BD}\,Q_{BD}^2/D_{BD}^5 \qquad (11.18)$$

Further simplifying the problem by assuming each loop to be the same length as the equivalent length

$$L_{BC} = L_{BD} = L_E \qquad (11.19)$$

We get

$$Q^2/D_E^5 = Q_{BC}^2/D_{BC}^5 = Q_{BD}^2/D_{BD}^5 \qquad (11.20)$$

Substituting for Q_{BD} in terms for Q_{BC} from Eqn (11.20), we get

$$Q^2/D_E^5 = Q_{BC}^2/D_{BC}^5 \qquad (11.21)$$

and

$$Q_{BC}^2/D_{BC}^5 = (Q - Q_{BC})^2/D_{BD}^5 \qquad (11.22)$$

From Eqns (11.21) and (11.22), we can solve for the two flows Q_{BC}, Q_{BD}, and the equivalent diameter D_E in terms of the known quantities Q, D_{BC}, and D_{BC}.

The following problem will illustrate the equivalent diameter approach in parallel piping systems.

Problem 11.4: Parallel Pipes in a Liquid Pipeline

A parallel pipe system, transporting water, similar to the one shown in Figure 11.3 is located in a horizontal plane with the following data.

Flow rate $Q = 2000$ gal/min

Pipe segment AB = 15.5 in inside diameter, 4000 ft

Pipe segment BCE = 12 in inside diameter, 8000 ft

Pipe segment BDE = 10 in inside diameter, 6500 ft

Pipe segment EF = 15.5 in inside diameter, 3000 ft

1. Calculate the flow rate through each parallel pipe and the equivalent diameter of a single pipe 5000-ft-long between B and E to that will replace the two parallel pipes.
2. Determine the pressure required at the origin A to provide a delivery pressure of 50 psig at the terminus F. Use the Hazen–Williams equation with a C factor = 120.

Solution

$$\text{(a)} \quad Q_1 + Q_2 = 2000$$

$$Q_1^2 L_1/D_1^5 = Q_2^2 L_2/D_2^5$$

where suffix 1 and 2 refer to the two branches BCE and BDE, respectively.

$$(Q_2/Q_1)^2 = (D_2/D_1)^5(L_1/L_2)$$
$$= (10/12)^5 \times (8000/6500)$$

$$Q_2/Q_1 = 0.7033$$

Solving, we get

$$Q_1 = 1174 \text{ gal/min}$$

$$Q_2 = 826 \text{ gal/min}$$

The equivalent pipe diameter for a single pipe 5000 ft long is calculated as follows

$$(2000)^2(5000)/D_E^5 = (1174)^2 \times 8000/(12)^5$$

$$\text{or} \quad D_E = 13.52 \text{ in}$$

Therefore, a 13.52-in-diameter pipe, 5000 ft long between B and E will replace the two parallel pipes.

(b) To determine the pressure required at the origin A, we will first calculate the pressure required at E for the pipe segment EF to provide a delivery pressure of 50 psig at the terminus F.

2.2 Parallel Pipes in Gas Pipelines

Similar to the parallel pipes in liquid pipelines discussed in the previous section, we can perform an analysis of the parallel pipes or loops in a gas pipeline system. In calculating the pressure drop through each parallel pipe in a gas pipeline we use a slightly different approach.

According to the second principle of parallel pipes, the pressure drop in pipe branch BCE must equal the pressure drop in pipe branch BDE. This is because both pipe branches have a common starting point (B) and common ending point (E). Therefore, the pressure drop in the branch pipe BCE and branch pipe BDE are each equal to $(P_B - P_E)$ where P_B and P_E are the pressures at junctions B and E, respectively.

Therefore, we can write

$$\Delta P_{BCE} = \Delta P_{BDE} = P_B - P_E \qquad (11.23)$$

ΔP represents pressure drop and ΔP_{BCE} is a function of the diameter and length of branch BCE and the flow rate Q_1. Similarly, ΔP_{BDE} is a function of the diameter and length of branch BCE and the flow rate Q_2.

To calculate the pressure drop in parallel pipes, we must first determine the flow split at junction B. We know that the sum of the two flow rates Q_1 and Q_2 must equal the given inlet flow rate Q. If both pipe loops BCE and BDE are equal in lengths and pipe inside diameters, we can infer that the flow rate will be split equally between the two branches.

Thus, for identical pipe loops:

$$Q_1 = Q_2 = \frac{Q}{2} \tag{11.24}$$

In this case, the pressure drop from B to E can be calculated assuming a flow rate of $\frac{Q}{2}$ flowing through one of the pipe loops.

To illustrate this further, suppose we are interested in determining the pressure at A for the given flow rate Q and a specified delivery pressure (P_F) at the pipe terminus F. We start with the last pipe segment EF and calculate the pressure required at E for a flow rate of Q to deliver gas at F at a pressure P_F. We could use the general flow equation for this and substitute P_E for upstream pressure P_1 and P_F for downstream pressure P_2. Having calculated P_E, we can now consider one of the pipe loops such as BCE and calculate the upstream pressure P_B required for a flow rate of $\frac{Q}{2}$ through BCE for a downstream pressure of P_E. In the general flow equation, the upstream pressure $P_1 = P_B$ and the downstream pressure $P_2 = P_E$.

This is correct only for identical pipe loops. Otherwise, the flow rate Q_1 and Q_2 through the pipe branches BCE and BDE will be unequal. From the calculated value of P_E, we can now determine the pressure required at A by applying the general flow equation to pipe segment AB that has a gas flow rate of Q. The upstream pressure P_1 will be calculated for a downstream pressure $P_2 = P_E$.

Consider now a situation in which the pipe loops are not identical. This means that the pipes BCE and BDE may have different lengths and different diameters. In this case, we must determine the flow split between these two branches by equating the pressure drops through each of the branches. Because Q_1 and Q_2 are two unknowns, we will use the flow conservation principle and the common pressure drop principle to determine the values of Q_1 and Q_2. From the general flow equation, we can state the following.

The pressure drop from friction in branch BCE can be calculated from

$$\left(P_B^2 - P_E^2\right) = \frac{K_1 L_1 Q_1^2}{D_1^5} \tag{11.25}$$

where
K_1 = a parameter that depends on gas properties, gas temperature, etc.
L_1 = length of pipe branch BCE

D_1 = inside diameter of pipe branch BCE
Q_1 = flow rate through pipe branch BCE
Other symbols are as defined previously.

K_1 is a parameter that depends on the gas properties, gas temperature, base pressure, and base temperature that will be the same for both pipe branches BCE and BDE in a parallel pipeline system. Hence we regard this as a constant from branch to branch.

Similarly, the pressure drop because of friction in branch BDE is calculated from

$$(P_B^2 - P_E^2) = \frac{K_2 L_2 Q_2^2}{D_2^5} \qquad (11.26)$$

where
K_2 = a constant like K_1
L_2 = length of pipe branch BDE
D_2 = inside diameter of pipe branch BDE
Q_2 = flow rate through pipe branch BDE
Other symbols are as defined earlier.

In Eqns (11.25) and (11.26), the constants K_1 and K_2 are equal because of they do not depend on the diameter or length of the branch pipes BCE and BDE. Combining both equations, we can state the following for common pressure drop through each branch.

$$\frac{L_1 Q_1^2}{D_1^5} = \frac{L_2 Q_2^2}{D_2^5} \qquad (11.27)$$

Simplifying further, we get the following relationship between the two flow rates Q_1 and Q_2.

$$\frac{Q_1}{Q_2} = \left(\frac{L_2}{L_1}\right)^{0.5} \left(\frac{D_1}{D_2}\right)^{2.5} \qquad (11.28)$$

Combining Eqn (11.27) with Eqn (11.28), we can solve for the flow rates Q_1 and Q_2.

To illustrate this, consider the inlet flow $Q = 100$ MMSCFD and the pipe branches as follows

$$L_1 = 10 \text{ mi} \quad D_1 = 15.5 \text{ in.} \quad \text{for branch BCE}$$

$$L_2 = 15 \text{ mi} \quad D_2 = 13.5 \text{ in.} \quad \text{for branch BDE}$$

From Eqn (11.24) for flow conservation, we get

$$Q_1 + Q_2 = 100$$

From Eqn (11.28), we get the ratio of flow rates as

$$\frac{Q_1}{Q_2} = \left(\frac{15}{10}\right)^{0.5} \left(\frac{15.5}{13.5}\right)^{2.5} = 1.73$$

Solving these two equations in Q_1 and Q_2, we get

$$Q_1 = 63.37 \text{ MMSCFD}$$

$$Q_2 = 36.63 \text{ MMSCFD}$$

Once we know the values of Q_1 and Q_2, we can easily calculate the common pressure drop in the branch pipes BCE and BDE. A problem will be used to illustrate this method.

Another method of calculating pressure drops in parallel pipes is using the equivalent diameter. In this method, we replace the pipe loops BCE and BDE with a certain length of an equivalent diameter pipe that has the same pressure drop as one of the branch pipes. The equivalent diameter pipe can be calculated using the general flow equation as explained next. The equivalent pipe with the same ΔP that will replace both branches will have a diameter D_e and a length equal to one of the branch pipes, say L_1.

Because of the pressure drop in the equivalent diameter pipe, which flows the full volume Q, is the same as that in any of the branch pipes, we can state the following:

$$\left(P_B^2 - P_E^2\right) = \frac{k_e L_e Q^2}{D_e^5} \tag{11.29}$$

where $Q = Q_1 + Q_2$ from Eqn (11.24) and K_e represents the constant for the equivalent diameter pipe of length L_e flowing the full volume Q. Equating the value of $(P_B^2 - P_E^2)$ to the corresponding values considering each branch separately, we get

$$\frac{K_1 L_1 Q_1^2}{D_1^5} = \frac{K_2 L_2 Q_2^2}{D_2^5} = \frac{K_e L_e Q^2}{D_e^5} \tag{11.30}$$

Also setting $K_1 = K_2 = K_e$ and $L_e = L_1$, we simplify Eqn (11.30) as follows.

$$\frac{L_1 Q_1^2}{D_1^5} = \frac{L_2 Q_2^2}{D_2^5} = \frac{L_1 Q^2}{D_e^5} \tag{11.31}$$

Using Eqn (11.30) in conjunction with Eqn (11.31), we solve for the equivalent diameter D_e as

$$De = D_1 \left[\left(\frac{1 + \text{Const1}}{\text{Const1}}\right)^2\right]^{1/5} \tag{11.32}$$

where

$$\text{Const1} = \sqrt{\left(\frac{D_1}{D_2}\right)^5 \left(\frac{L_2}{L_1}\right)} \tag{11.33}$$

And the individual flow rates Q_1 and Q_2 are calculated from

$$Q_1 = \frac{Q\,\text{Const1}}{1 + \text{Const1}} \tag{11.34}$$

and

$$Q_2 = \frac{Q}{1 + \text{Const1}} \tag{11.35}$$

To illustrate the equivalent diameter method, consider the inlet flow $Q = 100$ MMSCFD and the pipe loops as follows

$L_1 = 10$ mi $D_1 = 15.5$ in.   for branch BCE

$L_2 = 15$ mi $D_2 = 13.5$ in.   for branch BDE

From Eqn (11.35)

$$\text{Const1} = \sqrt{\left(\frac{15.5}{13.5}\right)^5 \left(\frac{15}{10}\right)} = 1.73$$

Using Eqn (11.32), the equivalent diameter is

$$De = 15.5 \left[\left(\frac{1 + 1.73}{1.73}\right)^2\right]^{1/5} = 18.60 \text{ in}$$

Thus the NPS 16 and NPS 14 pipes in parallel can be replaced with an equivalent pipe having an inside diameter of 18.6 in.

Next we calculate the flow rates in the two parallel pipes as follows

$$Q_1 = \frac{100 \times 1.73}{1 + 1.73} = 63.37 \text{ MMSCFD}$$

and

$$Q_2 = 36.63 \text{ MMSCFD}$$

Having calculated an equivalent diameter De, we can now calculate the common pressure drop in the parallel branches by considering the entire flow Q flowing through the equivalent diameter pipe.

Problem 11.5: Gas Pipeline

A gas pipeline consists of two parallel pipes, as shown in Figure 11.3. It is operated at a flow rate of 100 MMSCFD. The first pipe segment AB is 12 miles long and consists of NPS 16, 0.250-in wall thickness pipe. The loop BCE is 24 miles long and consists of NPS 14, 0.250-in wall thickness pipe. The loop BDE is 16 miles long and consists of NPS 12, 0.250-in wall thickness pipe. The last segment EF is 20 miles long, NPS 16, 0.250-in wall thickness pipe. Assuming a gas gravity of 0.6, calculate the outlet pressure at F and the pressures at the beginning and the end of the pipe loops and the flow rates through them. The inlet pressure at A = 1200 psig. Gas flowing temperature = 80 °F, base temperature = 60 °F, and base pressure = 14.73 psia. Compressibility factor Z = 0.92. Use the general flow equation with Colebrook friction factor $f = 0.015$.

Solution

From Eqn (11.28), the ratio of the flow rates through the two pipe loops is given by

$$\frac{Q_1}{Q_2} = \left(\frac{16}{24}\right)^{0.5} \left(\frac{14 - 2 \times 0.25}{12.75 - 2 \times 0.25}\right)^{2.5} = 1.041$$

And from Eqn (11.24)

$$Q_1 + Q_2 = 100$$

Solving for Q_1 and Q_2, we get
$Q_1 = 51.0$ MMSCFD and $Q_2 = 49.0$ MMSCFD.

Next considering the first pipe segment AB, we will calculate the pressure at B based on the inlet pressure of 1200 psig at A, using the general flow equation as follows.

$$100 \times 10^6 = 77.54 \left(\frac{1}{\sqrt{0.015}}\right) \left(\frac{520}{14.73}\right) \left[\frac{(1214.73^2 - P_2^2)}{0.6 \times 540 \times 12 \times 0.92}\right]^{0.5} 15.5^{2.5}$$

Solving for the pressure at B, we get

$$P_2 = 1181.33 \text{ psia} = 1166.6 \text{ psig}$$

This is the pressure at the beginning of the looped section at B. Next we calculate the outlet pressure at E of pipe branch BCE considering a flow rate of 51 MMSCFD through the NPS 14 pipe, starting at a pressure of 1181.33 psia at B.

Using the general flow equation, we get

$$51 \times 10^6 = 77.54 \left(\frac{1}{\sqrt{0.015}}\right)\left(\frac{520}{14.73}\right)\left[\frac{(1181.33^2 - P_2^2)}{0.6 \times 540 \times 24 \times 0.92}\right]^{0.5} 13.5^{2.5}$$

Solving for the pressure at E, we get

$$P_2 = 1145.63 \text{ psia} = 1130.9 \text{ psig}$$

We will now calculate the pressures using the equivalent diameter method.

From Eqn (11.35)

$$\text{Const1} = \sqrt{\left(\frac{13.5}{12.25}\right)^5 \left(\frac{16}{24}\right)} = 1.041$$

From Eqn (11.32), the equivalent diameter is

$$De = 13.5 \left[\left(\frac{1 + 1.041}{1.041}\right)^2\right]^{1/5} = 17.67 \text{ in.}$$

Thus we can replace the two branch pipes between B and E with a single piece of pipe 24 miles long having an inside diameter of 17.67 in flowing 100 MMSCFD.

The pressure at B was calculated earlier as

$$P_B = 1181.33 \text{ psia}$$

Using this pressure, we can calculate the downstream pressure at E for the equivalent pipe diameter as follows

$$100 \times 10^6 = 77.54 \left(\frac{1}{\sqrt{0.015}}\right)\left(\frac{520}{14.73}\right)\left[\frac{(1181.33^2 - P_2^2)}{0.6 \times 540 \times 24 \times 0.92}\right]^{0.5} 17.67^{2.5}$$

Solving for the outlet pressure at E, we get

$$P_2 = 1145.60 \text{ psia}$$

which is almost the same as what we calculate before.

The pressure at F will therefore be the same as what we calculated before.

Therefore, using the equivalent diameter method the parallel pipes BCE and BDE can be replaced with a single pipe 24 miles long having an inside diameter of 17.67 in.

Problem 11.6: Gas Pipeline (SI Units)

A natural gas pipeline DN 500 with a 12-mm wall thickness is 60 km long. The gas flow rate is 5.0 Mm^3/day at 20 °C. Calculate the inlet pressure required for a delivery pressure of 4 MPa (absolute), using the general flow equation with the modified Colebrook–White friction factor. Pipe roughness = 0.015 mm. To increase the flow rate through the pipeline, the entire line is looped with a DN 500 pipeline, 12-mm-wall thickness. Assuming the same delivery pressure, calculate the inlet pressure at the new flow rate of 8 Mm^3/day. Gas gravity = 0.65. Viscosity = 0.000119 poise. Compressibility factor Z = 0.88. Base temperature = 15 °C and base pressure = 101 kPa. If the inlet and outlet pressures are held the same as before, what length of the pipe should be looped to achieve the increased flow?

Solution

Pipe inside diameter D = 500 − 2 × 12 = 476 mm.
  Flow rate Q = 5.0 × 10^6 m^3/day.
  Base temperature T_b = 15 + 273 = 288 K.
  Gas flow temperature T_f = 20 + 273 = 293 K.
  Delivery pressure P_2 = 4 MPa.
  Calculate the Reynolds number as follows:

$$R = 0.5134 \left(\frac{101}{288}\right)\left(\frac{0.65 \times 5 \times 10^6}{0.000119 \times 476}\right) = 10,330,330$$

From the modified Colebrook–White equation, the transmission factor is

$$F = -4 Log_{10}\left(\frac{0.015}{3.7 \times 476} + \frac{1.4125F}{10,330,330}\right)$$

Solving by successive iteration, we get

$$F = 19.80$$

Using the general flow equation, the inlet pressure is calculated next.

$5 \times 10^6 = 5.747 \times 10^{-4}$

$$\times\; 19.80 \left(\frac{273+15}{101}\right)\left[\frac{P_1^2 - 4000^2}{0.65 \times 293 \times 60 \times 0.88}\right]^{0.5} \times (476)^{2.5}$$

Solving for the inlet pressure, we get

$$P_1 = 5077 \text{ kPa (absolute)} = 5.08 \text{ MPa (absolute)}$$

Therefore, the inlet pressure required at 5 Mm^3/day flow rate is 5.08 MPa.

Next, at 8 Mm^3/day flow rate, we calculate the new inlet pressure with the entire 60-km length looped with an identical DN 500 pipe. Because the loop is the same size as the main line, each parallel branch will carry half the total flow rate or 4 Mm³/day.

We calculate the Reynolds number for flow through one of the loops.

$$R = 0.5134 \left(\frac{101}{288}\right)\left(\frac{0.65 \times 4 \times 10^6}{0.000119 \times 476}\right) = 8,264,264$$

From the modified Colebrook–White equation, the transmission factor is

$$F = -4 Log_{10}\left(\frac{0.015}{3.7 \times 476} + \frac{1.4125 F}{8,264,264}\right)$$

Solving by successive iteration, we get

$$F = 19.70$$

Keeping the delivery pressure the same as before (4 MPa), using the general flow equation, we calculate the inlet pressure required as follows.

$$4 \times 10^6 = 5.747 \times 10^{-4}$$

$$\times\, 19.70 \left(\frac{273+15}{101}\right)\left[\frac{P_1^2 - 4000^2}{0.65 \times 293 \times 60 \times 0.88}\right]^{0.5} \times (476)^{2.5}$$

Solving for the inlet pressure, we get

$$P_1 = 4724 \text{ kPa (absolute)} = 4.72 \text{ MPa (absolute)}$$

Therefore, for the fully looped pipeline at 8 Mm³/day flow rate the inlet pressure required is

$$4.72 \text{ MPa.}$$

Next, keeping the inlet and outlet pressures the same at 5077 and 4000 kPa, respectively, at the new flow rate of 8 Mm³/day we assume L km of the pipe from the inlet is looped. We will calculate the value of L by first calculating the pressure at the point where the loop ends. Because each parallel pipe carries 4 Mm³/day, we use the Reynolds number and transmission factor calculated earlier.

$$R = 8,264,264 \quad \text{and} \quad F = 19.70$$

Using the general flow equation, we calculate the outlet pressure at the end of the loop of length L km as follows.

$$4 \times 10^6 = 5.747 \times 10^{-4}$$

$$\times\, 19.70 \left(\frac{273+15}{101}\right)\left[\frac{5077^2 - P_2^2}{0.65 \times 293 \times L \times 0.88}\right]^{0.5} \times (476)^{2.5}$$

Solving for pressure in terms of the loop length L, we get

$$P_2^2 = 5077^2 - 105,291.13 L \tag{11.36}$$

Next we apply the general flow equation for the pipe segment of length $(60 - L)$ km, that carries the full 8 Mm^3/day flow rate. The inlet pressure is P_2 and the outlet pressure is 4000 kPa.

The Reynolds number at 8 Mm^3/day is

$$R = 0.5134 \left(\frac{101}{288}\right)\left(\frac{0.65 \times 8 \times 10^6}{0.000119 \times 476}\right) = 16,528,528$$

From the modified Colebrook–White equation, the transmission factor is

$$F = -4Log_{10}\left(\frac{0.015}{3.7 \times 476} + \frac{1.4125F}{16,528,528}\right)$$

Solving by successive iteration, we get

$$F = 19.96$$

Using the general flow equation, we calculate the inlet pressure for the pipe segment of length (60-L) km as follows.

$$8 \times 10^6 = 5.747 \times 10^{-4}$$
$$\times 19.96 \left(\frac{273+15}{101}\right)\left[\frac{P_2^2 - 4000^2}{0.65 \times 293 \times (60-L) \times 0.88}\right]^{0.5}$$
$$\times (476)^{2.5}$$

Simplifying, we get

$$P_2^2 = 4000^2 + 410,263.77(60-L) \qquad (11.37)$$

From Eqns (11.36) and (11.37), eliminating P_2, we solve for L as follows.

$$5077^2 - 105,291.13L = 4000^2 + 410,263.77(60-L)$$

Therefore

$$L = 48.66 \text{ km}$$

Thus 48.66 km of the 60-km pipeline length will have to be looped starting at the pipe inlet so that at 8 Mm^3/day both inlet and outlet pressures will be the same as before at 5 Mm^3/day.

What will be the effect if the loop was installed starting at the downstream end of the pipeline and proceeding towards the upstream end? Will the results be the same? In the next section, we will explore the best location to install the pipe loop.

3. LOCATING PIPE LOOP: GAS PIPELINES

In the preceding example, we looked at looping an entire pipeline to reduce pressure drop and increase the flow rate. We also explored looping a portion of the pipe, beginning at the upstream end. How do we determine where the loop should be placed for optimum results? Should it be located upstream, downstream, or in a mid-section of the pipe? We will analyze this as follows.

Three looping scenarios are presented in Figure 11.4.

In case (1), a pipeline of length L is shown looped with X miles of pipe, beginning at the upstream end A. In case (2), the same length X of pipe is looped, but it is located on the downstream end B. Case (3) shows the mid-section of the pipeline being looped. For most practical purposes, we can say that the cost of all three loops will be the same as long as the loop length is the same.

To determine which of these cases are optimum, we must analyze how the pressure drop in the pipeline varies with distance from the pipe inlet to outlet. It is found that, if the gas temperature is constant throughout, at locations near the upstream end the pressure drops at a slower rate than at the downstream end. Therefore, there is more pressure drop in the downstream section compared to that in the upstream section. Hence, to reduce the overall pressure drop, the loop must be installed towards the downstream end of the pipe. This argument is valid only if the gas temperature is constant

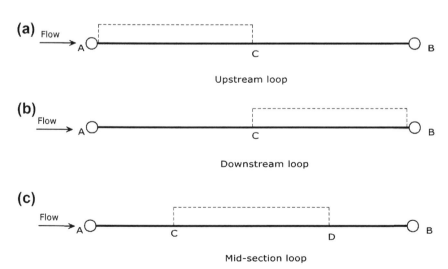

Figure 11.4 Different looping scenarios.

throughout the pipeline. In reality, because of heat transfer between the flowing gas and the surrounding soil (buried pipe) or the outside air (above-ground pipe), the gas temperature will change along the length of the pipeline. If the gas temperature at the pipe inlet is higher than that of the surrounding soil (buried pipe), the gas will lose heat to the soil and the temperature will drop from the pipe inlet to the pipe outlet. If the gas is compressed at the inlet using a compressor, then the gas temperature will be a lot higher than that of the soil immediately downstream of the compressor. The hotter gas will cause higher pressure drops (examine the general flow equation and see how the pressure varies with the gas flow temperature). Hence, in this case, the upstream segment will have a larger pressure drop compared with the downstream segment. Therefore, considering heat transfer effects, the pipe loop should be installed in the upstream portion for maximum benefit. The installation of the pipe loop in the mid-section of the pipeline as in case (3) in Figure 11.4 will not be the optimum location based on the preceding discussion. It can therefore be concluded that if the gas temperature is fairly constant along the pipeline, the loop should be installed toward the downstream end as in case (2). If heat transfer is taken into account and the gas temperature varies along the pipeline, with the hotter gas being upstream, the better location for the pipe loop will be on the upstream end as in case (1).

Looping pipes will be explored more in Chapter 14, where we discuss several case studies and pipeline economics.

CHAPTER TWELVE

Meters and Valves

In this chapter, we will discuss the various methods and instruments used in the measurement of liquid that flows through a pipeline. The formulas used for calculating the liquid velocities, flow rates, etc., from the pressure readings, their limitations, and the degree of accuracy attainable will be covered for some of the more commonly used instruments. These days, considerable work is being done in this field of flow measurement to improve the accuracy of instruments, particularly when custody transfer of products is involved.

1. HISTORY

Measurement of flow of liquids and solids has been going on for centuries. In the early days, the Romans used some form of flow measurement to allocate water from the aqueducts to the houses in their cities. This was necessary to control the quantity of water used by the citizens and prevent waste. Similarly, it is reported that the Chinese used to measure salt water in brine pots that were used for salt production. In later years, a commodity had to be measured so that it may be properly allocated and the ultimate user charged for it appropriately. Today, gasoline is dispensed from meters at gas stations and the recipient is billed according to the volume of gasoline so measured. Water companies measure water consumed by a household using water meters, whereas natural gas for residential and industrial consumers are measured by gas meters. In all these instances, the objective is to ensure that the supplier gets paid for the commodity and the user of the commodity is billed for the product at the agreed-upon price. In addition, in industrial processes that involve the use of liquids and gases to perform a specific function require accurate quantities to be dispensed so that the desired effect of the processes may be realized. In most cases involving consumers, certain regulatory or public agencies (for example, the department of weights and measures) periodically check flow measurement devices to ensure that they performing accurately and if necessary to calibrate them against a very accurate master device.

2. FLOW METERS

Several types of instruments are available to measure the flow rate of a liquid in a pipeline. Some measure the velocity of flow, whereas others directly measure the volume flow rate or the mass flow rate. The following flow meters are used in the pipeline industry:
- Venturi meter
- Flow nozzle
- Orifice meter
- Flow tube
- Rotameter
- Turbine meter
- Vortex flow meter
- Magnetic flow meter
- Ultrasonic flow meter
- Positive displacement meter
- Mass flow meter
- Pitot tube

The first four items listed are called *variable head meters* because the flow rate measured is based on the pressure drop because of a restriction in the meter, which varies with the flow rate. The last item, the Pitot tube, is also called a velocity probe because it actually measures the velocity of the liquid. From the measured velocity, we can calculate the flow rate using the conservation of mass equation, discussed in an earlier chapter.

$$\text{Mass flow} = (\text{Flow rate}) \times (\text{density})$$
$$= (\text{Area}) \times (\text{Velocity}) \times (\text{density}) \quad (12.1)$$
$$\text{Or } M = Q\rho = AV\rho$$

Because the liquid density is practically constant, we can state that:

$$Q = AV \quad (12.2)$$

where
 A: cross-sectional area of flow
 V: velocity of flow
 ρ: liquid density

We will discuss the principle of operation and the formulas used for some of the more common meters described here.

3. VENTURI METER

The venturi meter, also known as venturi tube, belongs to the category of variable head flowmeters. The principle of a venturi meter is depicted in Figure 12.1. This type of a venturi meter is also known as the Herschel type and it consists of a smooth gradual contraction from the main pipe size to the throat section, followed by a smooth, gradual enlargement from the throat section to the original pipe diameter.

The included angle from the main pipe to the throat section in the gradual contraction is generally in the range of $21° \pm 2°$. Similarly, the gradual expansion from the throat to the main pipe section is limited to a range of $5°$ to $15°$ in this design of venturi meter. This construction results in minimum energy loss, causing the discharge coefficient (discussed later) to approach the value of 1.0. This type of a venturi meter is generally rough cast with a pipe diameter range of 4.0–48.0 in. The throat diameter may vary quite a bit, but the ratio of the throat diameter to the main pipe diameter (d/D), also known as the *beta ratio*, represented by the symbol β, should range between 0.30 and 0.75.

The venturi meter consists of a main piece of pipe that decreases in size to a section called the throat, followed by a gradually increasing size back to the original pipe size. The liquid pressure at the main pipe section 1 is denoted by P_1 and that at the throat section 2 is represented by P_2. As the liquid flows

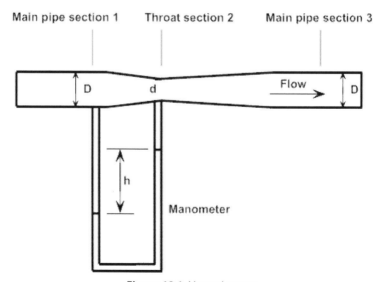

Figure 12.1 Venturi meter.

through the narrow throat section, it accelerates to compensate for the reduction in area, because the volume flow rate is constant ($Q = AV$ from Eqn (12.2)). As the velocity increases in the throat section, the pressure decreases (Bernoulli's equation). As the flow continues past the throat, the gradual increase in the flow area results in reduction of flow velocity back to the level at section 1, correspondingly the liquid pressure will increase to some value higher than at the throat section.

We can calculate the flow rate using the Bernoulli's equation, along with the continuity equation, based on the conservation of mass as shown in Eqn (12.1) previously.

If we consider the main pipe section 1 and the throat section 2 as a reference, and apply Bernoulli's equation for these two sections, we get

$$P_1/\gamma + V_1^2/(2g) + Z_1 = P_2/\gamma + V_2^2/(2g) + Z_2 + h_L \qquad (12.3)$$

And from the continuity equation:

$$Q = A_1 V_1 = A_2 V_2 \qquad (12.4)$$

where γ is the specific weight of liquid and h_L represents the pressure drop because of friction between section 1 and section 2.

Simplifying these equations yields the following:

$$V_1 = \sqrt{[2g(P_1 - P_2)/\gamma + (Z_1 - Z_2) - h_L]/[(A_1/A_2)^2 - 1]} \qquad (12.5)$$

The elevation difference $Z_1 - Z_2$ is negligible even if the venturi meter is positioned vertically. We will also drop the friction loss term h_L and include it in a coefficient C, called the discharge coefficient, and rewrite Eqn (12.5) as follows:

$$V_1 = C\sqrt{[2g(P_1 - P_2)/\gamma]/[(A_1/A_2)^2 - 1]} \qquad (12.6)$$

This equation gives us the velocity of the liquid in the main pipe section 1. Similarly, the velocity in the throat section, V_2, can be calculated using Eqn (12.4) as follows:

Velocity in throat.

$$V_2 = C\sqrt{[2g(P_1 - P_2)/\gamma]/[1 - (A_2/A_1)^2]} \qquad (12.7)$$

The volume flow rate Q can now be calculated using Eqn (12.4) as follows:

$$Q = A_1 V_1$$

Therefore, $\quad Q = CA_1 \sqrt{[2g(P_1 - P_2)/\gamma]/[(A_1/A_2)^2 - 1]}$ (12.8)

Because the beta ratio $\beta = d/D$ and $A_1/A_2 = (D/d)^2$.

We can write the above equations in terms of the beta ratio β as follows:

$$Q = CA_1 \sqrt{[2g(P_1 - P_2)/\gamma]/[(1/\beta)^4 - 1]} \quad (12.9)$$

It can be seen by examining Eqns (12.5), 12.6, and 12.7 that the discharge coefficient C actually represents the ratio of the velocity of liquid through the venturi to the ideal velocity when the energy loss (h_L) is zero. C is therefore a number less than 1.0. The value of C depends on the Reynolds number in the main pipe section 1 and is shown graphically in Figure 12.2. The Reynolds number greater than 2×10^5 the value of C remains constant at 0.984.

For smaller size piping, 2–10 in size, venturi meters are machined and hence have a better surface finish than the larger, rough cast meters. These smaller venturi meters have a C value of 0.995 for Reynolds number greater than 2×10^5.

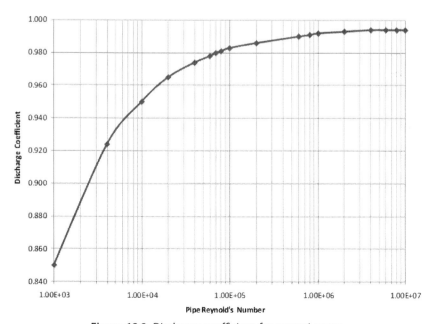

Figure 12.2 Discharge coefficient for venturi meter.

4. FLOW NOZZLE

A typical *flow nozzle* is illustrated in Figure 12.3. It consists of the main pipe section 1, followed by a gradual reduction in flow area and a subsequent short cylindrical section, and finally an expansion to the main pipe section 3 as shown in the figure.

The American Society of Mechanical Engineers (ASME) and the International Standards Organization have defined the geometries of these flow nozzles and published equations to be used with the flow nozzles. Because the smooth gradual contraction from the main pipe diameter D to the nozzle diameter d, the energy loss between the sections 1 and 2 is very small. We can apply the same Eqns 12.6 through 12.8 for the venturi meter as well as the flow nozzle. The discharge coefficient C for the flow nozzle is found to be 0.99 or better, for Reynolds numbers above 10^6. At lower Reynolds numbers, there is greater energy loss immediately following the nozzle throat, because of sudden expansion, and hence C values are lower.

Depending on the beta ratio and the Reynolds number, the discharge coefficient C can be calculated from the equation.

$$C = 0.9975 - 6.53\sqrt{(\beta/R)} \tag{12.10}$$

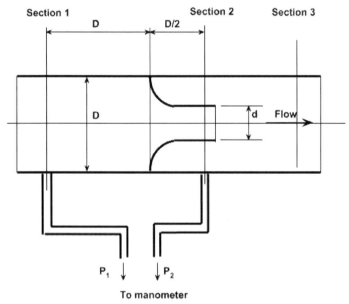

Figure 12.3 Flow nozzle.

where

$\beta = d/D$

and R is the Reynolds number based on the pipe diameter D.

Compared with the venturi meter, the flow nozzle is more compact in size because it does not require the length for gradual decrease in diameter at the throat, and the additional length for the smooth, gradual expansion from the throat to the main pipe size. However, there is more energy loss (and therefore, pressure head loss) in the flow nozzle because of the sudden expansion from the nozzle diameter to the main pipe diameter. The latter causes greater turbulence and eddies compared to the gradual expansion in the venturi meter.

5. ORIFICE METER

An *orifice meter* consists of a flat plate that has a sharp-edged hole accurately machined in it and placed concentrically in a pipe as shown in Figure 12.4. As liquid flows through the pipe, the flow suddenly contracts as it approaches the orifice and then suddenly expands after the orifice back to the full pipe diameter. This forms a *vena contracta* or a throat immediately past the orifice. This reduction in flow pattern at the vena contracta causes increased velocity and hence lower pressure at the throat, similar to the venturi meter, discussed previously.

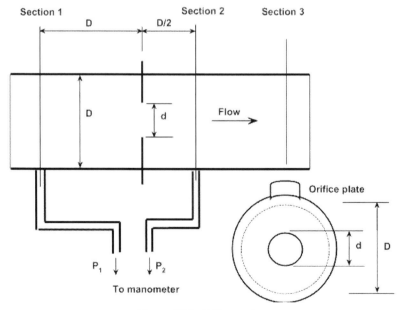

Figure 12.4 Orifice meter.

Table 12.1 Pressure taps for orifice

Inlet pressure tap, P1	Outlet pressure tap, P2
1. One pipe diameter upstream from plate face of plate	One-half pipe diameter downstream of inlet
2. One pipe diameter upstream from plate	At vena contracta
3. Flange taps, 1 in upstream from plate outlet face of plate	Flange taps, 1 in downstream from face of plate

The pressure difference between section 1, with the full flow and section 2 at the throat, can then be used to measure the liquid flow rate, using equations developed earlier for the venturi meter and the flow nozzle. Because of the sudden contraction at the orifice and the subsequent sudden expansion after the orifice, the coefficient of discharge C for the orifice meter is much lower than that of a venturi meter or a flow nozzle. In addition, depending on the pressure tap locations, section 1 and section 2, the value of C is different for orifices.

There are three possible pressure tap locations for an orifice meter as listed in Table 12.1.

Figure 12.5 shows the variation of C with the beta ratio d/D for various values of pipe Reynolds number.

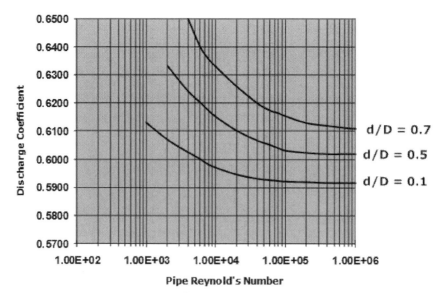

Figure 12.5 Orifice meter discharge coefficient.

Comparing the three types of flow meters discussed previously, we can conclude that the orifice plate has the highest energy loss due to the sudden contraction followed by the sudden expansion. On the other hand, the venturi meter has a lower energy loss compared with a flow nozzle because of the smooth, gradual reduction at the throat, followed by the smooth, gradual expansion after the throat.

Flow tubes are proprietary variable head flow meters that are streamlined in design to cause the least amount of energy loss. The flow tube is also a variable head type flow meter, which is manufactured by different companies with their own special, proprietary designs.

6. TURBINE METER

Turbine meters are used in a wide variety of applications. The food industry uses turbine meters for measurement of milk, cheese, cream, syrups, vegetable oils, etc. Turbine meters are also used in the oil industry.

Basically the turbine meter is a velocity measuring instrument. Liquid flows through a free-turning rotor that is mounted coaxially inside the meter. Upstream and downstream of the meter, a certain amount of straight piping lengths is required to ensure smooth velocities through the meter. The liquid striking the rotor causes it to rotate, the velocity of rotation being proportional to the flow rate. From the rotor speed measured and the flow area, we can compute the flow rate through the meter. The turbine meter must be calibrated because flow depends on the friction, turbulence and the manufacturing tolerance of the rotor parts.

For liquids with viscosity close to that of water the range of flow rates is 10:1. With higher and lower viscosity liquids it drops to 3:1. Density effect is similar to viscosity.

A turbine meter requires straight length of pipe before and after the meter. The straightening vanes located in the straight lengths of pipe helps eliminate swirl in flow. The upstream length of straight pipe has to be 10 D in length, where D is the pipe diameter. After the meter, the straight length of pipe is 5 D long. Turbine meters may be two-section type or three-section types, depending upon the number of pieces the meter assembly is composed of. For liquid flow measurement, the American Petroleum Institute (API) Manual of Petroleum Measurement Standard must be followed. In addition, bypass piping must also be installed to isolate the meter for maintenance and repair. To maintain flow measurements on a continuous basis for custody transfer, an entire spare meter unit will be required on the bypass piping, when the main meter is taken out of service for testing and maintenance.

From the turbine meter reading at flowing temperature, the flow rate at some base temperature (such as 60 °F) is calculated using the following equation:

$$Q_b = Q_f \times M_f \times F_t \times F_p \qquad (12.11)$$

where

Q_b: flow rate at base conditions, such as 60 °F and 14.7 psi
Q_f: measured flow rate at operating conditions, such as 80 °F and 350 psi
M_f: meter factor for correcting meter reading, based on meter calibration data.
F_t: temperature correction factor for correcting from flowing temperature to the base temperature.
F_p: pressure correction factor for correcting from flowing pressure to the base pressure.

7. POSITIVE DISPLACEMENT METER

The positive displacement meter, also known as PD meter, is most commonly used in residential applications for measuring water and natural gas consumption. PD meters can measure small flows with good accuracy. These meters are used when consistently high accuracy is required under steady flow through a pipeline. PD meters are accurate within $\pm 1\%$ over a flow range of 20:1. They are suitable for batch operations, mixing and blending of liquids that are clean, without deposits and with noncorrosive liquids.

PD meters basically operate by measuring fixed quantities of the liquid in *containers* that are alternatively filled and emptied. The individual volume are then totaled and displayed. As the liquid flows through the meter, the force from the flowing stream causes the pressure drop in the meter, which is a function of the internal geometry. The PD meter does not require the straightening vanes as with a turbine meter and hence is a more compact design. However, PD meters can be large and heavy compared with an equivalent turbine meter. Also, meter jamming can occur when stopped and restarted. This necessitates some form of bypass piping with valves to prevent pressure rise and damage to meter.

The accuracy of a PD meter depends on clearance between the moving and stationary components. Hence precision machined parts are required to maintain accuracy. They are not suitable for slurries or liquids with suspended particles that could jam the components and cause damage to the meter. PD meters include several types, such as the reciprocating piston type, rotating disk, rotary piston, sliding vane, and rotary vane type.

Many modern petroleum installations employ PD meters for crude oil and refined products. These have to be calibrated periodically to ensure accurate flow measurement. Most PD meters used in the oil industry are tested at regular intervals, such as daily or weekly, using a master meter, called a *meter prover* system that is installed as a fixed unit along with the PD meter.

Example 12.1

A venturi meter is used for measuring the flow rate of water at 70 °F. The flow enters a 6.625-in pipe, 0.250-in wall thickness and the throat diameter is 2.4 in. The venturi is of the Herschel type and is rough cast with a mercury manometer. The manometer reading shows a pressure difference of 12.2 in of mercury. Calculate the flow velocity in the pipe and the volume flow rate. Use specific gravity of 13.54 for mercury.

Solution

At 70 °F, the specific weight and viscosity are

$$\gamma = 62.3 \text{ lb/ft}^3 \text{ and } v = 1.05 \times 10^{-5} \text{ ft}^2/\text{s}$$

Assume first that the Reynolds number is greater than 2×10^5.

Therefore, the discharge coefficient C will be 0.984 from Figure 12.2.

The beta ratio $\beta = 2.4/(6.625 - 0.5) = 0.3918$ and therefore β is between 0.3 and 0.75.

$$A_1/A_2 = (1/0.3918)^2 = 6.5144$$

The manometer reading will give us the pressure difference by equating the pressures at the two points in the manometer.

$$P_1 + \gamma_w(y + h) = P_2 + \gamma_w y + \gamma_m h$$

where y is depth of the higher mercury level below the center line of the venture and γ_w and γ_m are the specific weights of water and mercury, respectively.

Simplifying this, we get

$$(P_1 - P_2)/\gamma_w = (\gamma_m/\gamma_w - 1)h = (13.54 * 62.4/62.3 - 1) \times 12.2/12$$
$$= 12.77 \text{ ft}$$

And using Eqn (12.6), we get

$$V_1 = 0.984\sqrt{[(64.4 \times 12.77)/(6.5144^2 - 1)]} = 4.38 \text{ ft/s}$$

Now we will check the value of Reynolds number:

$$R = 4.38 \times (6.125/12)/1.05 \times 10^{-5} = 2.13 \times 10^5$$

Because R is greater than 2×10^5, our assumption for C is correct.
The volume flow rate is $Q = A_1 V_1$.
Flow rate $= 0.7854 \ (6.125/12)^2 \times 4.38 = 0.8962 \text{ ft}^3/\text{s}$.
Or flow rate $= 0.8962 \times (1728/231) \times 60 = 402.25 \text{ gal/min}$.

Example 12.2
An orifice meter is used to measure the flow rate of kerosene in a 2-in schedule 40 pipe at 70 °F. The orifice is 1.0-in diameter. Calculate the volume flow rate if the pressure difference across the orifice is 0.54 psi. Specific gravity of kerosene $= 0.815$. Viscosity $= 2.14 \times 10^{-5} \text{ ft}^2/\text{s}$.

Solution
Because the flow rate depends on the discharge coefficient C, which in turn depends on the beta ratio and the Reynolds number, we will have to solve this example by trial and error.

The 2-in schedule 40 pipe is 2.375-in outside diameter and 0.154-in wall thickness.

The beta ratio $\beta = 1.0/(2.375 - 2 \times 0.154) = 0.4838$.

First, we assume a value for the discharge coefficient C (say 0.61) and calculate the flow rate from Eqn (12.6). We will then calculate the Reynolds number and get a more correct value of C from Figure 12.5. Repeating the method a couple of times will yield a more accurate value of C and finally the flow rate.

The density of kerosene $= 0.815 \times 62.3 = 50.77 \text{ lb/ft}^3$.
Ratio of areas $A_1/A_2 = (1/0.4838)^2 = 4.2717$.
Using Eqn (12.6), we get

$$V_1 = 0.61 \text{ sqrt}\left[(64.4 \times 0.54 \times 144/(50.77)/(4.2717^2 - 1)\right]$$

Reynolds number $= 1.4588 \times (2.067/12)/(2.14 \times 10^{-5}) = 11,742$

Using this value of Reynolds number, we get, from Figure 12.5, the discharge coefficient $C = 0.612$.

Because we earlier assumed C = 0.61, we were not too far off. Recalculating based on the new value of C:

$$V_1 = 0.612 \text{ sqrt}[$$

Reynolds number $= 1.46 \times (2.067/12)/(2.14 \times 10^{-5}) = 11{,}751$

This is quite close to the previous value of the Reynolds number. Therefore we will use the last calculated value for velocity. The volume flow rate is $Q = A_1 V_1$.

$$\text{Flow rate} = 0.7854(2.067/12)^2 \times 1.46 = 0.034 \text{ ft}^3/\text{s}$$

Or flow rate $= 0.034 \times (1728/231) \times 60 = 15.27$ gal/min.

7.1 Gas Transmission Pipeline

In this chapter, we will discuss the various types of valves and flow measurements used on gas pipeline. The design and construction codes for valves, materials of construction, and application of the different types of valves and their performance characteristics will be explained. The importance of flow measurement in a gas pipeline, the accuracy of available instruments, codes, and standards used will be discussed. Various American National Standards Institute (ANSI), API, and American Gas Association (AGA) formulas used in connection with orifice meters will be reviewed. Because a small error in measurement in gas flow in a pipeline can translate to several thousand dollar loss of revenue, it is important that industry strive to improve upon measurement methods. Accordingly, gas transportation companies and related industries have been researching for better ways to improve flow measurement accuracy. For a detailed discussion of gas flow measurement, the reader is referred to the publications listed in the Bibliography section.

8. PURPOSE OF VALVES

Valves are installed on pipelines and piping system to isolate sections of piping for maintenance, for directing the fluid from one location to another, shut down flow through pipe sections, and for protecting pipe and prevent loss of fluid in the event of a rupture. On long-distance pipelines transporting natural gas and other compressible fluids, design codes and regulatory requirements dictate that sections of pipeline be isolated by installing mainline block valves at certain fixed spacing. For example, Department of Transportation

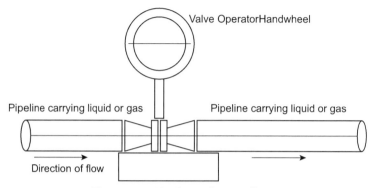

Figure 12.6 Mainline valve installation.

49 CFR, Part 192, requires that in class 1 locations mainline valves be installed 20 miles apart. Class locations were discussed in Chapter 6.

A typical mainline block valve installation on a gas transmission pipeline is illustrated in Figure 12.6.

9. TYPES OF VALVES

The various types of valves used in the gas pipeline industry include the following:
- Gate valve
- Ball valve
- Plug valve
- Butterfly valve
- Globe valve
- Check valve
- Control valve
- Relief valve
- Pressure-regulating valve

Each of these valves listed will be discussed in detail in the following sections.

Valves may be of screwed design, weld ends, or flanged ends. In the gas industry, large valves are generally of the welded type where the valve is attached to the pipe on either side by a welded joint to prevent gas leakage to the atmosphere. In smaller sizes, screwed valves are used. A typical weld end mainline valve along with smaller valves on either side is shown in Figure 12.7.

Valves may be operated manually using a hand wheel or using an electric, pneumatic, or gas operator as shown in Figure 12.8.

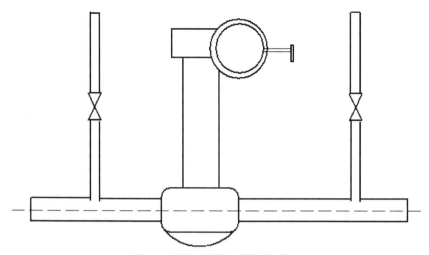

Figure 12.7 Mainline block valve.

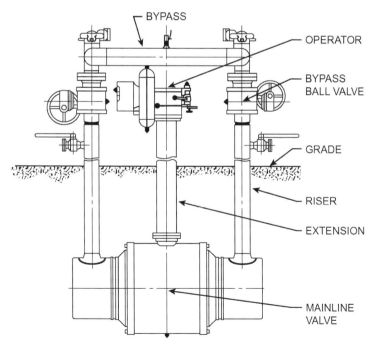

Figure 12.8 Valve with motor operator.

10. MATERIAL OF CONSTRUCTION

Most valves used in gas pipelines are constructed of steel and conform to specification such as API, ASME, and ANSI standards. For certain gases that are corrosive and require certain special properties, some exotic materials may be used. The next section lists applicable standards and codes used in the design and construction of valves and fittings on gas pipeline.

The valve trim material which refers to the various working parts of a valve such as the stem, wedge, disc, etc., are constructed of many different materials depending upon the pressure rating and service. Valve manufacturers designate their products using some form of a proprietary numbering system. However, the purchaser of the valve must specify the type of material and operating conditions required. A typical gate valve specification may be as follows: NPS 12, ANSI 600 gate valve, cast steel flanged ends rising stem 13% CR, single wedge CS, stellite faced, seat rings SS 304, ABC company #2308.

Valve operators may consist of hand wheel or lever that is attached to the stem of the valve. Gear systems are used for larger valves. Electric motor operated valves are quite commonly used in gas pipeline systems, as are gas and pneumatic operators. Many valves may be buried resulting in a portion of the valve and the operator above ground.

The pressure rating of a valve represents the internal pressure that the valve can be subject to under normal operating conditions. For example, ANSI 600 rating refers to a valve that may be safely operated at pressures up to 1440 psig. Most gas pipelines are operated around this pressure rating. Table 12.2 shows the ANSI pressure rating for valves and pipes.

If a valve is designated as an ANSI 600–rated valve, the manufacturer of the valve must hydrostatically test the valve at a higher pressure for a specified period

Table 12.2 ANSI pressure rating for valves

Class	Allowable Pressure psi
150	275
300	720
400	960
600	1440
900	2160
1500	3600

as required by the design code. Generally, the hydrotest pressure is 150% of the valve rating. This compares with a hydrotest pressure of 125% of maximum operating pressure (MOP) for pipelines as discussed in Chapter 6.

11. CODES FOR DESIGN AND CONSTRUCTION

The following is a list of applicable standards and codes used in the design and construction of valves and fittings on gas pipeline.

ASME B31.8: Gas Transmission and Distribution Piping Systems
ASME B16.3: Malleable Iron Threaded Fittings
ASME B16.5: Pipe Flanges and Flanged Fittings
ASME B16.9: Factory Made Wrought Steel Butt Welding Fittings
ASME B16.10: Face to Face and End to End Dimensions of Valves
ASME B16.11: Forged Steel Fittings, Socket Welding and Threaded Fittings
ASME B16.14: Ferrous Pipe Plugs, Bushing etc.
ASME B16.20: Metallic Gaskets
ASME B16.21: Non-Metallic Gaskets
ASME B16.25: Butt Welding Ends
ASME B16.28: Wrought Steel, Butt Welding, Short Radius Elbows and Returns
ASME B16.36: Orifice Flanges
ANSI B16.5: Steel pipe flanges, flanged valves and fittings
ANSI/ASTM A182: Forged or rolled alloy-steel pipe flanges, forged fittings and valves and parts for high temperature service.
API 593: Ductile iron plug valves
API 594: Wafer type check valves
API 595: Cast iron gate valves
API 597: Steel venturi gate valves
API 599: Steel plug valves
API 600: Steel gate valves
API 602: Compact cast-steel gate valves
API 603: Class 150 corrosion-resistant gate valves
API 604: Ductile iron gate valves
API 606: Compact carbon-steel gate valves (extended bodies)
API 609: Butterfly valves to 150 psig and 150 °F
API 6D: Pipeline valves
MSS DS-13: Corrosion resistant cast flanged valves
MSS SP-25: Standard marking system for valves, fittings, and flanges

12. GATE VALVE

A gate valve is generally used to completely shut off fluid flow or, in the fully open position, provide full flow in a pipeline. Thus it is used either in the fully closed or fully open positions. A gate valve consists of a valve body, seat and disc, a spindle, gland, and a wheel for operating the valve. The seat and the gate together perform the function of shutting off the flow of fluid. A typical gate valve is shown in Figure 12.9.

Gate valves are generally not suitable for regulating flow or pressure or operating in a partially open condition. For this service, a plug valve or a control valve should be used. It must be noted that because of the type of construction a gate valve requires many turns of the hand wheel to completely open or close the valve. When fully opened, gate valves offer little resistance to flow and its equivalent length to diameter ratio (L/D) is

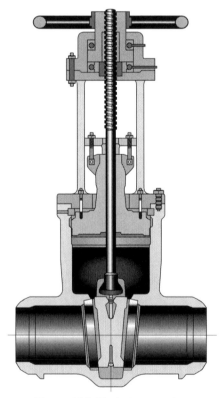

Figure 12.9 Typical gate valve.

Table 12.3 Equivalent lengths of valves and fittings

Description	L/D
Gate valve	8
Globe valve	340
Angle valve	55
Ball valve	3
Plug valve straightway	18
Plug valve 3-way thru-flo	30
Plug valve branch flo	90
Swing check valve	50
Lift check valve	600
Standard elbow: 90°	30
Standard elbow: 45°	16
Standard elbow long radius 90°	16
Standard tee thru-flo	20
Standard tee thru-branch	60
Mitre bends: $\alpha = 0$	2
Mitre bends: $\alpha = 30$	8
Mitre bends: $\alpha = 60$	25
Mitre bends: $\alpha = 90$	60

approximately 8. The equivalent L/D for commonly used valves and fittings is listed in Table 12.3.

The gate valves used in the mainlines carrying oil or gas must be of full bore or through conduit design to enable smooth passage of scrapers or pigs used for cleaning or monitoring pipelines. Such gate valves are referred to as full bore or through conduit gate valves.

13. BALL VALVE

A ball valve consists of a valve body in which a large sphere with a central hole equal to the inside diameter of the pipe is mounted. As the ball is rotated, in the fully open position the valve provides the through conduit or full bore required for unrestricted flow of the fluid and scrapers or pigs. Compared with a gate valve, a ball valve has very little resistance to flow in the fully open position. When fully open, the L/D ratio for a ball valve is approximately 3.0. The ball valve, like the gate valve, is generally used in the fully open or fully closed positions. A typical ball valve is shown in Figure 12.10.

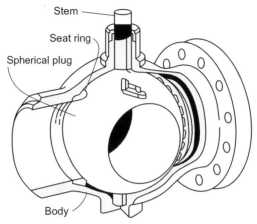

Figure 12.10 Typical ball valve.

Unlike a gate valve, a ball valve requires a one-quarter turn of the hand wheel to go from the fully open to the fully closed positions. Such quick opening and closing of a ball valve may be of importance in some installations where isolating pipe sections quickly is needed in the event of emergency.

14. PLUG VALVE

The plug valve traces its origin to the beginnings of the valve industry. It is a simple device for shutting off or allowing the flow of a fluid in a pipe by a simple quarter turn of the handle. In this sense it is similar to the ball valve. Plug valves are generally used in screwed piping and in small pipe sizes. Plug valves may be hand wheel operated or operated using a wrench or gearing mechanism. The L/D ratio for this type of valve ranges from 18 to 90 depending upon the design. A typical plug valve is shown in Figure 12.11.

15. BUTTERFLY VALVE

The butterfly valve was originally used where a tight closure was not absolutely necessary. However, over the years, these valves have been manufactured with fairly tight seals made of rubber or elastomeric materials that provide good shut off similar to other types of valves. Butterfly valves are used where space is limited. Unlike gate valves, butterfly valves can be

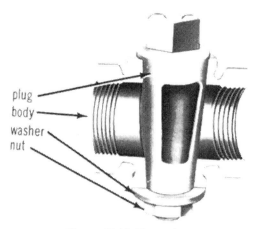

Figure 12.11 Plug valve.

used for throttling or regulating flow as well as in the full open and fully closed position. The pressure loss through a butterfly valve is small in comparison with the gate valve. The L/D ratio for this type of valve is approximately one-third of that of a gate valve. Butterfly valves are used in large and small sizes. They may be hand wheel–operated or operated using a wrench or gearing mechanism. A typical butterfly valve is shown in Figure 12.12.

Figure 12.12 Butterfly valve.

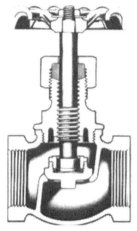

Figure 12.13 Globe valve.

16. GLOBE VALVE

Globe valves, so-called because of their outside shape, are widely used in plant piping. They are suitable for manual and automatic operation. Unlike the gate valve, globe valve can be used for regulating flow or pressures as well as complete shutoff of flow. It may also be used sometime as a pressure relief valve or as a check valve. Compared with a gate valve or ball valve, the globe valve has considerably higher pressure loss in the fully open position. This is because the flow of fluid changes direction as it goes through the valve. The L/D ratio for this type of valve is approximately 340. Globe valves are manufactured in sizes up to NPS 16. They are generally hand wheel–operated. A typical globe valve is shown in Figure 12.13.

17. CHECK VALVE

Check valves are normally in the close position and open when the fluid flows through it. It also has the capability of shutting off the flow in the event the pressure downstream exceeds the upstream pressure. In this respect, it is used for flow in one direction only. Thus it prevents backflow through the valve. Because flow of the fluid through the valve is allowed to be in one direction only, check valves must be installed properly by noting the normal direction of flow. An arrow stamped on the outside of the valve body indicates the direction of flow. Check valves may be classified as swing check valve and lift check valve. The L/D ratio for check valves range from

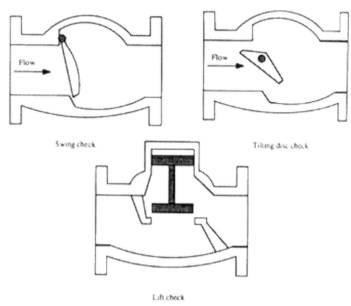

Figure 12.14 Check valve.

50 for the swing check valve to as high as 600 for lift check valves. Examples of typical check valves are shown in Figure 12.14.

18. PRESSURE CONTROL VALVE

A pressure control valve is used to automatically control the pressure at a certain point in a pipeline. In this respect, it is similar to a pressure regulator discussed next. Whereas the pressure regulator is generally used to maintain a constant downstream pressure, a pressure control valve is used to control the upstream pressure. The upstream and downstream are relative to the location of the valve on the pipeline. Generally, a bypass piping system around the control valve is installed to isolate the control valve in the event of an emergency or for maintenance work on the control valve. This is illustrated in Figure 12.15.

19. PRESSURE REGULATOR

A pressure regulator is a valve that is similar to a control valve. Its function is to control or regulate the pressure in a certain section of a pipeline system. For example, on a lateral piping that comes off of a main pipeline,

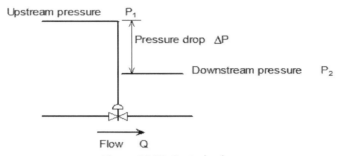

Figure 12.15 Control valve.

used for delivering gas to a customer, a lower pressure may be required on the customer side. If the main pipeline pressure at the point of connection to the lateral pipeline is 800 psig, whereas the customers piping is limited to 600 psig, a pressure regulator is used to reduce the pressure by 200 psig as shown in Figure 12.16.

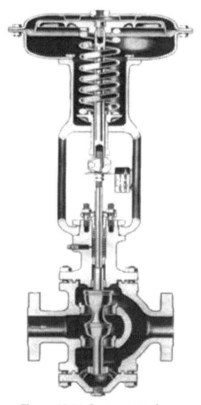

Figure 12.16 Pressure regulator.

20. PRESSURE RELIEF VALVE

The pressure relief valve is used to protect a section of piping by relieving the pipeline pressure when it reaches a certain value. For example, if the MOP of a pipeline system is 1400 psig, a pressure relief valve may be set at 1450 psig. Any upset conditions that cause the pipeline pressure to exceed the normal 1400 psig will cause the relief valve to open at the set point of 1450 and expel the gas to the atmosphere or to a relief vessel thereby protecting the pipeline from overpressure and eventually rupture. The difference between the relief valve set point (1450 psig) and the pipeline MOP (1400 psig) will depend upon the actual application, the valve type, and expected fluctuations in pressure. Generally, the difference will range between 20 and 50 psig. Too close a difference will result in frequent operation of the relief valve, which will be a nuisance and in many cases waste of valuable gas. A large difference between a relief valve set point and the pipeline MOP may render the valve ineffective.

21. FLOW MEASUREMENT

Gas flow measurement in a pipeline is necessary for properly accounting for the amount of gas transported from one point to another along a gas pipeline. The owner of the gas and the customer who purchases the gas both require that the correct amount of gas be delivered for the agreed-upon price. Even a very small error in flow measurement on large-capacity pipelines can result in huge losses to either the owner or customer of gas. For example, consider a gas pipeline transporting 300 Million standard cubic feet per day (MMSCFD) at a tariff of 50 cents per One thousand cubic Feet (MCF). An error of 1% in the gas flow measurement can translate to a loss of more than $500,000 per year to either the seller or the buyer. Hence it is easy to appreciate the importance of good, accurate flow measurement in gas pipelines. Over the years, gas flow measurement technology have improved considerably. Many organizations have jointly developed standards and procedures for measurement of natural gas through orifice meters installed in pipelines. AGA, API, ANSI, and ASME have together endorsed standards for orifice metering of natural gas. The AGA Measurement Committee Report No. 3 is considered to be the leading publication in this regard. This standard is also endorsed by ANSI and API and is referred to as ANSI/API 2530 standard. We will refer to sections of this standard when discussing orifice meters.

22. FLOW METERS

Because the orifice meter is the main flow measurement instrument used in the gas industry, we will discuss this first.

22.1 Orifice Meter

The orifice meter is a flat steel plate that has a concentric machined hole with a sharp edge and positioned inside the pipe as shown in Figure 12.4.

As the gas flows through the pipeline and then through the orifice plate, because of the reduction in a cross-sectional area as the gas approaches the orifice, the velocity of flow increases and, correspondingly, the pressure drops. After the orifice the cross-sectional area increases again back to the full pipe diameter, which results in expansion of gas and decrease in flow velocity. This process of accelerating flow through the orifice and subsequent expansion forms a vena contracta or a throat immediately past the orifice as shown in the figure. Three different types of orifice meters are illustrated in Figure 12.17.

The different types of orifice meters shown have different crest shapes that affect the extent of contraction of the jet of gas as it flows through the orifice. The contraction coefficient C_c is defined in terms of the area of cross-section of the vena contracta compared with the cross-sectional area of the orifice as defined as follows:

$$C_c = \frac{A_c}{A_o} \qquad (12.12)$$

where

C_c: contraction coefficient, dimensionless
A_c: cross-sectional area of the vena contracta, in^2
A_o: cross-sectional area of the orifice, in^2

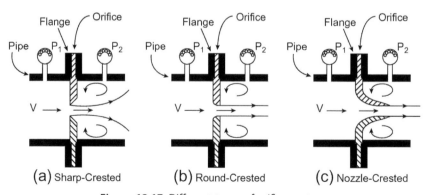

Figure 12.17 Different types of orifice meters.

The discharge through the orifice meter is represented by the following basic equation

$$Q = C_c C_v A_o \sqrt{\frac{2[(p_1 - p_2)/\rho + g(z_1 - z_2)]}{1 - C_c^2(A_o/A)^2}} \quad (12.13)$$

where
- Q: flow rate, ft^3/s
- C_c: contraction coefficient, dimensionless
- C_v: discharge coefficient, dimensionless
- A_o: cross-sectional area of the orifice, in^2
- A: cross-sectional area of pipe containing the orifice, in^2
- p_1: upstream pressure, psig
- p_2: downstream pressure, psig
- ρ: density of gas, lb/ft^3
- z_1: upstream elevation, ft
- z_2: downstream elevation, ft
- g: acceleration due to gravity.

When the elevation difference between the upstream and downstream pressure taps is negligible, the discharge equation for the orifice meter can be simplified to

$$Q = C_c C_v A_o \sqrt{\frac{2(p_1 - p_2)/\rho}{1 - C_c^2 \left(\frac{A_o}{A}\right)^2}} \quad (12.14)$$

where all symbols are as defined earlier.

For round crested and nozzle crested orifice meters, shown in Figure 12.13, the value of C_c may be taken as 1.0. This indicates an absence of vena contracta for these types of orifices. For the sharp crested orifice at high Reynolds numbers or for turbulent flow, C_c is calculated from the equation:

$$C_c = 0.595 + 0.29 \left(\frac{A_o}{A}\right)^{\frac{5}{2}} \quad (12.15)$$

where all symbols are as defined earlier.

There are basically two types of pressure measurements in orifice meters. These are called flange taps and pipe taps. They relate to the locations where the pressure measurements are taken. A flange tap requires that the upstream

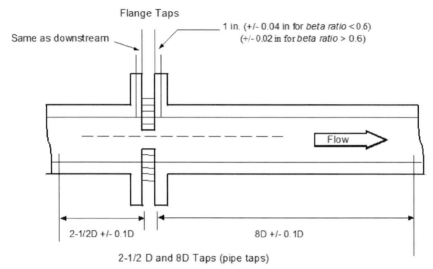

Figure 12.18 Flange taps and pipe taps.

tap be located at a distance of 1 in upstream of the nearest plate face and the downstream tap is located 1 in downstream of the nearest plate face. Pipe taps are such that the upstream tap be located at a distance of 2.5 times the inside diameter of the pipe upstream of the nearest plate face and the downstream tap is located at a distance of 8 times the inside diameter of the pipe, downstream of the nearest plate face.

See Figure 12.18, which illustrates the location of flange taps and pipe taps.

Several terms used in the calculation of the orifice flow must be explained first. The differential pressure for an orifice is the pressure difference between the upstream and downstream taps. The orifice diameter is defined as the arithmetic average of four or more inside diameter measurements evenly spaced. Strict tolerances for the orifice diameters are specified in the AGA3/ANSI 2530 standard. Table 12.4 shows these tolerances taken from the standard.

22.2 Meter Tube

The meter tube is the piece of pipe in which the orifice plate is installed along with straightening vanes as needed. A typical meter tube consisting of the orifice plate and straightening vanes is illustrated in Figure 12.19.

Table 12.4 Orifice plate diameter tolerances

Orifice diameter, in	Tolerance ±, in
0.250	0.0003
0.375	0.0004
0.500	0.0005
0.625	0.0005
0.750	0.0005
0.875	0.0005
1.000	0.0005
Above 1.000	0.0005 (Per inch diameter)

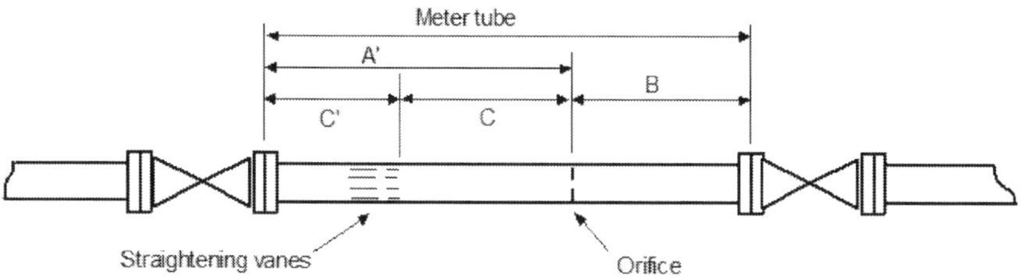

Figure 12.19 Meter tube installation.

The dimensions of the meter tube such as A, B, C, and C′ depend upon the orifice to pipe diameter ratio, also known as beta ratio β and are specified in the AGA Report No. 3. For example, for beta = 0.5:

$$A = 25 \quad A' = 10 \quad B = 4 \quad C = 5 \quad C' = 5.5$$

These numbers are actually multiples of the pipe or meter tube diameter. The requirements of straightening vanes before the orifice plate depend upon the specific installation. The main reason for straightening vanes is to reduce flow disturbance at the orifice plate from upstream fittings. Refer to AGA Report No. 3 for various meter tube configuration.

The orifice flow rate is the mass flow rate or volume flow rate of gas per unit of time.

The density is the mass per unit volume of gas at a specific temperature and pressure.

22.3 Expansion Factor

This is a dimensionless factor used to correct the calculated flow rate to take into account the reduction in gas density as it flows through an orifice, which is caused by the increased velocity and corresponding reduced static pressure. Methods of calculating the expansion factor Y will be discussed in subsequent sections.

The beta ratio is defined as the ratio of the orifice diameter to the meter tube diameter as follows:

$$\beta = \frac{d}{D} \qquad (12.16)$$

For orifice meters with flange taps beta ratio ranges between 0.15 and 0.70. For orifice meters with pipe taps beta ratio ranges between 0.20 and 0.67.where

β: beta ratio, dimensionless
d: orifice diameter, in
D: meter tube diameter, in

The fundamental orifice meter flow equation described in the ANSI 2530/AGA Report No. 3 is as follows:

$$q_m = \frac{C}{(1-\beta^4)^{0.5}} Y \frac{\pi}{4} d^2 (2g\rho_f \Delta P)^{0.5} \qquad (12.17)$$

$$\text{or} \quad q_m = KY \frac{\pi}{4} d^2 (2g\rho_f \Delta P)^{0.5} \qquad (12.18)$$

$$\beta = \frac{d}{D} \qquad (12.19)$$

$$K = \frac{C}{(1-\beta^4)^{0.5}} = \frac{CD^2}{(D^4 - d^4)^{0.5}} \qquad (12.20)$$

where

q_m: mass flow rate of gas, lb/s
ρ_f: density of gas, lb/ft^3
C: discharge coefficient
β: beta ratio, dimensionless
d: orifice diameter, in
D: meter tube diameter, in
Y: expansion factor, dimensionless
g: acceleration due to gravity, ft/s^2
ΔP: pressure drop across the orifice, psi
K: flow coefficient, dimensionless

These equations were arrived at using the conservation of energy and mass equations with thermodynamics and equation of state for the gas in question.

Essentially, these formulas give the mass flow rate of gas. We need to convert these to volume flow rate using the density. The coefficient of discharge C in the preceding equation is approximately 0.6 and the flow coefficient K is a value that is between 0.6 and 0.7. Both the flow coefficient K and the expansion factor Y are determined using test data. The volume flow rate at standard (base conditions) is calculated from the mass flow rate as follows:

$$q_v = \frac{q_m}{\rho_b} \tag{12.21}$$

where

q_v: volume flow rate, ft^3/s
q_m: mass flow rate, lb/s
ρ_b: gas density at base temperature, lb/ft^3

The expansion factor Y for low compressibility fluids such as water at 60 °F and 1 atm pressure is taken as 1.0. For gases, Y can be calculated as explained in the next section. The flow coefficient K is found to vary with D, the diameter of the meter tube, orifice diameter d, mass flow rate q_m, fluid density, and viscosity at flowing temperature. For gases, K also varies with the ratio of the differential pressure to the static pressure and k, the ratio of specific heat of the gas. In many cases, the flow coefficient K is considered to be a function of the Reynolds number, acoustic ratio, meter tube diameter, and the beta ratio. Rearranging Eqn (12.7), we get

$$KY = \frac{4q_m}{\pi d^2 \left[2g_c \rho_f \Delta P\right]^{0.5}} \tag{12.22}$$

Several empirical equations are available to calculate the flow coefficient K. The following equation by Buckingham and Bean is endorsed by the National Bureau of Standards and listed in AGA Report No. 3.

For flange taps:

$$K_e = 0.5993 + \frac{0.007}{D} + \left(0.364 + \frac{0.076}{D^{0.5}}\right)\beta^4$$

$$+ 0.4\left(1.6 - \frac{1}{D}\right)^5 \left[\left(0.07 + \frac{0.5}{D}\right) - \beta\right]^{2.5}$$

$$- \left(0.009 + \frac{0.034}{D}\right)(0.5 - \beta)^{1.5} + \left(\frac{65}{D^2} + 3\right)(\beta - 0.7)^{2.5}$$

$$\tag{12.23}$$

where
  K_e: flow coefficient for Reynolds number $R_d = d(10^6/15)$, dimensionless
  D: meter tube diameter, in
  d: orifice diameter, in
  β: beta ratio, dimensionless
  For pipe taps:

$$K_e = 0.5925 + \frac{0.0182}{D} + \left(0.440 - \frac{0.06}{D}\right)\beta^2 + \left(0.935 + \frac{0.225}{D}\right)\beta^5$$
$$+ 1.35\beta^{14} + \frac{1.43}{D^{0.5}}(0.25 - \beta)^{2.5}$$
(12.24)

where all symbols are as defined before.

For flange taps and pipe taps, the value of K_o is calculated from:

$$K_o = \frac{K_e}{1 + \frac{15 \times 10^{-6} E}{d}} \tag{12.25}$$

where the parameter E in Eqn (12.14) is found from:

$$E = d(830 - 5000\beta + 9000\beta^2 - 4200\beta^3 + B) \tag{12.26}$$

The value of parameter B in Eqn (12.26) is defined as follows:

$$\text{For flange taps} \quad B = \frac{530}{D^{0.5}} \tag{12.27}$$

$$\text{For pipe taps} \quad B = \frac{875}{D} + 75 \tag{12.28}$$

and finally the flow coefficient K is calculated from:

$$K = K_o\left(1 + \frac{E}{R_d}\right) \tag{12.29}$$

where
  K_o: flow coefficient for infinitely large orifice Reynolds number, dimensionless
  R_d: Reynolds number at the inlet of orifice, dimensionless

The Reynolds number used in the preceding equations is calculated from

$$R_d = \frac{V_f d \rho_f}{\mu} \tag{12.30}$$

where

R_d: Reynolds number at the inlet of orifice, dimensionless
V_f: velocity of fluid at inlet of orifice, ft/s
d: orifice diameter, ft
ρ_f: fluid density at flowing conditions, lb/ft^3
μ: dynamic viscosity of fluid, lb/ft.s

The values of flow coefficient K calculated using the preceding equations apply to orifice meters manufactured and installed in accordance with AGA Report No. 3 as long as the meter tube is greater than 1.6 in inside diameter and the beta ratio is between 0.10 and 0.75.

The uncertainties in flow coefficient K are as follows in accordance with AGA Report No. 3.

For flange taps the uncertainty is ±0.5% for $0.15 < \beta < 0.70$.
The uncertainty is greater than ±1.0% for $0.15 > \beta > 0.70$.
For pipe taps the uncertainty is ±0.75% for $0.20 < \beta < 0.67$.
The uncertainty is greater than ±1.5% for $0.20 > \beta > 0.67$.

22.4 Expansion Factor

The expansion factor Y is calculated in two ways. In the first method, it is calculated using the upstream pressure and in the second method using the downstream pressure. The following equation is used for the expansion factor Y_1 with reference to upstream pressure.

For flange taps:

$$Y_1 = 1 - \left(0.41 + 0.35\beta^4\right)\frac{x_1}{k} \tag{12.31}$$

For pipe taps:

$$Y_1 = 1 - \left[0.333 + 1.145\left(\beta^2 + 0.7\beta^5 + 12\beta^{13}\right)\right]\frac{x_1}{k} \tag{12.32}$$

And the pressure ratio x_1 is

$$x_1 = \frac{P_{f1} - P_{f2}}{P_{f1}} = \frac{h_w}{27.707 P_{f1}} \tag{12.33}$$

where

Y_1: expansion factor based on upstream pressure
x_1: ratio of differential pressure to absolute upstream static pressure
h_w: differential pressure between upstream and downstream taps in inches of water at 60 °F

P_{f1}: static pressure at upstream tap, psia
P_{f2}: static pressure at downstream tap, psia
x_1/k: acoustic ratio, dimensionless
k: ratio of specific heats of gas, dimensionless

The value of Y_1 calculated using these equations is subject to a tolerance from 0 to $\pm 0.5\%$ for the range of $x = 0$ to 0.20. For larger values of x, the uncertainty is larger. For flange taps, the values of Y_1 are valid for beta ratio range of 0.10–0.80. For pipe taps, the beta ratio range is 0.10–0.70.

With reference to the downstream pressure the expansion factor Y_2 is calculated using the following equations.

For flange taps:

$$Y_2 = Y_1 \left(\frac{1}{1 - x_1} \right)^{0.5} \tag{12.34}$$

$$Y_2 = (1 + x_2)^{0.5} - (0.41 + 0.35\beta^4) \frac{x_2}{k(1 + x_2)^{0.5}} \tag{12.35}$$

For pipe taps:

$$Y_2 = (1 + x_2)^{0.5} - \left[0.333 + 1.145 \left(\beta^2 + 0.7\beta^5 + 12\beta^{13} \right) \right] \frac{x_2}{k(1 + x_2)^{0.5}} \tag{12.36}$$

And the pressure ratio x_2 is

$$x_2 = \frac{P_{f1} - P_{f2}}{P_{f2}} = \frac{h_w}{27.707 P_{f2}} \tag{12.37}$$

where all symbols are as defined before.

The density of the flowing gas used in Eqn (12.6) must be obtained from equation of state or from tables. It is important to use the correct density in the flow equations. Otherwise the uncertainty in flow measurement could be as large as 10%. Generally, the density of the gas can be calculated from the perfect gas law discussed in Chapter 1 with the modification using the compressibility factor. The following equation is obtained by rearranging the real gas equation and using the gravity of gas. See Chapter 1 for details.

$$\rho_f = \frac{m}{V} = \frac{G_i M P_f}{Z_f R T_f} \tag{12.38}$$

$$\rho_{f1} = \frac{G_i M P_{f1}}{Z_{f1} R T_f} \qquad (12.39)$$

$$\rho_b = \frac{M G_i P_b}{R Z_b T_b} \qquad (12.40)$$

where

m: mass of gas
V: volume of gas
G_i: gravity of gas (air = 1.00)
M: molecular weight of gas
P_f: absolute gas pressure
Z_f: compressibility factor at flowing temperature
R: gas constant
T_f: absolute flowing temperature
subscript f_1 refers to upstream tap flowing conditions
subscript f_2 refers to downstream tap flowing conditions
subscript b refers to base conditions

Two other equations based on real gas specific gravity and taking the base conditions of 14.73 psia and 60 °F results in the gas densities at the upstream tap and at the base conditions as follows:

$$\rho_{f1} = \frac{M Z_b G P_{f1}}{0.99949 R Z_{f1} T_f} \qquad (12.41)$$

$$\rho_b = \frac{M G P_b}{0.99949 R T_b} \qquad (12.42)$$

Knowing the density at the upstream tap and at base condition, the following equation is used for the volume flow rate. This equation is derived from the equations listed in the preceding sections.

$$q_v = \frac{\pi}{4} \frac{\sqrt{2g}(KY_1 d^2)\sqrt{(\rho_{f1}\Delta P)}}{\rho_b} \qquad (12.43)$$

Combining all equations we have reviewed so far, the AGA Report No. 3 shows a compact equation for the flow of gas through an orifice meter as follows.

$$Q_v = C\sqrt{h_w P_f} \qquad (12.44)$$

where

Q_v: gas flow rate at base conditions, ft³/h
h_w: differential pressure between upstream and downstream taps in inches of water at 60 °F
P_f: absolute static pressure, psia
C: orifice flow constant

For P_f, subscript 1 is used for upstream and subscript 2 for downstream pressure.

The orifice flow constant C consists of the product of several factors that depend upon the Reynolds number, expansion factor, base pressure, base temperature, flowing temperature, gas gravity, and super compressibility factor of gas. It is given by the following equation

$$C = F_b F_r F_{pb} F_{tb} F_{tf} F_{gr} F_{pv} Y \qquad (12.45)$$

where the dimensionless factors are

F_b: basic orifice factor
F_r: Reynolds number factor
F_{pb}: pressure base factor
F_{tb}: temperature base factor
F_{tf}: flowing temperature factor
F_{gr}: gas relative density factor
F_{pv}: supercompressibility factor
Y: expansion factor.

These values of the factors that constitute the orifice flow constant C are defined in AGA Report No. 3 and listed in Appendix B of that publication. However, each of these factors may be calculated as follows:

The basic orifice factor is

$$F_b = 338.178 d^2 K_o \qquad (12.46)$$

where K_o is calculated using Eqn (12.14).

The Reynolds number factor is

$$F_r = 1 + \frac{E}{R_d} \qquad (12.47)$$

$$K = K_o F_r \qquad (12.48)$$

The pressure base factor is

$$F_{pb} = \frac{14.73}{P_b} \qquad (12.49)$$

The temperature base factor is

$$F_{tb} = \frac{T_b}{519.67} \qquad (12.50)$$

The flowing temperature factor is

$$F_{tf} = \left(\frac{519.67}{T_f}\right)^{0.5} \qquad (12.51)$$

The gas relative density factor is

$$F_{gr} = \left(\frac{1}{G_r}\right)^{0.5} \qquad (12.52)$$

where all symbols in the preceding equations are as defined before.

23. VENTURI METER

The venturi meter is based upon the Bernoulli equation. It consists of a smooth gradual contraction from the main pipe size to a reduced section known as the throat and finally expanding back gradually to the original pipe diameter.

This type of venturi meter is called Herschel type. The angle of contraction from the main pipe to the throat section is in the range of 21° ± 2°. The gradual expansion from the throat to the main pipe section is in the range of 5° to 15°. This design causes the least energy loss such that the discharge coefficient may be assumed at 1.0. Venturi meters range in size from 4.0 to 48.0 in. The beta ratio, equal to d/D, generally ranges between 0.30 and 0.75.

The gas pressure in the main pipe section is represented by P_1 and that at the throat it is represented by P_2. As gas flows through a venturi meter, it increases in flow velocity in the narrow throat section. Correspondingly, the pressure reduces in the throat section according to Bernoulli equation. After gas leaves the throat section, it reduces in flow velocity because of the increase in pipe cross-sectional area and it reaches the original flow velocity.

The flow velocity in the main pipe section before the throat is calculated from the known pressures P_1 and P_2.

$$V_1 = \frac{\sqrt{\left[\frac{2g(P_1-P_2)}{\rho} + (Z_1 - Z_2) - h_L\right]}}{\left(\frac{A_1}{A_2}\right)^2 - 1} \qquad (12.53)$$

Neglecting the elevation difference $Z_1 - Z_2$ and the friction loss h_L, this equation reduces to the following:

$$V_1 = C \frac{\sqrt{\left[\frac{2g(P_1 - P_2)}{\rho}\right]}}{\left(\frac{A_1}{A_2}\right)^2 - 1} \tag{12.54}$$

where
  V_1: velocity of gas in the main pipe section before the throat
  ρ: the average gas density
  A_1: cross-sectional area of pipe
  A_2: cross-sectional area of throat
  C: discharge coefficient, dimensionless

The volume flow rate is then calculated by multiplying the velocity by the cross-sectional area resulting in the following equation:

$$Q = CA_1 \frac{\sqrt{\left[\frac{2g(P_1 - P_2)}{\rho}\right]}}{\left(\frac{A_1}{A_2}\right)^2 - 1} \tag{12.55}$$

Using the beta ratio, we simplify the previous equation as follows:

$$Q = CA_1 \frac{\sqrt{\left[\frac{2g(P_1 - P_2)}{\rho}\right]}}{\left(\frac{1}{\beta}\right)^4 - 1} \tag{12.56}$$

The discharge coefficient C is a number less than 1.0 and it depends on the Reynolds number in the main pipe section. For Reynolds number greater than 2×10^5, the value of C remains constant at 0.984.

In smaller pipe sizes such as 2–10 in, venturi meters are machined and therefore have a better surface finish than the larger, rough cast meters. Smaller venturi meters have a C value of 0.995 for Reynolds numbers larger than 2×10^5.

24. FLOW NOZZLE

The flow nozzle is another device for measuring flow rate. It consists of a main pipe section, followed by a gradual reduction in cross-section area and a short cylindrical section, ending in a gradual expansion to the original pipe size.

The discharge coefficient C for a flow nozzle is approximately 0.99 for Reynolds numbers greater than 10^6. At lower Reynolds numbers, because of greater energy loss subsequent to the nozzle throat, C values are lower.

The discharge coefficient C depends on the beta ratio and Reynolds number. It is calculated using the following equation:

$$C = 0.9975 - 6.53\sqrt{\frac{\beta}{R}} \qquad (12.57)$$

where

β: d/D
R: Reynolds number based on the pipe diameter D.

Example 12.3
A 4-in diameter orifice meter is installed in a pipe with an inside diameter of 12.09 in. The differential pressure is measured at 30 in of water and the static pressure upstream is 600 psig. Gas gravity = 0.6, gas flowing temperature = 70 °F. The base temperature and the base pressure are 60 °F and 14.7 psia, respectively. Assuming flange taps, calculate the flow rate in standard ft^3/h.

The barometric pressure is 14.5 psia.

Solution
The basic orifice factor F_b is calculated from the AGA 3 appendix as follows:

$$F_b = 3258.5$$

$$(hP)^{0.5} = [30 \times (600 + 14.5)]^{0.5} = 135.78$$

$$F_r = 1 + \frac{0.0207}{135.78} = 1.0002$$

$$F_{pb} = \frac{14.73}{14.7} = 1.002$$

$$F_{tb} = \frac{60 + 460}{519.67} = 1.006$$

$$F_{tf} = \left(\frac{519.67}{70+460}\right)^{0.5} = 0.9902$$

$$F_{gr} = \left(\frac{1}{0.6}\right)^{0.5} = 1.291$$

$$F_{pv} = 1.0463$$

$$\frac{h}{P} = \frac{30}{614.5} = 0.0488$$

$$\beta = \frac{4}{12.09} = 0.3309$$

$$Y = 0.9995$$

$$C = 3258.5 \times 1.0002 \times 1.002 \times 1.006 \times 0.9902 \times 1.291 \times 1.0463 \times 0.9995$$
$$= 4391.96$$

Using Eqn (12.33), the flow rate is

$$Q_v = 4391.96 \times 135.78 = 596{,}340 \ \text{ft}^3/\text{h}$$

25. SUMMARY

In this chapter, we discussed the more commonly used devices for the measurement of liquid pipeline flow rates and the different types of valves used in pipelines. Several models of ball valves, gate valves, plug valves, and butterfly valves were reviewed. Among measuring devices, variable head flow meters, such as the venturi tube, flow nozzle, orifice meter, and the equations for calculating the velocities and flow rate from the pressure drop were explained. The importance of the discharge coefficient and how it varies with the Reynolds number and the beta ratio were also discussed. A trial-and-error method for calculating the flow rate through an orifice meter was illustrated using an example. For a more detailed description and analysis of flow meters, the reader should refer to any of the standard texts used in the industry.

In the gas pipeline section, we covered the topics of valves and flow measurement as it related to gas pipeline transportation. The various types of valves used and their functions were reviewed. The importance of flow measurement in natural gas pipeline transaction was explained. The predominantly used measuring device known as orifice meter was discussed in detail. The calculation methodology based on AGA Report No. 3 was reviewed. Also the venturi meter and the flow nozzle were discussed.

CHAPTER THIRTEEN

Pipeline Economics

In this chapter, we will discuss cost of a pipeline and the various components that contribute to the economics of pipelines. This will include the major components of the initial capital costs and that of the recurring annual costs. We will also examine how the transportation charge is established based on throughput rates, project life, interest rate, and financing scenarios.

1. ECONOMIC ANALYSIS

In any pipeline investment project, we must perform an economic analysis of the pipeline system to ensure that we have the right equipment and materials at the right cost to perform the necessary service and provide a profitable income for the venture. The previous chapters helped determine the pipe size, pipe material, pumping equipment, etc., necessary to transport a given volume of a product. In this chapter, we will analyze the cost implications and how to decide on the economic pipe size and pumping equipment required to provide the optimum rate of return (ROR) on our investment.

The major capital components of a pipeline system consist of the pipe, pump stations, storage tanks, valves, fittings, and meter stations. Once this capital is expended and the pipeline is installed, and pump station and other facilities built, we will incur annual operating and maintenance (O&M) cost for these facilities. Annual costs will also include general and administrative (G&A) costs including payroll costs, rental and lease costs, and other recurring costs necessary for the safe and efficient operation of the pipeline system. The revenue for this operation will be in the form of pipeline tariffs collected from companies that ship products through this pipeline. The capital necessary for building this pipeline system may be partly owner equity and partly borrowed money. There will be investment hurdles and ROR requirements imposed by equity owners and financial institutions that lend the capital for the project. Regulatory requirements will also dictate the maximum revenue that may be collected and the ROR that may be realized as transportation services. An economic analysis must be performed for the project taking into account all of these factors and a reasonable project life of 20–25 years or more in some cases.

These concepts will be illustrated by examples in subsequent sections of this chapter. Before we discuss the details of each cost component, it will be instructive and beneficial to calculate the transportation tariff and cost of service using a simple example.

Problem 13.1

Consider a new pipeline that is being built for transporting crude oil from a tank farm to a refinery. For the first phase (10 years), it is estimated that shipping volumes will be 100,000 barrels (bbl)/day. Calculations indicate that the 16-in pipeline, 100 miles long with two pump stations will be required. The capital costs for all facilities are estimated to be $72 million. The annual operating cost including electric power, O&M, G&A, etc., is estimated to be $5 million. The project is financed at a debt equity ratio 80/20. The interest rate on debt is 8% and the ROR allowed by regulators is 12%. Consider a project life of 20 years and overall tax rate of 40%.

1. What is the annual cost of service for this pipeline?
2. Based on the fixed throughput rate of 100,000 bbl/day and a load factor of 95%, what tariff rate can be charged within the regulatory guidelines?
3. During the second phase, volumes are expected to increase by 20% (years 11–20). Estimate the revised tariff rate for the second phase assuming no capital cost changes to pump stations and other facilities. Use an increased annual operating cost of $7 million and same load factor as before.

Solution

1. The total capital cost of all facilities is $72 million. We start by calculating the debt capital and equity capital based on the given 80/20 debt equity ratio.

$$\text{Debt capital} = 0.80 \times 72 = \$57.6 \text{ million}$$

$$\text{Equity capital} = 72 - 57.6 = \$14.4 \text{ million}$$

The debt capital of $57.6 million is borrowed from a bank or financial institution at the 8% annual interest rate. To retire this debt over the project life of 20 years, we must account for this interest payment in our annual costs.

In a similar manner, the equity investment of $14.4 million may earn 12% ROR. Because the tax rate is 40%, the annual cost component of the equity will be adjusted to compensate for the tax rate.

$$\text{Interest cost per year} = 57.6 \times 0.08 = \$4.61 \text{ million}$$

$$\text{Equity cost per year} = 14.4 \times 0.12/(1-0.4) = \$2.88 \text{ million}$$

Assuming a straight line depreciation for 20 years, our yearly depreciation cost for $72 million capital is calculated as follows:

$$\text{Depreciation per year} = 72/20 = \$3.6 \text{ million}$$

By adding annual interest expense, annual equity cost, and annual depreciation cost to the annual O&M cost, we get the total annual service cost to operate the pipeline.

Total cost of service $= 4.61 + 2.88 + 3.60 + 5.00 = \$16.09$ million/yr

2. Based on the total cost of service of the pipeline, we can now calculate the tariff rate to be charged, spread over a flow rate of 100,000 bbl/day and a load factor of 95%:

Tariff $= (16.09 \times 10^6)/(365 \times 100000 \times 0.95)$ or $\$0.4640$/bbl

3. During phase 2, the new flow rate is 120,000 bbl/day and the total cost of service becomes:

Total cost of service $= 4.61 + 2.88 + 3.60 + 7.00 = \$18.09$ million/yr

And the revised tariff becomes:

Tariff $= (18.09 \times 10^6)/(365 \times 120000 \times 0.95)$ or $\$0.4348$/bbl

2. CAPITAL COSTS

The capital cost of a pipeline project consists of the following major components:
- Pipeline
- Pump stations
- Tanks and manifold piping
- Valves, fittings, etc.
- Meter stations
- Supervisory control and data acquisition (SCADA) and telecommunications
- Engineering and construction management
- Environmental and permitting
- Right-of-way acquisition cost
- Other project costs such as allowance for funds used during construction (AFUDC) and contingency

2.1 Pipeline Costs

The capital cost of pipeline consists of material and labor for installation. To estimate the material cost, we will use the following method:

Pipe material cost $= 10.68\,(D-t)\,t \times 2.64 \times L \times$ cost per ton

or

$$\text{PMC} = 28.1952\, L(D-t)\, t(Cpt) \tag{13.1}$$

where
   PMC: pipe material cost, $
   L: pipe length, miles
   D: Pipe outside diameter, in
   t: pipe wall thickness, in
   Cpt: pipe cost, $/ton

In SI units:

$$\text{PMC} = 0.02463\, L(D-t)\, t(Cpt) \tag{13.2}$$

where
   PMC: pipe material cost, $
   L: pipe length, km
   D: pipe outside diameter, mm
   T: pipe wall thickness, mm
   Cpt: pipe cost, $/metric ton

Because the pipe will be coated, wrapped, and delivered to the site, we will have to increase the material cost by some factor to account for these items or add the actual cost of these items to the pipe material cost.

Pipe installation cost or labor cost is generally stated in $/ft or $/mile of pipe. It may also be stated based on an inch-diameter-mile of pipe. Construction contractors will estimate the labor cost of installing a given pipeline, based on detailed analysis of terrain, construction conditions, difficulty of access, and other factors. Historical data are available for estimating labor costs of various size pipelines. In this section, we will use approximate methods that should be verified with contractors taking into account current labor rates and geographic and terrain issues. A good approach is to express the labor cost in terms of $/inch diameter per mile of pipe. Thus we can say that a particular 16-in pipeline can be installed at a cost of $15,000 per inch-diameter-mile. Therefore for a 100-mile, 16-in pipeline we can estimate the labor cost to be

$$\text{Pipe Labor cost} = \$15000 \times 16 \times 100 = \$24 \text{ million}$$

Based on $/ft cost, this works out to be

$$\frac{24 \times 10^6}{100 \times 5280} = \$45.50 \text{ per ft}$$

In addition to labor costs for installing straight pipe, there may be other construction costs such as road crossings, railroad crossings, river crossings, etc., These are generally estimated as a lump sum for each item and added to the total pipe installed costs. For example there may be 10 highway and road crossings totaling $2 million and one major river crossing that may cost $500,000. For simplicity, however, we will ignore these items for the present.

2.2 Pump Station Costs

To estimate the pump station cost, a detailed analysis would consist of preparing a material take-off from the pump station drawings and getting vendor quotes on major equipment such as pumps, drivers, switchgear, valves, instrumentation, etc., and estimating the station labor costs.

An approximate cost for pump stations can be estimated using a value for cost in dollars per installed horsepower (HP). This is an all-inclusive number considering all facilities associated with the pump station. For example, we can use an installed cost of $1500 per HP and estimate that a pump station with 5000 HP will cost:

$$\$1500 \times 5000 = \$7.5 \text{ million}.$$

Here we used an all-inclusive number of $ 1500 per installed HP. This figure takes into account all material and equipment cost and construction labor. Such values of installed cost per HP can be obtained from historical data on pump stations constructed in the recent past. Larger HP pump stations will have smaller dollar $/HP costs, whereas smaller pump stations with less HP will have a higher $/HP cost, reflecting economies of scale.

2.3 Tanks and Manifold Piping

Tanks and manifold piping can be estimated fairly accurately by detailed material take-offs from construction drawings and from vendor quotes.

Generally, tank vendors quote installed tank costs in $/bbl. Thus if we have a 50,000-bbl tank, it can be estimated at:

$50000 \times \$10/\text{bbl} = \$500000$ based on an installed cost of $10/bbl

We would of course increase the total tankage cost by a factor of 10–20% to account for other ancillary piping and equipment.

As with installed HP costs, the unit cost for tanks decreases with tank size. For example, a 300,000-bbl tank may be based on $6 or $8 per bbl compared with the $10/bbl cost for the smaller 50,000-bbl tank.

2.4 Valves and Fittings

This category of items may also be estimated as a percentage of the total pipe cost. However, if there are several mainline block valve locations that can be estimated as a lump sum cost, we can estimate the total cost of valves and fittings as follows:

A typical 16-in mainline block valve installation may cost $100,000 per site including material and labor costs. If there are 10 such installations spaced 10 miles apart on a pipeline, we would estimate cost of valves and fittings to be $1.0 million.

2.5 Meter Stations

Meter stations may be estimated as a lump sum fixed price for a complete site. For example a 10 inch meter station with meter, valves, piping instrumentation may be priced at $250,000 per site including material and labor cost. If there are two such meter stations on the pipeline, we would estimate total meter costs at $500,000.

2.6 SCADA and Telecommunication System

This category covers costs associated with SCADA, telephone, microwave, etc. SCADA system costs include the facilities for remote monitoring, operation, and control of the pipeline from a central control center. Depending upon the length of the pipeline, number of pump stations, valve stations, etc., the cost of these facilities may range from $2 million to $5 million or more. An estimate based on the total project cost may range from 2% to 5%.

2.7 Engineering and Construction Management

Engineering and construction management consists of preliminary and detail engineering design costs and personnel costs associated with management and inspection of the construction effort for pipelines, pump stations, and other facilities. This category usually ranges from 15% to 20% of total pipeline project costs.

2.8 Environmental and Permitting

In the past, environmental and permitting costs used to be a small percentage of the total pipeline system costs. In recent times, because of stricter environmental and regulatory requirements, this category now includes items such as environmental impact reports, environmental studies pertaining to the flora and fauna, fish and game, endangered species, sensitive areas such as Native American burial sites, and allowance for habitat mitigations. The latter cost includes the acquisition of new acreage to compensate for areas disturbed by the pipeline route. This new acreage will then be allocated for parks, wildlife preserves, etc.

Permitting costs would include pipeline construction permits such as road crossings, railroad crossings, river and stream crossings, and permitting for antipollution devices for pump stations and tank farms.

Environmental and permitting costs may be as high as 10–15% of total project costs.

2.9 Right-of-Way Acquisitions Cost

Right-of-way (ROW) must be acquired for building a pipeline along private lands, farms, public roads, and railroads. In addition to initial acquisition costs, there may be annual lease costs that the pipeline company will have to pay railroads, agencies, and private parties for pipeline easement and maintenance. The annual ROW costs would be considered an expense and would be included in the operating costs of the pipeline. For example, the ROW acquisitions costs for a pipeline project may be $20 million, which would be included in the total capital costs of the pipeline project. In addition, annual lease payments for ROW acquired may be a total of $500,000 a year, which would be included with other operating costs such as pipeline O&M, G&A costs, etc.

Historically, ROW costs have been in the range of 6–8% of total project costs for pipelines.

2.10 Other Project Costs

Other project costs would include AFUDC, legal and regulatory costs, and contingency costs. Contingency costs cover unforeseen circumstances and design changes including pipeline rerouting to bypass sensitive areas, pump stations, and facilities modifications not originally anticipated at the start of the project. AFUDC and contingency costs will range between 15% and 20% of the total project cost.

3. OPERATING COSTS

The annual operating cost of a pipeline consists mainly of the following:
- Pump station energy cost (electric or natural gas)
- Pump station equipment maintenance costs (equipment overhaul, repairs, etc.)
- Pipeline maintenance cost including line rider, aerial patrol, pipe replacements, relocations, etc.
- SCADA and telecommunication costs
- Valve and meter station maintenance
- Tank farm operation and maintenance
- Utility costs: water, natural gas, etc.
- Ongoing environmental and permitting costs
- ROW lease costs
- Rentals and lease costs
- G&A costs, including payroll

In this list, pump station costs include electrical energy and equipment maintenance costs, which can be substantial. Consider two pump stations of 5000 HP each operating 24 h a day, 350 days a year, with a two-week shutdown for maintenance. This can result in annual O&M costs of $6–7 million based on electricity costs of 9–10 cents per kWh. In addition to the power cost, other components of O&M costs include annual maintenance and overhead, which can range from $0.50 million to $1.0 million depending on the equipment involved.

4. FEASIBILITY STUDIES AND ECONOMIC PIPE SIZE

In many instances, we have to investigate the technical and economic feasibility of building a new pipeline system to provide transportation services for liquids from a storage facility to a refinery or from a refinery to a tank farm. Other types of studies may include technical and economic feasibility studies for expanding the capacity of an existing pipeline system to handle additional throughput volumes because of increased market demand or refinery expansion.

Grassroots pipeline projects in which a brand new pipeline system needs to be designed from scratch involve analysis of the best pipeline route, optimum pipe size, and pumping equipment required to transport a given volume of liquid. In this section, we will learn how an economic pipe size is determined for a pipeline system based on an analysis of capital and operating costs.

Consider a project in which a 100-mile pipeline is to be built to transport 8000 bbl/h of refined products from a refinery to a storage facility. The question before us is: What pipe diameter and pump stations are the most optimum for handling this volume?

Let us assume that we selected a 16-in diameter pipe to handle the designated volume and we calculated that this system needs two pump stations of 2500 HP each. The total cost of this system of pipe and pump stations can be calculated and we will call this the 16-in cost option. If we chose a 20-in diameter pipeline, the design would require one 2000-HP pump station. In the first case, more HP and less pipe will be required. The 16-in pipeline system would require approximately 20% less pipe than the 20-in option. However, the 16-in option requires 2.5 times the HP required in the 20-in case. Therefore, the annual pump station operating costs for the 16-in system would be higher than the 20-in case because the electric utility cost for the 5000-HP pump stations will be higher than that for the 2000-HP station required for the 20-in system. Therefore, to determine the optimum pipe size required, we must analyze the capital costs and the annual operating costs to determine the scenario that gives us the least total cost, taking into account a reasonable project life. We would perform these calculations considering time value of money, and select the option that results in the lowest PV of investment.

Generally, in any situation, we must evaluate at least three or four different pipe diameters and calculate the total capital costs and operating costs for each pipe size. As indicated in the previous paragraph, the optimum pipe size and pump station configuration will be the alternative that minimizes total investment, after taking into account the time value of money over the life of the project. We will illustrate this using an example.

Problem 13.2

A city is proposing to build a 24-mile-long water pipeline to transport 14.4 million gal/day flow rate. There is static elevation head of 250 ft from the originating pump station to the delivery terminus. A minimum delivery pressure of 50 psi is required at the pipeline terminus.

The pipeline operating pressure must be limited to 1000 psi using steel pipe with a yield strength of 52,000 psi. Determine the optimum pipe diameter and the HP required for pumping this volume on a continuous basis, assuming 350 days of operation, 24 h a day. Electric costs for driving the pumps will be based on 8 cents/kWh. The interest rate on borrowed money is 8% per year. Use the Hazen–Williams equation for pressure drop with a C factor of 100.

Assume $700/ton for pipe material cost and $20,000/inch-diameter-mile for pipeline construction cost. For pump stations, assume a total installed cost of $1500 per HP. To account for items other than pipe and pump stations in the total cost, use a 25% factor.

Solution

First we have to bracket the pipe diameter range. If we consider 20-in diameter pipe, 0.250-in wall thickness, the average water velocity will be:

$$V = 0.4085 \times 10000/19.52 = 10.7 \text{ ft/s}$$

where 10,000 gal/min is the flow rate based on 14.4 million gal/day.

This is not a very high velocity; therefore, a 20-in pipe can be considered as one of the options. We will compare this with two other pipe sizes: 22 and 24 in nominal diameter.

Initially, we will assume 0.500-in pipe wall thickness for 22 in and 24 in. Later we will calculate the actual required wall thickness for the given maximum allowable operating pressure (MAOP). Using ratios, the velocity in the 22-in pipe will be approximately:

$$10.7 \times (19.5/21)^2 \quad \text{or} \quad 9.2 \text{ ft/s}$$

and the velocity in the 24-in pipe will be

$$10.7 \times (19.5/23)^2 \quad \text{or} \quad 7.7 \text{ ft/s}$$

Thus the selected pipe sizes 20, 22, and 24 in will result in a water velocity between 7.7 ft/s and 10.7 ft/s, which is an acceptable range of velocities in a pipe.

Next we need to choose a suitable wall thickness for each pipe size to limit the operating pressure to 1000 psi.

Using the internal design pressure Eqn (4.3), we calculate the pipe wall thickness required as follows.

For a 20-in pipe, the wall thickness is

$$T = 1000 \times 20/(2 \times 52000 \times 0.72) = 0.267 \text{ in}$$

Similarly, for the other two pipe sizes, we calculate:
For a 22-in pipe, the wall thickness is

$$T = 1000 \times 22/(2 \times 52000 \times 0.72) = 0.294 \text{ in}$$

And for a 24-in pipe, the wall thickness is

$$T = 1000 \times 24/(2 \times 52000 \times 0.72) = 0.321 \text{ in}$$

Using the closest commercially available pipe wall thicknesses, we choose the following three sizes:

20 inch, 0.281 inch wall thickness (MAOP = 1052 psi)

22 inch, 0.312 inch wall thickness (MAOP = 1061 psi)

24 inch, 0.344 inch wall thickness (MAOP = 1072 psi)

The revised MAOP values for each pipe size, with the slightly higher than required minimum wall thickness, were calculated as shown within parentheses. With the revised pipe wall thickness, the velocity calculated earlier will be corrected to:

$$V_{20} = 10.81 \text{ ft/s}$$

$$V_{22} = 8.94 \text{ ft/s}$$

and

$$V_{24} = 7.52 \text{ ft/s}$$

Next, we will calculate the pressure drop due to friction in each pipe size at the given flow rate of 10,000 gal/min, using the Hazen–Williams equation with a C factor of 100.

For the 20-in pipeline:

$$10000 \times 60 \times 24/42 = 0.1482 \times 100 \ (20 - 2 \times 0.281)^{2.63} (Pm/1.0)^{0.54}$$

rearranging and solving for the pressure drop, Pm, we get:

$$Pm = 63.94 \text{ psi/mi} \quad \text{for the 20 inch pipe}$$

Similarly, we get the following for the pressure drop in the 22 and 24-in pipelines:

$$Pm = 40.25 \text{ psi/mi} \quad \text{for the 22 inch pipe}$$

and

$$Pm = 26.39 \text{ psi/mi} \quad \text{for the 24 inch pipe}$$

We can now calculate the total pressure required for each pipe size, taking into account the friction drop in the 24-mile pipeline and the elevation head of 250 ft along with a minimum delivery pressure of 50 psi at the pipeline terminus.

Total pressure required at the origin pump station is

$$(63.94 \times 24) + 250 \times 1.0/2.31 + 50 = 1692.79 \text{ psi for the 20 inch pipe}$$

and

$$(40.25 \times 24) + 250 \times 1.0/2.31 + 50 = 1124.23 \text{ psi for the 22 inch pipe}$$

and

$$(26.39 \times 24) + 250 \times 1.0/2.31 + 50 = 791.59 \text{ psi for the 24 inch pipe}$$

Because the MAOP of the pipeline is limited to 1000 psi, it is clear that we would need two pump stations for the 20 and 22-in pipeline cases, whereas one pump station will suffice for the 24-in pipeline case.

The total brake HP (BHP) required for each case will be calculated from the previous total pressure and the flow rate of 10,000 gal/min assuming a pump efficiency of 80%. We will also assume that the pumps require a minimum suction pressure of 50 psi.

$$\text{BHP} = 10000 \times (1693 - 50)/(0.8 \times 1714) = 11,983 \text{ for 20 inch}$$

$$\text{BHP} = 10000 \times (1124 - 50)/(0.8 \times 1714) = 7,833 \text{ for 22 inch}$$

$$\text{BHP} = 10000 \times (792 - 50)/(0.8 \times 1714) = 5,412 \text{ for 24 inch}$$

Increasing the previous BHP values by 10% for installed HP and choosing the nearest motor size, we will use 14,000 HP for the 20-in pipeline system, 9000 HP for the 22-in system, and 6000 HP for the 24-in pipeline system.

If we had factored in a 95% efficiency for the electric motor and picked the next nearest size motor, we would have arrived at the same HP motors.

To calculate the capital cost of facilities, we will use $700 per ton for steel pipe, delivered to the construction site. The labor cost for installing the pipe will be based on $20,000 per inch-diameter-mile.

The installed cost for pump stations will be assumed at $1500/HP.

To account for other cost items discussed earlier in this chapter, we will add 25% to the subtotal of pipeline and pump station cost.

The summary of the estimated capital cost for the three pipe sizes are listed in the following table:

Based on total capital costs alone, it can be seen that the 24-in system is the best. However, we will have to look at the operating costs as well before making a decision on the optimum pipe size.

Next, we calculate the operating cost for each scenario using electrical energy costs for pumping. As discussed in an earlier section of this chapter, many other items enter into the calculation of annual operating costs, such as O&M, G&A, etc. For simplicity, we will increase the electrical cost of the pump stations by a factor to account for all other operating costs.

Using the BHP calculated at each pump station for the three cases and 8 cents/kWh for electricity cost, we find that the annual operating cost for 24 h operation per day, 350 days per year, the annual costs are:

$$11983 \times 0.746 \times 24 \times 350 \times 0.08 = \$6.0 \text{ million/yr for 20 inch}$$

$$7833 \times 0.746 \times 24 \times 350 \times 0.08 = \$9.93 \text{ million/yr for 22 inch}$$

$$5412 \times 0.746 \times 24 \times 350 \times 0.08 = \$2.71 \text{ million/yr for 24 inch}$$

Strictly speaking, these costs will have to be increased to account for the demand charge for starting and stopping electric motors. The utility company may charge based on the kW rating of the motor. It will range from $4 to $6 per kW/month. Using an average demand charge of $5/kW/mo, we get the following demand charges for the pump station in a 12-month period.

$$14000 \times 5 \times 12 = \$840{,}000$$

$$9000 \times 5 \times 12 = \$540{,}000$$

$$6000 \times 5 \times 12 = \$360{,}000$$

Adding the demand charges to the previously calculated electric power cost, we get the total annual electricity costs as

$$\$6.84 \text{ million/yr for 20 in}$$

$$\$4.47 \text{ million/yr for 22 in}$$

$$\$3.07 \text{ million/yr for 24 in}$$

Increasing these numbers by a 50% factor to account for other operating costs, such as O&M, G&A, etc., we get the following for total annual costs for each scenario:

$$\$10.26 \text{ million/yr for 20 inch}$$

$$\$6.7 \text{ million/yr for 22 inch}$$

$$\$4.6 \text{ million/yr for 24 inch}$$

Next, we will use a project life of 20 years and interest rate of 8% to perform a discounted cash flow analysis, to obtain the PV of these annual operating costs. Then the total capital cost calculated earlier listed in Table 13.1 will be added to

Table 13.1 Capital cost of three different pipe diameters

Capital cost	20 in	22 in	24 in
Million $	—	—	—
Pipeline	2.62	3.21	3.85
Pump stations	21.00	13.50	9.00
Other (25%)	5.91	4.18	3.21
Total	29.53	20.88	16.07

the PVs of the annual operating costs. The PV will then be obtained for each of the three scenarios.

PV of 20 inch system = \$29.53 + present value of \$10.26 million/yr at 8% for 20 years

or

$$PV_{20} = 29.53 + 100.74 = \$130.27 \text{ million}$$

Similarly, for the 22 and 24-in systems, we get

$$PV_{22} = 20.88 + 65.78 = \$86.66 \text{ million}$$

$$PV_{24} = 16.07 + 45.16 = \$61.23 \text{ million}$$

Thus, based on the net PV of investment, we can conclude that the 24-in pipeline system with one 6000-HP pump station is the preferred choice.

In the preceding calculations, for the sake of simplicity, we made several assumptions. We considered major cost components, such as pipeline and pump station costs, and added a percentage of the subtotal to account for other costs. Also in calculating the PV of the annual costs, we used constant numbers for each year. A more rigorous approach would require the annual costs be inflated by some percentage every year to account for inflation and cost of living adjustments. The Consumer Price Index could be used in this regard. As far as capital costs go, we can get more accurate results if we perform a more detailed analysis of the cost of valves, meters, and tanks instead of using a flat percentage of the pipeline and pump station costs. The objective in this chapter was to introduce the reader to the importance of economic analysis and to outline a simple approach to selecting the economical pipe size.

In addition, the earlier section on cost of services and tariff calculations provided an insight into how transportation companies finance a project and collect revenues for their services.

5. GAS PIPELINE

5.1 Components of Cost

In a gas pipeline system, the major components that contribute to the initial capital cost are the pipeline, compressor stations, mainline valve stations and metering facilities, telecommunications, and SCADA. Other costs include environmental and permitting costs, ROW, acquisition cost, engineering and construction management, legal and regulatory costs, contingency, and AFUDC.

The recurring annual costs will include O&M costs, fuel, energy and utility costs, rental, permitting, and annual ROW costs. The O&M costs will include payroll and G&A costs.

In any pipeline system constructed to provide transportation of gas, there will be capital costs and annual operating costs. If we decide on a useful life of the pipeline (say 20 or 30 years), we can annualize all costs and also determine the revenue stream necessary to amortize the total investment in the pipeline project. The revenue earned after expenses and taxes plus a percentage for profit divided by the volume transported will give the transportation tariff necessary. The calculation of capital cost, operating cost, and transportation tariff will be illustrated using an example.

Throughout this chapter, we will need to convert annual cash flows or expenses into PV and vice versa. A useful equation relating the PV of a series of annual payments over a number of years at a specified interest rate is as follows.

$$PV = \frac{R}{i}\left(1 - \frac{1}{(1+i)^n}\right) \qquad (13.3)$$

where

PV: present value, $
R: series of cash flows, $
i: interest rate, decimal value
n: number of periods, years

For example, $10,000 annual payments for 20 years at an annual interest rate of 10% results in a PV of

$$PV = \frac{10,000}{0.10}\left(1 - \frac{1}{(1+0.10)^{20}}\right) = \$85,136$$

Similarly, we can convert PV of $10 million into an annualized cost based on 8% interest for 30 years as follows. From Eqn (13.3):

$$10,000,000 = \frac{R}{0.08}\left(1 - \frac{1}{(1.08)^{30}}\right)$$

Solving for the annual cost R, we get

$$R = \$888,274$$

Next we will calculate the cost of service and transportation tariff using a simple example.

Problem 13.3
A natural gas pipeline transports 100 million standard cubic feet per day (MMSCFD) at a load factor of 95%. The capital cost is estimated at $60 million and the annual operating cost is $5 million. Amortizing the capital at 10% for a project life of 25 years, calculate the cost of service and transportation tariff for this pipeline.

Solution
All costs will be converted to annualized values for a 25$132#-year project life and 10% interest rate. This will be the cost of service on an annual basis. When this cost is divided by the annual pipeline throughput, we obtain the transportation tariff.

The capital cost of $60 million is first converted to annual cash flow at 10% interest rate for a period of 25 years. Using Eqn (10.1):

$$\text{Annualized capital cost} = \frac{60 \times 0.10}{\left[1 - \frac{1}{(1.10)^{25}}\right]} = \$6.61 \text{ million}$$

This assumes zero salvage value at the end of the 25-year useful life of the pipeline.

Therefore, for a project life of 25 years and a discount rate of 10%, the capital cost of $60 is equivalent to annual cost of $6.61 million. Adding the annual operating cost of $5 million.the total annual cost is $6.61 + \$5 = \$11.61$ million per year.

This annual cost is defined as the cost of service incurred each year. Actually, to be accurate, we should take into account several other factors such as the tax rate, depreciation of assets, and profit margin to arrive at a true cost of service.

The transportation tariff is defined as the cost of service divided by the annual volume transported. At 95% load factor and flow rate of 100 MMSCFD the transportation tariff is

$$\text{Tariff} = \frac{\$11.61 \times 10^6 \times 10^3}{100 \times 10^6 \times 365 \times 0.95} = \$0.3348 \text{ per MCF.}$$

In other words, for this pipeline, every MCF of gas transported requires a payment of approximately 33.5 cents to the pipeline owner that provides the transportation. This is a very rough and simplistic calculation of an example of tariff. In reality, we must take into account many other factors to arrive at an accurate cost of service. For example, the annual operating cost will vary from year to year over the life of the pipeline from inflation and other reasons. Taxes, depreciation of assets, and salvage value at the end of the life of the pipeline must also be considered. Nevertheless, the preceding analysis gives a quick overview of the approach used to calculate a rough value of the transportation cost.

6. CAPITAL COSTS

The capital cost of a pipeline project consists of the following major components.

- Pipeline
- Compressor stations
- Mainline valves stations
- Meter stations
- Pressure regulator stations
- SCADA and telecommunications
- Environmental and permitting
- Right of way acquisition
- Engineering and construction management.

In addition, there are other costs such as AFUDC and contingency. Each of the preceding major categories of capital cost will be discussed next.

6.1 Pipeline

The pipeline cost consists of those costs associated with the pipe material, coating, pipe fittings, and the actual installation or labor cost. In chapter 6, we introduced a simple formula to calculate the weight of pipe per unit length. From this and the pipe length, the total tonnage of pipe can be calculated. Given the cost per ton of pipe material, the total pipe material cost can be calculated. Knowing the construction cost per unit length of pipe we can also calculate the labor cost for installing the pipeline. The sum of these two costs is the pipeline capital cost.

For pipe weight, the cost of pipe required for a given pipeline length is found from:

$$PMC = \frac{10.68(D-T)TLC \times 5280}{2000} \quad (13.4)$$

where
PMC: pipe material cost, $
L: length of pipe, mi
D: pipe outside diameter, in
T: pipe wall thickness, in
C: pipe material cost, $/ton

In SI units:

$$PMC = 0.02463(D-T)TLC \quad (13.5)$$

where
PMC: pipe material cost, $
L: length of pipe, km
D: pipe outside diameter, mm
T: pipe wall thickness, mm
C: pipe material cost, $/metric ton

Generally, pipe will be supplied externally coated and wrapped. Therefore, we must add this cost or a percentage to the bare pipe cost to account for these extra costs and the delivery cost to the construction site. In the absence of actual cost, we may increase the bare pipe cost by a small percentage such as 5%. For example, using Eqn (10.2) for a 100-mile pipeline NPS 20 with 0.500-in wall thickness, the total pipe cost based on $800 per ton is

$$PMC = \frac{10.68(20-0.5)0.5 \times 100 \times 800 \times 5280}{2000} = \$21.99 \text{ million}$$

If the pipe is externally coated and wrapped and delivered to the field at an extra cost of $5 per foot, this cost can be added to the bare pipe cost as follows:

Pipe coating and wrapping cost = $5 \times 5280 \times 100 = \$2.64$ million

Therefore, the total pipe cost becomes

$$\$21.99 + \$2.64 = \$24.63 \text{ million}$$

The labor cost to install the pipeline may be represented in dollars per unit length of pipe. For example, the labor cost may be $60 per ft or

$316,800 per mile of pipe for a particular size pipe in a certain construction environment. This number will depend upon whether the pipeline is installed in open country, fields, or city streets. Such numbers are generally obtained from contractors who will take into consideration the difficulty of trenching, installing pipe, and backfilling in the area of construction. For estimation purposes, there is a wealth of historical data available for construction cost for various pipe sizes. Sometimes, the pipe installation cost is expressed in terms of dollars per inch diameter per mile of pipe. For example, an NPS 16 pipe may have an installation cost of $15,000 per inch-diameter-mile. Thus, if 20 miles of NPS 16 pipe is to be installed we estimate the labor cost as follows:

$$\text{Pipe installation cost} = \$15{,}000 \times 16 \times 20 = \$4.8 \text{ million}$$

If we convert this cost on a unit length basis, we get

$$\text{Pipe installation cost} = \frac{4.8 \times 10^6}{20 \times 5280} = \$45.45 \text{ per ft}$$

Table 13.2 shows typical installation costs for pipelines. It must be remembered that these numbers must be verified by discussions with construction contractors who are familiar with the construction location.

Several other construction costs must be added to the installation costs for straight pipe. These costs include road, highway and railroad crossings, and stream and river crossings. These costs may be provided as lump sum numbers that can be added to the pipeline installation costs to come up with a total pipeline construction costs. For example, a pipeline may include two road and highway crossings that total $300,000 in addition to a couple of river crossings costing $1 million. Compared with the installation cost of a

Table 13.2 Typical pipeline installation costs

Pipe diameter, in	Average cost $/in-dia/mi
8	18,000
10	20,000
12	22,000
16	14,900
20	20,100
24	33,950
30	34,600
36	40,750

long-distance pipeline, the road and river crossings total may be a small percentage.

6.2 Compressor Station Costs

To provide transportation of gas through a pipeline we have to install one or more compressor stations to provide the necessary gas pressure. Once we decide on the details of the compressor station equipment and piping, a detail bill of materials may be developed from the engineering drawings. Based upon quotation from equipment vendors, a detailed cost estimate of the compressor stations may be developed. In the absence of vendor data and in situations where a rough order of magnitude costs for compressor stations are desired, we may use an all-inclusive price of dollars per installed HP. For example, using an installed cost of $2000 per HP, for a 5000-HP compressor station the capital cost will be estimated as follows:

$$\text{Compressor station cost} = 2000 \times 5000 = \$10 \text{ million}$$

In these calculations, the all-inclusive number of $2000 per installed HP is expected to include material and equipment cost and the labor cost for installing the compressor equipment, piping, valves, instrumentation, and controls within the compressor stations. Generally, the $/HP number decreases as the size of the compressor HP increases. Thus a 5000-HP compressor station may be estimated on the basis of $2000 per HP whereas a 20,000-HP compressor stations will be estimated at an installed cost of $1500 per HP. These numbers are mentioned for illustration purposes only. Actual $/HP values must be obtained from historical pipeline cost data and in consultation with compressor station construction contractors and compressor station equipment vendors. Generally, the pipeline and compressor station costs constitute the bulk of the total pipeline project cost.

6.3 Mainline Valve Stations

Mainline block valves are installed to isolate sections of a pipeline for safety reasons and maintenance and repair. In the event of a pipeline rupture, the damaged pipeline section may be isolated by closing off the mainline valves on either side of the rupture location. For mainline valve stations installed at specified intervals along the pipeline, the cost of facilities may be specified as a lump sum figure that includes the mainline valve and operator, blowdown valves, and piping and other pipe and fittings that comprise the entire block valve installation. Generally, a lump sum figure may be obtained for a typical mainline block valve installation from a construction contractor. For example, an NPS 16 mainline valve installation may be estimated at $100,000 per site. In a 100-mile, NPS 16 pipeline, Department of Transportation code requirements may dictate that a mainline valve be installed every 20 miles. Therefore, in this case there would be six mainline valves for a 100-mile pipeline. At $100,000 per site, the total installed cost of all mainline valve stations will be $600,000. This will be added to the capital cost of the pipeline facilities.

6.4 Meter Stations and Regulators

Meter stations are installed for measuring the gas flow rate through the pipeline. These meter stations will consist of meters, valves, fittings, instrumentation, and controls. Meter stations may also be estimated as a fixed price including material and labor for a particular site. For example, a 10-in meter station may cost $300,000 lump sum. If there are four such meter stations on a 100-mile gas pipeline, the total meter station cost will be $1.2 million. The meter station costs, like the mainline valve station costs will be added to the pipeline cost.

Pressure regulating stations are installed at some locations on a gas pipeline to reduce the pressure for delivery to a customer or to protect a section of a pipeline with a lower maximum operating pressure. Such pressure regulating stations may also be estimated as a lump sum per site and added to the capital cost of the pipeline.

6.5 SCADA and Telecommunication Systems

Typically, on a gas pipeline, the pressures, flow rates, and temperatures are monitored along the pipeline by means of electronic signals sent from remote terminal units on various valves and meters to a central control center via telephone lines, microwave, or satellite communication system. The term SCADA is used to refer to these facilities. SCADA is used to remotely monitor, operate, and control a gas pipeline system from a central control center. In addition to monitoring valve status, flows, temperatures, and pressures along a pipeline, the compressor stations are also monitored. In many cases, starting and stopping of compressor units are performed remotely using SCADA. The cost of SCADA facilities range from $2 million to $5 million or more, depending upon the pipeline length, number of compressor stations, number of mainline valves, and meter stations.

Sometimes this category is estimate as a percentage of the total project cost such as 2 to 5%.

6.6 Environmental and Permitting

The environmental and permitting costs are those costs that are associated with the modifications to pipeline, compressor stations, and valve and meter stations to ensure that these facilities do not pollute the atmosphere, streams, and rivers, damage ecosystems including the flora and fauna, fish and game, and endangered species. Many sensitive areas such as Native American religious and burial sites must be considered and allowances made for mitigation of habitat in certain areas. Permitting costs may include those costs associated with changes needed to compression equipment, pipeline alignment such that toxic emissions from pipeline facilities do not endanger the environment, humans, and plant and animal life. In many cases, these costs include acquisition of land to compensate for the areas that were disturbed because of pipeline construction. Such lands acquired will be allocated for public use such as parks and wildlife preserves. Permitting costs will also include an environmental study, the preparation of an environmental impact report and permits for road crossings, railroad crossings, and stream and river crossings. These environmental and permitting costs on a gas pipeline project may range between 10% and 15% of total project costs.

6.7 Right-of-Way Acquisitions

ROW for a pipeline is acquired from private parties, state and local governments, and federal agencies for a fee. This fee may be a lump sum payment at the time of acquisition and additional annual fees to be paid for a certain duration. For example, the ROW may be acquired from private farms, cooperatives, Bureau of Land Management, and railroads. The initial cost for acquiring the ROW will be included in the capital cost of the pipeline. The annual rent or lease payment for land will be considered an expense. The latter will be included in the annual cost such as operating costs. These also include payroll and energy costs. As an example, the ROW acquisition costs for a gas pipeline may be $30 million. This cost would be added to the total capital costs of the gas pipeline. Also, there may be annual ROW lease payments of $300,000 a year, which would be added to other annual costs such as O&M cost and administrative costs. For most gas pipelines, the initial ROW costs will be in the range of 6% to 10% of total project costs.

6.8 Engineering and Construction Management

Engineering costs are those costs that pertain to the design and preparation of drawings for the pipeline, compressor stations, and other facilities. This will include both preliminary and detailed engineering design costs including development of specifications, manuals, purchase documents, equipment inspection, and other costs associated with materials and equipment acquisition for the project. The construction management costs include field personnel cost, rental facilities, office equipment, transportation, and other costs associated with overseeing and managing the construction effort for the pipeline and facilities. On a typical pipeline project, engineering and construction management costs ranges from 15% to 20% of total pipeline project costs.

6.9 Other Project Costs

In addition to the major cost categories discussed in the preceding sections, there are other costs that should be included in the total pipeline project cost. These include legal and regulatory costs necessary for filing an application with the Federal Energy Regulatory Commission (FERC) and state agencies that have jurisdiction over interstate and intrastate transportation of natural gas as well as a contingency cost intended to cover categories not considered or not envisioned when the project was conceptualized. As the project is engineered, new issues and problems may surface that require additional funds. These are generally included in the category of contingency cost. The final category of cost referred to as AFUDC, which is intended to cover the costs associated with financing the project during various stages of construction. Contingency and AFUDC costs may range between 15% and 20% of the total project cost (Table 13.3).

7. OPERATING COSTS

Once the pipeline, compressor stations, and ancillary facilities are constructed and the pipeline put into operation, there will be annual operating costs over the useful life of the pipeline, which may be 20–30 years or more. These annual costs consist of the following major categories:
- Compressor station fuel or electrical energy cost
- Compressor station equipment maintenance and repair costs
- Pipeline maintenance cost such as pipe repair, relocation, aerial patrol and monitoring
- SCADA and telecommunications

Table 13.3 A cost breakdown for a typical natural gas pipeline project

Description		Million $
Pipeline		160.00
Compressor stations		20.00
Mainline valve stations		1.20
Meter stations		1.20
Pressure regulator stations		0.10
SCADA and telecommunications	2–5%	5.48
Environmental and permitting	10–15%	21.90
Right-of-way acquisition	6–10%	14.60
Engineering and construction management	15–20%	36.50
Contingency	10%	26.10
Subtotal		287.08
Working capital		5.00
AFUDC	5%	14.35
Total		306.43

- Valve, regulator, and meter station maintenance
- Utility costs such as water, natural gas, etc.
- Annual or periodic environmental and permitting costs
- Lease, rental and other recurring ROW costs
- Administrative and payroll costs.

Compressor station costs include periodic equipment maintenance and overhaul costs. For example, a gas turbine–driven compressor unit may have to be overhauled every 18–24 months (Table 13.4).

Table 13.4 The annual operating cost of a typical gas pipeline

Description	$ per year
Salaries	860,000
Payroll overhead (20%)	172,000
Administrative and general (50%)	516,000
Vehicle expense	72,800
Office expenses (6%)	92,880
Misc materials and tools	100,000
Compressor station maintenance	—
Consumable materials	50,000
Periodic maintenance	150,000
ROW payments	350,000
Utilities	150,000
Gas control	100,000
SCADA contract install and maintenance	200,000
Internal corrosion inspection ($750,000/3 years)	250,000
Cathodic protection survey	100,000
Total O&M	3,163,680

Problem 13.4

A new pipeline is being constructed to transport natural gas from a gas processing plant to a power plant 100 miles away. Two project phases are envisioned. During the first phase, lasting 10 years, the amount of gas shipped is expected to be a constant volume of 120 MMSCFD at 95% load factor. A pipe size of NPS 16, 0.250-in wall thickness is required to handle the volumes with two compressor stations with a total of 5000 HP. The total pipeline cost may be estimated at $800,000 per mile and compressor station cost at $2000 per HP installed. The annual operating costs are estimated at $8 million. The pipeline construction project will be financed by borrowing 80% of the required capital at an interest rate of 6%. The regulatory ROR allowed on equity capital is 14%. Consider a project life of 20 years and overall tax rate of 40%.

1. Calculate the annual cost of service for this pipeline and the transportation tariff in $/MCF.
2. The second phase, lasting the next 10 years, is projected to increase throughput to 150 MMSCFD. Calculate the transportation tariff for the second phase considering the capital cost to increase by $20 million and the annual cost increases to $10 million, with the same load factor as phase 1.

Solution

1. First calculate the total capital cost of facilities of phase 1.

$$\text{Pipeline cost} = \$800{,}000 \times 100 = \$80 \text{ million}$$

$$\text{Compressor station cost} = \$2000 \times 5000 = \$10 \text{ million}$$

$$\text{Total capital cost} = \$80 + \$10 = \$90 \text{ million.}$$

A total of 80% of this capital of $90 million will be borrowed at 6% interest for 20 years.

From Eqn (13.3), the annual cost to amortize the loan is

$$\text{Loan amortization cost} = \frac{90 \times 0.8 \times 0.06}{1 - \left(\frac{1}{1.06}\right)^{20}} = \$6.28 \text{ million}$$

Therefore, we need to build into the cost of service the annual payment of $6.28 million to retire the debt of $72 million (80% of $90 million) over the project life of 20 years. On the remaining capital (equity) of ($90 − $72) or $18 million, 14% ROR per year is allowed.

Therefore, 14% of $18 million can be included in the cost of service to account for the equity capital.

Annual revenue on equity capital = 0.14 × $18 million = $2.52 million.

Because the tax rate is 40%, the adjusted annual revenue on equity capital is

$$\frac{\$2.52 \text{ million}}{1 - 0.4} = \$4.2 \text{ million}$$

Next add the operating cost of $8 million per year to the annual costs for debt and equity just calculated to arrive at the annual cost of service as follows:

$$\text{Annual payment to retire debt} = \$6.28 \text{ million}$$

$$\text{Annual revenue on equity capital} = \$4.2 \text{ million}$$

$$\text{Annual operating cost} = \$8 \text{ million}$$

Therefore,

$$\text{Annual cost of service} = 6.28 + 4.2 + 8 = \$18.48 \text{ million}.$$

The transportation tariff at 120 MMSCFD and 95% load factor is

$$\text{Tariff} = \frac{18.48 \times 10^6 \times 10^3}{120 \times 10^6 \times 365 \times 0.95} = \$0.4441 \text{ per MCF}.$$

2. In the second phase that lasts 10 years, the capital cost increases by $20 million. The extra $20 million will be assumed to be financed by 80% debt and 20% equity as before. The annual cost to amortize the debt is

$$\text{Loan amortization cost} = \frac{20 \times 0.8 \times 0.06}{1 - \left(\frac{1}{1.06}\right)^{10}} = \$2.17 \text{ million}$$

The remaining capital of ($20 − $16) or $4 million is equity that according to regulatory guidelines can earn at 14% interest. It must be noted that the interest rate and ROR used in this example are approximate and only for the purpose of illustration. The actual ROR allowed on a particular pipeline will depend on various factors such as the state of the economy, current FERC regulations, or state laws and may range from as low as 8% to as high as 16% or more. Similarly the interest rate of 6% used for debt amortization is also an illustrative number. The actual interest rate on debt will depend on various factors such as the state of the economy, money supply and the federal interest rate charged by Federal Reserve (prime rate). This rate will vary with the country where the pipeline is built and the multinational bank that may finance the pipeline project.

For phase 2, the annual revenue on equity capital is

$$4 \times 0.14 = \$0.56 \text{ million}$$

Accounting for 40% tax rate, the adjusted annual revenue on equity capital is

$$\frac{\$0.56}{1 - 0.4} = \$0.93 \text{ million}$$

Therefore for phase 2 the increase in capital of \$20 million and operating cost of \$2 million will result in an increase in cost of service of

Annual cost of service = $\$2.17 + 0.93 + 2 = \$5.1$ million.

In summary, for phase 2, the total cost of services is

$$\$18.48 + \$5.1 = \$23.58 \text{ million}$$

At a flow rate of 150 MMSCFD and 95% load factor the tariff for phase 2 is

$$\frac{23.58 \times 10^6 \times 10^3}{150 \times 10^6 \times 365 \times 0.95} = \$0.4534 \text{ per MCF}$$

8. DETERMINING ECONOMIC PIPE SIZE

For a particular pipeline transportation application, there is an economic or optimum pipe diameter that will result in the least cost of facilities. For example, a pipeline that requires 100 MMSCFD gas to be transported from a source location to a destination location may be constructed of a wide range of pipe materials and diameters. We may choose to use NPS 14, NPS 16, or NPS 18 pipe or any other pipe size for this application. Using the smallest diameter pipe will cause the greatest pressure drop and the highest HP requirement for a given volume flow rate. The largest pipe size will result in the lowest pressure drop and hence require the least HP. Therefore, the NPS 14 system will be the least in pipe material cost and highest in HP required. On the other hand, the NPS 18 system will require the least HP but considerably more pipe material cost because of the difference in pipe weight per unit length. Determining the optimum pipe size for an application will be illustrated in the next example.

Problem 13.5

A gas pipeline is to be constructed to transport 150 MMSCFD of natural gas from Dixie to Florence, 120 miles away. Consider three pipe sizes—NPS 14, NPS 16, and NPS 18—all having 0.250-in wall thickness. Determine the most economical pipe diameter taking into account the pipe material cost, cost of compressor stations, and fuel costs. The selection of pipe size may be based on a 20-year project life and a PV of discounted cash flow at 8% per year. Use $800 per ton for pipe material and $2000 per installed HP for compressor station cost. Fuel gas may be estimated at $3 per MCF.

The following information from hydraulic analysis is available:

NPS 14 pipeline: Two compressor stations, 8196 HP total. Fuel consumption is 1.64 MMSCFD

NPS 16 pipeline: One compressor station, 3875 HP. Fuel consumption is 0.78 MMSCFD

NPS 18 pipeline: One compressor station, 2060 HP. Fuel consumption is 0.41 MMSCFD

Solution

First calculate the capital cost of 120 miles of pipe for each case.

From Eqn (10.2), the cost of NPS 14 pipe is

$$PMC = \frac{10.68(14 - 0.250) \times 0.250 \times 120 \times 800 \times 5280}{2000} = \$9.3 \text{ million}$$

Similarly,

The cost of NPS 16 pipe is

$$PMC = \frac{10.68(16 - 0.250) \times 0.250 \times 120 \times 800 \times 5280}{2000}$$

$$= \$10.66 \text{ million}$$

and the cost of NPS 18 pipe is

$$PMC = \frac{10.68(18 - 0.250) \times 0.250 \times 120 \times 800 \times 5280}{2000}$$

$$= \$12.01 \text{ million}$$

Next, calculate the installed cost of compressor stations for each pipe size.

For NPS 14 pipe, compressor station cost is

$$8196 \times 2000 = \$16.39 \text{ million}$$

For NPS 16 pipe, compressor station cost is

$$3875 \times 2000 = \$7.75 \text{ million}$$

For NPS 18 pipe, compressor station cost is

$$2060 \times 2000 = \$4.12 \text{ million}$$

The operating fuel cost for each case will be calculated next considering fuel gas at $3 per MCF and 24 h a day operation for 350 days a year. A shutdown for 15 days per year is allowed for maintenance and any operational upset conditions.

For NPS 14 pipe, the fuel cost is

$$1.64 \times 10^3 \times 350 \times 3 = \$1.72 \text{ million per year}$$

For NPS 16 pipe, the fuel cost is

$$0.78 \times 10^3 \times 350 \times 3 = \$0.82 \text{ million per year}$$

For NPS 18 pipe, the fuel cost is

$$0.41 \times 10^3 \times 350 \times 3 = \$0.43 \text{ million per year}$$

The actual operating cost includes many other items besides the fuel costs. For simplicity, in this example we will only consider the fuel cost. The annual fuel cost for the project life of 20 years will be discounted at 8% in each case. This will then be added to the sum of the pipeline and compressor station capital cost to arrive at a PV.

The PV of a series of cash flows each equal to R for a period of n years at an interest rate of i percent is given by Eqn (10.1)

The PV of NPS 14 fuel cost is, from Eqn (10.1):

$$PV = \frac{1.72}{0.08}\left(1 - \frac{1}{(1+0.08)^{20}}\right) = 1.72 \times 9.8181 = \$16.89 \text{ million}$$

The PV of NPS 16 fuel cost is

$$PV = 0.82 \times 9.8181 = \$8.05 \text{ million}$$

The PV of NPS 18 fuel cost is

$$PV = 0.43 \times 9.8181 = \$4.22 \text{ million}$$

Therefore adding up all costs, the PV for NPS 14 is

$$PV_{14} = 9.3 + 16.39 + 16.89 = \$42.58 \text{ million}$$

Adding up all costs, the PV for NPS 16 is

$$PV_{16} = 10.66 + 7.75 + 8.05 = \$26.46 \text{ million}$$

And adding up all costs, the PV for NPS 18 is

$$PV_{18} = 12.01 + 4.12 + 4.22 = \$20.35 \text{ million}$$

Therefore, we see that the least cost option is NPS 18 pipeline with a PV of $20.35 million.

In the preceding example, if the flow rate had been lower or higher, the result may be different. For each pipe size if we were to calculate the HP required at various flow rates and the corresponding fuel consumption, we could generate a graph showing the variation of total cost with flow rate. Obviously as flow rate is increased the HP required and fuel consumption also increases. Performing these calculations for different pipe sizes will yield a graph similar to that shown in Figure 13.1. In the next example,

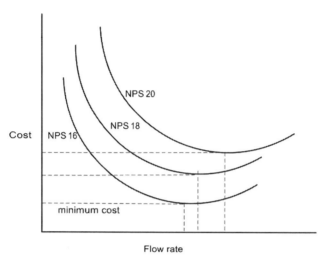

Figure 13.1 Pipeline cost versus flow rates for various pipe sizes.

we will consider three pipe sizes (NPS 16, NPS 18, and NPS 20) and calculate the capital cost and O&M cost for a range of flow rates to develop curves similar to those shown in Figure 13.1.

Problem 13.6

For a natural gas pipeline 120 miles long, three pipe sizes were analyzed for flow rate ranges of 50 MMSCFD to 500 MMSCFD, using the hydraulic simulation software GASMOD (www.systek.us). The following are the pipe sizes and flow rates studied.

NPS 16 pipe: flow rates — 50 MMSCFD to 200 MMSCFD.

NPS 18 pipe: flow rates — 50 MMSCFD to 300 MMSCFD.

NPS 20 pipe: flow rates — 100 MMSCFD to 500 MMSCFD.

The wall thickness was 0.250 in for NPS 16 and NPS 18 and 0.500 in for NPS 20.

From the hydraulic simulation, the number of compressor stations required, HP, and fuel consumption were obtained as shown in Table 13.5.

Table 13.5 Hydraulic simulation results for three pipe sizes

Flow rate MMSCFD	Compressor stations	Total HP	Fuel MMSCFD
NPS 16			
50	1	49	0.01
100	1	1072	0.21
150	1	3875	0.78
175	2	5705	1.14
200	2	9203	1.84
NPS 18			
50	1	49	0.01
100	1	209	0.04
150	1	2060	0.41
175	1	3394	0.68
200	1	4954	1
250	2	9348	1.87
300	2	17902	3.58
NPS 20			
100	1	98	0.02
150	1	1053	0.21
175	1	2057	0.41
200	1	3281	0.66
250	1	6312	1.26
300	2	10,519	2.1
400	2	31,401	6.28
500	2	73,207	14.64

Develop annualized costs for each pipe size and flow rate using the following assumptions.

The capital cost of the pipe material is based on $800 per ton.
For pipe installation cost, use the following:

NPS 16	$50 per foot
NPS 18	$60 per foot
NPS 20	$80 per foot

For compressor station capital cost, use $2000 per installed HP.
Fuel gas may be assumed at $3 per MCF.

The project life is 20 years and the interest rate for discounting cash flow is 8%.

Add 40% to the pipe and compressor capital costs to account for miscellaneous cost such as meter stations, valves, ROW, environmental, engineering and construction management, and contingency. The pipeline is assumed to be operational 350 days a year.

Solution

From the given hydraulic simulation data using the assumptions listed, we develop the total capital cost of pipe, compressor station, and miscellaneous cost.

The pipe material cost is calculated from Eqn (10.2) and using $800 per ton for pipe material cost as follows.

For NPS 16 pipe:

$$\text{Pipe material cost} = \frac{10.68(16 - 0.25)0.25 \times 120 \times \$800 \times 5280}{2000}$$

$$= \$10.66 \text{ million}$$

Similarly, for NPS 18 pipe:

$$\text{Pipe material cost} = \frac{10.68(18 - 0.25)0.25 \times 120 \times \$800 \times 5280}{2000}$$

$$= \$12.01 \text{ million}$$

and NPS 20 pipe material cost is

$$\text{Pipe material cost} = \frac{10.68(20 - 0.5)0.5 \times 120 \times \$800 \times 5280}{2000}$$

$$= \$26.39 \text{ million}$$

These costs are shown in Tables 13.6–13.8.
The labor cost for installing pipe is calculated as follows.
NPS 16 pipe:
Pipe installation cost = $50 × 5280 × 120 = $31.68 million.
Similarly, for NPS 18 pipe:
Pipe installation cost = $60 × 5280 × 120 = $38.02 million.
And for NPS 20 pipe:
Pipe installation cost = $80 × 5280 × 120 = $50.69 million.
Next we calculate the installed cost of compressor stations using $2000 per installed HP.

For the NPS 16 pipe at 100 MMSCFD flow rate, the HP required is 1072 and the installed cost is
$2000 × 1072 = $2.14 million.

Table 13.6 NPS 16 pipe cost summary

NPS 16

Flow rate	Number of compressor stations	Total HP	Fuel MMSCFD	Fuel $/yr	Pipe material, $	Pipe labor, $	Total pipe cost, $	Compressor station cost, $	Miscellaneous cost, $	Total capital, $	O&M cost, $/yr	Annualized capital, $/yr	Total annual cost, $	Annual cost, $/MCF
50	1	49	0.01	0.01	10.66	31.68	42.34	0.098	16.97	59.41	2.00	6.05	8.06	0.4607
100	1	1072	0.21	0.22	10.66	31.68	42.34	2.144	17.79	62.27	2.00	6.34	8.56	0.2447
150	1	3875	0.78	0.82	10.66	31.68	42.34	7.75	20.04	70.12	2.00	7.14	9.96	0.1897
175	2	5705	1.14	1.20	10.66	31.68	42.34	11.41	21.50	75.25	3.00	7.66	11.86	0.1937
200	2	9203	1.84	1.93	10.66	31.68	42.34	18.406	24.30	85.04	3.00	8.66	13.59	0.1942

Notes: Pipe material cost = $800/ton.
Pipe labor cost = $50/ft for NPS 16.
Compressor station cost = $2000 per installed HP.
Miscellaneous cost is 40% of pipe and compressor station cost.
Operating cost based on 350 days per year.
Fuel cost is $3 per MCF.
Capital cost is annualized at 8% interest for 20-year project life.
Table values in millions of dollars.

Table 13.7 NPS 18 pipe cost summary

NPS 18

Flow rate	Number of compressor stations	Total HP	Fuel MMSCFD	Fuel $/yr	Pipe material, $	Pipe labor, $	Total Pipe cost, $	Compressor station cost, $	Miscellaneous cost, $	Total capital, $	O&M cost, $/yr	Annualized capital, $/yr	Total annual cost, $/yr	Annual cost, $/MCF
50	1	49	0.01	0.01	12.01	38.02	50.03	0.098	20.05	70.18	2.00	7.15	9.16	0.5233
100	1	209	0.04	0.04	12.01	38.02	50.03	0.418	20.18	70.62	2.00	7.19	9.24	0.2639
150	1	2060	0.41	0.43	12.01	38.02	50.03	4.12	21.66	75.81	2.00	7.72	10.15	0.1934
175	1	3394	0.68	0.71	12.01	38.02	50.03	6.788	22.73	79.54	2.00	8.10	10.82	0.1766
200	1	4954	1.00	1.05	12.01	38.02	50.03	9.908	23.97	83.91	2.00	8.55	11.60	0.1657
250	2	9348	1.87	1.96	12.01	38.02	50.03	18.696	27.49	96.21	3.00	9.80	14.76	0.1687
300	2	17902	3.58	3.76	12.01	38.02	50.03	35.804	34.33	120.16	3.00	12.24	19.00	0.1809

Notes: Pipe material cost = $800/ton.
Pipe labor cost = $60/ft for NPS 18.
Compressor station cost = $2000 per installed HP.
Miscellaneous cost is 40% of pipe and compressor station cost.
Operating cost based on 350 days per year.
Fuel cost is $3 per MCF.
Capital cost is annualized at 8% interest for 20-year project life.
Table values in millions of dollars.

Table 13.8 NPS 20 pipe cost summary

NPS 20

Flow rate	Number of compressor stations	Total HP	Fuel MMSCFD	Fuel $/yr	Pipe material, $	Pipe labor, $	Total pipe cost, $	Compressor station cost, $	Miscellaneous cost, $	Total capital, $	O&M cost, $/yr	Annualized capital, $/yr	Total annual cost, $	Annual cost, $/MCF
100	1	98	0.02	0.02	26.39	50.69	77.08	0.196	30.91	108.18	2.00	11.02	13.04	0.3726
150	1	1053	0.21	0.22	26.39	50.69	77.08	2.106	31.67	110.86	2.00	11.29	13.51	0.2574
175	1	2057	0.41	0.43	26.39	50.69	77.08	4.114	32.48	113.67	2.00	11.58	14.01	0.2287
200	1	3281	0.66	0.69	26.39	50.69	77.08	6.562	33.46	117.10	2.00	11.93	14.62	0.2089
250	1	6312	1.26	1.32	26.39	50.69	77.08	12.624	35.88	125.58	2.00	12.79	16.11	0.1842
300	2	10519	2.1	2.21	26.39	50.69	77.08	21.038	39.25	137.38	3.00	13.99	19.20	0.1828
400	2	31401	6.28	6.59	26.39	50.69	77.08	62.802	55.95	195.83	3.00	19.95	29.54	0.2110
500	2	73207	14.64	15.37	26.39	50.69	77.08	146.414	89.40	312.89	3.00	31.87	50.24	0.2871

Notes: Pipe material cost = $800/ton.
Pipe labor cost = $80/ft for NPS 20.
Compressor station cost = $2000 per installed HP.
Miscellaneous cost is 40% of pipe and compressor station cost.
Operating cost based on 350 days per year.
Fuel cost is $3 per MCF.
Capital cost is annualized at 8% interest for 20-year project life.
Table values in millions of dollars.

Similarly, the installed cost of each compressor station for all cases are calculated and tabulated as shown in Table 10.5—10.7.

The miscellaneous cost is 40% of the sum of the pipe cost and compressor station cost as follows.

Pipe material cost = $10.66 million.
Pipe installation cost = $31.68 million.
Compressor station cost = $2.14 million.
Thus, for NPS 16 pipe at 100 MMSCFD, the miscellaneous cost is.

$$0.40 \times (10.66 + 31.68 + 2.14) = \$17.79 \text{ million}$$

The O&M cost is added to the annual fuel cost to obtain the total annual cost. The total capital cost is annualized at 8% interest for 20 years and added to the O&M and fuel costs. For example, for NPS 16 pipe at 100 MMSCFD flow rate, the total capital cost of $62.27 million is annualized at $6.34 million and added to the O&M and fuel cost to obtain the total annual cost of $8.56 million. Dividing this annual cost by the gas transported per year, we obtain the annual cost per MCF as follows:

$$\text{Annual cost per MCF} = \frac{8.56 \times 10^6 \times 10^3}{100 \times 10^6 \times 350} = \$0.2447$$

Similarly, for each pipe size and flow rate, the values are tabulated as shown in Table 10.5—10.7.

Upon reviewing Table 13.6 for NPS 16 pipe, we see that the annual cost per MCF decreases from $0.4607 to $0.1897 as the flow rate increases from 50 MMSCFD to 150 MMSCFD. After that, it increases with flow rate and reaches a value of $0.1942 at 200 MMSCFD. Therefore, for NPS 16 pipe, 150 MMSCFD is the optimum flow rate that results in the least transportation cost.

Similarly, from Table 13.7 for NPS 18 pipe, the annual cost per MCF decreases from $0.5233 to $0.1657 as the flow rate increases from 50 MMSCFD to 200 MMSCFD. After that it increases with flow rate and reaches a value of $0.1809 at 300 MMSCFD. Therefore, for NPS 18 pipe, 200 MMSCFD is the optimum flow rate that results in the least transportation cost.

Finally, from Table 13.8 for NPS 20 pipe, the annual cost per MCF decreases from $0.3726 to $0.1828 as the flow rate increases from 100 MMSCFD to 300 MMSCFD. After that it increases with flow rate and reaches a value of $0.2871 at 500 MMSCFD. Therefore, for NPS 20 pipe, 300 MMSCFD is the optimum flow rate that results in the least transportation cost.

A plot of the annual cost per MCF versus flow rate for the three pipe sizes is shown in Figure 13.2.

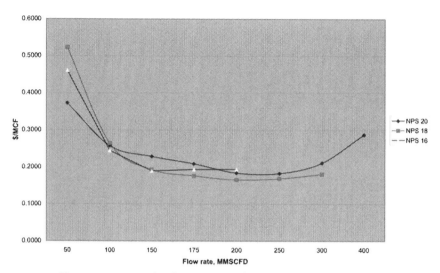

Figure 13.2 Annualized costs versus flow rate for three pipe sizes.

In the preceding calculations, to simplify matters, several assumptions were made. Miscellaneous costs were estimated as a percentage of the pipeline and compressor station costs. Also we considered the annual costs to be constant from year to year. A more accurate calculation would be to escalate the annual costs by a percentage each year to account for inflation, using the Consumer Price Index. Nevertheless, the preceding calculations illustrated a methodology of economic analysis to determine the most optimum pipe size.

Problem 13.7

In Chapter 5, we compared expanding the capacity of gas pipeline from Windsor to Cardiff using two options: Installing intermediate compressor station or installing pipe loop. Using the results of Example 5.1, compare the two options taking into account the capital cost, operating cost, fuel cost, and considering project life of 25 years. The capital will be financed with 70% debt at 8% interest. The regulatory return allowed on the 30% equity is 12%. The tax rate may be assumed at 35%. Fuel consumption is 0.2 MCF per day per HP and fuel gas cost is $3 per MCF. Assume a 350-day operation per year. Calculate annualized cost of service and transportation tariff both options. It is expected that the annual O&M cost will increase by $2 million for phase 1 and an additional $3 million for phase 2 compressor station options. For the looping option the incremental O&M cost is $0.5 million for phase 1 and $0.75 million for phase 2.

Solution

Phase 1 expansion.

This expansion results in a flow rate of 238.41 MMSCFD and the compressor station option requires installing the following HP:

Windsor compressor station: 8468 HP
Avon compressor station: 3659 HP

$$\text{Total HP} = 8468 + 3659 = 12{,}127 \text{ HP}$$

Incremental HP for phase 1 was calculated as

$$\Delta HP = 12{,}127 - 7064 = 5063 \text{ HP}$$

The cost of this incremental HP based on $2000 per installed HP is

$$\Delta \text{Capital cost} = 5063 \times 2000 = \$10.13 \text{ million}$$

The incremental fuel cost for 5063 HP is

$$\Delta \text{Fuel cost} = 5063 \times 0.2 \times \$3 \times 350 = \$1.06 \text{ million per year.}$$

The incremental capital cost of $10.13 million will be funded by 70% debt and 30% equity. Debt capital = $10.13 \times 0.7 = \$7.09$ million.

$$\text{Loan amortization cost} = \frac{10.13 \times 0.7 \times 0.08}{1 - \left(\frac{1}{1.08}\right)^{25}} = \$0.66 \text{ million per year.}$$

The remaining capital of ($10.13 − $7.09) or $3.04 million is equity that according to regulatory guidelines can earn at 12% interest.

The annual revenue allowed on equity capital is

$$3.04 \times 0.12 = \$0.36 \text{ million}$$

Accounting for 35% tax rate, the adjusted annual revenue on equity capital is

$$\frac{\$0.36}{1 - 0.35} = \$0.55 \text{ million}$$

Next add the O&M cost increase of $2 million per year and the fuel cost of $1.06 million to the annual costs for debt and equity just calculated to arrive at the annual cost of service as follows:

$$\text{Annual payment to retire debt} = \$0.66 \text{ million}$$

$$\text{Annual revenue on equity capital} = \$0.36 \text{ million}$$

$$\text{Annual operating cost} = \$2.0 \text{ million}$$

$$\text{Annual fuel cost} = \$1.06 \text{ million}$$

Therefore,

Incremental annual cost of service for phase 1 expansion compressor station option is

$$\$0.66 + 0.36 + 2.0 + 1.06 = \$4.08 \text{ million.}$$

This amount is the incremental annual cost of service over and above the cost of service for the initial flow rate of 188.41 MMSCFD.

The incremental tariff for an incremental flow rate of 50 MMSCFD for phase 1 expansion is

$$\text{Incremental tariff} = \frac{4.08 \times 10^6 \times 10^3}{50 \times 10^6 \times 350} = \$0.2331 \text{ per MCF.}$$

Next we calculate the cost of service and tariff considering the looping option.

In Example 5.1 for phase 1 expansion, we required installation of 50.03 miles of loop at a cost of $25.02 million. In addition to this cost of pipe loop, we must also include the cost of the increased HP requirement at Windsor for phase 1 flow rate, which was calculated at 1404 HP.

At $2000 per installed HP, the extra cost for incremental HP is $2.81 million.

Thus for phase 1, the total cost of looping pipe upstream of Cardiff and increased HP cost at Windsor compressor station was calculated as

$$\$25.02 + \$2.81 = \$27.83 \text{ million}$$

The incremental fuel cost for the extra 1404 HP is

$$\Delta \text{Fuel cost} = 1404 \times 0.2 \times \$3 \times 350 = \$0.30 \text{ million per year.}$$

The incremental capital of $27.83 million for the looping option would also be funded by 70% debt and 30% equity.

$$\text{Debt capital} = 27.83 \times 0.7 = \$19.48 \text{ million}$$

$$\text{Loan amortization cost} = \frac{19.48 \times 0.08}{1 - \left(\frac{1}{1.08}\right)^{25}} = \$1.82 \text{ million per year.}$$

The remaining capital of ($27.83 − $19.48) or $8.35 million is equity that according to regulatory guidelines can earn at 12% interest.

The annual revenue allowed on equity capital is

$$8.35 \times 0.12 = \$1.0 \text{ million}$$

Accounting for 35% tax rate, the adjusted annual revenue on equity capital is

$$\frac{\$1.0}{1-0.35} = \$1.54 \text{ million}$$

Next add the O&M cost increase of $0.5 million per year and the fuel cost of $0.30 million to the annual costs for debt and equity just calculated to arrive at the incremental annual cost of service as follows:

Annual payment to retire debt = $1.82 million

Annual revenue on equity capital = $1.0 million

Annual operating cost = $0.5 million

Annual fuel cost = $0.3 million

Therefore,

Incremental annual cost of service for phase 1 expansion looping option is

$$\$1.82 + 1.0 + 0.5 + 0.3 = \$3.62 \text{ million.}$$

This amount is the incremental annual cost of service over and above the cost of service for the initial flow rate of 188.41 MMSCFD.

The incremental tariff for an incremental flow rate of 50 MMSCFD for phase 1 expansion looping option is

$$\text{Incremental tariff} = \frac{3.62 \times 10^6 \times 10^3}{50 \times 10^6 \times 350} = \$0.2069 \text{ per MCF}$$

We can summarize the calculations as follows:
For phase 1 expansion:

Compressor station option:
Incremental annual cost of service = $4.08 million.
 Incremental tariff = $0.2331 per MCF.

Looping option:
Incremental annual cost of service = $3.62 million.
 Incremental tariff = $0.2069 per MCF.

It can be seen that the incremental annual cost of service and the incremental tariff for phase 1 expansion is less in the looping option than the compressor station option. Therefore, for phase 1 expansion, the looping option is the preferred choice.

For phase 2 expansion, the throughput increase of 50 MMSCFD will be on top of phase 1 expansion. Because the preferred choice for phase 1 expansion is the looping option, we must consider the increase in facilities required for phase 2 with 50.03 miles of pipe loop already installed. In Example 5.1 for phase 2, the loop required was calculated to be 76.26 miles. The incremental HP at Windsor was calculated as 1775 HP. Also, the incremental looping required and cost of increased HP at Windsor over the phase 1 values were calculated to be $16.66 million.

The incremental fuel cost for the extra 1775 HP is

$$\Delta \text{Fuel cost} = 1775 \times 0.2 \times \$3 \times 350 = \$0.37 \text{ million per year.}$$

The incremental capital of $16.66 million for the phase 2 looping option would also be funded by 70% debt and 30% equity.

$$\text{Debt capital} = 16.66 \times 0.7 = \$11.66 \text{ million}$$

$$\text{Loan amortization cost} = \frac{11.66 \times 0.08}{1 - \left(\frac{1}{1.08}\right)^{25}} = \$1.09 \text{ million per year.}$$

The remaining capital of ($16.66 − $11.66) or $5.0 million is equity that according to regulatory guidelines can earn at 12% interest.

The annual revenue allowed on equity capital is

$$\$5.0 \times 0.12 = \$0.6 \text{ million}$$

Accounting for 35% tax rate, the adjusted annual revenue on equity capital is

$$\frac{\$0.6}{1 - 0.35} = \$0.92 \text{ million}$$

Next add the O&M cost increase of $0.75 million per year and the fuel cost of $0.37 million to the annual costs for debt and equity just calculated to arrive at the incremental annual cost of service for phase 2 looping expansion as follows:

$$\text{Annual payment to retire debt} = \$1.09 \text{ million}$$

$$\text{Annual revenue on equity capital} = \$0.6 \text{ million}$$

$$\text{Annual operating cost} = \$0.75 \text{ million}$$

$$\text{Annual fuel cost} = \$0.37 \text{ million}$$

Therefore,

Incremental annual cost of service for phase 2 expansion looping option is

$$\$1.09 + 0.6 + 0.75 + 0.37 = \$2.81 \text{ million.}$$

This amount is the incremental annual cost of service over and above the cost of service for the phase 1 flow rate of 238.41 MMSCFD.

The incremental tariff for an incremental flow rate of 50 MMSCFD for phase 2 expansion looping option is

$$\text{Incremental tariff} = \frac{2.81 \times 10^6 \times 10^3}{50 \times 10^6 \times 350} = \$0.1606 \text{ per MCF}.$$

In summary,

For phase 2 expansion:

Looping option:

Incremental annual cost of service = $2.81 million.

Incremental tariff = $0.1606 per MCF.

These incremental costs are over and above the phase 1 numbers.

It must be noted that we did not consider a compressor station option for phase 2 expansion. This is because the preferred option for phase 1 expansion was installing loop. Because approximately 50 miles of pipe loop was already installed for phase 1, we simply looked at adding approximately 26 miles of extra loop for phase 2. For comparison, we could determine additional compressor station requirements for phase 2 instead of extending the loop. This is left as an exercise for the reader.

9. SUMMARY

We reviewed the major cost components of a pipeline system consisting of pipe, pump station, etc., and illustrated methods of estimating the capital costs of these items. The annual costs such as electrical energy, O&M etc., were also identified and calculated for a typical pipeline. Using the capital cost and operating cost, the annual cost of service was calculated based on specified project life, interest cost, etc. Thus we determined the transportation tariff that could be charged for shipments through the pipelines. Also a methodology of determining the optimum pipe size for a particular application using PV was explained. Considering three different pipe sizes, we determined the best option based on a comparison of PV of the three different cases.

In the previous chapters, we explored different scenarios of pipe sizes and pressures to transport natural gas through pipelines from one location to another. Various pressure drop formulas, compression requirements, and HP required were calculated without delving too much into costs of facilities. In this chapter, the economic aspects of pipelines will be reviewed. The economic pipe size required for a particular throughput will be arrived at considering the various costs that make up a pipeline system. The initial capital cost of pipeline and ancillary facilities will be discussed along with the annual O&M cost. Because pipelines are generally designed to transport gas belonging to one company by another company, a methodology for determining transportation cost or tariff will be analyzed.

A pipeline may be constructed to transport natural gas for the owner of the pipeline or for selling gas to another company or to transport some other company's gas. These three scenarios represent three major uses of pipeline transportation of natural gas. The economics involved in the selection of pipe diameter, compressor station, and related facilities will vary slightly for each scenario. As an owner company transporting its own gas, minimal facilities will probably be built. However, Department of Transportation codes and other regulatory requirements will still have to be met to ensure a safe pipeline operation that will not endanger humans or the environment. In the second scenario, in which a company builds a pipeline to transport its gas and sells the gas at the end of the pipeline to a customer, minimal facilities will be constructed without too much regulatory control. In the third scenario, a pipeline company constructs and operates a pipeline for the purpose of transporting gas belonging to other companies. This will be under the jurisdiction of FERC or a state agency such as Public Utilities Commission in California or the Texas Railroad Commission in Texas. An interstate pipeline in which the pipeline crosses one or more state boundaries will be regulated by FERC. A pipeline that is intrastate such as wholly within California will be subject to Public Utilities Commission rules and not FERC. Such regulatory requirements impose strict guidelines on the type and number of facilities and costs that may be passed on to the customer requesting gas transportation. These regulatory requirements will dictate that excessive capital facilities not be built and the amortized cost passed on to the customers. Whereas a private pipeline company transporting its own gas may build in extra compressor units as spares to ensure uninterrupted operation in the event of equipment failure, FERC regulated pipelines may not be able to do so. Thus pipeline economics will differ slightly from case to case.

In this chapter, we will not discuss other modes of transportation of gas such as truck transport of pressurized gas containers. The general economic principles discussed here are applicable to both private unregulated pipelines as well as FERC regulated pipelines used for interstate transportation of natural gas.

In this chapter, the economic aspects of a natural gas pipeline transportation were reviewed. A method for determining the optimum pipe size necessary to transport a certain flow rate was discussed. We introduced concepts of capital cost of pipeline and compressor stations and the annual O&M costs. The fuel consumption calculations were also explained. Taking into account time value of money and the ROR allowed on an equity investment in pipeline facilities we calculated an annual cost of transporting gas. From this annual cost, the transportation tariff was calculated. The economic pipe size for a particular application was illustrated using three different pipe sizes and estimating the initial capital cost and annual operating costs. A typical pipeline expansion scenario with option of installing compressors versus pipe loops was also explained using economic principles. Additionally, the major components of the capital cost of a typical pipeline system were reviewed.

10. PROBLEMS

1. A natural gas pipeline transports 120 MMSCFD at a load factor of 95%. The capital cost is estimated at $70 million and the annual operating cost is $6 million. Amortizing the capital at 8% for a project life of 20 years, calculate the cost of service and transportation tariff for this pipeline.
2. A new pipeline is being constructed to transport natural gas from a processing plant to a power plant 150 miles away. An initial phase and an expansion phase are contemplated. During the initial phase lasting 10 years, the amount of gas shipped is expected to be a constant volume of 100 MMSCFD at 95% load factor. A pipe size of NPS 18, 0.250-in wall thickness is required to handle the volumes with two compressor stations of 5000 HP total. The total pipeline cost may be estimated at $750,000 per mile and compressor station cost at $2000 per HP installed. The annual operating costs are estimated at $6 million. The construction project will be financed by borrowing 75% of the required capital at an interest rate of 6%. The Regulatory ROR allowed on equity is 13%. Consider a project life of 25 years and overall tax rate of 36%.
   a. Calculate the annual cost of service for this pipeline and the transportation tariff in $/MCF.
   b. The second phase lasting the next 10 years is projected to increase throughput to 150 MMSCFD. Calculate the transportation tariff for the expansion phase considering the capital cost to increase by

$30 million and the annual cost increases by $4 million, with the same load factor as before.

3. A gas pipeline is to be constructed to transport 200 MMSCFD of natural gas from Jackson to Columbus, 180 miles away. Consider three pipe sizes—NPS 18, NPS 20, and NPS 24—all constructed of API 5L-X52 pipe with suitable wall thickness for a maximum operating pressure of 1400 psig. Determine the most economical pipe diameter taking into account the pipe material cost, cost of compressor stations, and fuel costs. The selection of pipe size may be based on a 30-year project life and a PV of discounted cash flow at 6% per year. Use $750 per ton for pipe material and $2000 per installed HP for compressor station cost. Fuel gas may be estimated at $3 per MCF.

CHAPTER FOURTEEN

Case Studies

1. INTRODUCTION

In this chapter, we will discuss the number and size of compressor stations or pump stations required to move a liquid or gas through a transmission pipeline.

These studies require calculation of pipeline hydraulics, pump, or compressor horsepower (HP) and other cost analyses in order to determine the optimum pipe size, and power requirements that will produce the least cost option.

Several liquid and gas transmission pipelines will be discussed, including heated crude oil pipelines as well as refined products (gasoline, jet fuel, and diesel) some utizing drag reduction agent (DRA) to improve flow rate in liquids, batching, thermal and isothermal flows with constant and variable speed motor driven pumps or compressors.

Finally, we will also analyze transmission pipelines that are used to transport water for several small or medium-sized cities considering the popular Hazen–Williams method calculation.

2. CASE STUDY 1: REFINED PRODUCTS PIPELINE (ISOTHERMAL FLOW) PHOENIX TO LAS VEGAS PIPELINE

A pipeline is planned between Phoenix, AZ, and Las Vegas, NV, a distance of 420 mi as indicated in Figure 14.1. It is expected to deliver refined petroleum products (batched pipeline) consisting of gasoline, jet fuel, and diesel fuel. Deliveries are as listed here, with intermediate volume drop off at Kingman, Lake Havasu City (LHC) and Bullhead City (BHC), as indicated. The pipeline will run in a batched mode to satisfy the requirements of the delivery terminals. Appropriately sized storage tanks at both origin location (Phoenix) and the delivery terminus (Las Vegas) will be sized for a minimum of 10 days' storage, along with a transmix tank for the blended products because of contamination from individual batches.

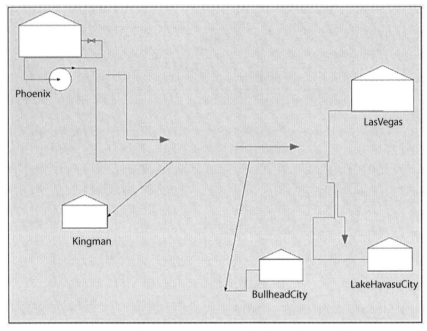

Figure 14.1 Phoenix to Las Vegas refined products pipeline.

Flow leaving Phoenix facility 400 million barrels per day (MBPD)
Kingman delivery: 80 MBPD
LHC delivery: 80 MBPD
BHC delivery: 70 MBPD
The remaining volume of 170 MBPD is delivered to the Las Vegas terminal.

The desired operating pressure is limited to 1400 psig.

The properties of the liquids transported are based on 70 °F average yearly temperature as shown below.

Product	Specific Gravity at 70 °F	Viscosity at 70 °F (cSt)	Batch Volume (Barrels (bbl))
Gasoline	0.746	1.2	50,000
Diesel	0.850	5.9	50,000
Jet fuel	0.815	2.5	50,000

Determine the optimum pipe size required based on the initial throughput of 400 MBPD without considering diameter change to account for reduction in volume resulting from drop offs at intermediate locations. Calculate the total capital cost of pump stations, pipelines, and major equipment such as scraper traps, main line valves, storage tanks, and inlet and outlet flow meters. Consider $1500 per ton for the pipe, coated and wrapped, $1800 per installed HP for each pump station and $500,000 for each flow meter assembly and $200,000 for mainline valve stations. Use a right-of-way and permitting cost of $15 million.

The project would be financed at 20% equity and 80% bank loan. The project life is 30 years. The allowable regulatory ROR (Rate of Return) on equity shall be 15% and interest on the loan amount shall be 10%. Assume 40% corporate income tax rate when calculating the allowable regulatory tariff.

2.1 Solution

The pipeline is 420 miles long and we will first study a 36-in pipeline 0.500-in wall pipeline as option 1. Next we will review the hydraulics and cost by increasing pipe size to 38 and 40 in as needed to reduce the velocity to nominal values around 3–5 ft/s and recommend the optimum least cost option.

At 400 MBD flow rate of the heaviest product, diesel, the average velocity and Reynolds number are

$$Vel = 0.0119 \times 400,000/(35 \times 35) = 3.89 \text{ ft/s}$$

$$R = 92.24 \times 400,000/(5.9 \times 35.0) = 178,673$$

The flow is therefore in the turbulent zone in the first segment of the pipeline.

The Colebrook–White friction factor is calculated next.

$$e/D = 0.002/35.0 = 0.0000571 \quad \text{assume } f = 0.01 \text{ initially}$$

A better approximation is then calculated from Eqn (5.21) for Colebrook–White friction as

$$1/\sqrt{f} = -2\text{Log}_{10}\left[(e/3.7D) + 2.51/(R\sqrt{f})\right]$$

or

$$= -2 \times \log\left[(0.0000571/3.7 + 2.51/(178673 \times f^{0.5})\right]$$

Solving by iteration $f = 0.0168$ approximately.

Next we calculate the psi/mi at this flow rate from Eqn (5.27):

$$Pm = 0.0605 \times 0.0168 \, (400,000)^2 \, (0.85/35^5) = 2.6275 \text{ psi/mi}$$

Therefore the total pressure drop for the first pipe segment from Phoenix to Kingman is

$$2.6275 \times 190 = 499.23 \text{ psig}$$

Similarly, next we calculate the R, f, and psi/mi for the second segment between Kingman and LHC at a flow rate of 320 MBD:

At 320 MBD flow rate of diesel, we get a Reynolds number of

$$R = 92.24 \times 320,000/(5.9 \times 35) = 142,938$$

The flow is therefore in the turbulent zone in the second segment of the pipeline.

The Colebrook–White friction factor is calculated next.

$$e/D = 0.002/35 = 0.000057 \quad \text{assume } f = 0.01 \text{ initially}$$

A better approximation is then calculated from Eqn (5.21) for Colebrook–White friction as

$$1/\sqrt{f} = -2\text{Log}_{10}\left[(e/3.7D) + 2.51/(R\sqrt{f})\right]$$

or

$$= -2 \times \text{Log}\left[(0.000057/3.7 + 2.51/(142,938 \times f^{0.5})\right]$$

Solving by iteration $f = 0.0175$ approximately.

Next we calculate the pressure drop in the second pipe segment from Kingman to LHC as

$$Pm = 0.0605 \times 0.0175 \, (320,000)^2 (0.85/35^5) = 1.7525 \text{ psi/mi}$$

Therefore the total pressure drop from Kingman to LHC =

$$1.7525 \times 60 = 105.2 \text{ psig}$$

Next, we calculate the Reynolds number, f, and Pm for the next pipe segment between LHC and BHC at 240 MBPD as:

At 240 MBD flow rate of diesel, we get a Reynolds number of

$$R = 92.24 \times 240,000/(5.9 \times 35) = 107,204$$

The Colebrook–White friction factor is calculated next.

$$e/D = 0.002/35 = 0.000057 \quad \text{assume } f = 0.01 \text{ initially}$$

A better approximation is then calculated from Eqn (5.21) for Colebrook–White friction as

$$1/\sqrt{f} = -2\text{Log}_{10}\left[(e/3.7D) + 2.51/(R\sqrt{f})\right]$$

or

$$= -2 \times \text{Log}\left[(0.000057/3.7 + 2.51/(107204 \times f^{0.5})\right]$$

Solving by iteration $f = 0.0185$ approximately.

$$Pm = 0.0605 \times 0.0185 \times (240,000)^2 (0.85/35^5) = 1.042 \text{ psi/mi}$$

The total pressure drop from LHC to BHC:

$$1.042 \times 85 = 88.6 \text{ psig}$$

And finally, we calculate the Reynolds number, f, and Pm for the last pipe segment between BHC and Las Vegas at 170 MBPD as:

At 170 MBD flow rate of diesel, we get a Reynolds number of

$$R = 92.24 \times 170,000/(5.9 \times 35) = 75,936$$

The Colebrook–White friction factor is calculated next.

$$e/D = 0.002/35 = 0.000057 \quad \text{assume } f = 0.01 \text{ initially}$$

A better approximation is then calculated from Eqn (5.21) for Colebrook–White friction as

$$1/\sqrt{f} = -2\text{Log}_{10}\left[(e/3.7D) + 2.51/(R\sqrt{f})\right]$$

or

$$= -2 \times \text{Log}\left[(0.000057/3.7 + 2.51/(75936 \times f^{0.5})\right]$$

Solving by iteration $f = 0.0198$ approximately

$$Pm = 0.0605 \times 0.0198 \times (170,000)^2 (0.85/355) = 0.5606 \text{ psi/mi}$$

The total pressure drop from BHC to Las Vegas:

$$0.5606 \times 85 = 47.7 \text{ psig}$$

Therefore the total pressure required to pump diesel from Phoenix to Las Vegas is as follows:

$$P_T = 500 + 106 + 89 + 48 + 50 = 793 \text{ psi}$$

Note that we have not reduced the pipe size with decreasing flow rate through each pipe segment. This was done deliberately to simplify calculations. In a subsequent problem, we will reduce the pipe size to compensate for reduction in flow rate after making deliveries at each location.

The material cost of the 420 miles of pipe at $1500 per ton is based on: weight of pipe = $420 \times 5280 \times 10.68 \times 0.5 (36 - 0.5)/2000 = 210{,}195$ tons.

Thus cost of pipe coated and wrapped is = $210{,}195 \times 1500 = \$315.3$ million.

Of course we must add a percentage for the transportation and taxes as well.

Adding 15% for this, we get $362.6 million.

The HP required is based on pressure at Phoenix.

$$\text{HP} = (793 - 50) \times 400000/(58776 \times 0.75)$$

$$= 6743 \text{ say } 8000 \text{ HP motor.}$$

At $1800/HP installed, total capital cost of pump station is $8000 \times 1800 = \$14.4$ million.

Additionally, we should add capital for valves, fittings, storage tanks, and delivery meters:

Flow meters = $500{,}000 \times 4 = \$2$ million.
Mainline valves = $200{,}000 \times 25 = \$5$ million.
Storage tanks at $5/bbl for four products = $\$5 \times 4 \times 4 = \$80$ million.
Miscellaneous pipe and fittings, scraper traps, etc. (10% of total) = $47 million.
Right-of-way and permitting cost = $15 million.
with 10% contingency.

Total capital required = $363 + 14.4 + 2 + 5 + 80 + 47 + 15 + 53 = \$580$ million.

Operating cost for pump and motors = $8000 \text{ HP} \times 0.746 \times 24 \times 365 \times 0.10 = \$5.28$ million/year.

Adding a general and administrative cost of $5 million/year.

Total annual operating cost = $10.3 million (Figure 14.2).

Using the software program LIQTHERM, the tariff rate = $0.72/bbl.

In the preceding analysis, we selected an NPS 36 pipe throughout the system. Next we will look at reducing the pipe size as product is delivered along the various delivery points such as Kingman, LHC, etc. We will first start with NPS 36 for the segment between Phoenix and Kingman and then reduce the pipe size to NPS 34 between Kingman and LHC, then NPS 32 between LHC and BHC, and finally NPS 30 all the way from the BHC to Las Vegas.

From Phoenix to Kingman remains the same as earlier calculations:

At 400 MBD flow rate of the heaviest product, diesel, the average velocity, and Reynolds number are.

Vel: 3.89 ft/s.

R: 178,673.

f: 0.0168 approximately.

Pm: 2.6275 psi/mi.

Therefore the total pressure drop for the first pipe segment from Phoenix to Kingman is

$$2.6275 \times 190 = 499.23 \text{ psig}$$

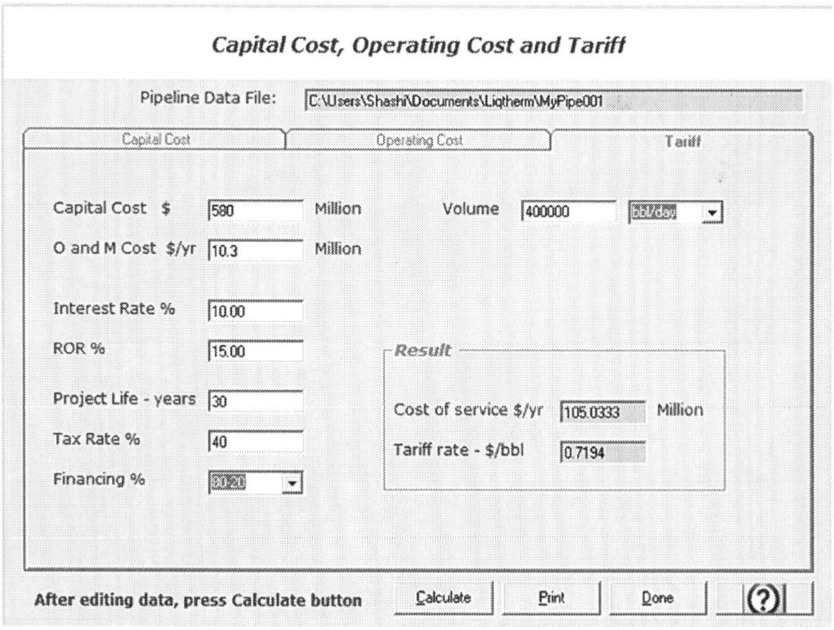

Figure 14.2 Tariff calculations using same size pipe.

Similarly, next we calculate the R, f, and psi/mi for the second between Kingman and LHC at a flow rate of 320 MBD and pipe size NPS 34 (segment using LIQTHERM software):
Vel: 3.44 ft/s.
R: 150,462.
f: 0.0173.
Pm: 2.2455 psi/mi.
Therefore the total pressure drop from Kingman to LHC = 2.2455 × 60 = 134.73 psig.
Next, we calculate the Reynolds number, f, and Pm for the next pipe segment between LHC and BHC at 240 MBPD and pipe size NPS 32.
Vel: 2.92 ft/s.
R: 120,069.
f: 0.0181.
Pm: 1.7998 psi/mi.
The total pressure drop from LHC to BHC = 1.7998 × 85 = 152.98 psig.
And finally, we calculate the Reynolds number, f, and Pm for the last pipe segment between BHC and Las Vegas at 170 MBPD and pipe size NPS 30.
Vel: 2.36 ft/s.
R: 90,864.
f: 0.0191.
Pm: 1.3291 psi/mi.
The total pressure drop from BHC to Las Vegas = 1.3291 × 85 = 112.98 psig
Therefore the total pressure required to pump diesel from Phoenix to Las Vegas is as follows:

$$P_T = 500 + 135 + 153 + 113 + 50 = 951 \text{ psi}$$

The weight of pipe per segment is
Phoenix to Kingman = 190 × 5280 × 10.68 × 0.5 (36 − 0.5)/2000 = 95,089 tons.
Kingman to LHC = 60 × 5280 × 10.68 × 0.5 (34 − 0.375)/2000 = 28,728 tons.
LHC to BHC = 85 × 5280 × 10.68 × 0.5 (32 − 0.375)/2000 = 38,301 tons.
BHC to Las Vegas = 85 × 5280 × 10.68 × 0.5 (30 − 0.375)/2000 = 35,904 tons.

The total pipe weight for 420 miles is

$$95,089 + 28,728 + 38,301 + 35,904 = 198,022 \text{ tons}$$

Thus cost of pipe coated and wrapped is = 198,022 × 1500 = $297.03 million.

Of course we must add a percentage for the transportation and taxes as well.

Adding 15% for this, we get $341.6 million.

The HP required is based on pressure at Phoenix.

$$HP = (951 - 50) \times 400,000/(58,776 \times 0.75)$$
$$= 8176 \text{ say } 10,000 \text{ HP motor.}$$

At $1800/HP installed, the total capital cost of pump station is 10,000 × 1800 = $18 million.

Additionally we should add capital for valves, fittings, storage tanks, and delivery meters:

Flow meters = $500,000 × 4 = $2 million.

$$\text{Mainline valves} = \$200,000 \times 25 = \$5 \text{ million}$$

Storage tanks @ $5/bbl for four products = $5 × 4 × 4 = $80 million.

Miscellaneous pipe and fittings, scraper traps, etc. (10% of total) = $47 million.

with 10% contingency.

Total capital required = 342 + 18 + 2 + 5 + 80 + 47 + 15 + 51 = $560 million.

Operating cost for pump and motors = 10,000 HP × 0.746 × 24 × 365 × 0.10 = $6.6 million/year.

Adding a general and administrative cost of $5 million/year:

Total annual operating cost = $11.6 million (Figure 14.3).

The tariff rate = $0.71.

3. CASE STUDY 2: HEAVY CRUDE OIL PIPELINE 2 MILES LONG WITHOUT HEATERS

A short 2-mile pipeline NPS 16 with 0.25-in wall thickness runs from one heavy oil terminal at Anaheim to a tank farm at Cuiaba and is used for transporting heavy crude oil. Assume the crude to have Sg = 0.895 and viscosity = 100 cSt at 60 °F and additionally Sg = 0.815 and viscosity of

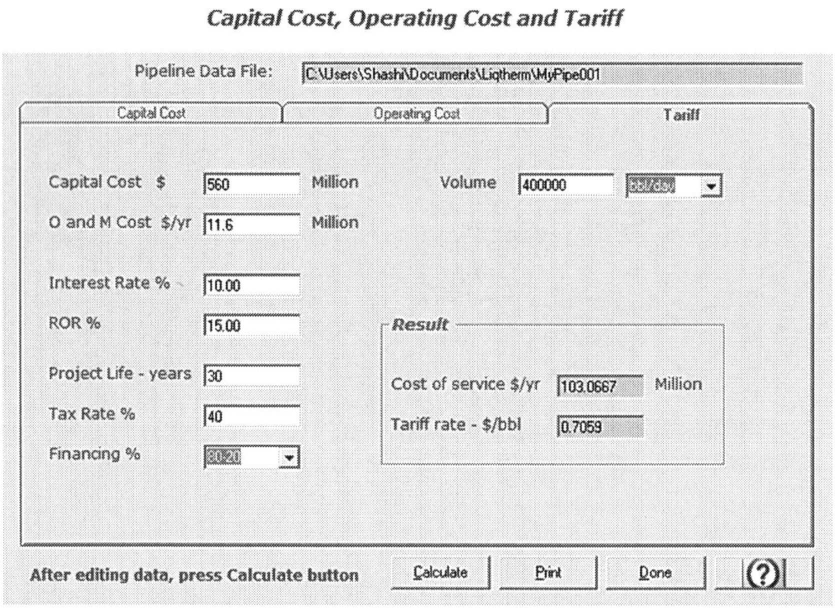

Capital Cost, Operating Cost and Tariff

Figure 14.3 Tariff calculation using variable size pipe.

25 cSt at 100 °F. The initial throughput is 5000 bbl/h. The maximum allowable operating pressure (MAOP) is 300 psig. The pipeline will transport heavy crude oil without additional heaters and will use a single pump at Anaheim.

The desired delivery pressure at Cuiaba is 100 psig with losses of 15 psig through a delivery meter. Subdivide the pipeline into short segments of 0.25 miles each and determine the temp variation along the pipeline starting from an inlet temp of 100 °F at Anaheim. Assume K subsoil = 0.6, Kins = 0.02 and soil temp is constant at 50 °F throughout.

(a) What is the pressure and HP required at Anaheim considering no viscosity correction for the constant speed pumps?

(b) Assuming two batches of heavy and light crudes are pumped through the pipeline in batches of 25,000 bbl (heavy) and 30,000 bbl (light crude). Determine the HP required and pressures at Anaheim, considering same subdivisions as part (a).

The light crude has following characteristic:

$$Sg = 0.830 \quad Visc = 15 \text{ cSt at } 60 \text{ °F}$$
$$Sg = 0.805 \quad Visc = 5 \text{ cSt at } 100 \text{ °F}$$

Use roughness = 0.002 in and Colebrook–White equation for pressure loss.

3.1 Solution

(a) Calculate the Reynolds number at inlet temperature first:

$$R = 92.24 \times 5000 \times 24/(5.0 \times 15.5) = 142,823$$

The Colebrook–White friction factor is calculated next:

$$e/D = 0.002/15.5 = .000129 \quad \text{assume } f = 0.01 \text{ initially}$$

A better approximation is then calculated from Eqn (5.21) for Colebrook–White friction as

$$1/\sqrt{f} = -2\text{Log}_{10}\left[(e/3.7D) + 2.51/(R\sqrt{f})\right]$$

or

$$= -2 \times \text{Log}\left[(0.000129/3.7 + 2.51/(403455 \times f^{0.5})\right]$$

Solving by iteration $f = 0.012$ approximately.

Next we calculate the psi/mi at this flow rate from Eqn (5.27):

$$Pm = 0.0605 \times 0.012 \times f(5,000)^2(0.815/15.5^5) = 61.57 \text{ psi/mi}$$

considering isothermal flow for 2 miles of pipe for now, the pressure at Anaheim for this flow rate is

$$P1 = 61.57 \times 2 + 100 + 15 + 400 \times 0.815/2.31 = 380 \text{ psig}$$

Because the elevation difference between Anaheim and Cuiba is $(500-100)$ ft.

If flow were isothermal the pressure required at Anaheim is approximately 380 psig. Therefore the HP required is

$$HP = (5000 \times 0.7) \times (380 - 20) \times (2.31/0.815) \times 1/(3960 \times 0.75)$$
$$= 1203.$$

Considering 20-psi pump suction pressure and 75% pump efficiency. However, the flow is not isothermal. Actually the temp of the crude oil drops as it reaches Cuiba and therefore its viscosity increases. This results in varying Reynolds number and therefore varying Pm.

We will therefore use subdivided pipeline to calculate the temp and pressure profile for thermal flow.

We will use LIQTHERM software (www.systek.us) to perform this simulation.

The result of the simulation is shown below along with the screens used (Figures 14.4–14.11).

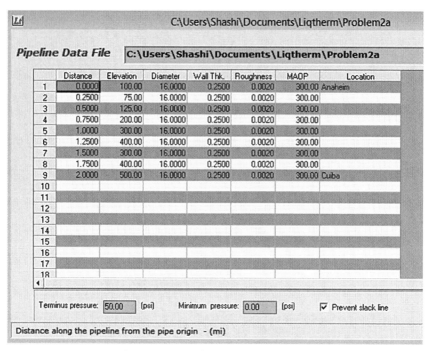

Figure 14.4 Pipeline data input screen.

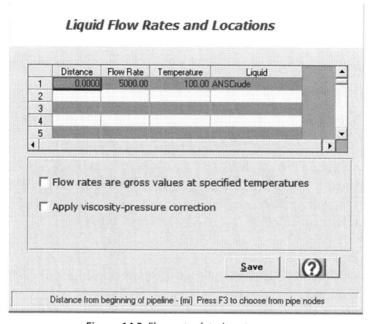

Figure 14.5 Flow rate data input screen.

Case Studies 531

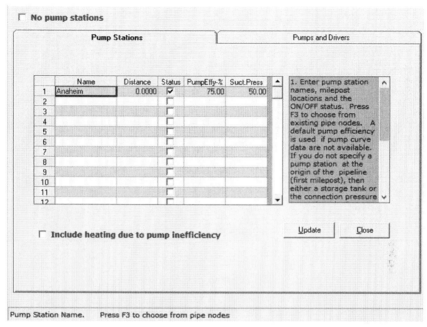

Figure 14.6 Pump station data input screen.

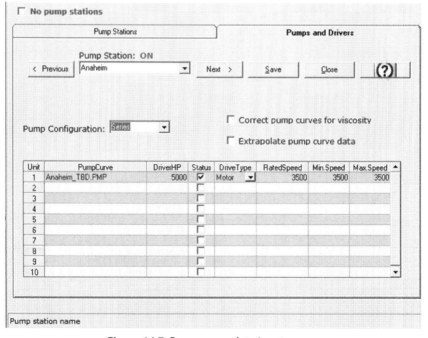

Figure 14.7 Pump curve data input screen.

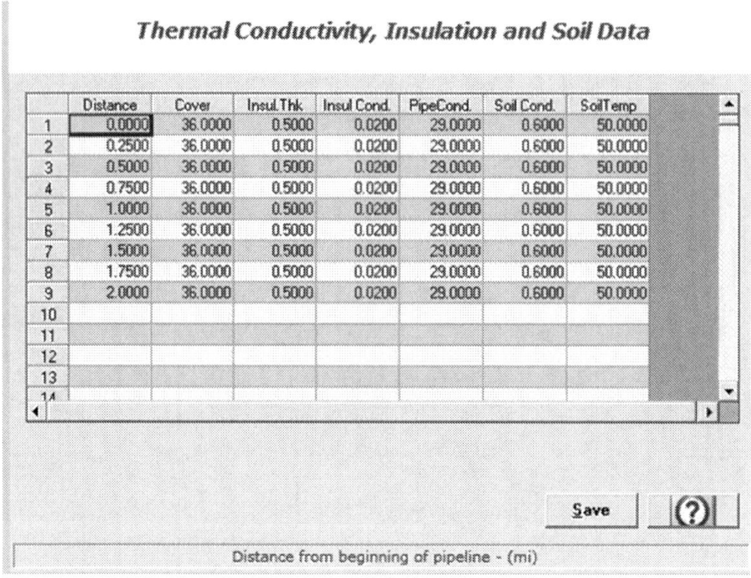

Figure 14.8 Conductivity input screen.

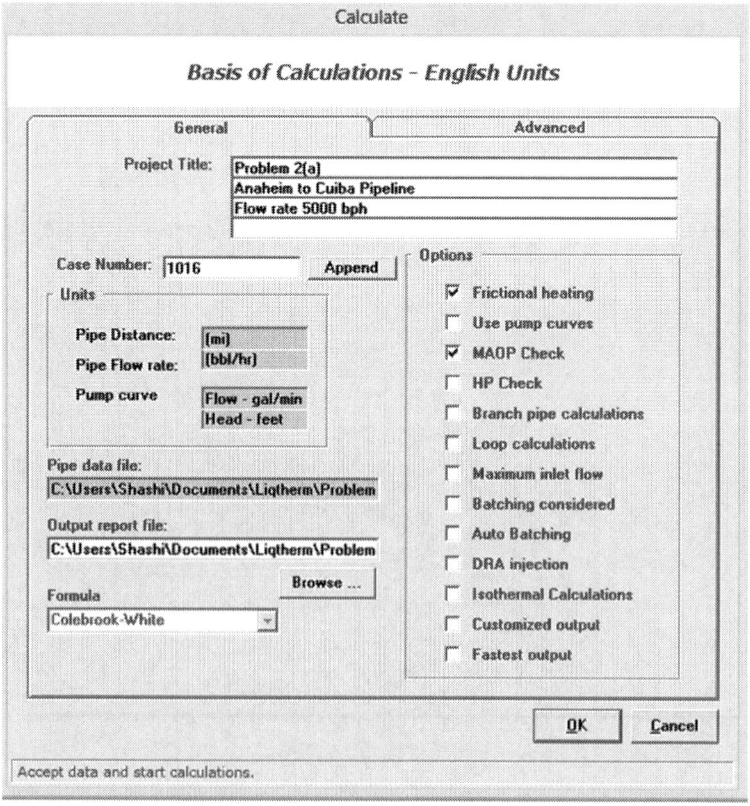

Figure 14.9 Start of calculation screen.

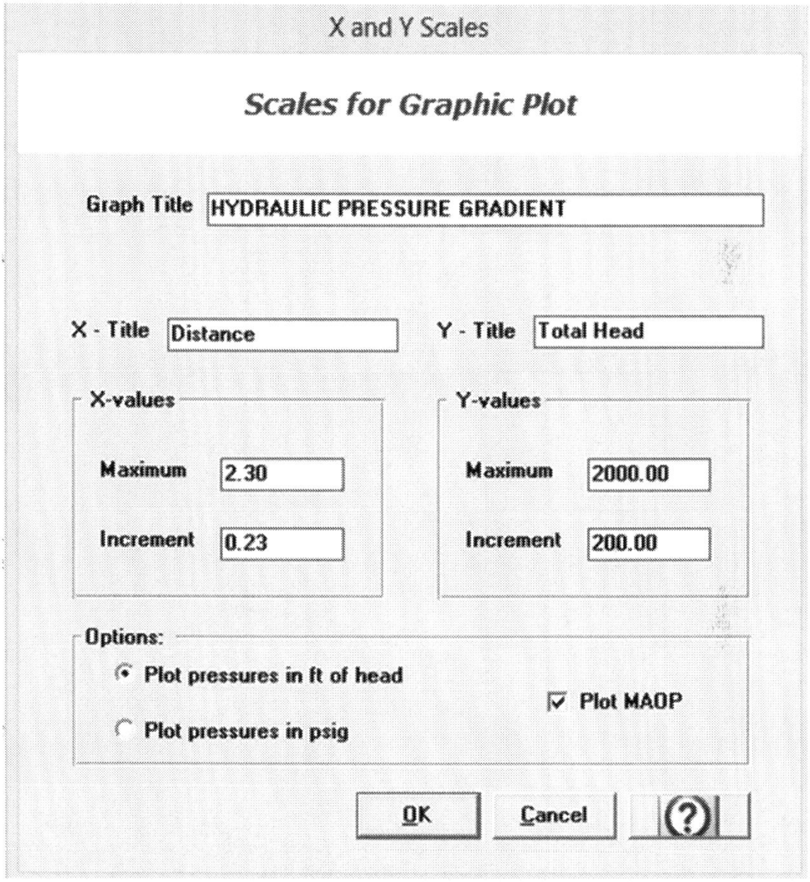

Figure 14.10 Graphic plot screen.

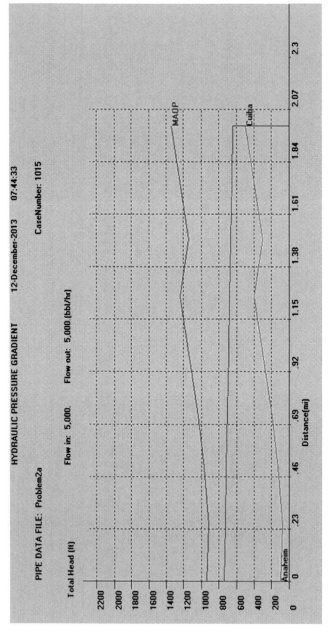

Figure 14.11 Anaheim to Cuiba hydraulic pressure gradient screen.

Case Studies

```
******* LIQTHERM  STEADY STATE PIPELINE HYDRAULIC SIMULATION REPORT *********
DATE: 12-December-2013     TIME:   07:44:19

PROJECT:                   Problem 2(a)
                           Anaheim to Cuiba Pipeline
                           Flow rate 5000 bph

Pipeline data file:        C:\Documents\Liqtherm\Problem2a

******* LIQTHERM - LIQUID PIPELINE STEADY STATE HYDRAULIC SIMULATION ********
****************** Version 6.00.820*************

CASE NUMBER:                              1015

CALCULATION OPTIONS:
Thermal Calculations:                     NO
Frictional Heating:                       YES
Use Pump Curves:                          NO
MAOP Check:                               YES
Horsepower Check:                         NO
Heating due to pump inefficiency:         NO
Valves/Fittings and Devices:              NO
Branch pipe calculations:                 NO
Loop pipe calculations:                   NO
Maximum Inlet Flow:                       NO
Batching Considered:                      NO
DRA Injection:                            NO
Correct volumes for temperature:          NO
Slack Line Calculations:                  NO
Customized Output:                        NO


Inlet flow rate:                          5,000.     (bbl/hr)
Outlet flow rate:                         5,000.     (bbl/hr)
Inlet flow temperature:                   100.00     (degF)
Outlet flow temperature:                  99.52      (degF)

Minimum Pipe pressure:                    0.00       (psi)
Pipe delivery pressure:                   50.00      (psi)


Pressure drop formula used:               Colebrook-White equation

Calculation sub-divisions:                2
Iteration Accuracy:                       MEDIUM



********** LIQUID PROPERTIES **********

Liquid properties file: C:\Users\Shashi\my documents\Liqtherm\Liquid Properties
Database

PRODUCT:                   ANSCrude
Specific gravity:          0.8950 at  60.0(degF)
                           0.8250 at  100.0(degF)

Viscosity:                 43.00 CST at 60.0(degF)
                           15.00 CST at 100.0(degF)


********** LIQUID FLOW RATES AND LOCATIONS **********

Location     Flow rate     Inlet Temp.     Product
(mi)         (bbl/hr)      (degF)
0.00         5,000.        100.0           ANSCrude
```

```
************ Pump Curve Data Not Considered ***********
Pump Sta.         Pump Efficiency(%)
Anaheim           75.00

NOTE: When not using pump curve data, an average pump efficiency
is used to calculate HP at each pump station.

********** PUMP STATIONS **********

Pump        Distance   Pump suct  Pump disch  Sta. disch  Throttled   BHP Reqd
TotHPinst.  KW
station                pressure   pressure    pressure    pressure    by pump
(Active)
            (mi)       (psi)      (psi)       (psi)       (psi)

Anaheim     0.00       50.00      228.52      228.52      0.00        486.
5000.       363.

       Total active pump stations: 1        TOTAL Power:          486.
5,000.      363.

NOTE: Throttle pressures are zero because pump curve data is not used.

Pump Station: Anaheim
Requires pump with following condition: Head : 499.86 (ft) at Flow : 3500.00(gal/min)

********* Heater Stations not Active **************

********** PIPELINE PROFILE DATA **********

Distance    Elevation  Diameter   Wall Thk.  Roughness   MAOP        Location
(mi)        (ft)       (in)       (in)       (in)        (psi)

0.0000      100.00     16.000     0.250      0.0020      300.        Anaheim
0.2500      75.00      16.000     0.250      0.0020      300.
0.5000      125.00     16.000     0.250      0.0020      300.
0.7500      200.00     16.000     0.250      0.0020      300.
1.0000      300.00     16.000     0.250      0.0020      300.
1.2500      400.00     16.000     0.250      0.0020      300.
1.5000      300.00     16.000     0.250      0.0020      300.
1.7500      400.00     16.000     0.250      0.0020      300.
2.0000      500.00     16.000     0.250      0.0020      300.        Cuiba

********** THERMAL CONDUCTIVITY PROFILE DATA **********

Distance    Burial depth  Insul.Thk    Thermal Conductivity       Soil Temp
Location
            (Cover)                    Insulation  Pipe   Soil
(mi)        (in)          (in)         (Btu/hr/ft/degF)           (degF)

0.00        36.00         0.500        0.02        29.00  0.60    50.00
0.25        36.00         0.500        0.02        29.00  0.60    50.00
0.50        36.00         0.500        0.02        29.00  0.60    50.00
0.75        36.00         0.500        0.02        29.00  0.60    50.00
1.00        36.00         0.500        0.02        29.00  0.60    50.00
1.25        36.00         0.500        0.02        29.00  0.60    50.00
1.50        36.00         0.500        0.02        29.00  0.60    50.00
1.75        36.00         0.500        0.02        29.00  0.60    50.00
2.00        36.00         0.500        0.02        29.00  0.60    50.00

********** VELOCITY, REYNOLD'S NUMBER AND PRESSURE DROP **********

Distance   Diameter.  FlowRate    Velocity  Reynolds   Press.drop   Location
(mi)       (in)       (bbl/hr)    (ft/sec)  number     (psi/mi)

0.0000     16.00      5,000.00    5.94      47,606.    17.77        Anaheim
0.2500     16.00      5,000.00    5.94      47,542.    17.78
0.5000     16.00      5,000.00    5.94      47,478.    17.78
0.7500     16.00      5,000.00    5.94      47,415.    17.79
```

```
1.0000     16.00    5,000.00    5.94    47,352.    17.80
1.2500     16.00    5,000.00    5.94    47,289.    17.80
1.5000     16.00    5,000.00    5.94    47,226.    17.81
1.7500     16.00    5,000.00    5.94    47,164.    17.82
2.0000     16.00    5,000.00    5.94    47,164.    17.82           Cuiba


********** TEMPERATURE AND PRESSURE PROFILE **********
Distance   Elevation FlowRate          Temp.    SpGrav   Viscosity  Pressure   MAOP
Location
  (mi)       (ft)    (bbl/hr)          (degF)              CST       (psi)    (psi)
Name

0.0000     100.00    5,000.00          100.00   0.8250    15.00      50.00    300.00
Anaheim

0.0000     100.00    5,000.00          100.00   0.8250    15.00     228.52    300.00
Anaheim
0.2500      75.00    5,000.00           99.94   0.8251    15.02     233.01    300.00
0.5000     125.00    5,000.00           99.88   0.8252    15.04     210.71    300.00
0.7500     200.00    5,000.00           99.82   0.8253    15.06     179.47    300.00
1.0000     300.00    5,000.00           99.76   0.8254    15.08     139.29    300.00
1.2500     400.00    5,000.00           99.70   0.8255    15.10      99.11    300.00
1.5000     300.00    5,000.00           99.64   0.8256    15.12     130.40    300.00
1.7500     400.00    5,000.00           99.58   0.8257    15.14      90.20    300.00
2.0000     500.00    5,000.00           99.52   0.8258    15.16      50.00    300.00
Cuiba
```

4. CASE STUDY 3: HEAVY CRUDE OIL PIPELINE FROM JOPLIN TO BEAUMONT (THERMAL FLOW WITH HEATERS AND NO BATCHING)

A 14-in, 0.250-in wall thickness pipeline is used for shipping heavy crude oil from Joplin to a delivery terminal at Beaumont. A pump station and a heater station are located at Joplin with the following data:

Joplin pump station:

    Suction pressure: 25 psig
    Total motor HP installed: 1800 HP

Joplin heater:

    Inlet temperature: 100 °F
    Outlet temperature: 150 °F
    Heater efficiency: 82%

Three identical pumps (600 HP each) are installed in series at Joplin pump station. Each pump curve is defined by JOPLIN.PMP as given below:

Flow Rate (gpm)	Head (ft)	Efficiency (%)
0.0	2020	0.0
400	2070	54.2
600	2060	68.2
800	2000	76.9
1100	1820	82.0
1200	1725	81.1
1400	1500	76.1

The crude oil properties are as follows:

Temperature (°F)	Specific Gravity	Viscosity (cp)
60	0.925	500
120	0.814	215

Pipe delivery pressure	75 psig
Pipe depth of cover	36 in
Insulation thickness	0.0 in (uninsulated)
Insulation conductivity	0.02 Btu/h/ft/°F
Pipe conductivity	29.0 Btu/h/ft/°F
Thermal conductivity of soil	0.54 Btu/hr/ft/°F
Soil temperature	60 °F

The pipeline elevation profile is as below:

Distance (mi)	Elevation (ft)
0.00	50.00
10.00	75.00
25.00	125.00
35.00	89.00
40.00	67.00
50.00	112.00
65.00	152.00
75.00	423.00
80.00	300.00
100.00	240.00

Determine the pressure and temperature profile and the HP required to transport the crude oil at 1500 bbl/h considering the pump curve data. Use a pipe absolute roughness of 0.002 in and constant MAOP for pipe equal to 1800 psig (Figures 14.12–14.14).

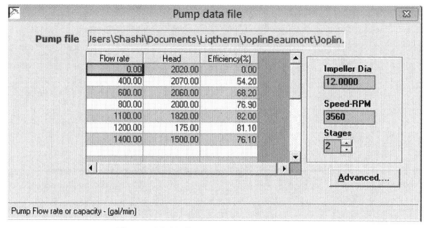

Figure 14.12 Pump curve input screen.

Case Studies 539

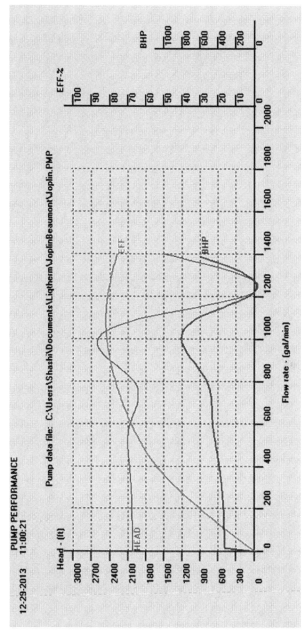

Figure 14.13 Pump curve plot.

```
******* LIQTHERM  STEADY STATE PIPELINE HYDRAULIC SIMULATION REPORT *********

DATE: 29-December-2013      TIME:  10:53:24

PROJECT:                    Joplin to Beaumont Pipeline with heavy crude
                            Problem 3
                            One pump station and one heater at Joplin

Pipeline data file:         \JoplinBeaumont\JoplintoBeaumont


******* LIQTHERM - LIQUID PIPELINE STEADY STATE HYDRAULIC SIMULATION ********
****************** Version 6.00.820**************

CASE NUMBER:                      1045

CALCULATION OPTIONS:
Thermal Calculations:             NO
Frictional Heating:               YES
Use Pump Curves:                  YES
Pump Curves Corrected for Viscosity:  NO
MAOP Check:                       YES
Horsepower Check:                 NO
Heating due to pump inefficiency: NO
Valves/Fittings and Devices:      NO
Branch pipe calculations:         NO
Loop pipe calculations:           NO
Maximum Inlet Flow:               NO
Batching Considered:              NO
DRA Injection:                    NO
Correct volumes for temperature:  NO
Slack Line Calculations:          NO
Customized Output:                NO


Inlet flow rate:                  1,500.     (bbl/hr)
Outlet flow rate:                 1,500.     (bbl/hr)
Inlet flow temperature:           100.00     (degF)
Outlet flow temperature:          65.42      (degF)

Minimum Pipe pressure:            0.00       (psi)
Pipe delivery pressure:           75.00      (psi)


Pressure drop formula used:       Colebrook-White equation
Calculation sub-divisions:        2
Iteration Accuracy:               MEDIUM


********** LIQUID PROPERTIES **********
Liquid properties file: \Liqtherm\Liquid Properties Database

PRODUCT:                  Product-A
Specific gravity:         0.9250 at   60.0(degF)
                          0.8140 at  120.0(degF)
Viscosity:                500.00 CP at  60.0(degF)
                          215.00 CP at 120.0(degF)

********** LIQUID FLOW RATES AND LOCATIONS **********

Location     Flow rate     Inlet Temp.    Product
(mi)         (bbl/hr)      (degF)

0.00         1,500.        100.0          Product-A
```

Case Studies

```
********** PUMP STATIONS **********

Pump         Distance   Pump suct   Pump disch  Sta. disch  Throttled   BHP Reqd
TotHPinst.   KW
station                 pressure    pressure    pressure    pressure    by pump
(Active)
             (mi)       (psi)       (psi)       (psi)       (psi)

Joplin       0.00       25.00       2364.84     1724.50     640.34      1749.
1800.        1305.

         Total active pump stations: 1         TOTAL Power:      1,749.
1,800.       1305.


********** PUMP AND DRIVER DATA **********

PumpSta.     Config.    Pump Curves       Status   Driver   RPM      Pump BHP
HPInstalled

Joplin       Series     JOPLIN.PMP        ON       Motor    3,560.   583.      600
                        JOPLIN.PMP        ON       Motor    3,560.   583.      600
                        JOPLIN.PMP        ON       Motor    3,560.   583.      600

Pump Station:     Joplin
Pump curve file: JOPLIN.PMP
Constant Speed Pump(s): 3,560. RPM
Pump curve: JOPLIN.PMP         Pump Status:ON
Pump impeller : 12.000(in)   Number of stages: 2
Operating point: 1050.00(gal/min)  2375.31(ft)  81.93(%)


Flow rate       Head       Efficiency     WaterHP
(gal/min)       (ft)       (%)            (HP)

 0              2020       0.01           0.00
 400            2070       54.2           385.78
 600            2060       68.2           457.66
 800            2000       76.9           525.41
 1100           1820       82             616.53
 1200           175        81.1           65.39
 1400           1500       76.1           696.85


Resultant Pump Curve:  Joplin  Pump station
Constant Speed Pump(s): 3,560. RPM
Operating point: 1050.00(gal/min)  7125.93(ft)  81.93(%)

Flow rate       Head       Efficiency     WaterHP
(gal/min)       (ft)       (%)            (HP)

 0.01           6060.00    0.01           153.03
 400.00         6210.00    54.20          1157.33
 600.00         6180.00    68.20          1372.97
 800.00         6000.00    76.90          1576.23
 1100.00        5460.00    82.00          1849.59
 1200.00        525.00     81.10          196.17
 1400.00        4500.00    76.10          2090.55


********** HEATER STATIONS **********

Heater       Distance    Heater Inlet   Heater Outlet   HeaterEffy
HeaterDuty   HeatingCost
Station      (mi)        Temp.          Temp.           (%)
(MMBtu/hr)   ($/MMBtu)

Joplin       0.00        100.00         150.00          82.00      11.87
5.00
```

```
********** PIPELINE PROFILE DATA **********
Distance    Elevation   Diameter    Wall Thk.   Roughness   MAOP
Location
(mi)        (ft)        (in)        (in)        (in)        (psi)

0.0000      50.00       14.000      0.250       0.0020      1800.       Joplin
10.0000     75.00       14.000      0.250       0.0020      1800.
25.0000     125.00      14.000      0.250       0.0020      1800.
35.0000     89.00       14.000      0.250       0.0020      1800.
40.0000     67.00       14.000      0.250       0.0020      1800.
50.0000     112.00      14.000      0.250       0.0020      1800.
65.0000     152.00      14.000      0.250       0.0020      1800.
75.0000     423.00      14.000      0.250       0.0020      1800.
80.0000     300.00      14.000      0.250       0.0020      1800.
100.0000    240.00      14.000      0.250       0.0020      1800.
Beaumont

********** THERMAL CONDUCTIVITY PROFILE DATA **********

Distance    Burial depth    Insul.Thk       Thermal Conductivity        Soil
Temp        Location
            (Cover)                         Insulation  Pipe    Soil
(mi)        (in)            (in)            (Btu/hr/ft/degF)                (degF)

0.00        36.00           0.000           0.02        29.00   0.54        60.00
10.00       36.00           0.000           0.02        29.00   0.54        60.00
25.00       36.00           0.000           0.02        29.00   0.54        60.00
35.00       36.00           0.000           0.02        29.00   0.54        60.00
40.00       36.00           0.000           0.02        29.00   0.54        60.00
50.00       36.00           0.000           0.02        29.00   0.54        60.00
65.00       36.00           0.000           0.02        29.00   0.54        60.00
75.00       36.00           0.000           0.02        29.00   0.54        60.00
80.00       36.00           0.000           0.02        29.00   0.54        60.00
100.00      36.00           0.000           0.02        29.00   0.54        60.00

********** VELOCITY, REYNOLD'S NUMBER AND PRESSURE DROP **********

Distance    Diameter.   FlowRate        Velocity    Reynolds    Press.drop
Location
(mi)        (in)        (bbl/hr)        (ft/sec)    number      (psi/mi)

0.0000      14.00       1,500.00        2.35        747.        12.76       Joplin
10.0000     14.00       1,500.00        2.35        929.        9.81
25.0000     14.00       1,500.00        2.35        697.        13.85
35.0000     14.00       1,500.00        2.35        619.        15.94
40.0000     14.00       1,500.00        2.35        591.        16.80
50.0000     14.00       1,500.00        2.35        552.        18.21
65.0000     14.00       1,500.00        2.35        518.        19.61
75.0000     14.00       1,500.00        2.35        505.        20.18
80.0000     14.00       1,500.00        2.35        501.        20.38
100.0000    14.00       1,500.00        2.35        501.        20.38       Beaumont

********** TEMPERATURE AND PRESSURE PROFILE **********

Distance    Elevation   FlowRate        Temp.       SpGrav      Viscosity   Pressure    MAOP
Location
(mi)        (ft)        (bbl/hr)        (degF)                  CP          (psi)       (psi)
Name

0.0000      50.00       1,500.00        100.00      0.8510      280.37      25.00       1800.00
Joplin

0.0000      50.00       1,500.00        150.00      0.7585      147.98      25.00       1800.00
Joplin

0.0000      50.00       1,500.00        150.00      0.7585      147.98      1724.50     1800.00
Joplin
10.0000     75.00       1,500.00        119.80      0.8144      215.58      1648.93     1800.00
25.0000     125.00      1,500.00        94.05       0.8620      304.28      1484.12     1800.00
35.0000     89.00       1,500.00        84.07       0.8805      350.12      1359.06     1800.00
40.0000     67.00       1,500.00        80.37       0.8873      369.19      1287.76     1800.00
```

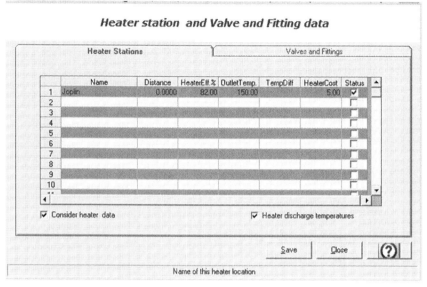

Figure 14.14 Heater station input screen.

```
 50.0000    112.00   1,500.00     74.87   0.8975   399.98   1102.44   1800.00
 65.0000    152.00   1,500.00     69.86   0.9068   430.73    813.81   1800.00
 75.0000    423.00   1,500.00     67.94   0.9103   443.27    511.38   1800.00
 80.0000    300.00   1,500.00     67.26   0.9116   447.82    458.98   1800.00
100.0000    240.00   1,500.00     65.42   0.9150   460.39     75.00   1800.00
Beaumont

Simulation started at:    10:53:09
Simulation completed at:  10:53:25
Total time taken: 16 seconds
Simulation Date:  29-December-2013
Output file: \JoplinBeaumont\JoplintoBeaumont.OUT
```

5. CASE STUDY 4: HEAVY CRUDE OIL PIPELINE (THERMAL FLOW WITH HEATERS AND DRA)

This Anaheim to Rio Grande pipeline will be transporting heated crude oil (160 °F discharge temperature) at certain locations to reduce oil viscosity and hence ease of pumpability.

Consider pipe sizes of 16- and 24-in NPS and determine the cost of facilities and operating costs for various pipe sizes and pump station HP (VFD motors) and heaters at strategic locations. DRA will also be considered to reduce costs (Figures 14.15–14.20).

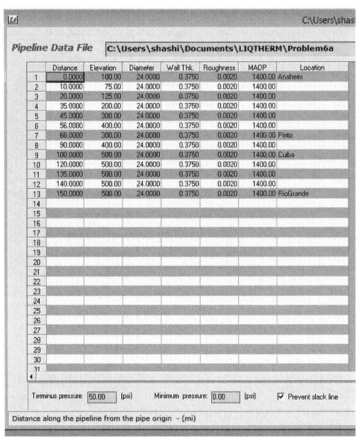

Figure 14.15 pipeline input screen.

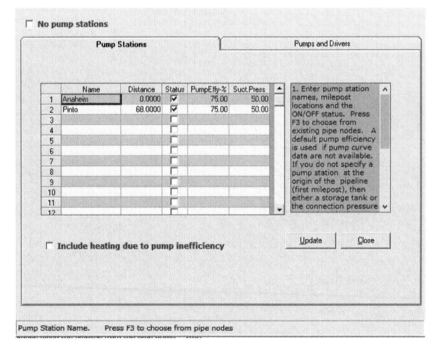

Figure 14.16 Pump station screen.

Case Studies

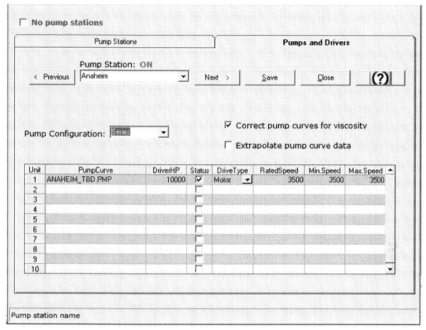

Figure 14.17 Anaheim pump details screen.

Thermal Conductivity, Insulation and Soil Data

	Distance	Cover	Insul.Thk	Insul.Cond	PipeCond	Soil Cond	SoilTemp
1	0.0000	36.0000	0.5000	0.0200	29.0000	0.6000	50.0000
2	10.0000	36.0000	0.5000	0.0200	29.0000	0.6000	50.0000
3	20.0000	36.0000	0.5000	0.0200	29.0000	0.6000	50.0000
4	35.0000	36.0000	0.5000	0.0200	29.0000	0.6000	50.0000
5	45.0000	36.0000	0.5000	0.0200	29.0000	0.6000	50.0000
6	56.0000	36.0000	0.5000	0.0200	29.0000	0.6000	50.0000
7	68.0000	36.0000	0.5000	0.0200	29.0000	0.6000	50.0000
8	90.0000	36.0000	0.5000	0.0200	29.0000	0.6000	50.0000
9	100.0000	36.0000	0.5000	0.0200	29.0000	0.6000	50.0000
10	120.0000	36.0000	0.5000	0.0200	29.0000	0.6000	50.0000
11	135.0000	36.0000	0.5000	0.0200	29.0000	0.6000	50.0000
12	140.0000	36.0000	0.5000	0.0200	29.0000	0.6000	50.0000
13	150.0000	36.0000	0.5000	0.0200	29.0000	0.6000	50.0000

Figure 14.18 Conductivity screen.

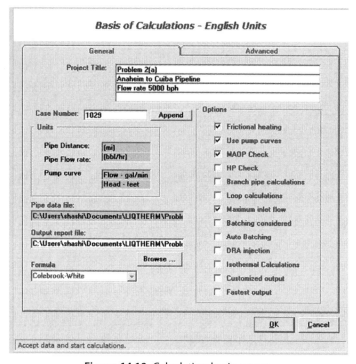

Figure 14.19 Calculation basis screen.

6. CASE STUDY 5: WATER PIPELINE FROM PAGE TO LAS CRUCES

A water pipeline is planned between a reservoir in Page, AZ, and the town of Las Cruces, NM, a distance of 295 miles to serve the residents of Las Cruces, NM. The estimated volume is 2000 million gal/year. Consider an average daily flow rate of 6 billion gal/year.

Select a steel pipeline of suitable pipe diameter and thickness based upon a max operating pressure of 720 psig to limit flow velocities to 3–5 ft/s As an alternative select a concrete pipeline and consider recommended Hazen–Williams C factors for the two materials. Estimate the power required, number of pump stations, and number of mainline valves and the capital and operating costs to accomplish this. Assume constant speed motor driven pumps and a storage tanks at the destination sufficient to satisfy a 7-day demand volume.

C factors for steel = 130.

for concrete C = 120.

Case Studies 547

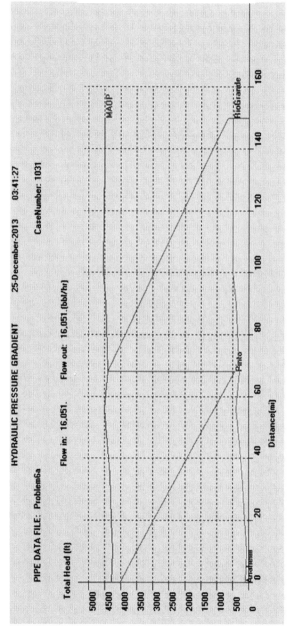

Figure 14.20 Hydraulic gradient.

6.1 Solution

Use the Hazen–Williams equation.

$$Q \text{ (bbl/day)} = 0.1482 \times C \times D^{2.63} (P_m/S_g)^{0.54}$$

$$Q \text{ (gpm)} = 6.7547 \times 10^{-3} \times C \times D^{2.63} (h_L)^{0.54}$$

h_L is in ft/1000 ft.

consider 36-in steel pipe, 0.500-in wall thickness.

$$\text{average velocity} = 0.0119 \times 6{,}000{,}000{,}000/(42 \times 365 \times 35.0 \times 35.0)$$
$$= 3.8 \text{ ft/s}$$

which is adequate.

$$6{,}000{,}000{,}000/(365 \times 24 \times 60) = 6.7547 \times 10^{-3} \times 130$$
$$\times (35.0)^{2.63} (h_L)^{0.54}$$

$h_L = 1.2538$ ft/1000 ft of pipe.

The total pressure required at origin is $1.2538 \times 5.28 \times 295 = 1953$ psi. Since MAOP is 720 psig, number of pump stations = 1953 psi/720 = 3 approximately.

Power required at the three pump stations is

$$3 \times (1953/3 - 50) \times (1/2.31) \times (6 \times 10^9)$$
$$\times (1/(365 \times 24 \times 60)/(3960 \times 0.75) = 3000 \text{ HP}$$

Cost of pumps and motors installed at $1800 per HP = $1800 × 3000 = $5.4 million.

$$\text{Storage tanks} = 7 \text{ days} = 7 \times 6{,}000{,}000{,}000/(365 \times 42)$$
$$= 2.8 \text{ million bbl} = \$20 \times 2.8 = \$56 \text{ million}$$

Mainline valves at 20 mile apart = 295/20 = 16 sets at $200,000 each site = $3.2 million.

Total capital = 5.4 + 56 + 3.2 + 10% for miscellaneous = 72 million.
Annual operating cost is estimated at $600/kWh plus labor about $5 million.

$$= 600 \times 3000 \times 0.746 \times 24 \times 365 + 5,000,000$$
$$= \$ 16.75 \text{ million per year}$$

7. CASE STUDY 6: GAS PIPELINE WITH MULTIPLE COMPRESSOR STATIONS FROM TAYLOR TO JENKS

A pipeline is planned between Taylor and a delivery point Jenks 220 miles away as shown in Figure 14.21. There are three compressor stations planned at Taylor, Trent, and Beaver as indicated. A flow rate of 500 million **standard cubic feet** per day (MMSCFD) of San Juan gas is delivered to Jenks at isothermal conditions of 70 °F.

Use the Colebrook–White equation for pressure drop and Standing–Katz for Z factor and consider Joule–Thompson effect for calculating the pressures and HP at each compressor station and also calculate the fuel consumed. Consider a delivery pressure of 500 psig at Jenks (Figures 14.22 and 14.23).

Holding delivery pressure.

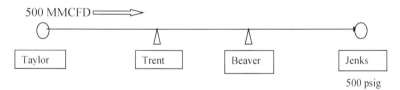

Figure 14.21 Gas pipeline with multiple compressor stations.

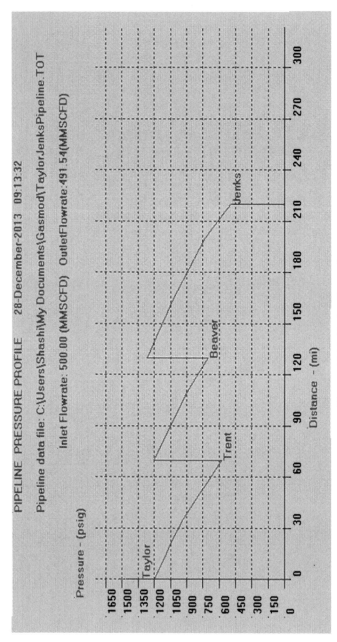

Figure 14.22 Hydraulic gradient.

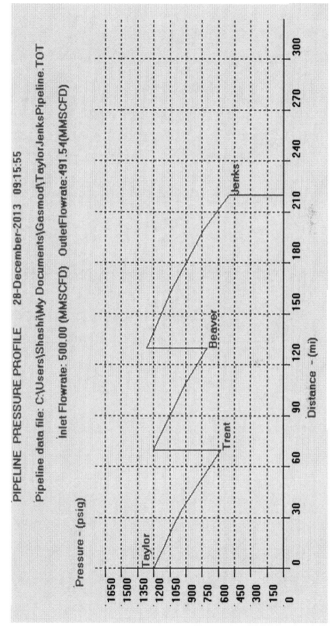

Figure 14.23 Hydraulic gradient with multiple compressor stations.

```
************ GASMOD - GAS PIPELINE HYDRAULIC SIMULATION ***********
************ Version 6.00.780 ************
DATE:                                28-December-2013    TIME:  09:13:26
PROJECT DESCRIPTION:
Pipeline from Taylor to Jenks
24 in pipeline - 220 miles long
3 compressor stations
Case Number:                         1055
Pipeline data file:                  C:\Users\Shashi\My
Documents\Gasmod\TaylorJenksPipeline.TOT

Pressure drop formula:               Colebrook-White
Pipeline efficiency:                 1.00
Compressibility Factor Method:       Standing-Katz

Inlet Gas Gravity(Air=1.0):          0.67883
Inlet Gas Viscosity:                 0.0000068(lb/ft-sec)
Gas specific heat ratio:             1.26
Polytropic compression index:        1.30


******** Calculations Based on Specified Thermal Conductivities of Pipe, Soil and
Insulation ********


Base temperature:                    60.00(degF)
Base pressure:                       14.70(psia)

Origin suction temperature:          70.00(degF)
Origin suction pressure:             800.00(psig)
Pipeline Terminus Delivery pressure: 495.32(psig)
Minimum pressure:                    400.0(psig)
Maximum gas velocity:                50.00(ft/sec)

Inlet  Flow rate:                    500.00(MMSCFD)
Outlet Flow rate:                    491.54(MMSCFD)

CALCULATION OPTIONS:
Polytropic compression considered:   YES
Branch pipe calculations:            NO
Loop pipe calculations:              NO
Compressor Fuel Calculated:          YES
Joule Thompson effect included :     YES
Customized Output:                   NO

ALL PRESSURES ARE GAUGE PRESSURES, UNLESS OTHERWISE SPECIFED AS ABSOLUTE PRESSURES


*************** PIPELINE PROFILE DATA ***********

Distance     Elevation     Diameter      Thickness     Roughness
(mi)         (ft)          (in)          (in)          (in)

0.00         620.00        24.000        0.500         0.000700
10.00        620.00        24.000        0.500         0.000700
15.00        620.00        24.000        0.500         0.000700
22.00        620.00        24.000        0.500         0.000700
35.00        620.00        24.000        0.500         0.000700
70.00        620.00        24.000        0.500         0.000700
80.00        620.00        24.000        0.500         0.000700
92.00        620.00        24.000        0.500         0.000700
110.00       620.00        24.000        0.500         0.000700
130.00       620.00        24.000        0.500         0.000700
142.00       620.00        24.000        0.500         0.000700
157.00       620.00        24.000        0.500         0.000700
163.00       620.00        24.000        0.500         0.000700
200.00       620.00        24.000        0.500         0.000700
220.00       500.00        16.000        0.375         0.000700
```

Case Studies

************** THERMAL CONDUCTIVITY AND INSULATION DATA ****************

Distance (mi)	Cover (in)	Thermal Conductivity (Btu/hr/ft/degF)			Insul.Thk (in)	Soil Temp (degF)
		Pipe	Soil	Insulation		
0.000	36.000	29.000	0.600	0.200	1.000	60.00
10.000	36.000	29.000	0.600	0.200	1.000	60.00
15.000	36.000	29.000	0.600	0.200	1.000	60.00
22.000	36.000	29.000	0.600	0.200	1.000	60.00
35.000	36.000	29.000	0.600	0.200	1.000	60.00
70.000	36.000	29.000	0.800	0.200	1.000	50.00
80.000	36.000	29.000	0.800	0.200	1.000	50.00
92.000	36.000	29.000	0.800	0.200	1.000	50.00
110.000	36.000	29.000	0.800	0.200	1.000	50.00
130.000	36.000	29.000	0.700	0.200	1.000	40.00
142.000	36.000	29.000	0.700	0.200	1.000	40.00
157.000	36.000	29.000	0.500	0.200	1.000	50.00
163.000	36.000	29.000	0.500	0.200	1.000	50.00
200.000	36.000	29.000	0.500	0.200	1.000	50.00
220.000	36.000	29.000	0.500	0.200	1.000	50.00

**************** LOCATIONS AND FLOW RATES ****************

Location GasName	Distance (mi)	Flow in/out (MMSCFD)	Gravity	Viscosity (lb/ft-sec)	Pressure (psig)	GasTemp. (degF)
Taylor SAN JUAN GAS	0.00	500.0000	0.6788	0.00000684	800.00	70.00
Jenks	220.00	-491.5423	0.6877	0.00000693	495.32	41.25

**************** COMPRESSOR STATION DATA **************

FLOW RATES, PRESSURES AND TEMPERATURES:

Name MaxPipe Temp (degF)	Flow Rate (MMSCFD)	Suct. Press. (psig)	Disch. Press. (psig)	Compr. Ratio	Suct. Loss. (psia)	Disch. Loss. (psia)	Suct. Temp. (degF)	Disch. Temp (degF)
Taylor 140.00	498.00	795.00	1210.00	1.5125	5.00	10.00	70.00	125.58
Trent 140.00	494.30	569.55	1210.00	2.0962	5.00	10.00	55.40	155.45
Beaver 140.00	491.54	700.73	1271.80	1.7982	5.00	10.00	52.90	130.57

Gas Cooling required at compressor station: Trent to limit station discharge temperature to 140 (degF)

************ COMPRESSOR EFFICIENCY, HP AND FUEL USED ****************

Name Installed (HP)	Distance (mi)	Compr Effy. (%)	Mech. Effy. (%)	Overall Effy. (%)	Horse Power	Fuel Factor (MCF/day/HP)	Fuel Used (MMSCFD)
Taylor 5000	0.00	85.00	98.00	83.30	10,010.49	0.2000	2.0021
Trent 5000	70.00	85.00	98.00	83.30	18,492.03	0.2000	3.6984
Beaver 5000	130.00	85.00	98.00	83.30	13,786.15	0.2000	2.7572

Total Compressor Station Horsepower: 42,288.67
15,000.

Total Fuel consumption: 8.4577 (MMSCFD)

WARNING!
Required HP exceeds the installed HP at compressor station: Taylor
Required HP exceeds the installed HP at compressor station: Trent
Required HP exceeds the installed HP at compressor station: Beaver

************** REYNOLD'S NUMBER AND HEAT TRANSFER COEFFICIENT **************

Distance (mi)	Reynold'sNum.	FrictFactor (Darcy)	Transmission Factor	HeatTransCoeff (Btu/hr/ft2/degF)	CompressibilityFactor (Standing-Katz)
0.000	29,024,499.	0.0100	19.97	0.2671	0.8490
10.000	29,024,499.	0.0100	19.97	0.2671	0.8459
15.000	29,024,499.	0.0100	19.97	0.2671	0.8431
22.000	29,024,499.	0.0100	19.97	0.2671	0.8421
35.000	29,024,499.	0.0100	19.97	0.2671	0.8515
70.000	29,024,499.	0.0100	19.97	0.3436	0.8556
80.000	29,024,499.	0.0100	19.97	0.3436	0.8444
92.000	29,024,499.	0.0100	19.97	0.3436	0.8339
110.000	29,024,499.	0.0100	19.97	0.3436	0.8355
130.000	29,024,499.	0.0100	19.97	0.3060	0.8392
142.000	29,024,499.	0.0100	19.97	0.3060	0.8235
157.000	29,024,499.	0.0100	19.97	0.2267	0.8153
163.000	29,024,499.	0.0100	19.97	0.2267	0.8199
200.000	29,024,499.	0.0100	19.97	0.2267	0.8553
220.000	29,024,499.	0.0100	19.97	0.2267	0.8553

******************* PIPELINE TEMPERATURE AND PRESSURE PROFILE *********************

Distance Location (mi)	Diameter (in)	Flow (MMSCFD)	Velocity (ft/sec)	Press. (psig)	GasTemp. (degF)	SoilTemp. (degF)	MAOP (psig)
0.00 Taylor	24.000	497.9979	24.22	1200.00	125.58	60.00	1440.00
10.00	24.000	497.9979	25.76	1127.38	111.03	60.00	1440.00
15.00	24.000	497.9979	26.63	1090.11	104.57	60.00	1440.00
22.00	24.000	497.9979	27.98	1036.62	96.38	60.00	1440.00
35.00	24.000	497.9979	31.09	931.59	83.41	60.00	1440.00
70.00 Trent	24.000	497.9979	49.55	574.55	55.40	50.00	1440.00
70.00 Trent	24.000	494.2995	24.04	1200.00	140.00	50.00	1440.00
80.00	24.000	494.2995	25.60	1126.05	117.31	50.00	1440.00
92.00	24.000	494.2995	27.81	1035.46	95.66	50.00	1440.00
110.00	24.000	494.2995	32.23	891.23	72.34	50.00	1440.00
130.00 Beaver	24.000	494.2995	40.31	705.73	52.90	40.00	1440.00
130.00 Beaver	24.000	491.5423	22.75	1261.80	130.57	40.00	1440.00
142.00	24.000	491.5423	24.29	1180.80	105.85	40.00	1440.00
157.00	24.000	491.5423	26.58	1077.85	83.53	50.00	1440.00
163.00	24.000	491.5423	27.65	1035.35	78.16	50.00	1440.00
200.00	24.000	491.5423	38.96	730.62	52.87	50.00	1440.00
220.00 Jenks	16.000	491.5423	129.51	495.32	41.25	50.00	1440.00

Gas velocity exceeds 50(ft/sec) @ location: 220.00(mi)

******************* LINE PACK VOLUMES AND PRESSURES *********************

Distance (mi)	Pressure (psig)	Line Pack (million std.cu.ft)
0.00	1200.00	0.0000
10.00	1127.38	12.9978
15.00	1090.11	6.3401
22.00	1036.62	8.6562
35.00	931.59	15.2169
70.00	574.55	32.8134
80.00	1126.05	9.2624
92.00	1035.46	14.8480
110.00	891.23	20.9852
130.00	705.73	20.1910
142.00	1180.80	12.6136
157.00	1077.85	20.2506
163.00	1035.35	7.8613
200.00	730.62	41.8639
220.00	495.32	15.7852

Total line pack in main pipeline = 239.6856(million std.cubic ft)

Started simulation at: 09:13:20
Finished simulation at: 09:13:26
Time elapsed   : 6 seconds
DATE: 28-December-2013

```
************ GASMOD - GAS PIPELINE HYDRAULIC SIMULATION ***********
************ Version 6.00.780 *************

DATE:                                    28-December-2013    TIME:  09:15:20
PROJECT DESCRIPTION:
Pipeline from Taylor to Jenks
24 in pipeline - 220 miles long
3 compressor stations
Holding delivery pressure

Pipeline data file:                      C:\Users\Shashi\My
Documents\Gasmod\TaylorJenksPipeline.TOT

Pressure drop formula:                   Colebrook-White
Pipeline efficiency:                     1.00
Compressibility Factor Method:           Standing-Katz

Inlet Gas Gravity(Air=1.0):              0.67883
Inlet Gas Viscosity:                     0.0000068(lb/ft-sec)
Gas specific heat ratio:                 1.26
Polytropic compression index:            1.30


******** Calculations Based on Specified Thermal Conductivities of Pipe, Soil and
Insulation ********


Base temperature:                        60.00(degF)
Base pressure:                           14.70(psia)

Origin suction temperature:              70.00(degF)
Origin suction pressure:                 800.00(psig)
Pipeline Terminus Delivery  pressure:    499.54(psig)
Minimum pressure:                        400.0(psig)
Maximum gas velocity:                    50.00(ft/sec)

Inlet  Flow rate:                        500.00(MMSCFD)
Outlet Flow rate:                        491.54(MMSCFD)

CALCULATION OPTIONS:
Polytropic compression considered:       YES
Branch pipe calculations:                NO
Loop pipe calculations:                  NO
Compressor Fuel Calculated:              YES
Joule Thompson effect included :         YES
Customized Output:                       NO
Holding Delivery Pressure at terminus

ALL PRESSURES ARE GAUGE PRESSURES, UNLESS OTHERWISE SPECIFED AS ABSOLUTE PRESSURES

***************** PIPELINE PROFILE DATA ***********

    Distance     Elevation      Diameter       Thickness       Roughness
    (mi)         (ft)           (in)           (in)            (in)

    0.00         620.00         24.000         0.500           0.000700
    10.00        620.00         24.000         0.500           0.000700
    15.00        620.00         24.000         0.500           0.000700
    22.00        620.00         24.000         0.500           0.000700
    35.00        620.00         24.000         0.500           0.000700
    70.00        620.00         24.000         0.500           0.000700
    80.00        620.00         24.000         0.500           0.000700
    92.00        620.00         24.000         0.500           0.000700
    110.00       620.00         24.000         0.500           0.000700
    130.00       620.00         24.000         0.500           0.000700
    142.00       620.00         24.000         0.500           0.000700
    157.00       620.00         24.000         0.500           0.000700
    163.00       620.00         24.000         0.500           0.000700
    200.00       620.00         24.000         0.500           0.000700
    220.00       500.00         16.000         0.375           0.000700
```

```
************** THERMAL CONDUCTIVITY AND INSULATION DATA ****************

Distance   Cover      Thermal Conductivity        Insul.Thk      Soil Temp
(mi)       (in)       (Btu/hr/ft/degF)            (in)           (degF)
                      Pipe    Soil    Insulation
0.000      36.000     29.000  0.600   0.200       1.000          60.00
10.000     36.000     29.000  0.600   0.200       1.000          60.00
15.000     36.000     29.000  0.600   0.200       1.000          60.00
22.000     36.000     29.000  0.600   0.200       1.000          60.00
35.000     36.000     29.000  0.600   0.200       1.000          60.00
70.000     36.000     29.000  0.800   0.200       1.000          50.00
80.000     36.000     29.000  0.800   0.200       1.000          50.00
92.000     36.000     29.000  0.800   0.200       1.000          50.00
110.000    36.000     29.000  0.800   0.200       1.000          50.00
130.000    36.000     29.000  0.700   0.200       1.000          40.00
142.000    36.000     29.000  0.700   0.200       1.000          40.00
157.000    36.000     29.000  0.500   0.200       1.000          50.00
163.000    36.000     29.000  0.500   0.200       1.000          50.00
200.000    36.000     29.000  0.500   0.200       1.000          50.00
220.000    36.000     29.000  0.500   0.200       1.000          50.00


**************** LOCATIONS AND FLOW RATES ****************

Location       Distance    Flow in/out   Gravity   Viscosity    Pressure   GasTemp.
GasName
               (mi)        (MMSCFD)                (lb/ft-sec)  (psig)     (degF)

Taylor         0.00        500.0000      0.6788    0.00000684   800.00     70.00
SAN JUAN GAS
Jenks          220.00      -491.5425     0.6877    0.00000693   499.54     41.35


**************** COMPRESSOR STATION DATA ***************


FLOW RATES, PRESSURES AND TEMPERATURES:

Name      Flow       Suct.      Disch.     Compr.    Suct.    Disch.    Suct.    Disch.
MaxPipe
          Rate       Press.     Press.     Ratio     Loss.    Loss.     Temp.    Temp
Temp
          (MMSCFD)   (psig)     (psig)               (psia)   (psia)    (degF)   (degF)
(degF)
Taylor    498.00     795.00     1210.00    1.5125    5.00     10.00     70.00    125.58
140.00
Trent     494.30     569.55     1210.00    2.0962    5.00     10.00     55.40    155.45
140.00
Beaver    491.54     700.73     1273.25    1.8003    5.00     10.00     52.90    130.52
140.00


Gas Cooling required at compressor station: Trent to limit station discharge
temperature to 140 (degF)


************* COMPRESSOR EFFICIENCY, HP AND FUEL USED ****************


Name      Distance   Compr    Mech.    Overall   Horse       Fuel         Fuel       Installed
                     Effy.    Effy.    Effy.     Power       Factor       Used
          (mi)       (%)      (%)      (%)                   (MCF/day/HP) (MMSCFD)   (HP)
Taylor    0.00       85.00    98.00    83.30     10,010.66   0.2000       2.0021     5000
Trent     70.00      85.00    98.00    83.30     18,500.49   0.2000       3.7001     5000
Beaver    130.00     85.00    98.00    83.30     13,776.30   0.2000       2.7553     5000

Total Compressor Station Horsepower:             42,287.45                15,000.

Total Fuel consumption:            8.4575 (MMSCFD)


WARNING!
Required HP exceeds the installed HP at compressor station: Taylor
Required HP exceeds the installed HP at compressor station: Trent
Required HP exceeds the installed HP at compressor station: Beaver
```

Case Studies

```
************** REYNOLD'S NUMBER  AND  HEAT TRANSFER COEFFICIENT **************
```

Distance (mi)	Reynold'sNum.	FrictFactor (Darcy)	Transmission Factor	HeatTransCoeff (Btu/hr/ft2/degF)	CompressibilityFactor (Standing-Katz)
0.000	29,024,499.	0.0100	19.97	0.2671	0.8490
10.000	29,024,499.	0.0100	19.97	0.2671	0.8459
15.000	29,024,499.	0.0100	19.97	0.2671	0.8431
22.000	29,024,499.	0.0100	19.97	0.2671	0.8421
35.000	29,024,499.	0.0100	19.97	0.2671	0.8515
70.000	29,024,499.	0.0100	19.97	0.3436	0.8556
80.000	29,024,499.	0.0100	19.97	0.3436	0.8444
92.000	29,024,499.	0.0100	19.97	0.3436	0.8339
110.000	29,024,499.	0.0100	19.97	0.3436	0.8355
130.000	29,024,499.	0.0100	19.97	0.3060	0.8390
142.000	29,024,499.	0.0100	19.97	0.3060	0.8233
157.000	29,024,499.	0.0100	19.97	0.2267	0.8150
163.000	29,024,499.	0.0100	19.97	0.2267	0.8195
200.000	29,024,499.	0.0100	19.97	0.2267	0.8546
220.000	29,024,499.	0.0100	19.97	0.2267	0.8546

```
******************* PIPELINE TEMPERATURE AND PRESSURE PROFILE ********************
```

Distance Location (mi)	Diameter (in)	Flow (MMSCFD)	Velocity (ft/sec)	Press. (psig)	GasTemp. (degF)	SoilTemp. (degF)	MAOP (psig)
0.00 Taylor	24.000	497.9979	24.22	1200.00	125.58	60.00	1440.00
10.00	24.000	497.9979	25.76	1127.38	111.03	60.00	1440.00
15.00	24.000	497.9979	26.63	1090.11	104.57	60.00	1440.00
22.00	24.000	497.9979	27.98	1036.62	96.38	60.00	1440.00
35.00	24.000	497.9979	31.09	931.59	83.41	60.00	1440.00
70.00 Trent	24.000	497.9979	49.55	574.55	55.40	50.00	1440.00
70.00 Trent	24.000	494.2978	24.04	1200.00	140.00	50.00	1440.00
80.00	24.000	494.2978	25.60	1126.05	117.31	50.00	1440.00
92.00	24.000	494.2978	27.81	1035.46	95.66	50.00	1440.00
110.00	24.000	494.2978	32.23	891.23	72.34	50.00	1440.00
130.00 Beaver	24.000	494.2978	40.31	705.73	52.90	40.00	1440.00
130.00 Beaver	24.000	491.5425	22.72	1263.25	130.52	40.00	1440.00
142.00	24.000	491.5425	24.26	1182.38	105.82	40.00	1440.00
157.00	24.000	491.5425	26.53	1079.61	83.52	50.00	1440.00
163.00	24.000	491.5425	27.60	1037.20	78.15	50.00	1440.00
200.00	24.000	491.5425	38.82	733.37	52.90	50.00	1440.00
220.00 Jenks	16.000	491.5425	128.44	499.54	41.35	50.00	1440.00

Gas velocity exceeds 50(ft/sec) @ location: 220.00(mi)

****************** LINE PACK VOLUMES AND PRESSURES ********************

Distance (mi)	Pressure (psig)	Line Pack (million std.cu.ft)
0.00	1200.00	0.0000
10.00	1127.38	12.9977
15.00	1090.11	6.3400
22.00	1036.62	8.6561
35.00	931.59	15.2168
70.00	574.55	32.8132
80.00	1126.05	9.2624
92.00	1035.46	14.8479
110.00	891.23	20.9851
130.00	705.73	20.1908
142.00	1182.38	12.6278
157.00	1079.61	20.2865
163.00	1037.20	7.8773
200.00	733.37	41.9847
220.00	499.54	15.8791

Total line pack in main pipeline = 239.9654 (million std.cubic ft)

8. CASE STUDY 7: GAS PIPELINE HYDRAULICS WITH INJECTIONS AND DELIVERIES

An 18-i/16-in-diameter, 420-mile-long buried pipeline defined here is used to transport 150 MMSCFD of natural gas from Compton to the delivery terminus at Harvey. There are three compressor stations located at Compton, Dimpton, and Plimpton, with a gas turbine–driven centrifugal compressors. The pipeline is *not* insulated and the maximum pipeline temperature is limited to 140 °F because of the material of the pipe external coating. The MAOP is 1440 psig. Determine the temperature and pressure profile and horsepower required at each compressor station (Figure 14.24).

The pipeline profile is defined below:

Distance (mi)	Elevation (ft)	Pipe Diameter (in)	Wall Thickness (in)	Roughness (in)
0.0	620	18.00	0.375	0.000700
45	620	18.00	0.375	0.000700
48	980	18.00	0.375	0.000700
85	1285	16.00	0.375	0.000700
160	1500	16.00	0.375	0.000700
200	2280	16.00	0.375	0.000700
238	950	16.00	0.375	0.000700
250	891	16.00	0.375	0.000700
295	670	16.00	0.375	0.000700
305	650	16.00	0.375	0.000700
310	500	16.00	0.375	0.000700
320	420	16.00	0.375	0.000700
330	380	16.00	0.375	0.000700
380	280	16.00	0.375	0.000700
420	500	16.00	0.375	0.000700

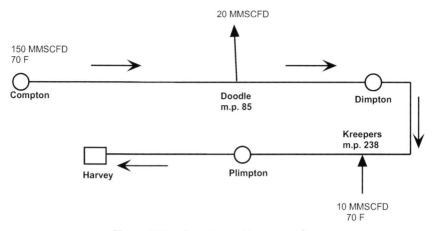

Figure 14.24 Compton to Harvey pipeline.

Gas specific heat (K) ratio	1.26
Maximum gas velocity	50 ft/s
Pipeline efficiency	1.00
Base temperature	60 °F
Base pressure	14.70 psia
Pressure drop formula	AGA fully turbulent
Compressibility factor	Standing-Katz
Polytropic index	1.3

A flow rate of 150 MMSCFD enters the pipeline at Compton (milepost 0.0) and at an intermediate location named Doodle (milepost 85), a delivery of 20 MMSCFD is made. Additionally an injection of 10 MMSCFD is made at Kreepers (milepost 238). The resulting flow then continues to the end of the pipeline. Gas inlet temperature is 70 °F at both locations.

Inlet gas specific gravity (air = 1.00)	0.600
Inlet gas viscosity	0.000008 lb/ft-sec

The compressor stations are as follows:

Compressor Station	Location (mi)	Discharge Pressure (psig)
Compton	0.00	1400
Dimpton	160.0	1400
Plimpton	295.0	1400

The installed HP at each compressor stations is 5000 HP.

Origin suction pressure	800 psig
Pipeline delivery pressure	500 psig
Minimum pipe pressure	400 psig
Station suction loss	5 psig
Station discharge loss	10 psig
Compressor adiabatic efficiency	85%
Compressor mechanical efficiency	98%
Fuel consumption	0.2 MCF/day/HP
Pipe burial depth	36 in
Pipe thermal conductivity	29 Btu/h/ft/°F
Soil thermal conductivity	0.800 Btu/h/ft/°F
Ambient soil temperature	65 °F
Origin suction temperature	70 °F

```
************ GASMOD - GAS PIPELINE HYDRAULIC SIMULATION ************
************ Version 6.00.780 *************
DATE:                                    4-January-2013      TIME:  07:08:01
PROJECT DESCRIPTION:
Problem 1
Pipeline from Compton to Harvey
18"/16" pipeline - 420 miles long
3 compressor stations

Pipeline data file:                      C:\Users\Shashi\My
Documents\Gasmod\Problem1.TOT

Pressure drop formula:                   AGA Turbulent
Pipeline efficiency:                     1.00
Compressibility Factor Method:           Standing-Katz

Inlet Gas Gravity(Air=1.0):              0.60000
Inlet Gas Viscosity:                     0.0000080(lb/ft-sec)
Gas specific heat ratio:                 1.26
Polytropic compression index:            1.30

**** Calculations Based on Specified Thermal Conductivities of Pipe, Soil and Insulation ******

Base temperature:                        60.00(degF)
Base pressure:                           14.70(psia)

Origin suction temperature:              70.00(degF)
Origin suction pressure:                 800.00(psig)
Pipeline Terminus Delivery  pressure:    851.27(psig)
Minimum pressure:                        400.0(psig)
Maximum gas velocity:                    50.00(ft/sec)

Inlet  Flow rate:                        150.00(MMSCFD)
Outlet Flow rate:                        137.82(MMSCFD)

CALCULATION OPTIONS:
Polytropic compression considered:       YES
Branch pipe calculations:                NO
Loop pipe calculations:                  NO
Compressor Fuel Calculated:              YES
Joule Thompson effect included :         NO
Customized Output:                       NO

ALL PRESSURES ARE GAUGE PRESSURES, UNLESS OTHERWISE SPECIFED AS ABSOLUTE PRESSURES

    **************** PIPELINE PROFILE DATA ************

    Distance      Elevation     Diameter      Thickness     Roughness
    (mi)          (ft)          (in)          (in)          (in)

    0.00          620.00        18.000        0.375         0.000700
   45.00          620.00        18.000        0.375         0.000700
   48.00          980.00        18.000        0.375         0.000700
   85.00         1285.00        16.000        0.375         0.000700
  160.00         1500.00        16.000        0.375         0.000700
  200.00         2280.00        16.000        0.375         0.000700
  238.00          950.00        16.000        0.375         0.000700
  250.00          891.00        16.000        0.375         0.000700
  295.00          670.00        16.000        0.375         0.000700
  305.00          650.00        16.000        0.375         0.000700
  310.00          500.00        16.000        0.375         0.000700
  320.00          420.00        16.000        0.375         0.000700
  330.00          380.00        16.000        0.375         0.000700
  380.00          280.00        16.000        0.375         0.000700
  420.00          500.00        16.000        0.375         0.000700

    *************** THERMAL CONDUCTIVITY AND INSULATION DATA ****************

    Distance  Cover      Thermal Conductivity       Insul.Thk      Soil Temp
    (mi)      (in)       (Btu/hr/ft/degF)           (in)           (degF)
                         Pipe    Soil   Insulation

    0.000     36.000     29.000  0.800  0.020       0.000          65.00
   45.000     36.000     29.000  0.800  0.020       0.000          65.00
```

```
 48.000   36.000   29.000  0.800   0.020       0.000       65.00
 85.000   36.000   29.000  0.800   0.020       0.000       65.00
160.000   36.000   29.000  0.800   0.020       0.000       65.00
200.000   36.000   29.000  0.800   0.020       0.000       65.00
238.000   36.000   29.000  0.800   0.020       0.000       65.00
250.000   36.000   29.000  0.800   0.020       0.000       65.00
295.000   36.000   29.000  0.800   0.020       0.000       65.00
305.000   36.000   29.000  0.800   0.020       0.000       65.00
310.000   36.000   29.000  0.800   0.020       0.000       65.00
320.000   36.000   29.000  0.800   0.020       0.000       65.00
330.000   36.000   29.000  0.800   0.020       0.000       65.00
380.000   36.000   29.000  0.800   0.020       0.000       65.00
420.000   36.000   29.000  0.800   0.020       0.000       65.00
```

**************** LOCATIONS AND FLOW RATES ****************

Location	Distance (mi)	Flow in/out (MMSCFD)	Gravity	Viscosity (lb/ft-sec)	Pressure (psig)	GasTemp. (degF)
Compton	0.00	150.0000	0.6000	0.00000800	800.00	70.00
Doodle	85.00	-20.0000	0.6000	0.00000800	1172.79	65.01
Kreepers	238.00	10.0000	0.6000	0.00000800	1142.82	65.01
Harvey	420.00	-137.8152	0.6057	0.00000808	851.27	65.00

**************** COMPRESSOR STATION DATA ***************

FLOW RATES, PRESSURES AND TEMPERATURES:

Name	Flow Rate (MMSCFD)	Suct. Press. (psig)	Disch. Press. (psig)	Compr. Ratio	Suct. Loss. (psia)	Disch. Loss. (psia)	Suct. Temp. (degF)	Disch. Temp (degF)	MaxPipe Temp (degF)
Compton	149.13	795.00	1410.00	1.7595	5.00	10.00	70.00	147.11	140.00
Dimpton	128.48	840.04	1410.00	1.6668	5.00	10.00	65.00	133.67	140.00
Plimpton	137.82	861.17	1410.00	1.6266	5.00	10.00	65.00	130.22	140.00

Gas Cooling required at compressor station: Compton to limit station discharge temperature to 140 (degF)

************* COMPRESSOR EFFICIENCY, HP AND FUEL USED ****************

Name	Distance (mi)	Compr Effy. (%)	Mech. Effy. (%)	Overall Effy. (%)	Horse Power	Fuel Factor (MCF/day/HP)	Fuel Used (MMSCFD)	Installed (HP)
Compton	0.00	85.00	98.00	83.30	4,329.48	0.2000	0.8659	5000
Dimpton	160.00	85.00	98.00	83.30	3,275.63	0.2000	0.6551	5000
Plimpton	295.00	85.00	98.00	83.30	3,318.70	0.2000	0.6637	5000

Total Compressor Station Horsepower: 10,923.81   15,000.

Total Fuel consumption: 2.1847 (MMSCFD)

************* REYNOLD'S NUMBER AND HEAT TRANSFER COEFFICIENT **************

Distance (mi)	Reynold'sNum.	FrictFactor (Darcy)	Transmission Factor	HeatTransCoeff (Btu/hr/ft2/degF)	CompressibilityFactor (Standing-Katz)
0.000	8,758,087.	0.0102	19.84	0.4624	0.8323
45.000	8,758,087.	0.0102	19.84	0.4624	0.8206
48.000	8,758,087.	0.0102	19.84	0.4624	0.8263
85.000	8,578,128.	0.0104	19.63	0.4992	0.8500
160.000	8,578,128.	0.0104	19.63	0.4992	0.8321
200.000	8,578,128.	0.0104	19.63	0.4992	0.8282
238.000	9,242,408.	0.0104	19.63	0.4993	0.8367
250.000	9,242,408.	0.0104	19.63	0.4993	0.8529
295.000	9,242,408.	0.0104	19.63	0.4993	0.8513
305.000	9,242,408.	0.0104	19.63	0.4993	0.8306
310.000	9,242,408.	0.0104	19.63	0.4993	0.8224
320.000	9,242,408.	0.0104	19.63	0.4993	0.8186
330.000	9,242,408.	0.0104	19.63	0.4993	0.8289
380.000	9,242,408.	0.0104	19.63	0.4993	0.8542
420.000	9,242,408.	0.0104	19.63	0.4993	0.8542

************* PIPELINE TEMPERATURE AND PRESSURE PROFILE *************

Distance (mi)	Diameter (in)	Flow (MMSCFD)	Velocity (ft/sec)	Press. (psig)	GasTemp. (degF)	SoilTemp. (degF)	MAOP (psig)	Location
0.00	18.000	149.1341	11.07	1400.00	140.00	65.00	1440.00	Compton
45.00	18.000	149.1341	11.96	1294.52	65.73	65.00	1440.00	
48.00	18.000	149.1341	12.14	1275.15	65.52	65.00	1440.00	
85.00	16.000	129.1341	14.61	1172.79	65.01	65.00	1440.00	Doodle
160.00	16.000	129.1341	20.08	845.04	65.00	65.00	1440.00	Dimpton
160.00	16.000	128.4790	12.20	1400.00	133.67	65.00	1440.00	Dimpton
200.00	16.000	128.4790	13.77	1238.91	65.70	65.00	1440.00	
238.00	16.000	138.4790	16.08	1142.82	65.01	65.00	1440.00	Kreepers
250.00	16.000	138.4790	16.83	1090.68	65.00	65.00	1440.00	
295.00	16.000	138.4790	21.02	866.17	65.00	65.00	1440.00	Plimpton
295.00	16.000	137.8152	13.09	1400.00	130.22	65.00	1440.00	Plimpton
305.00	16.000	137.8152	13.45	1361.65	87.96	65.00	1440.00	
310.00	16.000	137.8152	13.59	1347.75	78.21	65.00	1440.00	
320.00	16.000	137.8152	13.95	1312.39	69.27	65.00	1440.00	
330.00	16.000	137.8152	14.36	1274.95	66.36	65.00	1440.00	
380.00	16.000	137.8152	17.16	1064.61	65.00	65.00	1440.00	
420.00	16.000	137.8152	21.38	851.27	65.00	65.00	1440.00	Harvey

************* LINE PACK VOLUMES AND PRESSURES *************

Distance (mi)	Pressure (psig)	Line Pack (million std.cu.ft)
0.00	1400.00	0.0000
45.00	1294.52	38.8174
48.00	1275.15	2.7536
85.00	1172.79	32.1715
160.00	845.04	41.2678
200.00	1238.91	20.7075
238.00	1142.82	25.0119
250.00	1090.68	7.3423
295.00	866.17	23.7981
305.00	1361.65	5.4573
310.00	1347.75	3.5963
320.00	1312.39	7.2528
330.00	1274.95	7.1750
380.00	1064.61	32.2835
420.00	851.27	20.6483

Total line pack in main pipeline = 268.2833(million std.cubic ft)

9. CASE STUDY 8: GAS PIPELINE WITH TWO COMPRESSOR STATIONS AND TWO PIPE BRANCHES

A 12- /14-in-diameter, 180-mile-long natural gas pipeline defined here is used to transport natural gas from Davis to the delivery terminus at Harvey. There are two compressor stations located at Davis (milepost 0.0) and Frampton (milepost 82.0), respectively. Each compressor station operates at a maximum discharge of 1200 psig. At Davis, 100 MMSCFD (0.600 specific gravity) of gas enters the pipeline at 80 °F inlet temperature. An outgoing pipe branch (branch 1—NPS 8) is located at milepost 25.0 that is used to deliver 30 MMSCFD of gas from the main pipeline to a delivery location 32 miles away. At milepost 90.0, there is an incoming pipe branch (branch 2) that is used to inject an additional volume of 50 MMSCFD gas (0.615 specific gravity) into the main pipeline at 80 °F inlet temperature. The incoming branch pipe is NPS 10, 40 miles long. The pipelines are uninsulated and the maximum pipeline temperature is limited to 140 °F. The soil temperature is assumed to be 60 °F. Use an overall heat transfer coefficient (U factor) of 0.500 and the Colebrook–White equation for pressure drop. The pipe MAOP is 1440 psig. The base pressure and base temperature is 14.73 psia and 60 °F, respectively. The delivery pressure at Harvey is 400 psi. Minimum pressure required is 300 psig. Origin suction pressure is 850 psig.

Determine the temperature, pressure profile, and HP required at each compressor station and the gas fuel consumption. Use the data on the report to create the pipeline and branches (Figure 14.25).

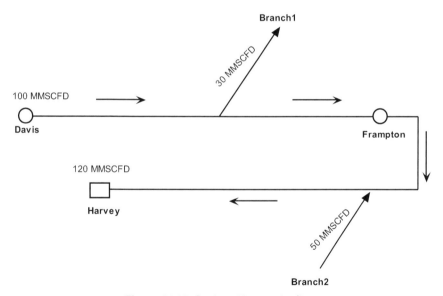

Figure 14.25 Davis to Harvey pipeline.

```
************ GASMOD - GAS PIPELINE HYDRAULIC SIMULATION ************
************ Version 6.00.780 ************

DATE:                                   4-January-2013      TIME:   09:23:25
PROJECT DESCRIPTION:
Problem 2
Pipeline from Davis to Harvey
12"/14" pipeline - 180 miles long
2 compressor stations
Branch in and Branch out

Pipeline data file:                     C:\Users\Shashi\MyDocuments\Gasmod\Problem2.TOT

Pressure drop formula:                  Colebrook-White
Pipeline efficiency:                    1.00
Compressibility Factor Method:          CNGA

Inlet Gas Gravity(Air=1.0):             0.60000
Inlet Gas Viscosity:                    0.0000080(lb/ft-sec)
Gas specific heat ratio:                1.29
Polytropic compression index:           1.30


**** Calculations Based on Specified Thermal Conductivities of Pipe, Soil and Insulation ****

Base temperature:                       60.00(degF)
Base pressure:                          14.73(psia)

Origin suction temperature:             80.00(degF)
Origin suction pressure:                850.00(psig)
Pipeline Terminus Delivery  pressure:   445.83(psig)
Minimum pressure:                       300.0(psig)
Maximum gas velocity:                   50.00(ft/sec)

Inlet   Flow rate:                      100.00(MMSCFD)
Outlet Flow rate:                       119.36(MMSCFD)

CALCULATION OPTIONS:
Polytropic compression considered:      YES
Branch pipe calculations:               YES
Loop pipe calculations:                 NO
Compressor Fuel Calculated:             YES
Joule Thompson effect included :        NO
Customized Output:                      NO
ALL PRESSURES ARE GAUGE PRESSURES, UNLESS OTHERWISE SPECIFED AS ABSOLUTE PRESSURES


**************** PIPELINE PROFILE DATA ************

   Distance      Elevation      Diameter       Thickness      Roughness
   (mi)          (ft)           (in)           (in)           (in)

   0.00          220.00         12.750         0.250          0.000700
   12.00         340.00         12.750         0.250          0.000700
   25.00         450.00         12.750         0.250          0.000700
   35.00         189.00         12.750         0.250          0.000700
   82.00         225.00         12.750         0.250          0.000700
   90.00         369.00         14.000         0.250          0.000700
   112.00        412.00         14.000         0.250          0.000700
   125.00        518.00         14.000         0.250          0.000700
   152.00        786.00         14.000         0.250          0.000700
   180.00        500.00         14.000         0.250          0.000700

************** THERMAL CONDUCTIVITY AND INSULATION DATA ****************

   Distance  Cover      Thermal Conductivity        Insul.Thk      Soil Temp
   (mi)      (in)       (Btu/hr/ft/degF)            (in)           (degF)
                        Pipe    Soil   Insulation
   0.000     36.000     29.000  0.800  0.020        0.000          60.00
   12.000    36.000     29.000  0.800  0.020        0.000          60.00
   25.000    36.000     29.000  0.800  0.020        0.000          60.00
   35.000    36.000     29.000  0.800  0.020        0.000          60.00
   82.000    36.000     29.000  0.800  0.020        0.000          60.00
   90.000    36.000     29.000  0.800  0.020        0.000          60.00
   112.000   36.000     29.000  0.800  0.020        0.000          60.00
   125.000   36.000     29.000  0.800  0.020        0.000          60.00
   152.000   36.000     29.000  0.800  0.020        0.000          60.00
   180.000   36.000     29.000  0.800  0.020        0.000          60.00
```

Case Studies

```
*************** LOCATIONS AND FLOW RATES ****************
```

Location	Distance (mi)	Flow in/out (MMSCFD)	Gravity	Viscosity (lb/ft-sec)	Pressure (psig)	GasTemp. (degF)
Davis	0.00	100.0000	0.6000	0.00000800	850.00	80.00
BranchOut	25.00	-30.0000	0.6000	0.00000800	1023.06	63.42
BranchIn	90.00	50.0000	0.6150	0.00000800	1170.30	77.05
Harvey	180.00	-119.3557	0.6077	0.00000802	445.83	60.00

```
*************** COMPRESSOR STATION DATA ***************
```
FLOW RATES, PRESSURES AND TEMPERATURES:

Name	Flow Rate (MMSCFD)	Suct. Press. (psig)	Disch. Press. (psig)	Compr. Ratio	Suct. Loss. (psia)	Disch. Loss. (psia)	Suct. Temp. (degF)	Disch. Temp (degF)	MaxPipe Temp (degF)
Davis	99.64	845.00	1210.00	1.4246	5.00	10.00	80.00	132.60	140.00
Frampton	69.36	800.57	1210.00	1.5022	5.00	10.00	60.00	118.60	140.00

```
************* COMPRESSOR EFFICIENCY, HP AND FUEL USED ****************
```

Name	Distance (mi)	Compr Effy. (%)	Mech. Effy. (%)	Overall Effy. (%)	Horse Power	Fuel Factor (MCF/day/HP)	Fuel Used (MMSCFD)	Installed (HP)
Davis	0.00	85.00	98.00	83.30	1,823.24	0.2000	0.3646	5000
Frampton	82.00	85.00	98.00	83.30	1,398.23	0.2000	0.2796	5000

Total Compressor Station Horsepower: 3,221.47   10,000.

Total Fuel consumption: 0.6442 (MMSCFD)

```
************** REYNOLD'S NUMBER  AND  HEAT TRANSFER COEFFICIENT **************
```

Distance (mi)	Reynold'sNum.	FrictFactor (Darcy)	Transmission Factor	HeatTransCoeff (Btu/hr/ft2/degF)	CompressibilityFactor (CNGA)
0.000	8,256,276.	0.0114	18.76	0.5000	0.8729
12.000	8,256,276.	0.0114	18.76	0.5000	0.8574
25.000	5,770,328.	0.0115	18.65	0.5000	0.8582
35.000	5,770,328.	0.0115	18.65	0.5000	0.8699
82.000	5,770,328.	0.0115	18.65	0.5000	0.8636
90.000	9,089,641.	0.0112	18.93	0.5452	0.8463
112.000	9,089,641.	0.0112	18.93	0.5452	0.8542
125.000	9,089,641.	0.0112	18.93	0.5452	0.8720
152.000	9,089,641.	0.0112	18.93	0.5452	0.9054
180.000	9,089,641.	0.0112	18.93	0.5452	0.9054

```
****************** PIPELINE TEMPERATURE AND PRESSURE PROFILE *********************
```

Distance (mi)	Diameter (in)	Flow (MMSCFD)	Velocity (ft/sec)	Press. (psig)	GasTemp. (degF)	SoilTemp. (degF)	MAOP (psig)	Location
0.00	12.750	99.6354	17.08	1200.00	132.60	60.00	1440.00	Davis
12.00	12.750	99.6354	18.37	1115.01	77.76	60.00	1440.00	
25.00	12.750	69.6354	13.97	1023.06	63.42	60.00	1440.00	BranchOut
35.00	12.750	69.6354	14.37	994.40	60.54	60.00	1440.00	
82.00	12.750	69.6354	17.61	805.57	60.00	60.00	1440.00	Frampton
82.00	12.750	69.3557	11.89	1200.00	118.60	60.00	1440.00	Frampton
90.00	14.000	119.3557	17.27	1170.30	77.05	60.00	1440.00	BranchIn
112.00	14.000	119.3557	19.34	1043.57	61.08	60.00	1440.00	
125.00	14.000	119.3557	21.02	959.00	60.20	60.00	1440.00	
152.00	14.000	119.3557	26.74	750.59	60.01	60.00	1440.00	
180.00	14.000	119.3557	44.44	445.83	60.00	60.00	1440.00	Harvey

****************** LINE PACK VOLUMES AND PRESSURES ********************

```
Distance   Pressure    Line Pack
(mi)       (psig)      (million std.cu.ft)

0.00       1200.00     0.0000
12.00      1115.01     4.3364
25.00      1023.06     4.7278
35.00      994.40      3.4912
82.00      805.57      14.5681
90.00      1170.30     2.5028
112.00     1043.57     10.1587
125.00     959.00      5.4912
152.00     750.59      9.6160
180.00     445.83      6.8828
```

Total line pack in main pipeline =   61.7750(million std.cubic ft)

************* PIPE BRANCH CALCULATION SUMMARY ************

Number of Pipe Branches = 2

BRANCH TEMPERATURE AND PRESSURE PROFILE:

Outgoing Branch File: C:\Users\Shashi\My Documents\Gasmod\BRANCHOUT.TOT

Branch Location: BranchOut at 25 (mi)
Minimum delivery pressure: 300 (psig)

```
Distance  Elevation  Diameter  Flow      Velocity  Press.   Gas Temp.  Amb Temp.  Location
(mi)      (ft)       (in)      (MMSCFD)  (ft/sec)  (psig)   (degF)     (degF)

0.00      450.00     8.625     30.000    13.68     1023.06  63.42      60.00      MP25
5.00      200.00     8.625     30.000    13.97     1002.05  60.38      60.00
8.00      278.00     8.625     30.000    14.23     983.27   60.10      60.00
12.00     292.00     8.625     30.000    14.57     959.99   60.02      60.00
20.00     358.00     8.625     30.000    15.35     910.55   60.00      60.00
32.00     420.00     8.625     30.000    16.78     831.48   60.00      60.00      End
```

Total line pack in branch pipeline C:\Users\Shashi\My Documents\Gasmod\BRANCHOUT.TOT = 4.5212(million std.cubic ft)

Incoming Branch File: C:\Users\Shashi\My Documents\Gasmod\BRANCHIN.TOT

Branch Location: BranchIn at 90 (mi)

```
Distance  Elevation  Diameter  Flow      Velocity  Press.   Gas Temp.  Amb Temp.  Location
(mi)      (ft)       (in)      (MMSCFD)  (ft/sec)  (psig)   (degF)     (degF)

0.00      250.00     10.750    50.000    11.05     1331.69  140.00     80.00      BranchIn
12.00     389.00     10.750    50.000    11.05     1280.75  83.12      80.00
23.00     465.00     10.750    50.000    11.89     1236.11  80.17      80.00
34.00     520.00     10.750    50.000    12.34     1190.53  80.01      80.00
40.00     369.00     10.750    50.000    12.55     1170.40  80.00      80.00      MP90
```

Total line pack in branch pipeline C:\Users\Shashi\My Documents\Gasmod\BRANCHIN.TOT = 11.5694(million std.cubic ft)

Compressor Power reqd. at the beginning of branch: 1,386.23 HP
Compression ratio: 1.65
Suction temperature: 80.00 (degF)
Suction pressure: 814.70 (psig)
Suction piping loss: 5.00 (psig)
Discharge piping loss: 10.00 (psig)

10. SAMPLE PROBLEM 9: A PIPELINE WITH TWO COMPRESSOR STATIONS, TWO PIPE BRANCHES, AND A PIPE LOOP IN THE SECOND SEGMENT OF THE PIPELINE TO HANDLE AN INCREASE IN FLOW

All data are identical to the previous problem, except that a pipe loop (NPS 14, 0.25-in wall thickness) has been added from milepost 112 to milepost 152 to help reduce the pressures and HP required for an injection from the second branch at milepost 90. The injection volume at milepost 90 is 80 MMSCFD.

The compressor station at Frampton will have to work harder to handle the increased volume from milepost 90 to the end of the pipeline. However, with 40 miles of pipe loop installed in the pipe segment from Frampton to Harvey, the pressures and HP are reduced because the flow of approximately 150 MMSCFD is split between the main pipeline and the loop (Figure 14.26).

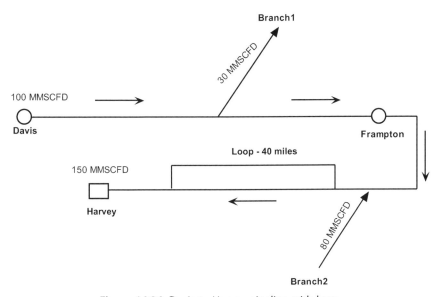

Figure 14.26 Davis to Harvey pipeline with loop.

```
************ GASMOD - GAS PIPELINE HYDRAULIC SIMULATION ***********
************ Version 6.00.780 ************
DATE:                               4-January-2013       TIME:    10:23:33
PROJECT DESCRIPTION:
Problem 3
Pipeline from Davis to Harvey
12"/14" pipeline - 180 miles long
2 compressor stations
Branch in and Branch out and Loop
Case Number:                        1004
Pipeline data file:                 C:\Users\Shashi\MyDocuments\Gasmod\Problem3.TOT

Pressure drop formula:              Colebrook-White
Pipeline efficiency:                1.00
Compressibility Factor Method:      CNGA

Inlet Gas Gravity(Air=1.0):         0.60000
Inlet Gas Viscosity:                0.0000080(lb/ft-sec)
Gas specific heat ratio:            1.29
Polytropic compression index:       1.30


******** Calculations Based on Specified Fixed Overall Heat Transfer Coefficient ********

Base temperature:                   60.00(degF)
Base pressure:                      14.73(psia)

Origin suction temperature:         80.00(degF)
Origin suction pressure:            850.00(psig)
Pipeline Terminus Delivery pressure:  375.77(psig)
Minimum pressure:                   300.0(psig)
Maximum gas velocity:               50.00(ft/sec)

Inlet  Flow rate:                   100.00(MMSCFD)
Outlet Flow rate:                   149.36(MMSCFD)

CALCULATION OPTIONS:
Polytropic compression considered:  YES
Branch pipe calculations:           YES
Loop pipe calculations:             YES
Compressor Fuel Calculated:         YES
Joule Thompson effect included :    NO
Customized Output:                  NO
ALL PRESSURES ARE GAUGE PRESSURES, UNLESS OTHERWISE SPECIFED AS ABSOLUTE PRESSURES


**************** PIPELINE PROFILE DATA ***********

Distance    Elevation     Diameter     Thickness     Roughness
(mi)        (ft)          (in)         (in)          (in)

0.00        220.00        12.750       0.250         0.000700
12.00       340.00        12.750       0.250         0.000700
25.00       450.00        12.750       0.250         0.000700
35.00       189.00        12.750       0.250         0.000700
82.00       225.00        12.750       0.250         0.000700
90.00       369.00        14.000       0.250         0.000700
112.00      412.00        14.000       0.250         0.000700
125.00      518.00        14.000       0.250         0.000700
152.00      786.00        14.000       0.250         0.000700
180.00      500.00        14.000       0.250         0.000700


**************** LOCATIONS AND FLOW RATES ****************

Location    Distance    Flow in/out    Gravity    Viscosity      Pressure    GasTemp.
            (mi)        (MMSCFD)                  (lb/ft-sec)    (psig)      (degF)

Davis       0.00        100.0000       0.6000     0.00000800     850.00      80.00
BranchOut   25.00       -30.0000       0.6000     0.00000800     1023.06     63.42
BranchIn    90.00       80.0000        0.6150     0.00000800     1170.30     77.70
Harvey      180.00      -149.3557      0.6092     0.00000801     375.77      60.00
```

Case Studies

```
*************** COMPRESSOR STATION DATA **************
FLOW RATES, PRESSURES AND TEMPERATURES:

Name      Flow      Suct.    Disch.    Compr.   Suct.    Disch.   Suct.    Disch.   MaxPipe
          Rate      Press.   Press.    Ratio    Loss.    Loss.    Temp.    Temp     Temp
          (MMSCFD)  (psig)   (psig)             (psia)   (psia)   (degF)   (degF)   (degF)

Davis     99.64     845.00   1210.00   1.4246   5.00     10.00    80.00    132.60   140.00
Frampton  69.36     800.57   1210.00   1.5022   5.00     10.00    60.00    118.60   140.00

************* COMPRESSOR EFFICIENCY, HP AND FUEL USED ****************
Name      Distance  Compr.   Mech.    Overall  Horse    Fuel              Fuel       Installed
                    Effy.    Effy.    Effy.    Power    Factor            Used       (HP)
          (mi)      (%)      (%)      (%)               (MCF/day/HP)      (MMSCFD)

Davis     0.00      85.00    98.00    83.30    1,823.24 0.2000            0.3646     5000
Frampton  82.00     85.00    98.00    83.30    1,398.23 0.2000            0.2796     5000

Total Compressor Station Horsepower:           3,221.47                              10,000.

Total Fuel consumption:               0.6442(MMSCFD)

************** REYNOLD'S NUMBER  AND  HEAT TRANSFER COEFFICIENT **************
Distance  Reynold'sNum.  FrictFactor  Transmission  HeatTransCoeff       CompressibilityFactor
(mi)                     (Darcy)      Factor        (Btu/hr/ft2/degF)    (CNGA)

0.000     8,256,276.     0.0114       18.76         0.5000               0.8729
12.000    8,256,276.     0.0114       18.76         0.5000               0.8574
25.000    5,770,328.     0.0115       18.65         0.5000               0.8582
35.000    5,770,328.     0.0115       18.65         0.5000               0.8699
82.000    5,770,328.     0.0115       18.65         0.5000               0.8636
90.000    11,401,803.    0.0111       18.98         0.5457               0.8512
112.000   5,700,901.     0.0113       18.78         0.5000               0.8607
125.000   5,700,901.     0.0113       18.78         0.5000               0.8678
152.000   11,401,803.    0.0111       18.98         0.5000               0.9010
180.000   11,401,803.    0.0111       18.98         0.5000               0.9010

****************** PIPELINE TEMPERATURE AND PRESSURE PROFILE ********************

Distance  Diameter  Flow      Velocity  Press.   GasTemp. SoilTemp. MAOP    Location
(mi)      (in)      (MMSCFD)  (ft/sec)  (psig)   (degF)   (degF)    (psig)

0.00      12.750    99.6354   17.08     1200.00  132.60   60.00     1440.00 Davis
12.00     12.750    99.6354   18.37     1115.01  77.76    60.00     1440.00
25.00     12.750    69.6354   13.97     1023.06  63.42    60.00     1440.00 BranchOut
35.00     12.750    69.6354   14.37     994.40   60.54    60.00     1440.00
82.00     12.750    69.6354   17.61     805.57   60.00    60.00     1440.00 Frampton

82.00     12.750    69.3557   11.89     1200.00  118.60   60.00     1440.00 Frampton
90.00     14.000    149.3557  21.61     1170.30  77.70    60.00     1440.00 BranchIn
112.00    14.000    74.6779   13.08     964.39   61.96    60.00     1440.00 LOOP

125.00    14.000    74.6779   13.60     926.81   60.17    60.00     1440.00
152.00    14.000    149.3557  29.87     842.64   60.00    60.00     1440.00 ENDLOOP
180.00    14.000    149.3557  65.58     375.77   60.00    60.00     1440.00 Harvey

NOTE: On looped portion of pipeline, the flow rate and velocity shown
above correspond to the portion of flow through the mainline only.
The remaining flow goes through the pipe loop.

Gas velocity exceeds  50(ft/sec)  @ location: 180.00(mi)

****************** LINE PACK VOLUMES AND PRESSURES ********************

Distance  Pressure  Line Pack
(mi)      (psig)    (million std.cu.ft)

0.00      1200.00   0.0000
12.00     1115.01   4.7855
25.00     1023.06   4.8517
35.00     994.40    3.5088
82.00     805.57    14.5793
90.00     1170.30   2.7077
112.00    964.39    9.7483
125.00    926.81    5.1399
152.00    842.64    9.9456
180.00    375.77    7.2152

Total line pack in main pipeline =   62.4820(million std.cubic ft)
```

************* PIPE LOOP CALCULATION SUMMARY ***********

Number of Pipe loops: 1

Pipe loop-1: C:\Users\Shashi\My Documents\Gasmod\LOOP1.TOT
Loop starts on main pipeline at: 112.00 (mi)
Loop ends on main pipeline at: 152.00 (mi)
Total mainline length looped: 40.00 (mi)

PIPE LOOP TEMPERATURE AND PRESSURE PROFILE:

| Distance | Elev. | Dia. | FlowRate | Velocity | Pressure | GasTemp | SoilTemp | MAOP | Location |
(mi)	(ft)	(in)	(MMSCFD)	(ft/sec)	(psig)	(degF)	(degF)	(psig)	
0.00	412.00	14.00	74.6779	13.08	964.39	61.96	60.00	1000.00	BeginLoop
10.00	500.00	14.00	74.6779	13.48	935.43	60.30	60.00	1000.00	
20.00	600.00	14.00	74.6779	13.92	905.39	60.05	60.00	1000.00	
30.00	700.00	14.00	74.6779	14.40	874.40	60.01	60.00	1000.00	
40.00	786.00	14.00	74.6779	14.94	842.64	60.00	60.00	1000.00	EndLoop

Total line pack in loop pipeline C:\Users\Shashi\My Documents\Gasmod\LOOP1.TOT = 15.1410(million std.cubic ft)

************* PIPE BRANCH CALCULATION SUMMARY ***********

Number of Pipe Branches = 2

BRANCH TEMPERATURE AND PRESSURE PROFILE:

Outgoing Branch File: C:\Users\Shashi\My Documents\Gasmod\BRANCHOUT.TOT

Branch Location: BranchOut   at   25 (mi)
Minimum delivery pressure: 300 (psig)

| Distance | Elevation | Diameter | Flow | Velocity | Press. | Gas Temp. | Amb Temp. | Location |
(mi)	(ft)	(in)	(MMSCFD)	(ft/sec)	(psig)	(degF)	(degF)	
0.00	450.00	8.625	30.000	13.68	1023.06	60.05	60.00	MP25
5.00	200.00	8.625	30.000	13.96	1002.19	60.01	60.00	
8.00	278.00	8.625	30.000	14.23	983.42	60.00	60.00	
12.00	292.00	8.625	30.000	14.57	960.15	60.00	60.00	
20.00	358.00	8.625	30.000	15.35	910.72	60.00	60.00	
32.00	420.00	8.625	30.000	16.78	831.66	60.00	60.00	End

Total line pack in branch pipeline C:\Users\Shashi\My Documents\Gasmod\BRANCHOUT.TOT = 4.5268(million std.cubic ft)

Incoming Branch File: C:\Users\Shashi\My Documents\Gasmod\BRANCHIN.TOT

Branch Location: BranchIn   at   90 (mi)

| Distance | Elevation | Diameter | Flow | Velocity | Press. | Gas Temp. | Amb Temp. | Location |
(mi)	(ft)	(in)	(MMSCFD)	(ft/sec)	(psig)	(degF)	(degF)	
0.00	250.00	10.750	80.000	15.30	1540.96	140.00	80.00	BranchIn
12.00	389.00	10.750	80.000	15.30	1432.35	89.54	80.00	
23.00	465.00	10.750	80.000	17.65	1333.14	81.58	80.00	
34.00	520.00	10.750	80.000	19.16	1227.01	80.26	80.00	
40.00	369.00	10.750	80.000	20.08	1170.39	80.10	80.00	MP90

Total line pack in branch pipeline C:\Users\Shashi\My Documents\Gasmod\BRANCHIN.TOT = 12.6391(million std.cubic ft)

Compressor Power reqd. at the beginning of branch: 2,904.54 HP
Compression ratio: 1.90
Suction temperature: 80.00 (degF)
Suction pressure: 814.70 (psig)
Suction piping loss: 5.00 (psig)
Discharge piping loss: 10.00 (psig)

11. SAMPLE PROBLEM 10: SAN JOSE TO PORTAS PIPELINE WITH INJECTION AND DELIVERY IN SI UNITS

A 450-/400-mm (10-mm wall thickness) diameter, 680-km-long buried pipeline defined below is used to transport 4.5 Mm³/day of natural gas from San Jose to the delivery terminus at Portas. There are three compressor stations located at San Jose, Tapas, and Campo, with gas turbine–driven centrifugal compressors. The pipeline is *not* insulated and the maximum pipeline temperature is limited to 60 °C because of the material of the pipe external coating. The maximum allowable operating pressure is 9900 kPa. Determine the temperature and pressure profile and horsepower required at each compressor station (Figure 14.27).

Gas specific heat ratio	1.26
Maximum gas velocity	15 m/s
Pipeline efficiency	1.0
Polytropic index	1.2
Base temperature	15 °C
Base pressure	101 kPa
Pressure drop formula	Colebrook–White
Compressibility factor	Standing-Katz

A flow rate of 4.5 Mm³/day enters the pipeline at San Jose (kilometer post 0.0) and at an intermediate location named Anaheim (kilometer post 135), a delivery of 0.5 Mm³/day is made. Additionally an injection of 0.25 Mm³/day is made at Grande (kilometer post 380). The resulting flow then

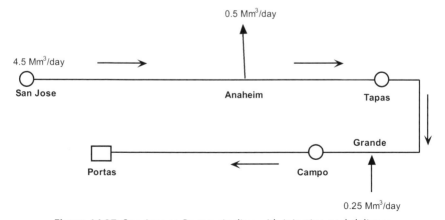

Figure 14.27 San Jose to Portas pipeline with injection and delivery.

continues to the end of the pipeline. Gas inlet temperature is 20 °C at both locations.

| Inlet gas specific gravity (air = 1.00) | 0.600 |
| Inlet gas viscosity | 0.000119 poise |

The pipeline profile is defined as follows:

Milepost (km)	Elevation (m)	Diameter (mm)	Wall Thickness (mm)	Roughness (mm)
0	200	450	10	0.02
72	200	450	10	0.02
77	300	400	10	0.02
135	392	400	10	0.02
260	457	400	10	0.02
320	695	400	10	0.02
380	290	400	10	0.02
402	272	400	10	0.02
475	204	400	10	0.02
490	198	400	10	0.02
500	150	400	10	0.02
515	128	400	10	0.02
532	116	400	10	0.02
612	85	400	10	0.02
680	152	400	10	0.02

The compressor stations are as follows:

Compressor Station	Location (km)	Discharge Pressure (kPa)
San Jose	0.00	9600
Tapas	260.0	9600
Campo	475.0	9000

The installed power at each compressor stations is 4000 KW.

Origin suction pressure	5500 kPa
Pipeline delivery pressure	3500 kPa
Minimum pipe pressure	2000 kPa
Station suction loss	35 kPa
Station discharge loss	70 kPa
Compressor adiabatic efficiency	85.0%
Compressor mechanical efficiency	98.0%
Fuel consumption	7.59 m^3/day/KW
Pipe burial depth	915 mm
Pipe thermal conductivity	50 W/m/°C
Soil thermal conductivity	1.4 W/m/°C
Ambient soil temperature	15 °C
Origin suction temperature	20 °C

Case Studies

```
************ GASMOD - GAS PIPELINE HYDRAULIC SIMULATION ***********
************ Version 6.00.780 ************

DATE:                                   4-January-2013      TIME:  11:37:19
PROJECT DESCRIPTION:
Problem 4
Pipeline from SanJose to Portas
3 compressor stations
Case Number:                            1004
Pipeline data file:                     C:\Users\Shashi\MyDocuments\Gasmod\Problem4.TOT

Pressure drop formula:                  Colebrook-White
Pipeline efficiency:                    1.00
Compressibility Factor Method:          Standing-Katz

Inlet Gas Gravity(Air=1.0):             0.60000
Inlet Gas Viscosity:                    0.0001190(Poise)
Gas specific heat ratio:                1.26
Polytropic compression index:           1.20


**** Calculations Based on Specified Thermal Conductivities of Pipe, Soil and Insulation ********


Base temperature:                       15.00(degC)
Base pressure:                          101.00(kPa)abs

Origin suction temperature:             20.00(degC)
Origin suction pressure:                5500.00(kPa)
Pipeline Terminus Delivery  pressure:   3578.25(kPa)
Minimum pressure:                       2000.0(kPa)
Maximum gas velocity:                   15.00(m/sec)

Inlet   Flow rate:                      4.50(Mm3/day)
Outlet  Flow rate:                      4.16(Mm3/day)

CALCULATION OPTIONS:
Polytropic compression considered:      YES
Branch pipe calculations:               NO
Loop pipe calculations:                 NO
Compressor Fuel Calculated:             YES
Joule Thompson effect included :        NO
Customized Output:                      NO

ALL PRESSURES ARE GAUGE PRESSURES, UNLESS OTHERWISE SPECIFED AS ABSOLUTE PRESSURES



              *************** PIPELINE PROFILE DATA ************


    Distance      Elevation     Diameter       Thickness      Roughness
     (km)         (meters)       (mm)            (mm)           (mm)

    0.00          200.00        450.000         10.000         0.020000
   72.00          200.00        450.000         10.000         0.020000
   77.00          300.00        400.000         10.000         0.020000
  135.00          392.00        400.000         10.000         0.020000
  260.00          457.00        400.000         10.000         0.020000
  320.00          695.00        400.000         10.000         0.020000
  380.00          290.00        400.000         10.000         0.020000
  402.00          272.00        400.000         10.000         0.020000
  475.00          204.00        400.000         10.000         0.020000
  490.00          198.00        400.000         10.000         0.020000
  500.00          150.00        400.000         10.000         0.020000
  515.00          128.00        400.000         10.000         0.020000
  532.00          116.00        400.000         10.000         0.020000
  612.00           85.00        400.000         10.000         0.020000
  680.00          152.00        400.000         10.000         0.020000
```

*************** THERMAL CONDUCTIVITY AND INSULATION DATA ****************

Distance (km)	Cover (mm)	Thermal Conductivity (W/m/degC) Pipe	Soil	Insulation	Insul.Thk (mm)	Soil Temp (degC)
0.000	915.000	50.000	1.400	0.030	0.000	15.00
72.000	915.000	50.000	1.400	0.030	0.000	15.00
77.000	915.000	50.000	1.400	0.030	0.000	15.00
135.000	915.000	50.000	1.400	0.030	0.000	15.00
260.000	915.000	50.000	1.400	0.030	0.000	15.00
320.000	915.000	50.000	1.400	0.030	0.000	15.00
380.000	915.000	50.000	1.400	0.030	0.000	15.00
402.000	915.000	50.000	1.400	0.030	0.000	15.00
475.000	915.000	50.000	1.400	0.030	0.000	15.00
490.000	915.000	50.000	1.400	0.030	0.000	15.00
500.000	915.000	50.000	1.400	0.030	0.000	15.00
515.000	915.000	50.000	1.400	0.030	0.000	15.00
532.000	915.000	50.000	1.400	0.030	0.000	15.00
612.000	915.000	50.000	1.400	0.030	0.000	15.00
680.000	915.000	50.000	1.400	0.030	0.000	15.00

**************** LOCATIONS AND FLOW RATES ****************

Location	Distance (km)	Flow in/out (Mm3/day)	Gravity	Viscosity (Poise)	Pressure (kPa)	GasTemp. (degC)
SanJose	0.00	4.5000	0.6000	0.00011900	5500.00	20.00
Anaheim	135.00	-0.5000	0.6000	0.00011900	7260.35	15.01
Grande	380.00	0.2500	0.6000	0.00011900	7619.17	15.01
Portas	680.00	-4.1633	0.6088	0.00012075	3578.25	15.00

**************** COMPRESSOR STATION DATA ***************

FLOW RATES, PRESSURES AND TEMPERATURES:

Name	Flow Rate (Mm3/day)	Suct. Press. (kPa)	Disch. Press. (kPa)	Compr. Ratio	Suct. Loss. (kPa)abs	Disch. Loss. (kPa)abs	Suct. Temp. (degC)	Disch. Temp (degC)	MaxPipe Temp (degC)
SanJose	4.47	5465.00	9670.00	1.7555	35.00	70.00	20.00	62.49	60.00
Tapas	3.94	3726.59	9670.00	2.5528	35.00	70.00	15.00	87.37	60.00
Campo	4.16	5143.02	9070.00	1.7488	35.00	70.00	15.00	56.47	60.00

Gas Cooling required at compressor station: SanJose to limit station discharge temperature to 60 (degC)
Gas Cooling required at compressor station: Tapas to limit station discharge temperature to 60 (degC)

************ COMPRESSOR EFFICIENCY, POWER AND FUEL USED ****************

Name	Distance (km)	Compr Effy. (%)	Mech. Effy. (%)	Overall Effy. (%)	Power (KW)	Fuel Factor (m3/day/KW)	Fuel Used (Mm3/day)	Installed (KW)
SanJose	0.00	85.00	98.00	83.30	3,321.91	7.5900	0.0252	4000
Tapas	260.00	85.00	98.00	83.30	5,096.19	7.5900	0.0387	4000
Campo	475.00	85.00	98.00	83.30	3,003.43	7.5900	0.0228	4000

Total Compressor Station Power: 11,421.53 (KW)   12,000. (KW)

Total Fuel consumption: 0.0867 (Mm3/day)

WARNING!
Required power exceeds the installed power at compressor station: Tapas

************* REYNOLD'S NUMBER AND HEAT TRANSFER COEFFICIENT **************

Distance (km)	Reynold'sNum.	FrictFactor (Darcy)	Transmission Factor	HeatTransCoeff (W/m2/degC)	CompressibilityFactor (Standing-Katz)
0.000	9,443,126.	0.0110	19.09	2.6824	0.8297
72.000	9,443,126.	0.0110	19.09	2.6824	0.8147
77.000	10,685,642.	0.0111	18.95	2.8976	0.8264
135.000	9,491,659.	0.0112	18.92	2.8959	0.8692
260.000	9,491,659.	0.0112	18.92	2.8959	0.8314
320.000	9,491,659.	0.0112	18.92	2.8959	0.8227
380.000	10,088,651.	0.0112	18.93	2.8971	0.8340
402.000	10,088,651.	0.0112	18.93	2.8971	0.8566
475.000	10,088,651.	0.0112	18.93	2.8971	0.8600
490.000	10,088,651.	0.0112	18.93	2.8971	0.8377
500.000	10,088,651.	0.0112	18.93	2.8971	0.8270
515.000	10,088,651.	0.0112	18.93	2.8971	0.8239
532.000	10,088,651.	0.0112	18.93	2.8971	0.8396
612.000	10,088,651.	0.0112	18.93	2.8971	0.8823
680.000	10,088,651.	0.0112	18.93	2.8971	0.8823

```
******************* PIPELINE TEMPERATURE AND PRESSURE PROFILE ********************

Distance  Diameter    Flow      Velocity  Press.    GasTemp.  SoilTemp.  MAOP      Location
(km)      (mm)        (Mm3/day) (m/sec)   (kPa)     (degC)    (degC)     (kPa)

0.00      450.000     4.4748    3.73      9600.00   60.00     15.00      9900.00   SanJose
72.00     450.000     4.4748    4.07      8801.88   15.56     15.00      9900.00
77.00     400.000     4.4748    5.29      8669.43   15.40     15.00      9900.00
135.00    400.000     3.9748    5.60      7260.35   15.01     15.00      9900.00   Anaheim
260.00    400.000     3.9748    10.56     3761.59   15.00     15.00      9900.00   Tapas

260.00    400.000     3.9361    4.21      9600.00   60.00     15.00      9900.00   Tapas
320.00    400.000     3.9361    4.79      8410.44   15.86     15.00      9900.00
380.00    400.000     4.1861    5.62      7619.17   15.01     15.00      9900.00   Grande
402.00    400.000     4.1861    5.99      7139.99   15.00     15.00      9900.00
475.00    400.000     4.1861    8.17      5178.02   15.00     15.00      9900.00   Campo

475.00    400.000     4.1633    4.74      9000.00   56.47     15.00      9900.00   Campo
490.00    400.000     4.1633    4.91      8690.06   31.72     15.00      9900.00
500.00    400.000     4.1633    5.00      8525.02   23.76     15.00      9900.00
515.00    400.000     4.1633    5.17      8244.51   18.23     15.00      9900.00
532.00    400.000     4.1633    5.39      7909.97   16.03     15.00      9900.00
612.00    400.000     4.1633    7.03      6038.71   15.00     15.00      9900.00
680.00    400.000     4.1633    11.73     3578.25   15.00     15.00      9900.00   Portas

******************* LINE PACK VOLUMES AND PRESSURES ********************
Distance  Pressure    Line Pack
(km)      (kPa)       (million std.cu.m)

0.00      9600.00     0.0000
72.00     8801.88     1.0538
77.00     8669.43     0.0784
135.00    7260.35     0.6409
260.00    3761.59     0.9399
320.00    8410.44     0.4561
380.00    7619.17     0.6655
402.00    7139.99     0.2222
475.00    5178.02     0.6053
490.00    8690.06     0.1241
500.00    8525.02     0.1118
515.00    8244.51     0.1692
532.00    7909.97     0.1881
612.00    6038.71     0.7599
680.00    3578.25     0.4340

Total line pack in main pipeline =   6.4492(million std.cubic m)
```

Appendix

A.1 UNITS AND CONVERSIONS

Item	English Units	SI Units	Conversion: English to SI
Mass	Slugs (slugs)	Kilograms (kg)	1 lb = 0.45359 kg
	Pound mass (lbm)		1 slug = 14.594 kg
	US tons	Metric tons (t)	1 US ton = 0.9072 t
	Long tons		1 long ton = 1.016 t
Length	Inches (in)	Millimeters (mm)	1 in = 25.4 mm
	Feet (ft)	Meters (m)	1 ft = 0.3048
	Miles (mi)	Kilometers (km)	1 mi = 1.609 km
Area	Square feet (ft^2)	Square meters (m^2)	1 ft^2 = 0.0929 m^2
Volume	Cubic inch (in^3)	Cubic millimeters (mm^3)	1 in^3 = 16,387.0 mm^3
	Cubic feet (ft^3)	Cubic meters (m^3)	1 ft^3 = 0.02832 mm^3
	US gallons (gal)	Liters (L)	1 gal = 3.785 L
	Barrel (bbl)		1 bbl = 42 US gal
Density	Slugs per cubic foot ($slug/ft^3$)	Kilograms per cubic meter (kg/m^3)	1 $slug/ft^3$ = 515.38 kg/m^3
Specific weight	Pound per cubic foot (lb/ft^3)	Newton per cubic meter (N/m^3)	1 lb/ft^3 = 157.09 N/m^3
Viscosity (kinematic)	ft^2/s	m^2/s	1 ft^2/s = 0.092903 m^2/s
Flow rate	Gallons/minute (gal/min)	Liters/minute (L/min)	1 gal/min = 3.7854 L/min
	Barrels/hour (bbl/hr)	Cubic meters/hour (m^3/hr)	1 bbl/hr = 0.159 m^3/h
	Barrels/day (bbl/day)		
Force	Pounds (lb)	Newton (N)	1 lb = 4.4482 N
Pressure	Pounds/square inch (psi)	kiloPascal (kPa)	1 psi = 6.895 kPa
	lb/in^2	kilograms/square centimeter (kg/cm^2)	1 psi = 0.0703 kg/cm^2
Velocity	Feet/second (ft/s)	meters/second (m/s)	1 ft/s = 0.3048 m/s
Work and energy	British thermal units (Btu)	Joule (J)	1 Btu = 1055.0 J

(Continued)

Item	English Units	SI Units	Conversion: English to SI
Power	Btu/hour	Watt (W) Joules/second (J/s)	1 Btu/hr = 0.2931 W
	Horsepower (HP)	Kilowatt (kW)	1 HP = 0.746 kW
Temperature	Degree Fahrenheit (°F)	Degree Celsius (°C)	1 °F = 9/5 °C + 32
	Degree Rankin (°R)	Degree Kelvin (°K)	1 °R = °F + 460
			1 °K = °C + 273
Thermal conductivity	Btu/hr/ft/°F	W/m/°C	1 Btu/hr/ft/°F = 1.7307 W/m/°C
Specific heat	Btu/lb/°F	kJ/kg/°C	1 Btu/lb/°F = 4.1869 kJ/kg/°C

A.2 COMMON PROPERTIES OF PETROLEUM FLUIDS

Product	Viscosity cSt @ 60 °F	°API Gravity	Specific Gravity @ 60 °F	Reid Vapor Pressure
Regular gasoline				
Summer grade	0.70	62.0	0.7313	9.5
Interseasonal grade	0.70	63.0	0.7275	11.5
Winter grade	0.70	65.0	0.7201	13.5
Premium gasoline				
Summer grade	0.70	57.0	0.7467	9.5
Interseasonal grade	0.70	58.0	0.7165	11.5
Winter grade	0.70	66.0	0.7711	13.5
No. 1 fuel oil	2.57	42.0	0.8155	
No. 2 fuel oil	3.90	37.0	0.8392	
Kerosene	2.17	50.0	0.7796	
Jet fuel JP-4	1.40	52.0	0.7711	2.7
Jet fuel JP-5	2.17	44.5	0.8040	

A.3 SPECIFIC GRAVITY AND API GRAVITY

Liquid	Specific Gravity @ 60 °F	API Gravity @ 60 °F
Propane	0.5118	N/A
Butane	0.5908	N/A
Gasoline	0.7272	63.0
Kerosene	0.7796	50.0
Diesel	0.8398	37.0
Light crude	0.8348	38.0
Heavy crude	0.8927	27.0
Very heavy crude	0.9218	22.0
Water	1.0000	10.0

A.4 VISCOSITY CONVERSIONS

Viscosity (SSU)	Viscosity (cSt)	Viscosity (SSF)
31.0	1.00	
35.0	2.56	
40.0	4.30	
50.0	7.40	
60.0	10.30	
70.0	13.10	12.95
80.0	15.70	13.70
90.0	18.20	14.44
100.0	20.60	15.24
150.0	32.10	19.30
200.0	43.20	23.50
250.0	54.00	28.0
300.0	65.00	32.5
400.0	87.60	41.9
500.0	110.00	51.6
600.0	132.00	61.4
700.0	154.00	71.1
800.0	176.00	81.0
900.0	198.00	91.0

(*Continued*)

Viscosity (SSU)	Viscosity (cSt)	Viscosity (SSF)
1000.0	220.00	100.7
1500.0	330.00	150
2000.0	440.00	200
2500.0	550.00	250
3000.0	660.00	300
4000.0	880.00	400
5000.0	1100.00	500
6000.0	1320.00	600
7000.0	1540.00	700
8000.0	1760.00	800
9000.0	1980.00	900
10,000.0	2200.00	1000
15,000.0	3300.00	1500
20,000.0	4400.00	2000

A.5 THERMAL CONDUCTIVITIES

Substance	Thermal Conductivity (Btu/hr/ft/°F)
Fire clay brick (burnt at 2426 °F)	0.60–0.63
Fire clay brick (burnt at 2642 °F)	0.74–0.81
Fire clay brick (Missouri)	0.58–1.02
Portland cement	0.17
Mortar cement	0.67
Concrete	0.47–0.81
Cinder concrete	0.44
Glass	0.44
Granite	1.0–2.3
Limestone	0.73–0.77
Marble	1.6
Sandstone	0.94–1.2
Corkboard	0.025
Fiber insulating board	0.028

Substance	Thermal Conductivity (Btu/hr/ft/°F)
Aerogel, silica	0.013
Coal, anthracite	0.15
Coal, powdered	0.067
Ice	1.28
Sandy soil, dry	0.25—0.40
Sandy soil, moist	0.50—0.60
Sandy soil, soaked	1.10—1.30
Clay soil, dry	0.20—0.30
Clay soil, moist	0.40—0.50
Clay soil, moist to wet	0.60—0.90
River water	2.00—2.50
Air	2.00

A.6 ABSOLUTE ROUGHNESS OF PIPE

Pipe Material	Roughness (mm)	Roughness (in)
Riveted steel	0.9—9.0	0.0354—0.354
Concrete	0.3—3.0	0.0118—0.118
Wood stave	0.18—0.9	0.0071—0.0354
Cast iron	0.26	0.0102
Galvanized iron	0.15	0.0059
Asphalted cast iron	0.12	0.0047
Commercial steel	0.045	0.0018
Wrought iron	0.045	0.0018
Drawn tubing	0.0015	0.000059

A.7 TYPICAL HAZEN–WILLIAMS C-FACTORS

Pipe Material	C-factor
Smooth pipes (all metals)	130–140
Smooth wood	120
Smooth masonry	120
Vitrified clay	110
Cast iron (old)	100
Iron (worn/pitted)	60–80
Polyvinyl chloride (PVC)	150
Brick	100

A.8 FRICTION LOSS IN VALVES
Resistance Coefficient K

Appendix 583

Description	L/D	Nominal Pipe Size: in										
		½	¾	1.0	1¼	1½	2	2½ to 3	4	6	8–10	12–16	18–24
Gate valve	8	0.22	0.20	0.18	0.18	0.15	0.15	0.14	0.14	0.12	0.11	0.10	0.10
Globe valve	340	9.2	8.5	7.8	7.5	7.1	6.5	6.1	5.8	5.1	4.8	4.4	4.1
Ball valve	3	0.08	0.08	0.07	0.07	0.06	0.06	0.05	0.05	0.05	0.04	0.04	0.04
Butterfly valve							0.86	0.81	0.77	0.68	0.63	0.35	0.30
Plug valve straightway	18	0.49	0.45	0.41	0.40	0.38	0.34	0.32	0.31	0.27	0.25	0.23	0.22
Plug valve 3-way thru-flo	30	0.81	0.75	0.69	0.66	0.63	0.57	0.54	0.51	0.45	0.42	0.39	0.36
Plug valve branch: flo	90	2.43	2.25	2.07	1.98	1.89	1.71	1.62	1.53	1.35	1.26	1.17	1.08

A.9 EQUIVALENT LENGTHS OF VALVES AND FITTINGS

Description	L/D
Gate valve	8
Globe valve	340
Ball valve	3
Swing check valve	50
Standard elbow: 90°	30
Standard elbow: 45°	16
Long radius elbow: 90°	16

Example: 14-in gate valve has L/D ratio = 8.
Equivalent length = 8 × 14 in = 112 in = 9.25 ft.

A.10 SEAM JOINT FACTORS FOR PIPES

Specification	Pipe Class	Seam Joint Factor (E)
ASTM A53	Seamless	1.00
	Electric resistance welded	1.00
	Furnace lap welded	0.80
	Furnace butt welded	0.60
ASTM A106	Seamless	1.00
ASTM A134	Electric fusion arc welded	0.80
ASTM A135	Electric resistance welded	1.00
ASTM A139	Electric fusion welded	0.80
ASTM A211	Spiral welded pipe	0.80
ASTM A333	Seamless	1.00
ASTM A333	Welded	1.00
ASTM A381	Double submerged arc welded	1.00
ASTM A671	Electric—fusion—welded	1.00
ASTM A672	Electric—fusion—welded	1.00
ASTM A691	Electric—fusion—welded	1.00
API 5L	Seamless	1.00
	Electric resistance welded	1.00
	Electric flash welded	1.00
	Submerged arc welded	1.00
	Furnace lap welded	0.80
	Furnace butt welded	0.60

Specification	Pipe Class	Seam Joint Factor (E)
API 5LX	Seamless	1.00
	Electric resistance welded	1.00
	Electric flash welded	1.00
	Submerged arc welded	1.00
API 5LS	Electric resistance welded	1.00
	Submerged arc welded	1.00

A.11 ANSI PRESSURE RATINGS

Class	Allowable Pressure (psi)
150	275
300	720
400	960
600	1440
900	2160
1500	3600

A.12 APPROXIMATE PIPELINE CONSTRUCTION COST

Pipe Diameter (in)	Average Cost ($/in-dia/mi)
8	18,000
10	20,000
12	22,000
16	14,900
20	20,100
24	33,950
30	34,600
36	40,750

REFERENCES

[1] Code of Federal Regulations, CFR 49, Part 192, transportation of natural or other gas by pipeline: minimum federal safety standards, US Government Printing Office, Washington, D.C.
[2] Code of Federal Regulations, CFR 49, Part 195, transportation of hazardous liquids by pipeline: minimum federal safety standards, US Government Printing Office, Washington, D.C.
[3] ASME 31.4-2006, 2006. Pipeline transportation system for liquid hydrocarbon and other liquids. American Society of Mechanical Engineers, New York, N.Y.
[4] ASME 31.8-2007, 2007. Gas transportation and distribution piping systems. American Society of Mechanical Engineers, New York, N.Y.
[5] Shashi Menon, E., 2005. Piping calculations manual. McGraw Hill, New York, NY.
[6] Mohitpour, M., Botros, K., Van Hardeveld, T., 2008. Pipeline pumping and compression systems, a practical approach. ASME Press, New York, NY.
[7] Shashi Menon, E., 2005. Gas pipeline hydraulics. CRC Press, Taylor & Francis, Boca Raton, FL.
[8] Guo, B., Song, S., Chacko, J., Ghalambor, A., 2005. Offshore pipelines. Gulf Professional Publishing, Burlington, MA.
[9] Silowash, B., 2010. Piping systems manual. McGraw Hill, New York, NY.
[10] Shashi Menon, E., Ozanne, H., Bubar, B., Bauer, W., Wininger, G., 2011. Pipeline planning and construction field manual. Elsevier Publishing Company, Waltham, MA.

Index

Note: Page numbers followed by "f" and "t" indicate figures and tables respectively

A

Above-ground pipeline, 283–284
Abrupt contraction, 177–178, 177f
Absolute pressure, 66
Absolute roughness of pipe, 581
Absolute viscosity of liquid, 41
ACI. *See* American Concrete Institute
Actual cubic feet per minute (ACFM), 398
Adiabatic bulk modulus, 52
Adiabatic compression, 378–380, 379f
Adjustment factor, 72
Affinity laws, 399
 for centrifugal pumps, 349–351
AFUDC. *See* Allowance for funds used during construction
AGA. *See* American Gas Association
AGA NB-13 Committee Report, 210–212
AGA NB-13 method, 209
AISC. *See* American Institute of Steel Construction
Allowable operating pressure, 83–85
Allowance for funds used during construction (AFUDC), 475
American Concrete Institute (ACI), 25
American Gas Association (AGA), 443
 equation, 184, 209–212
 method, 75–76
American Institute of Steel Construction (AISC), 24–25
American National Standards Institute (ANSI), 24, 443
American Petroleum Institute (API), 22, 24, 34–35, 439
 gravity, 7–8, 579
 standards and recommended practices, 22–23
American Society for Testing and Materials (ASTM), 24

American Society of Civil Engineers (ASCE), 15
American Society of Mechanical Engineers (ASME), 15, 24, 31, 436
 ASME B16 standards, 21–22
 ASME b31 standards, 17–19
American Water Works Association (AWWA), 24
ANSI. *See* American National Standards Institute
ANSI pressure ratings, 585
ANSI/API 2530 standard, 455
API. *See* American Petroleum Institute
Apparent molecular weight of gas mixture, 59
Approximate pipeline construction cost, 585
ARCO formulas, 52
ASCE. *See* American Society of Civil Engineers
ASME. *See* American Society of Mechanical Engineers
ASME B16 standards
 ASME/ANSI B16.5 pipe flanges and flanged fittings, 22
 metallic gaskets for pipe flanges, 21
 nonmetallic flat gaskets, 21–22
ASME b31 standards, 17–19
ASTM. *See* American Society for Testing and Materials
ASTM D341 chart method, 45f
Aude equation. *See* T.R. Aude equation
Aude K factor, 174
available NPSH ($NPSH_A$), 358–359
Average pipe segment pressure, 188–189
AWWA. *See* American Water Works Association
Axial stress. *See* Longitudinal stress

B

B&PVC. *See* Boiler and Pressure Vessel Code
Baku–Tbilisi–Erzurum pipeline (BTE pipeline). *See* South Caucasus pipeline
Balgzand Bacton pipeline, 9
Ball valve, 449–450, 450f
Barlow's equation, 96–97
 derivation of, 99–101
 for internal pressure, 85–89
Barometer, 151, 151f
Base pressure. *See* Gauge pressure
Base units, 31–32
Batches, 93–95
Batching, 336–338, 519
Baume scale, 38
Benjamin Miller formula. *See* Miller equation
BEP. *See* Best efficiency point
Bernoulli's equation, 181–183, 434, 467
Best efficiency point (BEP), 343–344
Beta ratio (β), 433, 459–460, 468
BHC. *See* Bullhead City
BHP. *See* Brake horsepower
Blending Index method, 48–51
Blowdown calculations, 144–145
Boiler and Pressure Vessel Code (B&PVC), 19–20
Bolivia–Brazil pipeline, 8–9
Boyle's law, 66
Brake horsepower (BHP), 286, 319–321, 340, 384
 flow rate *vs.*, 345–347
Bulk modulus of liquid, 51–53
 adiabatic bulk modulus, 52
 isothermal bulk modulus, 53
Bullhead City (BHC), 519–520
Buried pipeline, 282–283
Butterfly valve, 450–451, 451f

C

California Natural Gas Association method (CNGA method). *See* Canadian Natural Gas Association method (CNGA method)
Canadian Natural Gas Association method (CNGA method), 76–78
Capital costs, 475, 489
 compressor station costs, 492
 construction management, 478, 495
 engineering management, 478, 495
 environmental cost, 479, 494
 fittings, 478
 mainline valve stations, 492–493
 manifold piping, 477–478
 meter stations, 478, 493
 permitting cost, 479, 494
 pipeline costs, 475–477, 489–491
 project costs, 479–480, 495
 pump station costs, 477
 regulators, 493
 ROW acquisitions cost, 479, 494
 SCADA, 478, 493
 tanks, 477–478
 telecommunication system, 478, 493
 valves, 478
Carbon dioxide (CO_2), 1
case pressure (P_c), 362–363
Caspian pipeline, 12–13
Centimeter-gram-second system (CGS system), 30–31
centipoise (cP), 42
centistokes (cSt), 42
Centrifugal compressors, 396
Centrifugal displacement compressors, 397–398
Centrifugal pumps, 339f, 341f
 affinity laws for, 349–351
 head and efficiency *vs.* flow rate, 343–344
 NPSH, 358–359
 reciprocating pumps *vs.*, 338–342
CGS system. *See* Centimeter-gram-second system
Changing pipe delivery pressure effect, 241–242
Charles's law, 57
Check valves, 452–453, 453f
Churchill equation, 161
Circumferential stress. *See* Hoop stress
CNGA method. *See* Canadian Natural Gas Association method
CO_2. *See* Carbon dioxide

Coated pipes, internally, 179–181
Codes, 15–19. *See also* Standards
 B&PVC, 19–20
Colebrook equation. *See* Colebrook–White equation
Colebrook–White equation, 161, 167, 198–202. *See also* Hazen–Williams equation
Colebrook–White friction factor, 521–523
Composite rating chart, 341
Compressed gas discharge temperature, 382
Compressibility factor, 65, 69, 72–78
 AGA method, 75–76
 CNGA method, 76–78
 Dranchuk, Purvis, and Robinson method, 75
 Hall-Yarborough method, 73–75
 Standing-Katz method, 73
Compression power, 383–387
Compression ratio, 370–371, 377, 395
Compressor stations, 393f, 404, 519.
 See also Liquid-pump stations
 adiabatic compression, 378–380, 379f
 centrifugal displacement compressors, 397–398
 compressed gas discharge temperature, 382
 compression power, 383–387
 gas flow rate, 400–401
 gas pipeline
  with multiple stations, 549
  with three stations, 388f
  with two stations, 370f, 562–563
 hydraulic balance, 376
 isothermal compression, 376–378
 locations, 369–375
 optimum compressor locations, 387–393
 pipeline with two stations, 567
 piping losses, 401–403
 polytropic compression, 381
 positive displacement compressors, 397–398
Compressors
 head, 400–401
 in parallel, 393–396, 395f

performance curves, 398–400
 in series, 393–396, 393f
Conservation of mass, 57
Constant speed, 519, 546
Constant temperature, 66
Construction management, 478, 495
Continuity, 53–54
Control pressure, 363
cP. *See* centipoise
Critical flow, 196
Critical pressure of pure substance, 69
cSt. *See* centistokes

D

Darcy friction factor, 159
Darcy–Weisbach equation, 158–159
degree Kelvin (K), 66
degree Rankin (°R), 66
Density
 effect, 439
 of gas, 58
 of liquid, 36
 mass, 36
Department of Transportation (DOT), 85
 DOT 192, 20
 DOT 195, 20
Derived units, 32–34
Design factor, 87–88
Discharge coefficient, 434, 436–437, 468
Drag reduction, 179–181
Drag reduction agent (DRA), 6–7, 179–180, 519
Drag-reducing agent. *See* Drag reduction agent (DRA)
Drake Well, 2
Dranchuk, Purvis, and Robinson method, 75
Dynamic viscosity of liquid. *See* Absolute viscosity of liquid

E

E-Q curve, 340
Economic analysis, 473–475, 519
Economic pipe size, 480–486
Elevation adjustment parameter, 187–188
Energy equation, 54–56
Engineering management, 478, 495

English system of units, 30
"English units" of measurement, 30
Environmental costs, 479, 494
Equivalent diameters, 235
Equivalent length, 235
   pipe
      gas pipelines, 407–412, 408f
      liquid pipelines, 412–415
      of valves and fittings, 11
Erosional velocity, 192–194. *See also* Velocity
Expansion factor, 459–467

F

Fannings friction factor, 159, 164
FCI. *See* Fluid Control Institute
Federal and State laws, 20
Federal Energy Regulatory Commission (FERC), 20, 495
   filing, 20
Flange taps, 464
Flowing temperature factor, 467
Fluid Control Institute (FCI), 26
Fluid flow, 53
   AGA equation, 209–212
   average pipe segment pressure, 188–189
   Colebrook–White equation, 167, 198–202
   continuity, 53–54
   energy equation, 54–56
   equations, 183–184, 231–233
   erosional velocity, 192–194
   friction factor, 159–164, 197–198
   Fritzsche equation, 228–229
   gas pipelines, 181–183
   general flow equation, 184–187
   Hazen–Williams equation, 168–170
   IGT equation, 222–225
   liquid
      pressure, 149–154
      velocity, 154–156
   measurement, 431
   meters, 432, 456–467
   Miller equation, 172–173
   minor losses, 175–179
   modified Colebrook–White equation, 206–209
   Mueller equation, 227–228
   nozzle, 436–437, 436f, 469–470
   Panhandle A equation, 216–219
   Panhandle B equation, 219–222
   pipe elevations effect, 187–188
   pipe roughness effect, 229–231
   in pipes, 149
   pressure drop from friction, 165–167
   regimes, 158–159
   Reynolds number of flow, 194–197
   Shell-MIT equation, 170–171
   Spitzglass equation, 225–227
   T.R. Aude equation, 173–175
   transmission factor, 202–206
   velocity of gas in pipeline, 189–192
   Weymouth equation, 213–215
Foot-pound-second (FPS), 30
Foot-slug-second (FSS), 30
Fourier heat conduction formula. *See* Temperature gradient
FPS. *See* Foot-pound-second
Friction factor, 159–164, 197–198. *See also* Transmission factor
Friction loss in valves, 9–10
Frictional component, 237
Frictional heating, 284–285
Fritzsche equation, 228–229
FSS. *See* Foot-slug-second
Full bore gate valves, 449
Fully rough pipe flow, 196–197
Fully turbulent zone, 209
Fundamental flow equation. *See* General flow equation

G

G&A cost. *See* General and administrative cost
Gas deviation factor. *See* Compressibility factor
Gas flow measurement, 455
Gas flow rate, 400–401
Gas pipeline. *See also* Liquid pipelines
   hydraulics with injections and deliveries, 558–562
   modeling, 292–294
   with multiple compressor stations, 549
   parallel piping system, 419–428

Index 593

pipe loop location, 429–430
pipe wall thickness, 95–96
series piping system, 407–412
system, 487–489
with two compressor stations, 562–563
with two pipe branches, 562–563
Gas relative density factor, 467
Gas transmission pipeline, 89, 443
    Class 1, 89
    Class 2, 90
    Class 3, 90
    Class 4, 90–92
GasAndes Pipeline, 9
Gases, 322–327. *See also* Liquids
    high-pressure, 397
    properties, 29, 56
GASMOD software, 291–292, 294–297
Gasoline, 431
Gate valve, 448–449, 448f
Gauge pressure, 66
General and administrative cost (G&A cost), 473
General flow equation, 184–187
Globe valves, 452, 452f
Grade tapering, 335
Gradual enlargement, 176–177, 176f
Gravity effect, 321–322

H

H-Q curve. *See* Head-capacity curve
Hall-Yarborough method, 73–75
Hazen–Williams C-factors, 9
Hazen–Williams equation, 168–170, 548
Head loss, 178–179
Head-capacity curve, 340
Heat
    balance, 279–280
    entering and leaving pipe segment, 281–282
    exchangers, 397–398
Heat transfer
    above-ground pipeline, 283–284
    buried pipeline, 282–283
Heating value of gas, 79–80
Heavy crude oil pipeline, 527–529. *See also* Liquid pipelines
    flow rate data input screen, 530f
    graphic plot screen, 533f
    from Joplin to Beaumont, 537–538
    pump curve data input screen, 531f
    thermal flow with heaters, 543
Herschel type, 433, 467
HHP. *See* Hydraulic horsepower
HI method. *See* Hydraulic Institute method
High vapor pressure liquids transportation, 263–264
High-pressure gas, 397
Hoop stress, 84, 96
Horsepower (HP), 237, 317, 383, 477, 519
    requirement to liquid transportation, 317
    BHP, 319–321
    gas, 322–327
    gravity effect, 321–322
    HHP, 317–318
    viscosity effect, 321–322
100% BEP point, 351–354
Hydraulic balance, 330–333, 376
Hydraulic horsepower (HHP), 317–318
Hydraulic Institute method (HI method), 26–27, 351–354
Hydraulic pressure gradient
    gas pipeline, 264–267
    liquid pipeline, 258–263
Hydrocarbon gases, 60t–62t
Hydrostatics, 149
    test pressure, 83–85, 104–143
    testing, 85

I

Ideal gas(es), 64–68
    law, 65
IEEE. *See* Institute of Electronics and Electrical Engineers
IGT distribution equation. *See* Institute of Gas Technology equation (IGT equation)
Institute of Electronics and Electrical Engineers (IEEE), 15
Institute of Gas Technology equation (IGT equation), 222–225
Internal design pressure equation, 102–103

Intrastate pipelines, 20
Isentropic process, 378. *See also* Adiabatic compression
Isothermal
 bulk modulus, 53
 compression, 376–378
 flow, 273, 519–527
Isothermal condition. *See* Constant temperature

J

Joule–Thompson cooling effect, 290

K

kilowatts (kW), 317
Kinematic viscosity, 42

L

L/D ratio, 178–179
Lake Havasu City (LHC), 519–520
Laminar flow, 155
Line fill volume, 35, 93–95
Line pack, 235
LIQTHERM software, 529
Liquefied petroleum gas pipeline (LPG pipeline), 236
Liquid pipelines. *See also* Gas pipeline; Heavy crude oil pipeline; Refined products pipeline
 parallel piping system, 417–419
 series piping system, 412–415
Liquid-pump stations. *See* Compressor stations
 affinity laws for centrifugal pumps, 349–351
 batching, 336–338
 BHP *vs.* flow rate, 345–347
 centrifugal pump, 343–344
  reciprocating pumps *vs.*, 338–342
 control pressure, 363
 hydraulic balance, 330–333
 multiple pumps *vs.* system head curve, 358
 multipump station pipelines, 329–330
 NPSH, 358–361
 NPSH *vs.* flow rate, 347
 open channel flow, 335–336
 pipe grade change, 335
 pump
  configuration, 354–356
  head curve *vs.* system head curve, 357
  station configuration, 361–363
 requirement, 330–333
 slack line, 335–336
 specific gravity and viscosity, 351–354
 specific speed, 348–349
 telescoping pipe wall thickness, 334–335, 334f
 throttle pressure, 363
 variable speed pumps, 364
 viscosity correction chart, 352f
 VSD pump *vs.* control valve, 364–367
Liquid(s)
 density, 432
 pressure, 149–154
 properties, 29
 Reynolds number, 156–158
 velocity, 154–156
Logarithmic mean temperature difference (LMTD), 280–281
Long-distance pipelines, 2–5, 3t–5t
Longitudinal stress, 96
Looped piping system. *See* Parallel piping system
LPG pipeline. *See* Liquefied petroleum gas pipeline

M

Mainline valves, 103–104
Maintenance costs, 480
Manufacturers Standardization Society (MSS), 23–24
 SP-44 steel pipeline flanges, 24
MAOP. *See* Maximum allowable operating pressure
Mass, 34, 56–57. *See also* Volume density, 36
Maximum allowable operating pressure (MAOP), 83–84, 102–103, 180, 259, 329–330, 369
Maximum operating pressure (MOP), 102–103, 446–447
MBPD. *See* Million barrels per day

Meter prover system, 441–443
Meter-kilogram-second system (MKS system), 30–31
Meter(s)
  flow meters, 432
  flow nozzle, 436–437, 436f
  liquids and solids flow measurement, 431
  orifice meter, 437–439, 437f, 456–458, 456f
  PD meter, 440–443
  tube, 458–459, 459f
  turbine meter, 439–440
  venturi meter, 433–435, 433f
Metric Conversion Act, 31
Miller equation, 172–173
Million barrels per day (MBPD), 520
Million standard cubic feet per day (MMSCFD), 405–406
Minor losses, 175–179
  abrupt contraction, 177–178, 177f
  gradual enlargement, 176–177, 176f
  head loss and L/D ratio, 178–179
  internally coated pipes and drag reduction, 179–181
MIT equation. *See* Shell-MIT equation
MKS system. *See* Meter-kilogram-second system
MMSCFD. *See* Million standard cubic feet per day
Modified Colebrook–White equation, 206–209
Moody diagram, 162, 198
MOP. *See* Maximum operating pressure
MSS. *See* Manufacturers Standardization Society
Mueller equation, 227–228
Multipump station pipelines, 329–330

N

NACE. *See* National Association of Corrosion Engineers
National Association of Corrosion Engineers (NACE), 26
National Fire Protection Association (NFPA), 24
Natural gas, 64. *See also* Gases; Ideal gas(es)
  mixtures, 69–71

Net positive suction head (NPSH), 329
  flow rate *vs.*, 347, 347f
  requirement vs. availability, 358–361
Newton's law, 41
NFPA. *See* National Fire Protection Association
Nonhydrocarbon components adjustment, 72
Normal boiling point of liquid, 51
NPSH. *See* Net positive suction head
$NPSH_A$. *See* available NPSH
$NPSH_R$. *See* required NPSH

O

Open channel flow, 335–336
Operating and maintenance cost (O&M cost), 473
Optimum pipe size, 481
Orifice flow rate, 459
Orifice meter, 437–439, 437f, 456–458, 456f
Overall heat transfer coefficient, 278–279

P

Panhandle A equation, 216–219
Panhandle B equation, 219–222
Parallel pipeline, 235
Parallel piping system, 415–416, 415f. *See also* Series piping system
  gas pipelines, 419–428
  liquid pipelines, 417–419
Partially turbulent zone, 209
Pascal's law, 149
PD. *See* Positive displacement
Perfect gas equation. *See* Ideal gas law
Petroleum fluids, 7, 29t
PFI. *See* Pipe Fabrication Institute
Pipe, 1
  Balgzand Bacton pipeline, 9
  Bolivia–Brazil pipeline, 8–9
  Caspian pipeline, 12–13
  costs, 475–477, 489–491
  drag factor, 209–210
  elevations effect, 187–188, 237–241
  expansion, 516
  GasAndes pipeline, 9
  grade change, 335

Pipe (*Continued*)
  with injection and delivery, 571–575
  installation cost, 476
  with intermediate injections and
      deliveries, 242–255
  labor cost, 476
  long-distance, 2–5, 3t–5t
  loop location, 429–430
  material and grade, 101
  with pipe loop, 567
  Rockies express pipeline, 7–8
  roughness effect, 229–231
  seam joint factor, 11–12, 103t
  segment outlet temperature, 285–286
  South Caucasus pipeline, 11
  stress, 84f
  taps, 462
  Tennes, 7. *See also* Gas pipeline
  tonnage determination, 145–148
  Trans-Alaska pipeline, 5–7
  Trans-Mediterranean natural gas pipeline,
      9–10
  TransCanada pipeline, 8
  transmission, 1
  with two compressor stations, 567
  with two pipe branches, 567
  West–East natural gas pipeline, 11–12
  Yamal–Europe pipeline, 10
Pipe Fabrication Institute (PFI), 24
Pipe material cost (PMC), 476
Pipeline. *See* Pipe
Pipeline economics
  analysis, 473–475
  capital costs, 475–480, 489–495
  feasibility studies, 480–486
  gas pipeline system, 487–489
  NPS 18 pipe cost, 506t
  NPS 20 pipe cost, 507t
  operating costs, 480, 495–499
  pipe size, 480–486, 499–514
Pipeline stress design, 83
  allowable operating pressure, 83–85
  Barlow's equation, 85–89, 96–97
  batches, 93–95
  blowdown calculations, 144–145
  gas pipelines, 95–96
  gas transmission pipeline, 89–92

hydrostatic test pressure, 83–85,
    104–143
internal design pressure equation,
    102–103
line fill volume, 93–95
mainline valves, 103–104
pipe material and grade, 101
pipe tonnage determination, 145–148
thick wall pipes, 97–99
Plug valve, 450, 451f
PMC. *See* Pipe material cost
Polytropic compression, 381
Positive displacement (PD), 339
  meter, 440–443
  pumps, 339
Positive displacement compressors,
    397–398
Power curve, 340
Pressure, 149
  base factor, 466
  control valve, 453, 454f
  drop, 180
    from friction, 165–167
  hydraulic pressure gradient
    gas pipeline, 264–267
    liquid pipeline, 258–263
  limitation, 415–416
  piping, 17
  regulators, 268–271, 453–454, 454f
  relief valve, 455
  requirement for fluid transportation,
      235
    changing pipe delivery pressure effect,
        241–242
    frictional component, 237
    high vapor pressure liquids
        transportation, 263–264
    pipeline elevation effect, 237–241
    pipeline with intermediate injections
        and deliveries, 242–255
    system head curves, 255–257
    total pressure drop, 236
Project costs, 479–480
Pseudo critical
  pressure, 69
  properties, 71b
  temperature, 69

Index 597

Psig pressure, 66
Pump
  characteristic curves, 340
  liquid heating from pump inefficiency, 286–289
Pump head curve, system head curve *vs.*, 357
Pump stations, 329, 519, 521. *See also* Liquid-pump stations
  costs, 477

R
Radian, 32
Rate of return (ROR), 473
Real gases, 69
Reciprocating pumps, 340f
  centrifugal pumps *vs.*, 338–342
Reducer, 405
Refined products pipeline, 519–520. *See also* Gas pipeline; Heavy crude oil pipeline; Liquid pipelines
  Phoenix to Las Vegas, 520f
  solution, 521–527
  tariff calculations, 525f, 528f
Regulations, 15–19
Relative density, 37
Relief valves, 268–271
required NPSH (NPSH$_R$), 358–359
Revised Panhandle equation. *See* Panhandle B equation
Reynolds number, 59–63, 156–158, 462–463, 529
  factor, 466
  of flow, 194–197
Right-of-way (ROW), 479
  acquisitions cost, 479, 494
Rockies express pipeline, 7–8
ROR. *See* Rate of return
ROW. *See* Right-of-way

S
SAW pipes. *See* Submerged arc welded pipes
Saybolt Seconds Furol (SSF), 42
Saybolt Seconds Universal (SSU), 42
SCADA. *See* Supervisory control and data acquisition

SCFD. *See* Standard cubic feet per day
Seam joint factor, 11–12, 87–88
Series pipeline, 235
Series piping system, 405, 405f. *See also* Parallel piping system
  gas pipelines, 407–412, 408f
  liquid pipelines, 412–415
  pressures calculating, 406
Shah Deniz pipeline. *See* South Caucasus pipeline
Shell-MIT equation, 170–171
SI units. *See* "Systeme Internationale"; units
Simulation model reports review, 294–315
Slack line, 335–336
Slider supports, 5–6
Smooth pipe flow, 196–197
SMYS. *See* Specified minimum yield strength
Sour gas adjustment, 72
South Caucasus pipeline, 11
Specific gravity, 7–8
  of gas, 58–59
  of liquid, 37–41
    blended liquids, 40–41
    variation with temperature, 39
Specific speed, 348–349
Specific volume, 58
Specific weight
  of gas, 58
  of liquid, 36–37
Specified minimum yield strength (SMYS), 85
Spitzglass equation, 225–227
SSF. *See* Saybolt Seconds Furol
SSU. *See* Saybolt Seconds Universal
Standard cubic feet per day (SCFD), 184
Standards, 15–19
  FCI, 26
  HI pump standards, 26–27
  PFI, 24
Standing-Katz chart, 72
Standing-Katz method, 73
Static suction head, 359
Steradian angle, 33
Stokes, 42

Stonewall limit, 399
Streamline flow. *See* Laminar flow
Submerged arc welded pipes
 (SAW pipes), 88
Suction specific speed, 348
Supercompressibility factor, 72–73
Supervisory control and data acquisition
 (SCADA), 475, 478
Supplementary units, 32–33
Surge line, 399
Swamee–Jain equation, 161
System head curve, 255–257
 pump head curve *vs.*, 357
"Systeme Internationale" units
 (SI units), 31

T

T.R. Aude equation, 173–175
Tariff calculations, 514–515
Telecommunication system, 478, 493
Telescoping pipe wall thickness, 334–335, 334f
Temperature
 base factor, 467
 de-ration factor, 87–88
 gradient, 277
 temperature-dependent flow, 273–277
 variation, 292–294
Tennes, 7. *See* Gas pipeline
Thermal
 conductivities, 8–9, 277–278
 effects, 273
 flow with heaters, 537–538
  and DRA, 543
Thermal hydraulics, 273
 frictional heating, 284–285
 gas pipeline modeling, 292–294
 heat entering and leaving pipe segment, 281–282
 heat transfer
  above-ground pipeline, 283–284
  buried pipeline, 282–283
 isothermal flow, 273
 isothermal *vs.*, 289–292
 liquid heating from pump inefficiency, 286–289
 liquid pipelines, 277–289

LMTD, 280–281
 pipe segment outlet temperature, 285–286
 simulation model reports review, 294–315
 temperature variation, 292–294
 temperature-dependent flow, 273–277
Thick wall pipes, 97–99
Throttle pressure, 363
Through conduit gate valves, 449
Trans-Alaska Pipeline, 1, 5–7
Trans-Mediterranean natural gas pipeline, 9–10
TransCanada pipeline, 8
Transition flow, 196–197
Transmission factor, 165, 202–206
Transmission pipelines, 85
Transportation cost, 514–515
Trial-and-error approach, 161
Turbine meter, 439–440

U

Units of measurement, 30
 base units, 31–32
 CGS, 30–31
 classes, 31
 and conversions, 5–7
 derived units, 32–34
 English system of units, 30
 Metric Conversion Act, 31
 SI units, 31
 supplementary units, 32–33
Untreated *vs.* treated pressure drops. *See* Pressure—drop
US customary system (USCS), 56–57, 184–185, 336

V

Valves, 443
 ball valve, 449–450, 450f
 butterfly valve, 450–451, 451f
 check valves, 452–453, 453f
 codes for design and construction, 447
 gate valve, 448–449, 448f
 globe valves, 452, 452f
 mainline block valve, 445f
 mainline valve installation, 444f

material of construction, 446–447
with motor operator, 445f
plug valve, 450, 451f
pressure
control valve, 453, 454f
regulator, 453–454, 454f
relief valve, 455
types, 444
Vapor pressure of liquid, 51
Variable head meters, 432
Variable speed, 519
Variable speed drive (VSD), 363
VSD pump *vs.* control valve, 364–367
Velocity, 154–156
of gas in pipeline, 189–192
probe, 432
Venturi meter, 433–435, 433f, 467–468
discharge coefficient for, 435f
Venturi tube. *See* Venturi meter
Viscosity
conversions, 8
effect, 321–322
of gas, 59–64
of liquid, 41–51
blended products, 48–51
blending chart, 50f
variation with temperature, 44–47
Viscous flow. *See* Laminar flow
Volume
flow rate, 468
of gas, 57–58
of liquid, 34–36
Von Karman rough pipe flow equation, 210
Von Karman smooth pipe transmission factor, 210
VSD. *See* Variable speed drive

W

Water pipeline, 546–549
Weight density, 36
West–East natural gas pipeline, 11–12
Weymouth equation, 213–215

Y

Yamal–Europe pipeline, 10

Made in the USA
Coppell, TX
29 July 2021